Lecture Notes in Computer Science 6663

Commenced Publication in 1973
Founding and Former Series Editors:
Gerhard Goos, Juris Hartmanis, and Jan van Leeuwen

Harrie de Swart (Ed.)

Relational and Algebraic Methods in Computer Science

12th International Conference, RAMICS 2011
Rotterdam, The Netherlands, May 30 – June 3, 2011
Proceedings

 Springer

Volume Editor

Harrie de Swart
Erasmus University Rotterdam
Faculty of Philosophy
P.O. Box 1738, 3000 DR Rotterdam, The Netherlands
E-mail: deswart@fwb.eur.nl

ISSN 0302-9743   e-ISSN 1611-3349
ISBN 978-3-642-21069-3   ISBN 978-3-642-21070-9 (eBook)
DOI 10.1007/978-3-642-21070-9
Springer Heidelberg Dordrecht London New York

Library of Congress Control Number: Applied for

CR Subject Classification (1998): F.4, I.1, I.2.3, D.2.4, D.3.4

LNCS Sublibrary: SL 1 – Theoretical Computer Science and General Issues

Typesetting: Camera-ready by author, data conversion by Scientific Publishing Services, Chennai, India

Printed on acid-free paper

Springer is part of Springer Science+Business Media (www.springer.com)

Preface

This volume contains the proceedings of the 12th International Conference on Relational and Algebraic Methods in Computer Science (RAMiCS 2011) with a special track on Computational Social Choice and Social Software. The conference took place in Rotterdam, The Netherlands, from May 30 to June 3, 2011. Over the past 20 years, the RelMiCS (Relational Methods in Computer Science) and AKA (Applications of Kleene Algebra) conferences have been a main forum for researchers who use the calculus of relations and similar algebraic formalisms as methodological and conceptual tools. At the last of these conferences it was decided that the two series should be united under the new title "Relational and Algebraic Methods in Computer Science" (RAMiCS). This year, special attention was paid to the fact that the meetings started 20 years ago at the Banach Center in Warsaw. It was commemorated with an invited lecture by Chris Brink, who, together with Ewa Orlowska and Gunther Schmidt, was one of the originators of this series.

Relational and algebraic methods and software tools like RELVIEW turn out to be useful for solving problems in social choice and game theory. For that reason this conference included a special track on Computational Social Choice and Social Software, organized by the CFSC (Computational Foundations of Social Choice) and SSEAC (Social Software for Elections, the Allocation of tenders and Coalition formation) projects of the ESF LogiCCC programme.

Each submission was reviewed by three Programme Committee members. The committee decided to accept 18 papers. The programme also included five invited talks, of which three were on relational and algebraic methods, by Chris Brink, Bernhard Möller (included) and Renate Schmidt (included), and two on social choice theory, by Donald Saari and Agnieszka Rusinowska (included). In addition, there were two tutorials on relational and algebraic methods, by Georg Struth (included) and Michael Winter (included), and two on social choice theory, by Donald Saari (included) and Felix Brandt (included). These tutorials were part of a special PhD programme, where PhD students also had the opportunity to present their work in progress.

I am very grateful to the members of the Programme Committee and the external referees for their care and diligence in reviewing the submitted papers. I would also like to thank the Faculty of Philosophy of the Erasmus University in Rotterdam for having accepted to host this conference, in particular Willy Ophelders, Amanda Koopman, Linda Degener and Lizzy Patilaya for their assistance. I also gratefully appreciate the excellent facilities offered by the Easy-Chair conference administration system. Last but not least I would like to thank the European Science Foundation (ESF) and the Erasmus Trust Fund for their generous financial support.

March 2011   Harrie de Swart

Conference Organization

Programme Chair

Harrie de Swart   Rotterdam, The Netherlands

Programme Committee

Rudolf Berghammer	University of Kiel, Germany
Felix Brandt	Technical University Munich, Germany
Jules Desharnais	Laval University, Canada
Ulle Endriss	University of Amsterdam, The Netherlands
Marcelo Frias	University of Buenos Aires, Argentina
Hitoshi Furusawa	University of Kagoshima, Japan
Peter Höfner	University of Augsburg, Germany
Ali Jaoua	University of Qatar, Qatar
Peter Jipsen	Chapman University, USA
Wolfram Kahl	McMaster University, Canada
Larissa Meinicke	Macquarie University, Sydney, Australia
Bernhard Möller	University of Augsburg, Germany
Ewa Orlowska	National Institute of Telecommunications, Warsaw, Poland
Agnieszka Rusinowska	University of Paris 1, France
Gunther Schmidt	UniBw Munich, Germany
Renate Schmidt	University of Manchester, UK
Georg Struth	University of Sheffield, UK
Michael Winter	Brock University, Canada

External Reviewers

Bernd Brassel	Martin Eric Mueller
Han-Hing Dang	Koki Nishizawa
Guillaume Feuillade	Ingrid Rewitzky
Roland Glück	Patrick Roocks
Timothy Griffin	Kim Solin
Annabelle McIver	Toshinori Takai
Roger Maddux	Dmitry Tishkovsky

Sponsors

LogiCCC programme of the European Science Foundation (ESF)
Erasmus Trust Fund, Rotterdam
Faculty of Philosophy, Erasmus University, Rotterdam

Table of Contents

Building Structured Theories
(Invited Paper)

Bernhard Möller

Institut für Informatik, Universität Augsburg, D-86135 Augsburg, Germany
bernhard.moeller@informatik.uni-augsburg.de

Abstract. We provide a set of syntactic tools for structuring large collections of logical theories. Their use is demonstrated by a formalisation of algebras that are used in describing the semantics of concepts in programming languages, but also of more general systems.

1 Introduction

Within the series of RelMiCS, AKA and now RAMiCS conferences we have seen many algebraic theories, starting with relation and Kleene algebras, which have diversified considerably to cover more and more application areas. Still, many of them share a significant common core, and hence it seems adequate to think about their connections in a systematic way. At the same time, some of the theories are quite complex. This is similar to the situation in programming, where one tries to cope with that using suitable structuring mechanisms, such as inheritance and encapsulation.

In the present paper we attempt a similar structured presentation of some essential RAMiCS theories. While there is already some work in that direction in connection with treating these theories with automatic theorem provers [6, 15, 29, 30], we try to modularise further in a number of new and perhaps unusual ways to pinpoint more clearly which parts of the theories depend on which others.

Of course, there is already a lot of work on structuring larger formal theories. There is the long series of languages designed in the field of *algebraic specification*, like CLEAR [3], CIP-L [2], ASL [67], ACT ONE [14] and CASL [4]. They all comprise some sort of structuring mechanism, and many show notational similarity to what we will use in the present paper. However, by their nature they are mostly restricted to first-order equational logic, whereas we will be more liberal. There is also work on structuring specifications in Edinburgh LCF [40, 56]. General structured specification frameworks based on category theory appear in [12, 16, 20, 58, 60, 61]. And there is the interesting dependently typed functional language Agda [1] with proof assistant, which also allows expressing structured theories.

What we present here deviates from these approaches in that we introduce a number of additional construction mechanisms. Moreover, we forego the definition of a semantics in terms of operations on model classes or of pushouts/colimits. Rather, we view our structuring tools as syntactic devices that abbreviate certain

H. de Swart (Ed.): RAMICS 2011, LNCS 6663, pp. 1–21, 2011.

compounds of formulas and can be re-used and instantiated to exhibit common and recurring parts of specifications. For their meaning we rely on the standard semantics of first-order and higher-order logic.

In motivating the particular ingredients of the theories we present we frequently resort to their use in specifying the semantics of transition systems and the like. However, as it has been demonstrated in many excellent papers throughout this series of conferences, the theories have much wider applicability, and we hope that our methods of structuring will help in extending the algebraic treatment to many further areas.

2   Theories and Definitions

A *theory* has a name and may have an imports clause that specifies on which other theories it depends, a list of *sorts* (i.e., names for carrier sets), a list of operators and a list of predicates, each with their typing, a list of axioms (which should be independent) and a list of properties, starting with the keyword derives, that follow from the axioms. We will only write down the non-empty ones of these; list items are separated by the symbol | or line breaks, sometimes also by a horizontal line. For space limitations we usually list only a few of the more interesting/important derived properties. The operators and predicates are called the *constituents* of the theory. Occasionally we will mark certain constituents as hidden, since they only have auxiliary character for formulating certain axioms in a more convenient and generic way. All non-hidden constituents are visible to the outside and can be imported by other theories. A theory may also contain a list of typed variables that are used in the axioms or derived properties. We omit the explicit definition of variables whose type can be inferred from the typing of the operators and predicates that are applied to them. We use the standard convention that all free variables in a logical formula are implicitly universally quantified.

Definitions are similar to theories except that they do not contain axioms. Rather they give, following the keywords defined by, definitional equalities or equivalences for each of their constituents. The only exception are new constants that may be added without giving particular properties for them.

The distinction between theories and definitions is purely for documentation purposes. For brevity we will refer to them uniformly as *(building) blocks*. Blocks may be freely *imported* and/or *instantiated*, possibly under renaming. For the latter we use simple positional notation, listing the new names between parentheses after the block name. The meaning of an import is simple replacement of the block name by its body (with renaming if specified). If no renaming list is given, the block is imported with its original names. Hence upon import of several blocks into another one, identical names mean identical constituents.

An instantiated block may also be used in the axioms, defined by or derives parts of other blocks; in this case its constituent information is ignored and only the logical formulas in its body are copied in (under renaming if specified). In this case the block serves as a function from constituent names to sets of formulas.

By this twofold use of blocks we achieve a certain notational economy, as will be seen in the examples.

Types may be simple identifiers or Cartesian product, function or power set types. Mostly, however, we will use the higher types only in auxiliary blocks to improve the structuring; they will then disappear again after instantiation of these blocks. Only at the very end of the paper, when we talk about quantales, some higher types persist. A unary predicate is identified with the subset of elements that satisfy it. Use of such a predicate in the position of a type then achieves subsorting. In particular, if variables are declared to be of such a subsort type, quantifiers involving them range over the subsort only.

As first examples to show our notation at work we specify some aspects of comparison, in particular, of preorders and partial orders. First we just introduce the type of the comparison predicate.

theory	COMPARE
sorts	S
predicates	$\leq \subseteq S \times S$

Next, even without any assumptions on the predicate $\leq$, we define the concepts of isotony and antitony. This already involves predicates of higher type that take functions as arguments.

definition	ISO_ANTI
imports	COMPARE
predicates	$isotone, antitone \subseteq S \to S$
defined by	$isotone(f) \Leftrightarrow_{df} \forall x, y . x \leq y \Rightarrow f(x) \leq f(y)$ $antitone(f) \Leftrightarrow_{df} \forall x, y . x \leq y \Rightarrow f(y) \leq f(x)$

We now introduce a general mechanism for propagating properties like isotony and antitony to binary operators. This again involves higher-order concepts.

theory	LEFT_ARG
sorts	S
operators	$g : S \times S \to S$
predicates	$P \subseteq S \to S$
hidden	$right\_const : S \to (S \to S)$
axioms	$right\_const(y)(x) = g(x, y)$ $\forall y . P(right\_const(y))$

Now for instance LEFT_ARG($T, \circ, isotone$) expresses that an operator $\circ : T \times T \to T$ on some set T is isotone in its left argument. A symmetrical theory RIGHT_ARG propagates a predicate to the right argument of a binary operator. Below we will also use this mechanism to express left and right distributivity of a binary operator in terms of distributivity of a unary one.

Next, we specify preorders and partial orders.

theory	PREORDER
imports	COMPARE
axioms	$x \leq x$
	$x \leq y \wedge y \leq z \Rightarrow x \leq z$

theory	POSET
imports	PREORDER
axioms	$x \leq y \wedge y \leq x \Rightarrow x = y$

The fact that a set T with a binary relation $\preceq$ on it forms a partial order can then be expressed as $\mathsf{POSET}(T, \preceq)$.

With a similar theory $\mathsf{BOOLEAN\_ALG}(S, \sqcup, \sqcap, \neg, \bot, \top)$ one specifies Boolean algebras; we omit the detailed axioms and properties. We also introduce some standard notation for Boolean algebras without listing derived properties:

definition	BOOLEAN_OPS
imports	BOOLEAN_ALG
operators	$-, \rightarrow : S \times S \rightarrow S$
defined by	$x - y =_{df} x \sqcap \neg y \mid x \rightarrow y =_{df} \neg x \sqcup y$

3   Sequential Composition

We start our treatment of semantic theories with sequential composition which occurs in many quite different contexts. Sequentiality can concern time and space, like in sequences of events or elements of a list or an array.

We will denote sequential composition abstractly by $\cdot$. Concrete instances are concatenation of formal languages (with finite or infinite words), relational composition or gluing of sets of trajectories or program sequencing.

For now we do not require any laws about sequential composition. This is captured by our first block:

theory	GROUPOID
sorts	S
operators	$\cdot : S \times S \rightarrow S$

Already with this extremely general structure we can describe interesting and important computational phenomena, as will be shown in the next section.

But first we specify commutativity and idempotence:

theory	COMMUTATIVE
imports	GROUPOID
axioms	$x \cdot y = y \cdot x$

theory	IDEMPOTENT
imports	GROUPOID
axioms	$x \cdot x = x$

4   Annihilation

An element x is a *left annihilator* w.r.t. to sequential composition if composition with any element on the right does not change it.

definition	LEFT_ANNI
imports	GROUPOID
predicates	$left\_anni \subseteq S$
defined by	$left\_anni(x) \Leftrightarrow_{df} \forall y . x \cdot y = x$

Left annihilation means absolute domination. It can be used to model catastrophic failure: after an annihilating element "nothing else can happen". Note, however, that sometimes left annihilation is a highly desirable property: when studying infinite computations, like the ones initiated by (hopefully) always continuing operating systems, we usually do not want to take "behaviour after infinity" into account, hence the desired sets of behaviours of such systems should be left annihilators. In general, there may be various left annihilators in a groupoid. For instance, in UTP [21] both the totally undefined and the totally unreliable process are left annihilators. Symmetrically one specifies a right annihilator using a definition RIGHT_ANNI with a predicate $right\_anni$.

5   Characterising Failure

Although it seems almost paradoxical, something useful can be achieved with annihilators. The ideas here are inspired by [11,48] and were generalised in [44].

We assume that there is a distinguished left annihilator. It is intended to represent systems about which nothing definitive can be said and which hence can be viewed as "failing" in some sense. Since we are denoting sequential composition by $\cdot$, a fitting notation for such an element is 0.

theory	FAILURE
imports	LEFT_ANNI
operators	$0 : S$
axioms	$left\_anni(0)$

We will now, in our diction, adopt the view that an element x is "failing" iff it represents a system that fails to terminate. Hence, as discussed above, x should be a left annihilator. As all the computations of such a system are infinite, we will call x *purely infinite*. Dually, an element x will be called *purely finite* if it "notices" subsequent nontermination, i.e, if $x \cdot 0 = 0$. Notice that 0 is both purely infinite and purely finite, but is the only such element. A semantic algebra is *strict* if all its elements are purely finite, i.e., iff 0 is also a right annihilator.

definition	FIN_INF
imports	FAILURE
predicates	$purely\_inf \subseteq S$ $purely\_fin \subseteq S$
defined by	$purely\_inf(x) \Leftrightarrow_{df} left\_anni(x)$ $purely\_fin(x) \Leftrightarrow_{df} x \cdot 0 = 0$

theory	STRICT_COMP	
imports	FAILURE	RIGHT_ANNI
axioms	$right\_anni(0)$	

6   Further Aspects of Sequential Composition

Typically one requires at least associativity of sequential composition. This is captured by our next blocks. We specify left associativity; a symmetric theory R_ASSOC provides the predicate *right_assoc* of right associativity.

theory	L_ASSOC
imports	GROUPOID
predicates	$left\_assoc \subseteq S$
axioms	$left\_assoc(x) \Leftrightarrow_{df} \forall\, y, z \,.\, x \cdot (y \cdot z) = (x \cdot y) \cdot z$

Left-associativity and pure infiniteness show interesting connections:

theory	L_ASSOC_INF
imports	FAILURE ¦ L_ASSOC
derives	$purely\_inf(x) \Rightarrow left\_assoc(x)$ $left\_assoc(x) \Rightarrow (purely\_inf(x) \Leftrightarrow x = x \cdot 0)$

Using the associativity predicates we can talk about semigroups.

theory	SEMIGROUP
imports	L_ASSOC ¦ R_ASSOC
axioms	$left\_assoc(x)$
derives	$right\_assoc(y)$

Frequently, one also assumes a unit 1 of composition. Concrete instances are the language ε consisting just of the empty word, the identity relation or the empty program skip. This leads to the next block, which can further be combined with pure finiteness.

theory	MONOID
imports	SEMIGROUP
operators	$1 : S$
axioms	$1 \cdot x = x = x \cdot 1$

theory	ONE_FIN
imports	MONOID ¦ FIN_INF
derives	$purely\_fin(1)$

7   Concurrency

A well studied algebraic framework for concurrency are the various process calculi (ACP, CCS, CSP, ...), with varying properties of choice and sequential composition. We will here treat some aspects of the recent approach of *concurrent Kleene algebras* [22]. Next to sequential composition · these offer a parallel composition ∥. A basic idea is that the elements of such an algebra abstractly represent sets of traces of some kind. These traces consist of events that are occurrences of certain primitive actions such as communications or assignments. One assumes that there are certain causal or temporal dependences between

events. Sequential composition has to respect these dependences, i.e., in a composition $x \cdot y$ no event in a trace in x may depend on a "future" event in a trace in y. Parallel composition is much more liberal in that it does not impose such a restriction (at the expense of allowing "hazardous" programs with race conditions on the resources involved).

To capture this algebraically, one introduces a comparison relation $\leq$ where $x \leq y$ expresses that y is more liberal than x. Let us now look at the terms $(x \parallel y) \cdot (z \parallel u)$ and $(x \cdot z) \parallel (y \cdot u)$. The first of these is a sequential composition of two parallel compositions. Therefore neither x nor y may depend on z or u. The second one is a parallel composition of two sequential compositions with the same basic constituents. This is more liberal than the first one, since only dependence of x on z and of y on u must be excluded. This fundamental property is the basis of the algebraic axiomatisation.

theory	CONCURRENT_BIGROUPOID		
imports	GROUPOID$(S, \cdot)$	GROUPOID$(S, \parallel)$	COMPARE
axioms	$(x \parallel y) \cdot (z \parallel u) \leq (x \cdot z) \parallel (y \cdot u)$		

Since one wants to construct longer derivation chains, one frequently requires $\leq$ to be a partial order to admit transitivity steps. Moreover, in the basic model of [22] these operators are associative and there is an idle process 1 which is a common unit of $\cdot$ and $\parallel$.

theory	CONCURRENT_BIMONOID
imports	CONCURRENT_BIGROUPOID
operators	$1 : S$
axioms	MONOID$(S, \cdot, 1)$	MONOID$(S, \parallel, 1)$

An extension of this basic theory admits, a.o., a simple algebraic treatment of the rely/guarantee calculus of [33]. The details would lead too far here. Further applications are under way.

8 The Frame Rule and Separation Logic

Over every ordered groupoid one can define a very general form of Hoare triple, for which the classical rules of sequencing and weakening hold:

definition	GROUPOID_HOARE_TRIPLE
imports	GROUPOID $\mid$ POSET
predicates	$\_\{\_\}\_ \subseteq S \times S \times S$
defined by	$x \{y\} z \Leftrightarrow_{df} x \cdot y \leq z$
derives	$x \{u \cdot v\} z \Leftrightarrow \exists y . x \{u\} y \wedge y \{v\} z$
	$x \leq u \wedge u \{y\} v \wedge v \leq z \Rightarrow x \{y\} z$

If the groupoid is even a monoid one can also infer the classical rule for skip, viz. $x \{1\} z \Leftrightarrow x \leq z$. In the presence of parallel composition one obtains

the concurrency rule, and the so-called frame allows modular reasoning where a disjoint context may be added in parallel to a program without invalidating the reasoning using triples:

theory	CONCURRENT_HOARE_TRIPLE	
imports	CONCURRENT_BIGROUPOID	GROUPOID_HOARE_TRIPLE
derives	$x\{y\}z \wedge u\{v\}w \Rightarrow (x\parallel u)\{y\parallel v\}(z\parallel w)$	
	$x\{y\}z \Rightarrow (u\parallel x)\{y\}(u\parallel z)$	

A variant of the frame rule is of particular interest in the so-called separation logic [54] which allows modular reasoning about shared and mutable data structures with pointers. Again, the details would lead too far.

9   Choice

The second fundamental concept that occurs in many circumstances is that of choosing — in a more or less biased way — between a number of possibilities. If the number of these possibilities is finite one speaks of *bounded choice*, otherwise of *unbounded choice*. Bounded choice is mostly represented by a binary operator for choosing between two alternatives, which under suitable assumptions allows an inductive definition of choosing between any positive number of possibilities. We will see below how to deal with zero possibilities. Frequent notations for that operator are $[\,]\,, \sqcup, \sqcap$ and $+$, of which we will use the latter. Its typical axioms are associativity, commutativity and idempotence, which are the axioms for a meet or join semilattice.

theory	SEMILATTICE		
sorts	S		
operators	$+ : S \times S \to S$		
axioms	SEMIGROUP$(S, +)$	COMMUTATIVE$(S, +)$	IDEMPOTENT$(S, +)$

It is well known that every semilattice induces an order. In this paper we interpret $+$ as a join operator and write $\leq$ for the induced order, which is a derived concept and hence not specified by a theory but by a definition. It develops its full power in combination with a semilattice.

definition	SUBSUMPTION
imports	GROUPOID$(S, +)$
predicates	$\leq \,\subseteq\, S \times S$
defined by	$x \leq y \Leftrightarrow_{df} x + y = y$

theory	SEMILATTICE_WITH_SUBSUMPTION		
imports	SEMILATTICE	SUBSUMPTION	ISO_ANTI
derives	POSET$(S, \leq)$		
	$x \leq x + y$	$x + y \leq z \Leftrightarrow x \leq z \wedge y \leq z$	
	LEFT_ARG$(S, +, isotone)$	RIGHT_ARG$(S, +, isotone)$	

In many cases, a semilattice of sets under union as join is used; the subsumption order there coincides with set inclusion.

Over semilattices we can specify the property of distributivity:

definition	DIST
imports	SEMILATTICE_WITH_SUBSUMPTION
predicates	$distributive, super\_distributive \subseteq S \to S$
defined by	$distributive(f) \Leftrightarrow_{df} \forall x,y \,.\, f(x+y) = f(x) + f(y)$ $super\_distributive(f) \Leftrightarrow_{df} \forall x,y \,.\, f(x) + f(y) \le f(x+y)$
derives	$distributive(f) \Rightarrow isotone(f)$ $super\_distributive(f) \Leftrightarrow isotone(f)$

Often a unit for choice is assumed. It represents the choice between zero possibilities. Its interpretation ranges from catastrophic error over failure to chaos. Since we denote choice by $+$, a fitting notation for its unit is 0. The definition of the subsumption order implies that 0 is its least element. Moreover, the fact that 0 is the unit of choice means that choice is angelic; a 0 branch will always be eliminated in favour of the other branch: $0 + x = x = x + 0$.

theory	SEMILATTICE_WITH_MIN
imports	SEMILATTICE
operators	$0 : S$
axioms	$MONOID(S, +, 0)$
derives	$x + y = 0 \Rightarrow x = 0 = y$

definition	SUBSUMPTION_WITH_MIN
imports	SEMILATTICE_WITH_MIN SUBSUMPTION
derives	$0 \le x$

10   Choice and Composition: Idempotent Semirings

An idempotent left semiring combines choice and composition such that composition is distributive in its left argument and isotone in its right one. As usual, we have composition bind tighter than choice.

theory	IL_SEMIRING				
imports	SEMILATTICE_WITH_MIN				
	MONOID	FAILURE	SUBSUMPTION	ISO_ANTI	DIST
predicates	$right\_dist \subseteq S$				
axioms	LEFT_ARG$(S, \cdot, distributive)$		RIGHT_ARG$(S, \cdot, isotone)$		

In a left idempotent semiring we can study pure infiniteness a bit further.

definition	IL_SEMIRING_INF
imports	IL_SEMIRING ¦ FIN_INF
operators	$\inf : S \to S$
defined by	$\inf x =_{df} x \cdot 0$
derives	$purely\_inf(x) \Rightarrow right\_dist(x)$ $\inf x \le x$ ¦ $purely\_inf(y) \wedge y \le x \Rightarrow y \le \inf x$

The last property means that inf x is the greatest purely infinite element below x. Therefore we call inf x the *purely infinite part* of x.

A number of applications use weak idempotent semirings, in which composition is right-distributive while 0 need not be a right annihilator:

theory	WEAK_I_SEMIRING
imports	IL_SEMIRING
axioms	RIGHT_ARG($S, \cdot, distributive$)

If additionally 0 is also a right annihilator and composition also distributes over choice in its right argument one speaks of an *idempotent semiring*.

theory	I_SEMIRING	
imports	WEAK_I_SEMIRING	STRICT_COMP

11   Converse and Relation Algebra

In many cases one is interested in reverting the "flow of control" as underlying sequential composition. To this end one uses the converse $x^\smile$ of an element x:

theory	CONVERSE
imports	GROUPOID
operators	$\_^\smile : S \to S$
axioms	$(x \cdot y)^\smile = y^\smile \cdot x^\smile$

Abstract relation algebra results from combining converse with a Boolean idempotent semiring:

theory	RELATION_ALGEBRA	
imports	BOOLEAN_ALG $\mid$ I_SEMIRING($S, \sqcup, \bot, \cdot, 1$) $\mid$ CONVERSE	
	SUBSUMPTION $\mid$ ISO_ANTI	
axioms	$(x \sqcup y)^\smile = x^\smile \sqcup y^\smile \mid x \cdot \overline{x^\smile \cdot \overline{y}} \leq y$	
derives	$isotone(\_^\smile)$	
	$x \cdot y \sqcap z = \bot \Leftrightarrow x^\smile \cdot z \sqcap y = \bot \Leftrightarrow z \cdot y^\smile \sqcap x = \bot$	

There is no need to tell the RAMiCS audience that there are many more interesting and useful consequences of the axioms.

12   Iteration

Following Kleene's seminal work [37], arbitrary finite iteration of an element x is denoted by x^*. The axiomatisation follows [38].

theory	LEFT_KLEENE_ALG
imports	IL_SEMIRING ┊ SUBSUMPTION ┊ ISO_ANTI
operators	$\_^* : S \to S$
axioms	$1 + x \cdot x^* \leq x^* \;\vert\; y + x \cdot z \leq z \Rightarrow x^* \cdot y \leq z$
derives	$isotone(\_^*)$ $x^* \cdot x^* = x^* = (x^*)^* \;\vert\; (x \cdot y)^* \cdot x = x \cdot (y \cdot x)^* \;\vert\; (x + y)^* = x^* \cdot (y \cdot x^*)^*$ $x \leq 1 \Rightarrow x^* = 1$

The symmetrical axioms that describe "iteration at the right" may need adjustments due to the application circumstances (e.g. in probabilistic algebras) [42, 64, 43]. Infinite iteration is added using seminal ideas from [53]; the axiomatisation follows [5].

theory	LEFT_OMEGA_ALG
imports	LEFT_KLEENE_ALG ┊ SUBSUMPTION ┊ ISO_ANTI
operators	$\_^\omega : S \to S$
axioms	$x^\omega = x \cdot x^\omega \;\vert\; z \leq y + x \cdot z \Rightarrow z \leq x^\omega + y \cdot x^*$
derives	$isotone(\_^\omega)$ $0^\omega = 0 \;\vert\; x^* \cdot x^\omega = x^\omega \;\vert\; x^\omega \cdot y \leq x^\omega \;\vert\; (x^\omega)^\omega \leq x^\omega$ $(x \cdot y)^\omega = x \cdot (y \cdot x)^\omega \;\vert\; (x + y)^\omega = (x^* \cdot b)^\omega + (x^* \cdot b)^* \cdot x^\omega$ $x \leq 1^\omega$

The last derived property motivates the following definition.

definition	OMEGA_TOP
imports	LEFT_OMEGA_ALG
operators	$\top : S$
defined by	$\top =_{df} 1^\omega$
derives	$x \leq x \cdot x \Rightarrow x^\omega = x \cdot \top \;\vert\; x^\omega = x^\omega \cdot \top$

The latter property makes x^ω, e.g., not adequate for the precise description of Zeno effects in hybrid systems. Hence again an adjustment may be needed.

13   Tests: Modelling Sets of States

Elements of semirings frequently represent sets of transitions. To represent sets of states one may use special transitions that abstract `assert` statements as known from programming. A statement `assert B` skips (i.e., leaves the state unchanged) if B holds and aborts otherwise. Considered as a relation, it is a subset of the identity relation on program states. Hence sets of program states or predicates characterising such sets are in one-to-one correspondence with subidentity relations. A central property is that for them intersection and composition coincide. All this lays the basis for an algebraic representation of general sets of states. Such an approach was presented, e.g., in [41] by distinguishing particular semiring elements which, following [39], we call *tests*. Since we want the tests, the algebraic counterparts of predicates, to form a Boolean algebra, we first specify complementation.

definition	IL_SEMIRING_WITH_COMPL
imports	IL_SEMIRING
predicates	$are\_complements \subseteq S \times S$ $test \subseteq S$
defined by	$are\_complements(p,q) \Leftrightarrow_{df} p + q = 1 \wedge p \cdot q = 0 \wedge q \cdot p = 0$ $test(p) \Leftrightarrow_{df} \exists q \,.\, are\_complements(p,q)$
derives	$are\_complements(p,q) \Leftrightarrow are\_complements(q,p)$ $are\_complements(p,q) \Rightarrow p \leq 1 \wedge q \leq 1$ $are\_complements(p,q) \wedge are\_complements(p,r) \Rightarrow q = r$ $are\_complements(0,1)$

Now we can specify the notion of an idempotent left semiring with tests.

theory	IL_SEMIRING_WITH_TESTS
imports	IL_SEMIRING_WITH_COMPL
operators	$\neg : test \rightarrow test$
axioms	$\mathrm{COMMUTATIVE}(test, \cdot)$ $\neg p = q \Leftrightarrow_{df} are\_complements(p,q)$
derives	$\mathrm{BOOLEAN\_ALG}(test, +, \cdot, \neg, 0, 1)$ $\mathrm{I\_SEMIRING} \Rightarrow \mathrm{COMMUTATIVE}(test, \cdot)$

The commutativity requirement for tests is equivalent to distributivity of composition over choice also in its right argument on the subset of tests. However, in concrete algebras it usually is more onerous to check distributivity, a property involving three variables, than commutativity, which only involves two.

The above specification is somewhat unsatisfactory in that it is not purely equational and involves subsorting. This makes automatic verification quite cumbersome or even excludes the use of some automatic verification systems. We will discuss alternative specifications below.

Using tests we can give algebraic semantics to a simple programming language. Composition and choice are already present in semirings. We can enrich this by case distinction:

definition	IFTHENELSE
imports	IL_SEMIRING_WITH_TESTS
operators	$if\_then\_else : test \times S \times S \rightarrow S$
defined by	$if\ p\ then\ x\ else\ y =_{df} p \cdot x + \neg p \cdot y$

Using finite iteration we can also define a while loop:

definition	WHILE	
imports	IL_SEMIRING_WITH_TESTS	LEFT_KLEENE_ALG
operators	$while\_do\_ : test \times S \rightarrow S$	
defined by	$while\ p\ do\ x =_{df} (p \cdot x)^{*} \cdot \neg p$	

Moreover, we can give an algebraic definition of standard Hoare triples; it appears in [39] and admits a simple algebraic soundness proof of Hoare logic.

definition	HOARE_TRIPLE
imports	IL_SEMIRING_WITH_TESTS
predicates	$\{\_\} \_ \{\_\} \subseteq \text{test} \times S \times \text{test}$
defined by	$\{p\}\ x\ \{q\} \Leftrightarrow_{df} p \cdot x \cdot \neg q = 0$

14   Domain and Antidomain

An important concept for transition systems is the set of *enabled* states, i.e., the set of states from which transitions are possible. For a transition relation R, this is the *domain* of R. We apply this nomenclature also to the general case. Using tests, a *predomain* operator can be characterised algebraically by quite simple equational axioms [9]. The domain operator shows a stronger interplay between predomain and composition, which can be used, e.g., for an algebraic proof of relative completeness of the Hoare calculus [46].

theory	PREDOMAIN
imports	IL_SEMIRING_WITH_TESTS
operators	$\ulcorner\_ : S \rightarrow \text{test}$
variables	$p : \text{test} \mid x, y : S$
axioms	$x \leq \ulcorner x \cdot x \mid \ulcorner(p \cdot x) \leq p$
derives	$\ulcorner(x + y) = \ulcorner x + \ulcorner y \mid \ulcorner(x \cdot y) \leq \ulcorner(x \cdot \ulcorner y)$
	$\ulcorner(p \cdot x) = p \cdot \ulcorner x \mid \ulcorner p = p$

theory	DOMAIN
imports	PREDOMAIN
axioms	$\ulcorner(x \cdot \ulcorner y) \leq \ulcorner(x \cdot y)$

In a similar fashion one can specify a codomain operator. If one assumes an idempotent semiring with tests, the axioms are just the mirror images of the ones for domain. In case of a general idempotent left semiring, however, distributivity of codomain over choice needs to be stated as an additional axiom [44].

Let us now briefly discuss the mentioned alternative axiomatisation of tests and the domain operator. This has been carried out in a form specific to relation algebra in [31] and in the general semiring setting in [8]. The idea is to avoid explicit subsorting and to characterise the tests implicitly as the image of an antidomain operator @ which yields the negation of the domain of its argument. Over a full idempotent semiring the axiomatisation is surprisingly simple — however, to fully grasp it, good knowledge about tests and domain is almost mandatory. This is why we introduced their theories beforehand. In using the antidomain theory, rather than quantifying over a variable $p : \text{test}$ one uses a variable $x : S$ and writes @x instead of p. For the case of general left semirings additional axioms are necessary; this is the subject of ongoing work.

theory	ANTIDOMAIN		definition	DERIVED_DOMAIN
imports	I_SEMIRING		imports	ANTIDOMAIN
operators	$@ : S \rightarrow S$		operators	$\ulcorner{} : S \rightarrow S \mid \neg : test \rightarrow test$
axioms	$@x \cdot x = 0 \mid @x + @@x = 1$		predicates	$test \subseteq S$
	$@(x \cdot y) \le @(x \cdot @@y)$		defined by	$test(x) \Leftrightarrow_{df} \exists\, y . x = @y$
				$\ulcorner x =_{df} @@x \mid \neg p = @@p$
			derives	DOMAIN

15   Modal Operators: Diamond and Box

Many properties of transition systems can be described by the modal operators diamond and box, which express existential and universal quantification over the successor or predecessor states of a given set of states. We show exemplarily how to define the forward modal operators in terms of domain; the backward ones can be defined analogously in terms of codomain. Given a transition system x, a state s satisfies the predicate $\langle x \rangle q$ iff s has a successor under x that satisfies q. This is equivalent to saying that s lies in the inverse image of q under x. The box operator $[x]q$ is the De Morgan dual of diamond. This is the basis of the following specification.

definition	FORWARDMODAL	
imports	DOMAIN	
operators	$\langle {}_- \rangle {}_-, [{}_-]{}_- : S \times test \rightarrow test$	
defined by	$\langle x \rangle q =_{df} \ulcorner(x \cdot q) \mid [x]q =_{df} \neg \langle x \rangle \neg q$	
variables	$u : purely\_fin \mid z : right\_dist \mid p, q : test \mid x, y : S$	
derives	$\langle u \rangle 0 = 0$	$[u]1 = 1$
	$\langle z \rangle(p + q) = \langle z \rangle p + \langle z \rangle q$	$[z](p \cdot q) = [z]p \cdot [z]q$
	$\langle z \rangle p - \langle z \rangle q \le \langle z \rangle(p - q)$	$[z](p \rightarrow q) \le [z]p \rightarrow [z]q$
	$\langle x + y \rangle p = \langle x \rangle p + \langle y \rangle p$	$[x + y]p = [x]p \cdot [y]p$
	$\langle x \cdot y \rangle p = \langle x \rangle(\langle y \rangle p)$	$[x \cdot y]p = [x]([y]p)$
	$\{p\}\ x\ \{q\} \Leftrightarrow p \le [x]q$	

The last property shows that $[x]q$ is the algebraic counterpart of the weakest liberal precondition operator wlp.x.q used in program correctness calculi [10].

Equivalently, one can axiomatise one of the modal operators directly and define the other one and domain in terms of it. For instance, one can use the last property above to axiomatise the box operator. Then the diamond operator is defined as the De Morgan dual of box: $\langle x \rangle q =_{df} \neg[x]\neg q$. Finally, domain can be retrieved as $\ulcorner x =_{df} \langle x \rangle 1$.

Interestingly, in presence of star no special axioms are needed to establish star induction for diamond and box:

theory	MODAL_STAR
imports	LEFT_KLEENE_ALG ¦ FORWARDMODAL
variables	$p, q : test$ ¦ $x : S$
derives	$p \leq q \wedge p \leq [x]p \;\Rightarrow\; p \leq [x^*]q$
	$(\{p \cdot r\}\, x\, \{p\}) \wedge p \cdot \neg r \leq q \;\Rightarrow\; (\{p\} \text{ while } p \text{ do } x \, \{q\})$

The second property is a special case of the first one and corresponds to the familiar inference rule for the while loop. More generally, box calculus does not only admit an algebraic soundness proof of Hoare logic, but also one of relative completeness [46].

Moreover, box can also be used to model total and general correctness and the wp operator. In fact, wp turns out to be the box operator in an algebra of commands [47]. Hence the abstract relative completeness result can immediately be re-used to show relative completeness of wp-based Hoare logic.

16   Logics of Knowledge and Belief

In this section we use some of our blocks to build algebraic theories of knowledge and belief. The idea is to abstract the access relations of Kripke models for multiagent systems to elements of an idempotent semiring with tests and to represent the knowledge and belief operators as instances of the general box operator with suitable additional axioms. The monomodal case is obtained by setting set $\Box p =_{df} [x]p$ for some fixed transition element x.

definition	MULTIAGENT$_n$
imports	MODAL_ISEMIRING ¦ MODAL_STAR
operators	$a_1, \ldots, a_n, a : S$ ¦ $K_1, \ldots, K_n, E, C : test \rightarrow test$
defined by	$K_1 p =_{df} [a_1]p$ ¦ $\cdots$ ¦ $K_n p =_{df} [a_n]p$
	$a =_{df} a_1 + \cdots + a_n$ ¦ $E p =_{df} [a]$ ¦ $C p =_{df} [a^+]$
derives	$C p \leq C(C p)$ ¦ $C p \cdot C q \leq C(C p \cdot C q)$ ¦ $C p \cdot C q \leq C(C p \cdot q)$

The positive and negative introspection axioms are captured as follows.

definition	INTROSPECTION
imports	IL_SEMIRING_WITH_TESTS
predicates	$sat\_posintro, sat\_negintro \subseteq test \rightarrow test$
axioms	$sat\_posintro(f) \Leftrightarrow_{df} \forall p . f(p) \leq f(f(p))$
	$sat\_negintro(f) \Leftrightarrow_{df} \forall p . \neg f(p) \leq f(\neg f(p))$

This allows specifying belief logic:

theory	MULTIBELIEF$_n$
imports	MULTIAGENT$_n$ ¦ INTROSPECTION
axioms	$sat\_posintro(K_1)$ ¦ $\cdots$ ¦ $sat\_posintro(K_n)$
	$sat\_negintro(K_1)$ ¦ $\cdots$ ¦ $sat\_negintro(K_n)$

The axiom of truth (or reflexivity of the access elements) is expressed by

definition	TRUTH
imports	IL_SEMIRING_WITH_TESTS
predicates	$sat\_truth \subseteq test \rightarrow test$
axioms	$sat\_truth(f) \Leftrightarrow_{df} \forall p \,.\, f(p) \leq p$

Then multiagent knowledge logic is specified by

theory	MULTIKNOW$_n$	
imports	MULTIBELIEF$_n$	TRUTH
axioms	$sat\_truth(\mathsf{K}_1)$ $\cdots$	$sat\_truth(\mathsf{K}_n)$
derives	$\mathsf{C}(p \rightarrow \mathsf{E}p) \leq p \rightarrow \mathsf{C}p$	$\mathsf{C}p \cdot \mathsf{C}q = \mathsf{CC}p \cdot \mathsf{CC}q = \mathsf{C}(\mathsf{C}p \cdot \mathsf{C}q)$

It should be clear how further special-purpose multimodal logics can be constructed along these lines.

17   Quantales and Temporal Logics

Now we really leave the first-order setting. For a number of applications it is important that the underlying left semiring is not only a semilattice, but even a complete lattice in which composition distributes over arbitrary/non-empty suprema in its left/right argument. Such structures are known as left quantales and, in case the underlying semiring is even full, quantales [49,55]. For systematic reasons we call the supremum operator lub (least upper bound) rather than sup.

theory	LQUANTALE	
imports	SEMIGROUP	POSET
operators	$\mathsf{lub}\_ : \mathcal{P}(S) \rightarrow S$	
axioms	$\mathsf{lub}\,T \leq y \Leftrightarrow \forall x \,.\, x \in T \Rightarrow x \leq y$	
	$\mathsf{lub}\,T \cdot y = \mathsf{lub}\,\{x \cdot y \mid x \in T\}$	
	$T \neq \emptyset \Rightarrow y \cdot \mathsf{lub}\,T = \mathsf{lub}\,\{y \cdot x \mid x \in T\}$	

Again, frequently a unit of composition is useful.

theory	UL_QUANTALE	
imports	LQUANTALE	MONOID

From a unital left quantale we can derive an idempotent left semiring.

definition	UL_QUANTALE_AS_SEMIRING	
imports	UL_QUANTALE	
operators	$0 : S$	$+ : S \times S \rightarrow S$
defined by	$0 =_{df} \mathsf{lub}\,\emptyset$	$x + y =_{df} \mathsf{lub}\,(\{x\} \cup \{y\})$
derives	IL_SEMIRING	

In a unital left quantale, star and omega can be *defined* as least/greatest fixpoints. To this end we first enrich left quantales by an infimum operation glb (greatest lower bound).

definition	UL_QUANTALE_WITH_GLB
imports	UL_QUANTALE
operators	$\text{glb}\,\_ : \mathcal{P}(S) \to S$
defined by	$\text{glb}\,T \ =_{df} \ \text{lub}\,\{x \in S \mid \forall y \in T \,.\, x \leq y\}$

Least and greatest fixpoints are defined using the Tarski/Knaster theorem.

definition	FIXPOINTS	
imports	UL_QUANTALE_WITH_GLB	ISO_ANTI
operators	$\mu, \nu : isotone \ \to \ S$	
defined by	$\mu f \ =_{df} \ \text{glb}\,\{z \in S \mid f(z) \leq z\}$	
	$\nu f \ =_{df} \ \text{lub}\,\{z \in S \mid z \leq f(z)\}$	

Now we can specify iteration in a quantale.

definition	UL_QUANTALE_WITH_ITERATION	
imports	UL_QUANTALE_AS_SEMIRING	FIXPOINTS
operators	$^*, {}^{\omega} : S \to S$	
hidden	$f : S \times S \to (S \to S)$	
defined by	$f(x,y)(z) \ =_{df} \ x + y \cdot z$	
	$y^* \ =_{df} \ \mu f(1,y)$	$y^{\omega} \ =_{df} \ \nu f(0,y)$
derives	LEFT_KLEENE_ALG	
	FULLY_DIST $\Rightarrow$ LEFT_OMEGA_ALG	

where

theory	FULLY_DIST	
imports	UL_QUANTALE_AS_SEMIRING	FIXPOINTS
axioms	$\text{glb}\,\{x + y \mid y \in T\} \ = \ x + \text{glb}\,T$	

Further applications of left quantales concern, e.g., hybrid systems and the various temporal logics. For instance, to capture CTL* one can interpret the quantale elements as abstracting sets of computation paths (the semantics of path formulas) and tests as abstracting sets of states (the semantics of state formulas); a distinguished element n abstracts the single-step transition relation. Then we can define an until operator U as $x \, \mathsf{U} \, y \ =_{df} \ \mu z \,.\, y + (x \sqcap \mathsf{n} \cdot z)$. This admits proving all standard CTL laws purely algebraically [45]. Moreover, for the sublogics CTL and LTL the general CTL* semantics can be transformed into simplified versions in ω-regular form. These do no longer use the full power of quantales but just star and omega. Finally, for LTL even just star is used. This provides interesting connections between μ-calculus representations and star/omega-algebra. Other logics like ITL, IL, DC or NL can also be captured in this setting. For lack of space we cannot spread out the details in form of structured theories here.

18   Conclusion and Outlook

The algebraic structures presented form a comprehensive and flexible framework. They cover various semantic models in a uniform algebraic fashion. Further applications have concerned residuals (e.g., to define generalised modal operators as in [65,57]), predicate transformer semantics (e.g., as demonic refinement algebra [66,59] or command/design algebra [47,18,19]), probabilistic programs [42, 64,43], game algebra [51,52], hybrid systems [26,27], neighbourhood logic [25] and linked object structures and separation logic [13,7]. There is even a greater variety of applications outside the realm of program semantics. Many of them use standard relation algebra; for these the ideas in the present paper are not as useful, since the underlying theory is fixed. However, we mention a few that use variants of the semiring setting, for which the idea of building blocks of theories may have some profit. For instance [17, 24] provide algebraic descriptions of some aspects of routing systems. But also the cardinality operator in Dedekind categories and allegories as used for flow problems [35,36], or collagories [34] could be organised in the form of structure theories as proposed here.

As mentioned in the introduction, machine support for this type of theories has been intensively studied. This has resulted in large, modularised collections of theories and automatic proofs on the web [28,23,32,50,62]. Moreover, there are strong links with the TPTP project [63].

As a continuation of that and the ideas in the present paper we envisage a system for composing and analysing structured theories, of course with check for syntactic well-formedness including type constraints. Moreover, the system should perform a normalisation of a structured theory into a "flattened" unstructured one and then determine which fragment of logic is actually used, in particular, whether the overall theory is equivalent to a first-order one. It should then make suggestions which of the existing automatic theorem provers look most promising for use with that theory.

We are convinced that there is much more potential in the algebraic approach. What needs to be done is to explore further areas to see whether the structuring mechanisms we have proposed in the present paper are sufficient and maybe can notationally be streamlined further.

Acknowledgement. I am grateful for valuable comments by H.-H. Dang, R. Glück, P. Höfner, P. Roocks and A. Zelend.

References

1. Agda, http://wiki.portal.chalmers.se/agda/pmwiki.php
2. Bauer, F.L., et al.: The Munich Project CIP. LNCS, vol. 183. Springer, Heidelberg (1985)

3. Burstall, R., Goguen, J.: The Semantics of CLEAR, A Specification Language. In: Bjorner, D. (ed.) Abstract Software Specifications. LNCS, vol. 86, pp. 292–332. Springer, Heidelberg (1980)
4. CoFI (The Common Framework Initiative): CASL Reference Manual. LNCS, vol. 2960 (IFIP Series). Springer, Heidelberg (2004)
5. Cohen, E.: Separation and Reduction. In: Backhouse, R., Oliveira, J. (eds.) MPC 2000. LNCS, vol. 1837, pp. 45–59. Springer, Heidelberg (2000)
6. Dang, H.-H., Höfner, P.: Automated Higher-Order Reasoning about Quantales. In: Konev, B., Schmidt, R., Schulz, S. (eds.) Proc. Workshop on Practical Aspects of Automated Reasoning (2010)
7. Dang, H.-H., Höfner, P., Möller, B.: Algebraic Separation Logic. J. Log. Algebr. Program. (forthcoming, 2011)
8. Desharnais, J., Struth, G.: Modal Semirings Revisited. In: Audebaud, P., Paulin-Mohring, C. (eds.) MPC 2008. LNCS, vol. 5133, pp. 360–387. Springer, Heidelberg (2008)
9. Desharnais, J., Möller, B., Struth, G.: Kleene Algebra with Domain. ACM Transactions on Computational Logic 7, 798–833 (2006)
10. Dijkstra, E.: A Discipline of Programming. Prentice-Hall, Englewood Cliffs (1976)
11. Dijkstra, E.: Computation Calculus Bridging a Formalization Gap. Sci. Comput. Program. 37, 3–36 (2000)
12. Durán, F., Meseguer, J.: Structured Theories and Institutions. Theoretical Computer Science 309, 357–380 (2003)
13. Ehm, T.: Pointer Kleene Algebra. In: Berghammer, R., Möller, B., Struth, G. (eds.) RelMiCS 2003. LNCS, vol. 3051, pp. 99–111. Springer, Heidelberg (2004)
14. Ehrig, H., Mahr, B.: Fundamentals of Algebraic Specification I: Equations and Initial Semantics. EATCS-Monographs in Theor. Comp. Sci., vol. 6. Springer, Heidelberg (1985)
15. Foster, S., Struth, G., Weber, T.: Automated Engineering of Relational and Algebraic Methods in Isabelle/HOL. In: de Swart, H. (ed.) RAMiCS 2011. LNCS, vol. 6663, pp. 52–67. Springer, Heidelberg (2011)
16. Giunchiglia, F., Pecchiari, P., Talcott, C.: Reasoning Theories. J. Autom. Reason. 26, 291–331 (2001)
17. Gurney, A., Griffin, T.: Pathfinding Through Congruences. In: de Swart, H. (ed.) RAMiCS 2011. LNCS, vol. 6663, pp. 180–195. Springer, Heidelberg (2011)
18. Guttmann, W., Möller, B.: Modal design algebra. In: Dunne, S., Stoddart, B. (eds.) UTP 2006. LNCS, vol. 4010, pp. 236–256. Springer, Heidelberg (2006)
19. Guttmann, W., Möller, B.: Normal design algebra. J. Log. Algebr. Program. 79, 144–173 (2010)
20. Harper, R., Sannella, D., Tarlecki, A.: Structured Theory Presentations and Logic Representations. Ann. Pure Appl. Logic 67, 113–160 (1994)
21. He, J., Hoare, T.: Unifying Theories of Programming. Prentice-Hall, Englewood Cliffs (1998)
22. Hoare, T., Möller, B., Struth, G., Wehrman, I.: Concurrent Kleene Algebra and its Foundations. J. Log. Algebr. Program. (forthcoming, 2011)
23. Höfner, P.: Algebraic Reasoning with Prover9, http://www.kleenealgebra.de
24. Höfner, P., McIver, A.: Towards an Algebra of Routing Tables. In: de Swart, H. (ed.) RAMiCS 2011. LNCS, vol. 6663, pp. 212–229. Springer, Heidelberg (2011)
25. Höfner, P., Möller, B.: Algebraic Neighbourhood Logic. J. Log. Algebr. Program. 76, 35–59 (2008)
26. Höfner, P., Möller, B.: An Algebra of Hybrid Systems. J. Log. Algebr. Program. 78, 74–97 (2009)

27. Höfner, P., Möller, B.: Fixing Zeno Gaps. Theor. Comp. Sci. (forthcoming, 2011)
28. Höfner, P., Müller, M., Zeissler, S.: ATPPortal: A User-friendly Web Based Interface for Automated Theorem Provers and for Automatically Generated Proofs. University of Augsburg, Institute of Computer Science, Report TR1010-08, http://opus.bibliothek.uni-augsburg.de/volltexte/2010/1673/
29. Höfner, P., Struth, G.: Automated Reasoning in Kleene Algebra. In: Pfenning, F. (ed.) CADE 2007. LNCS (LNAI), vol. 4603, pp. 279–294. Springer, Heidelberg (2007)
30. Höfner, P., Struth, G., Sutcliffe, G.: Automated Verification of Refinement Laws. Annals of Mathematics and Artificial Intelligence 35, 35–62 (2009)
31. Hollenberg, M.: An Equational Axiomatization of Dynamic Negation and Relational Composition. Journal of Logic, Language and Information 6, 381–401 (1997)
32. Hurd, J.: OpenTheory: Package Management for Higher Order Logic Theories. In: Dos Reis, G., Théry, L. (eds.) PLMMS 2009 — Proc. ACM SIGSAM 2009 International Workshop on Programming Languages for Mechanized Mathematics Systems, pp. 31–37. ACM, New York (2009)
33. Jones, C.: Development Methods for Computer Programs Including a Notion of Interference. PhD Thesis, University of Oxford. Programming Research Group, Technical Monograph 25 (1981)
34. Kahl, W.: Collagories — Relation-Algebraic Reasoning for Gluing Constructions. J. Log. Algebr. Program. (forthcoming, 2011)
35. Kawahara, Y.: On the Cardinality of Relations. In: Schmidt, R. (ed.) RelMiCS/AKA 2006. LNCS, vol. 4136, pp. 251–265. Springer, Heidelberg (2006)
36. Kawahara, Y., Winter, M.: Cardinal Addition in Distributive Allegories. In: Berghammer, R., Jaoua, A.M., Möller, B. (eds.) RelMiCS 2009. LNCS, vol. 5827, pp. 227–241. Springer, Heidelberg (2009)
37. Kleene, S.: Representation of Events in Nerve Nets and Finite Automata. In: Shannon, C., McCarthy, J. (eds.) Automata Studies, pp. 3–42. Princeton University Press, Princeton (1956)
38. Kozen, D.: A Completeness Theorem for Kleene Algebras and the Algebra of Regular Events. Information and Computation 110, 366–390 (1994)
39. Kozen, D.: Kleene Algebra with Tests. Trans. Programming Languages and Systems 19, 427–443 (1997)
40. Luo, Z., Burstall, R.: A Set-theoretic Setting for Structuring Theories in Proof Development. University of Edinburgh, Laboratory for Foundations of Computer Science, Report ECS-LFCS-92-206 (1992)
41. Manes, E., Benson, D.: The Inverse Semigroup of a Sum-Ordered Semiring. Semigroup Forum 31, 129–152 (1985)
42. McIver, A., Cohen, E., Morgan, C.: Using Probabilistic Kleene Algebra for Protocol Verification. In: Schmidt, R. (ed.) RelMiCS/AKA 2006. LNCS, vol. 4136, pp. 296–310. Springer, Heidelberg (2006)
43. Meinicke, L., Solin, K.: Refinement Algebra for Probabilistic Programs. Electr. Notes Theor. Comput. Sci. 201, 177–195 (2008)
44. Möller, B.: Kleene Getting Lazy. Sci. Comput. Program. 65, 195–214 (2007)
45. Möller, B., Höfner, P., Struth, G.: Quantales and Temporal Logics. In: Johnson, M., Vene, V. (eds.) AMAST 2006. LNCS, vol. 4019, pp. 263–277. Springer, Heidelberg (2006)
46. Möller, B., Struth, G.: Algebras of Modal Operators and Partial Correctness. Theor. Comp. Sci. 351, 221–239 (2006)
47. Möller, B., Struth, G.: wp is wlp. In: MacCaull, W., Winter, M., Düntsch, I. (eds.) RelMiCS 2005. LNCS, vol. 3929, pp. 200–211. Springer, Heidelberg (2006)

48. Moszkowski, B.: A Complete Axiomatization of Interval Temporal Logic with Infinite Time. In: LICS 2000, pp. 241–252 (2000)
49. Mulvey, C.: &. Rendiconti del Circolo Matematico di Palermo 12, 99–104 (1986)
50. Paulson, L., Nipkow, T., Wenzel, M.: Isabelle,
    http://www.cl.cam.ac.uk/research/hvg/Isabelle/
51. Parikh, R.: Propositional Logics of Programs: New Directions. In: Karpinski, M. (ed.) FCT 1983. LNCS, vol. 158, pp. 347–359. Springer, Heidelberg (1983)
52. Pauly, M., Parikh, R.: Game Logic; An Overview. Studia Logica 75, 165–182 (2003)
53. Park, D.: On the Semantics of Fair Parallelism. In: Bjørner, D. (ed.) Abstract Software Specifications. LNCS, vol. 86, pp. 504–526. Springer, Heidelberg (1980)
54. Reynolds, J.: An Introduction to Separation Logic. In: Broy, M., Sitou, W., Hoare, T. (eds.) Engineering Methods and Tools for Software Safety and Security, pp. 285–310. IOS Press, Amsterdam (2009)
55. Rosenthal, K.: Quantales and their Applications. Pitman Research Notes in Math, vol. 234. Longman Scientific and Technical (1990)
56. Sannella, D., Burstall, R.M.: Structured Theories in LCF. In: Ausiello, G., Protasi, M. (eds.) CAAP 1983. LNCS, vol. 159, pp. 377–391. Springer, Heidelberg (1983)
57. Sintzoff, M.: Iterative Synthesis of Control Guards Ensuring Invariance and Inevitability in Discrete-Decision Games. In: Owe, O., Krogdahl, S., Lyche, T. (eds.) From Object-Orientation to Formal Methods. LNCS, vol. 2635, pp. 272–301. Springer, Heidelberg (2004)
58. Smith, D.: Automating the Design of Algorithms. In: Möller, B., Schuman, S., Partsch, H. (eds.) Formal Program Development. LNCS, vol. 755, pp. 324–354. Springer, Heidelberg (1993)
59. Solin, K., von Wright, J.: Enabledness and Termination in Refinement Algebra. Sci. Comput. Program. 74, 654–668 (2009)
60. SPECWARE, http://www.specware.org/
61. Srinivas, Y., Jüllig, R.: Specware: Formal Support for Composing Software. In: Möller, B. (ed.) MPC 1995. LNCS, vol. 947, pp. 399–422. Springer, Heidelberg (1995)
62. Struth, G.: Isabelle Repository for Relational and Algebraic Methods,
    http://staffwww.dcs.shef.ac.uk/people/G.Struth/isa/
63. Sutcliffe, G.: The TPTP Problem Library and Associated Infrastructure: The FOF and CNF Parts, v3.5.0. J. Autom. Reas. 43, 337–362 (2009)
64. Takai, T., Furusawa, H.: Monadic Tree Kleene Algebra. In: Schmidt, R. (ed.) RelMiCS/AKA 2006. LNCS, vol. 4136, pp. 402–416. Springer, Heidelberg (2006), http://www.sci.kagoshima-u.ac.jp/~furusawa/person/Papers/correct_monadic_kleene_algebra.pdf
65. von Karger, B.: Temporal Algebra. Mathematical Structures in Computer Science 8, 277–320 (1998)
66. von Wright, J.: Towards a Refinement Algebra. Sci. Comput. Program. 51, 23–45 (2004)
67. Wirsing, M.: Structured Algebraic Specifications: A Kernel Language. Theor. Comput. Sci. 42, 123–249 (1986)

Social Networks:
Prestige, Centrality, and Influence
(Invited Paper)

Agnieszka Rusinowska[1], Rudolf Berghammer[2], Harrie De Swart[3],
and Michel Grabisch[1]

[1] Centre d'Economie de la Sorbonne, Université Paris I Panthéon-Sorbonne
106-112 Bd de l'Hôpital, 75647 Paris Cedex 13, France
{agnieszka.rusinowska,michel.grabisch}@univ-paris1.fr
[2] Institut für Informatik, Universität Kiel, Olshausenstraße 40, 24098 Kiel, Germany
rub@informatik.uni-kiel.de
[3] Department of Philosophy, Erasmus University Rotterdam
P.O. Box 1738, 3000 DR Rotterdam, The Netherlands
deSwart@fwb.eur.nl

Abstract. We deliver a short overview of different centrality measures
and influence concepts in social networks, and present the relation-algebraic
approach to the concepts of power and influence. First, we briefly discuss
four kinds of measures of centrality: the ones based on degree, closeness, betweenness, and the eigenvector-related measures. We consider centrality of
a node and of a network. Moreover, we give a classification of the centrality
measures based on a topology of network flows. Furthermore, we present a
certain model of influence in a social network and discuss some applications
of relation algebra and RELVIEW to this model.

Keywords: social network, centrality, prestige, influence, relation algebra, RELVIEW.

1 Introduction

Social networks play a central role in our activities, in social phenomena, in
economic and political life. It is therefore crucial to provide an exhaustive analysis of social network structures and to study the impact they may have on
human's behavior. Many scholars are particularly interested in measures that
allow to compare networks. Also measures that compare nodes (representing
agents) within a network and show how a node relates to the network are of
interest. The question appears how central a node is and what its position and
prestige in a network are. The concept of centrality as applied to human communication was introduced already in the late 1940's, and since then many different
measures of centrality have been developed. They usually capture complementary aspects of a node's position, any hence a particular measure can be more
appropriate for some applications and less for others.

H. de Swart (Ed.): RAMICS 2011, LNCS 6663, pp. 22–39, 2011.

One of the aims of this paper is to deliver a brief overview of the main central-
ity measures. Four kinds of measures are presented: degree centrality, closeness
centrality, betweenness centrality, Katz prestige and Bonacich centrality. We
also briefly discuss a categorization of centrality measures based on a topology
of network flows.

Social networks are particularly important in studying all kinds of influence
phenomena. They are very useful for analyzing the diffusion of information and
the formation of opinions and beliefs. It is therefore not surprising that there
are numerous works in different scientific fields on the 'network approach' to
interaction and influence.

One of the leading dynamic models on information transmission, opinion and
consensus formation in networks is introduced by DeGroot [14]. Individuals start
with initial opinions on a subject and put some weights on the current beliefs
of other agents in forming their own beliefs for the next period. These beliefs
are updated over time. Several variations and generalizations of the DeGroot
model are presented e.g. in [15,20,21,22,36]. Surveys of models of influence and
different approaches to this phenomenon can be found e.g. in [27,29,36,38].

Another framework of influence in networks is introduced in [33]. In the original one-step model, agents have to make their acceptance-rejection decision on
a specific issue. Each agent has an inclination to say either 'yes' or 'no', but due
to possible influence of the other agents, his final decision ('yes' or 'no') may be
different from his initial inclination. This framework is extensively investigated
e.g. in [24,25,26,28,29,30,39].

Relation algebra is used very successfully for formal problem specification,
prototyping, and algorithm development. For details on relations and relational
algebra, see e.g. [13,16,17,40]. RELVIEW is a BDD-based tool for the visualization
and manipulation of relations and for prototyping and relational programming. It
has been developed at Kiel University. The tool is written in the C programming
language and makes full use of the X-windows graphical user interface. Details
and applications can be found e.g. in [3,4,9].

Several of our works are devoted to applications of relation algebra and REL-
VIEW to Game Theory and Social Choice Theory. In [5] we present such an
application to coalition formation, where with the help of relation algebra and
RELVIEW the set of all feasible stable governments is determined. A stable government is by definition not dominated by any other government. In [6] we deal
with the case where all governments are dominated. By using notions from relation algebra, graph theory and social choice theory, and by using RELVIEW we
can compute a government that is as close as possible to being non-dominated.
In [7] we apply relation algebra and RELVIEW to networks, i.e., to compute some
measures of agents' strength in a network, like power, success, and influence. In
[8] we present relation-algebraic models of simple games and develop relational
specifications for solving some basic game-theoretic problems. We test fundamental properties of simple games, compute specific players and coalitions, and
apply relation algebra to determine power indices.

In this paper we also aim at presenting a relation-algebraic approach to the concepts of influence in a social network. We recapitulate relation-algebraic specifications (presented in [7]) of the following concepts of the model of influence ([25,33,39]): the inclination and decision vectors, the group decision, the Hoede-Bakker index, the inclination vectors of potential and observed influence, and the set of followers.

The paper is structured as follows. In Section 2 the basic concepts in network theory are recalled. In Section 3 we discuss the main centrality measures. Section 4 concerns the model of influence in a social network. In Section 5 the relation-algebraic preliminaries are presented. Section 6 is devoted to the relation-algebraic approach to the concepts of influence. In Section 7 we present some concluding remarks.

2   The Basic Concepts in Network Theory

In this section we present the preliminaries on networks. For textbooks on network theory, see e.g. [23,36,44].

Let $N = \{1, 2, ..., n\}$ be a (finite) set of nodes. By $g_{ij} \in \{0, 1\}$ we denote a relationship between nodes i and j, where

$$g_{ij} = \begin{cases} 1 & \text{if there is a link between } i \text{ and } j \\ 0 & \text{otherwise.} \end{cases} \tag{1}$$

In what follows we only consider undirected links, i.e., we assume that $g_{ij} = g_{ji}$.

A *network* g is defined as a set of nodes N with links between them. Let $\mathcal{G}$ denote the collection of all possible networks on n nodes.

By $N_i(g)$ we denote the *neighborhood (the set of neighbors)* of node i in network g, i.e., the set of nodes with which node i has a link:

$$N_i(g) = \{j \in N : g_{ij} = 1\}. \tag{2}$$

The *degree* $d_i(g)$ of a node i in g is the number of i's neighbors in g, i.e.,

$$d_i(g) = |N_i(g)|. \tag{3}$$

A network g is said to be *regular* if every node has the same number of neighbors, i.e., if for some $d \in \{0, 1, ..., n-1\}$, $d_i(g) = d$ for each $i \in N$.

A *complete network* is a regular network with $d = n - 1$. The *empty network* is a regular network with $d = 0$.

One of the concerns when analyzing a network is to check how one node may be reached from another one. We distinguish between the following definitions:

- A *walk* is a sequence of nodes in which two nodes have a link (they are neighbors), and a node or a link may appear more than once. Its length is simply the number of links in the walk.
- A *trail* is a walk in which all links are distinct.
- A *path* is a trail in which all nodes are distinct.

- A *cycle* is a trail with at least 3 nodes in which the initial node and the end node are the same.
- A *geodesic* between two nodes is a shortest path between them.

If there is a path between i and j in g, then the *geodesic distance* $d(i, j; g)$ between these two nodes i and j is therefore equal to

$$d(i, j; g) = the\ number\ of\ links\ in\ a\ shortest\ path\ between\ i\ and\ j. \qquad (4)$$

If there is no path between i and j in g, we set $d(i, j; g) = \infty$.

A *star* is a network in which there exists some node i (referred to as the *center* of the star) such that every link in the network involves node i.

Two nodes belong to the same *component* if and only if there exists a path between them. A network is *connected* if there exists a path between any pair of nodes $i, j \in N$. Consequently, a network is connected if and only if it consists of a single component.

The *adjacency matrix* $\mathbf{G}$ of a (undirected or directed) network g is defined as $\mathbf{G} = [g_{ij}]$ with g_{ij} as in (1). In other words, an entry in the matrix $\mathbf{G}$ corresponding to the pair $\{i, j\}$ signifies the presence or absence of a link between i and j. Let $\mathbf{G}^k$ denote the kth power of $\mathbf{G}$, i.e., $\mathbf{G}^k = [g_{ij}^k]$, where g_{ij}^k measures the number of walks of length k that exist between i and j in network g. We have $\mathbf{G}^0 = \mathbf{I}$, where $\mathbf{I}$ is the $n \times n$ identity matrix.

3   Different Measures of Centrality in Networks

The concept of *centrality* captures a kind of prominence of a node in a network. The economic and sociological literature offers several such concepts. For surveys of different notions of centrality, see e.g. [19,23,36]. In this paper, we recapitulate several well-known centrality measures. The presentation is based on the three references mentioned above.

As presented in [36], measures of centrality can be categorized into the following main groups:

(1) Degree centrality
(2) Closeness centrality
(3) Betweenness centrality
(4) Prestige- and eigenvector-related centrality.

3.1   Degree Centrality

The degree centrality indicates how well a node is connected in terms of direct connections, i.e., it keeps track of the degree of the node. This measure can be seen as an index of the node's *communication activity*.

The *degree centrality* $C_d(i; g)$ *of node* i *in network* g is given by

$$C_d(i; g) = \frac{d_i(g)}{n - 1} = \frac{|N_i(g)|}{n - 1} \qquad (5)$$

where $N_i(g)$ and $d_i(g)$ are defined in (2) and (3). Obviously, $0 \leq C_d(i;g) \leq 1$.

Let i^* be a node which attains the highest degree centrality $C_d(i^*;g)$ in g. The *degree centrality* $C_d(g)$ *of network* g is given by

$$C_d(g) = \frac{\sum_{i=1}^{n} [C_d(i^*;g) - C_d(i;g)]}{\max_{g' \in \mathcal{G}} [\sum_{i=1}^{n} [C_d(i^*;g') - C_d(i;g')]]}. \tag{6}$$

Since the minimum degree is 1 and the maximum degree is $(n-1)$, one can easily see that the denominator of (6) is equal to $\frac{(n-2)(n-1)}{(n-1)}$, and hence

$$C_d(g) = \frac{\sum_{i=1}^{n} [C_d(i^*;g) - C_d(i;g)]}{n-2}.$$

Note that $C_d(g) = 1$ if g is a star, and $C_d(g) = 0$ if g is a regular network.

3.2 Closeness Centrality

The *closeness centrality* is based on proximity and measures how easily a node can reach other nodes in a network. It is a kind of a measure of the node's *independence* or *efficiency*.

The *closeness centrality* $C_c(i;g)$ of node i in network g is defined as

$$C_c(i;g) = \frac{n-1}{\sum_{j \neq i} d(i,j;g)} \tag{7}$$

where $d(i,j;g)$ is the geodesic distance between i and j as defined in (4), and $(n-1)$ is the minimum possible total distance from i to all other nodes in g. There is a whole family of closeness measures [44] based on different conventions for dealing with non-connected networks and other possible measures of distance.

Let i^* be a node which attains the highest closeness centrality $C_c(i^*;g)$ in g. The *closeness centrality* $C_c(g)$ *of network* g is given by

$$C_c(g) = \frac{\sum_{i=1}^{n} [C_c(i^*;g) - C_c(i;g)]}{\max_{g' \in \mathcal{G}} [\sum_{i=1}^{n} [C_c(i^*;g') - C_c(i;g')]]}. \tag{8}$$

One can show (see e.g. [19]) that

$$C_c(g) = \frac{\sum_{i=1}^{n} [C_c(i^*;g) - C_c(i;g)]}{(n-2)(n-1)/(2n-3)}.$$

Note that $C_c(g) = 1$ if g is a star, and $C_c(g) = 0$ if g is a cycle. Obviously, although $C_d(g) = C_c(g)$ for g being a star or a cycle, in general $C_d(g) \neq C_c(g)$.

3.3 Betweenness Centrality

The betweenness centrality (introduced in [18]) is based on how important a node is in terms of connecting other nodes. It is useful as an index of the potential of a node for *control of communication*.

By $P_i(kj)$ and $P(kj)$ we denote the number of geodesics between k and j containing $i \notin \{k, j\}$, and the total number of geodesics between k and j, respectively.

The *betweenness centrality* $C_b(i; g)$ *of node* i *in network* g is defined as

$$C_b(i; g) = \frac{2}{(n-1)(n-2)} \sum_{k \neq j: i \notin \{k,j\}} \frac{P_i(kj)}{P(kj)}. \tag{9}$$

Note that $\frac{P_i(kj)}{P(kj)}$ is the probability that i falls on a randomly selected geodesic linking k and j, and the number of all pairs of nodes (different from i) is equal to $\binom{n-1}{2} = \frac{(n-1)(n-2)}{2}$. In particular, if g is a star, then $C_b(i; g) = 1$ for i being the center and $C_b(i; g) = 0$ otherwise.

Let i^* be a node which attains the highest betweenness centrality $C_b(i^*; g)$ in g. The *betweenness centrality* $C_b(g)$ *of network* g is given by

$$C_b(g) = \frac{\sum_{i=1}^{n} [C_b(i^*; g) - C_b(i; g)]}{n - 1}. \tag{10}$$

3.4 Prestige- and Eigenvector-Related Centrality Measures

There exist other measures of centrality that take into account a richer range of direct and indirect influences in networks. The measures developed e.g. in [10,11,37] are based on the idea that a node's importance is determined by the importance of its neighbors.

The *Katz prestige* $P_i^K(g)$ *of node* i *in* g is defined as

$$P_i^K(g) = \sum_{j \neq i} g_{ij} \frac{P_j^K(g)}{d_j(g)}. \tag{11}$$

This means that the Katz prestige of i is equal to the sum of the prestiges of i's neighbors divided by their respective degrees. In other words, the measure is corrected by the number of neighbors of node j (if j has more relationships, then i gets less prestige from being connected to j). Note that this definition is self-referential. (11) can be rewritten as

$$\mathbf{P}^K(g) = \mathbf{G}'\mathbf{P}^K(g)$$

$$(\mathbf{I} - \mathbf{G}')\mathbf{P}^K(g) = \mathbf{0}$$

where $\mathbf{P}^K(g)$ is the $n \times 1$ vector of $P_i^K(g)$, $i \in N$, $\mathbf{I}$ is the $n \times n$ identity matrix, and $\mathbf{G}' = [g_{ij}']$ is the normalized adjacency matrix with $g_{ij}' = \frac{g_{ij}}{d_j(g)}$. In other words, calculating the Katz prestige is reduced to finding the unit eigenvector of $\mathbf{G}'$. Obviously, $\mathbf{P}^K(g)$ is determined up to a scale factor.

Katz [37] introduced another measure of prestige, where the prestige of a node is a weighted sum of the walks that emanate from it, and a walk of length k is worth a^k, for some parameter $0 < a < 1$. The *second prestige measure of Katz* is given by

$$\mathbf{P}^{K2}(g, a) = (\mathbf{I} - a\mathbf{G})^{-1}a\mathbf{G}\mathbf{1} \tag{12}$$

where $\mathbf{1}$ is the $n \times 1$ vector of 1s, and a is sufficiently small.

The *Bonacich centrality* is an extension of the second prestige measure of Katz and is expressed by

$$\mathbf{C}^{B}(g, a, b) = (\mathbf{I} - b\mathbf{G})^{-1}a\mathbf{G}\mathbf{1} \tag{13}$$

where $a > 0$ and $b > 0$ are scalars, and b is sufficiently small.

3.5   Categorizing Centrality Measures by a Topology of Network Flows

The relation between the major centrality measures and different flow processes is extensively discussed in [12]. Centrality measures make implicit assumptions about network flow, and hence they are matched to the kinds of flows they are appropriate for.

The typology of network flows is based on two dimensions:

- the trajectory dimension - kinds of trajectories that traffic may follow: geodesics, paths, trails, walks;
- the transmission dimension - methods of spread: parallel (simultaneous) duplication, serial (once at a time) duplication, transfer.

Table 1 classifies different kinds of traffic based on these two dimensions.

Table 1. Topology of flow processes (see [12])

	parallel duplication	serial duplication	transfer
geodesics	-	mitotic reproduction	package delivery
paths	internet name-server	viral infection	mooch
trails	e-mail broadcast	gossip	used goods
walks	attitude influencing	emotional support	money exchange

Table 2 classifies the major centrality measures presented above, based on flow processes.

Since each centrality measure is appropriate for particular kinds of flows, applying these measures to other flow processes that they are not designed for leads to wrong results. For example, one can use the closeness and betweenness centrality measures for package delivery, but it is inappropriate to use them to indicate who will receive news early in a gossip. For a discussion on this classification, see [12].

Table 2. Flow processes and major centrality measures (see [12])

	parallel duplication	serial duplication	transfer
geodesics		closeness	closeness
			betweenness
paths	closeness, degree		
trails	closeness, degree		
walks	closeness, degree		
	Bonacich eigenvector		
	Katz prestige		

4 The Model of Influence in a Social Network

In this section we present a framework of influence originally introduced in [33] and refined in [25,39].

4.1 The Hoede-Bakker Index

We consider a social network with a set of agents (players, actors, voters) denoted by $N := \{1, 2, ..., n\}$ who are to make a certain acceptance-rejection decision on a specific proposal. Each agent $k \in N$ has an *inclination* i_k either to say 'yes' (denoted by $+1$) or 'no' (denoted by -1). Let $i = (i_1, i_2, ..., i_n)$ denote an *inclination vector* and $I := \{-1, +1\}^n$ be the set of all inclination vectors.

It is assumed that agents may influence each other, and due to the influences, the final decision of an agent may be different from his original inclination. Formally, each inclination vector $i \in I$ is transformed into a *decision vector* $Bi = ((Bi)_1, (Bi)_2, ..., (Bi)_n)$, where $B : I \rightarrow I, i \mapsto Bi$ is the *influence function*. Let $B(I)$ be the set of all decision vectors under B and let $\mathcal{B}$ denote the set of all influence functions.

We also assume a *group decision function* $gd : B(I) \rightarrow \{-1, +1\}$, having the value $+1$ if the group decision is 'yes', and the value -1 if the group decision is 'no'. The set of all group decision functions will be denoted by $\mathcal{G}$.

In [39] we introduce the following generalized index. Given $B \in \mathcal{B}$ and $gd \in \mathcal{G}$, the *generalized Hoede-Bakker index of player* $k \in N$ is defined as

$$\text{GHB}_k(B, gd) := \frac{|I_k^{++}| - |I_k^{+-}| + |I_k^{--}| - |I_k^{-+}|}{2^n} \tag{14}$$

where

$$I_k^{++} := \{i \in I \mid i_k = +1 \ \wedge \ gd(Bi) = +1\}$$
$$I_k^{+-} := \{i \in I \mid i_k = +1 \ \wedge \ gd(Bi) = -1\}$$
$$I_k^{--} := \{i \in I \mid i_k = -1 \ \wedge \ gd(Bi) = -1\}$$
$$I_k^{-+} := \{i \in I \mid i_k = -1 \ \wedge \ gd(Bi) = +1\}.$$

Obviously all the four sets depend on (B, gd), which has been skipped for convenience of notation.

Note that the generalized Hoede-Bakker index, although defined in the influence setup, does not measure any influence. As remarked in [39] the GHB index is a kind of 'net Success', i.e., 'Success - Failure'.

4.2   The Influence Indices

Measures of influence, the so called *influence indices*, are defined in [25]. Below we recall these definitions.

Concerning notation, for convenience we omit braces for sets, e.g., $N \setminus \{j\}$ is written as $N \setminus j$. For any $S \subseteq N$, $|S| \geq 2$, we introduce the set I_S of all inclination vectors in which all members of S have the same inclination

$$I_S := \{i \in I \mid \forall k, j \in S \ [i_k = i_j]\} \tag{15}$$

and $I_k := I$, for any $k \in N$. For $i \in I_S$ we denote by i_S the value i_k for some $k \in S$. Let for each $S \subseteq N$ and $j \in N \setminus S$, $I_{S \to j}$ denote the set of all inclination vectors of *potential influence of S on j*, that is,

$$I_{S \to j} := \{i \in I_S \mid i_j = -i_S\}. \tag{16}$$

Moreover, for each $B \in \mathcal{B}$, let $I^*_{S \to j}(B)$ denote the set of all inclination vectors of *observed influence of S on j under $B \in \mathcal{B}$*, that is,

$$I^*_{S \to j}(B) := \{i \in I_{S \to j} \mid (Bi)_j = i_S\}. \tag{17}$$

In [25] we introduce the *weighted influence indices*, whose main idea is to give a relative importance to the different inclination vectors. For each $S \subseteq N$, $j \in N \setminus S$ and $i \in I_S$, we introduce a *weight* $\alpha_i^{S \to j} \in [0, 1]$ *of influence of coalition S on $j \in N \setminus S$ under the inclination vector $i \in I_S$*. There is no normalization on the weights, but we assume that for each $S \subseteq N$ and $j \in N \setminus S$, there exists $i \in I_{S \to j}$ such that $\alpha_i^{S \to j} > 0$.

Given $B \in \mathcal{B}$, for each $S \subseteq N$, $j \in N \setminus S$, the *weighted influence index* of coalition S on player j is defined as

$$d_\alpha(B, S \to j) := \frac{\sum_{i \in I^*_{S \to j}(B)} \alpha_i^{S \to j}}{\sum_{i \in I_{S \to j}} \alpha_i^{S \to j}} \in [0, 1]. \tag{18}$$

It is the (weighted) proportion of situations of observed influence among all situations of potential influence. Two particular ways of weighting lead to the *possibility influence index* $\overline{d}(B, S \to j)$ and the *certainty influence index* $\underline{d}(B, S \to j)$. We have for each $S \subseteq N$, $j \in N \setminus S$ and $B \in \mathcal{B}$

$$\overline{d}(B, S \to j) = d_{\overline{\alpha}}(B, S \to j), \text{ where } \overline{\alpha}_i^{S \to j} = 1 \text{ for each } i \in I_S$$

and

$$\underline{d}(B, S \to j) = d_{\underline{\alpha}}(B, S \to j), \text{ where for each } i \in I_S$$

$$\underline{\alpha}_i^{S \to j} = \begin{cases} 1, & \text{if } \forall p \notin S \cup j, i_p = -i_S \\ 0, & \text{otherwise.} \end{cases}$$

Consequently, we have

$$\overline{d}(B, S \to j) = \frac{|I_{S \to j}^*(B)|}{|I_{S \to j}|} \in [0, 1] \qquad (19)$$

$$\underline{d}(B, S \to j) = \frac{|\{i \in I_{S \to j}^*(B) \mid \forall p \notin S \ [i_p = -i_S]\}|}{2} \in \{0, \frac{1}{2}, 1\}. \qquad (20)$$

The possibility influence index gives therefore the fraction of potential influence situations that happen to be situations of observed influence indeed. The certainty influence index measures also such a fraction, except that it focuses only on situations in which the coalition in question is the only one which influences the agent.

4.3 Followers and Kernel

The key concept of the influence framework is the concept of *follower* of a given coalition, that is, an agent who always follows the inclination of that coalition when all members of the coalition have the same inclination. The *follower function* of $B \in \mathcal{B}$ is a mapping $F_B : 2^N \to 2^N$ defined as

$$F_B(S) := \{k \in N \mid \forall i \in I_S, (Bi)_k = i_S\}, \quad \forall S \subseteq N, S \neq \emptyset \qquad (21)$$

and $F_B(\emptyset) := \emptyset$. We say that $F_B(S)$ is the *set of followers of S under B*. The set of all follower functions is denoted by $\mathcal{F}$. In [25] it is shown that

$$d_\alpha(B, S \to j) = 1, \quad \forall j \in F_B(S) \setminus S.$$

Another important concept of the influence model is the concept of *kernel* of an influence function, which is the set of 'truly' influential coalitions. Assume F_B is not identical to the empty set. The *kernel* of B is defined as

$$\mathcal{K}(B) := \{S \in 2^N \mid F_B(S) \neq \emptyset, \text{ and } S' \subset S \Rightarrow F_B(S') = \emptyset\}. \qquad (22)$$

In [25] we also define some specific influence functions and study their properties, e.g., the sets of followers and kernels of these functions.

4.4 Further Research on Influence

The model of influence presented above, i.e., the model of initial inclinations and final decisions, is studied extensively in several other works:

- In [26] we generalize the basic yes-no model of influence to a framework in which every agent has a totally ordered set of possible actions, the same for each player, and he has an inclination to choose a particular action. We investigate the generalized influence indices, different influence functions, and other tools related to the influence in the multi-choice model.

- In [28] we consider the influence model with a continuum of actions. In this generalized framework we introduce and study measures of positive and negative influence and other tools for analyzing influence. Also the set of fixed points under a given influence function is analyzed. Furthermore, we study linear influence functions.
- The results presented in [24] concern a comparison of the influence model with the framework of command games [34,35]. We show that the framework of influence is more general than the framework of the command games. In particular, we define several influence functions which capture the command structure. For some influence functions we define the equivalent command games.
- In [30] we establish the exact relations between the key concepts of the influence model and the framework of command games. We deliver sufficient and necessary conditions for a function to be a follower function, and describe the structure of the set of all influence functions that lead to a given follower function. We also deliver sufficient and necessary conditions for a function to be a command function, and describe the minimal sets generating a normal command game. In addition, we study the relation between command games and influence functions.
- We also study the dynamics of influence. In [29] the yes-no model with a single step of mutual influence is generalized to a framework with iterated influence. We analyze the decision process in which the mutual influence does not stop after one step but iterates, and we study the convergence of an influence function. In particular, we investigate stochastic influence functions and apply the theory of Markov chains to the analysis of such functions. Moreover, we propose a general framework of influence based on aggregation functions.

5   Relation-Algebraic Preliminaries

In this section we present the basics of relation algebra.

If X and Y are sets, then a subset R of the Cartesian product $X \times Y$ is called a (binary) relation with *domain* X and *range* Y. We denote the set (also called type) of all relations with domain X and range Y by $[X \leftrightarrow Y]$ and write $R : X \leftrightarrow Y$ instead of $R \in [X \leftrightarrow Y]$. If X and Y are finite sets of size m and n respectively, then we may consider a relation $R : X \leftrightarrow Y$ as a Boolean matrix with m rows and n columns and entries from $\{0, 1\}$. The Boolean matrix interpretation of relations is used as one of the graphical representations of relations within the RELVIEW tool. We can speak about rows, columns and entries of a relation and write $R_{x,y}$ instead of $\langle x, y \rangle \in R$ or $x\,R\,y$.

The basic operations on relations are R^{T} (*transposition, conversion*), $\overline{R}$ (*complement, negation*), $R \cup S$ (*union, join*), $R \cap S$ (*intersection, meet*), RS (*composition, multiplication*), and the special relations O (*empty relation*), L (*universal relation*), and I (*identity relation*). If R is included in S we write $R \subseteq S$, and equality of R and S is denoted as $R = S$.

A *membership relation* $\mathsf{E} : X \leftrightarrow 2^X$ relates $x \in X$ and $Y \in 2^X$ iff $x \in Y$.

The expression $\mathrm{syq}(R, S) := \overline{R^\mathsf{T} \overline{S}} \cap \overline{\overline{R}^\mathsf{T} S}$ is by definition the *symmetric quotient* $\mathrm{syq}(R, S) : Y \leftrightarrow Z$ of two relations $R : X \leftrightarrow Y$ and $S : X \leftrightarrow Z$. Many properties of this construct can be found e.g. in [40]. In particular, for all $y \in Y$ and $z \in Z$ the relationship $\mathrm{syq}(R, S)_{y,z}$ holds iff for all $x \in X$ the equivalence $R_{x,y} \leftrightarrow S_{x,z}$ is true, i.e., if the y-column of R and the z-column of S coincide.

Given a Cartesian product $X \times Y$ of two sets X and Y, there are two projection functions which decompose a pair $u = (u_1, u_2)$ into its first component u_1 and its second component u_2. For a relation-algebraic approach it is useful to consider the corresponding *projection relations* $\pi : X \times Y \leftrightarrow X$ and $\rho : X \times Y \leftrightarrow Y$ such that for all pairs $u \in X \times Y$ and elements $x \in X$ and $y \in Y$ we have $\pi_{u,x}$ iff $u_1 = x$ and $\rho_{u,y}$ iff $u_2 = y$.

Projection relations enable us to describe the well-known pairing operation of functional programming relation-algebraically as follows: For relations $R : Z \leftrightarrow X$ and $S : Z \leftrightarrow Y$ we define their *pairing* (frequently also called *fork* or *tupling*) $[R, S] : Z \leftrightarrow X \times Y$ by $[R, S] := R\pi^\mathsf{T} \cap S\rho^\mathsf{T}$. Then for all $z \in Z$ and pairs $u = (u_1, u_2) \in X \times Y$ a simple reflection shows that $[R, S]_{z,u}$ iff R_{z,u_1} and S_{z,u_2}.

Column vectors are relations v with $v = v\mathsf{L}$. As for a column vector the range is irrelevant, we consider only vectors $v : X \leftrightarrow \mathbf{1}$ with a specific singleton set $\mathbf{1} := \{\bot\}$ as range. A column vector $v : X \leftrightarrow \mathbf{1}$ can be considered as a Boolean matrix with exactly one column, i.e., as a Boolean column vector, and it describes the subset $\{x \in X \mid v_{x,\bot}\}$ of its domain X. If $v : X \leftrightarrow \mathbf{1}$ describes the subset S of X in the sense above, then the injective mapping $inj(v) : S \leftrightarrow X$ is obtained from the identity relation $\mathsf{I} : X \leftrightarrow X$ by removing all rows which correspond to a 0-entry in v. Hence, we have $inj(v)_{j,k}$ iff $j = k$.

A non-empty column vector v is a *column point* if $vv^\mathsf{T} \subseteq \mathsf{I}$, i.e., it is injective in the relational sense. In the Boolean matrix model, a column point $v : X \leftrightarrow \mathbf{1}$ is a Boolean column vector in which exactly one entry is 1.

Vectors also allow to formalize the notions of y-columns and x-rows. For a relation $R : X \leftrightarrow Y$ and $y \in Y$, the column vector $v : X \leftrightarrow \mathbf{1}$ equals the y-column of R if for all $x \in X$ we have $v_{x,\bot}$ iff $R_{x,y}$.

Row vectors are relations defined as the transposes of column vectors. We only need row vectors v of the specific type $[\mathbf{1} \leftrightarrow Y]$ that correspond to Boolean row vectors. Then v describes the subset $\{y \in Y \mid v_{\bot,y}\}$ of its range Y.

If $v : 2^M \leftrightarrow \mathbf{1}$ represents the subset $\mathcal{S}$ of 2^M and the size of the domain of $w : W \leftrightarrow \mathbf{1}$ is at most $|M| + 1$, then for all $X \in 2^M$ we have $cardfilter(v, w)_{X,\bot}$ iff $X \in \mathcal{S}$ and $|X| < |W|$. Hence, the complement of $cardfilter(\mathsf{L}, w)$ represents the subset of 2^M whose elements have at least size $|W|$.

6   Applying Relation Algebra to the Model of Influence

In this section we deal with the relation-algebraic approach to the model of influence in a social network. We recall some selected results presented in [7].

6.1   Modeling the Inclination and Decision Vectors

For modeling inclination vectors and decision vectors, we use *column vectors*. For modeling subsets of the sets I and $B(I)$, we use *row vectors*.

We assume a social network with a set N of players. Let $D : N \leftrightarrow N$ be the relation of the dependency graph of the network. This means that there is an arc from an agent $j \in N$ to an agent $k \in N$ iff $D_{j,k}$ holds. Then the set of the dependent agents is described relation-algebraically by the column vector

$$depend(D) := D^{\mathsf{T}}\mathsf{L} \tag{23}$$

of type $[N \leftrightarrow \mathbf{1}]$, where L has type $[N \leftrightarrow \mathbf{1}]$ as well.

The set I of all inclination vectors can immediately be modeled by the columns of the membership relation $\mathsf{E} : N \leftrightarrow 2^N$. Hence, we regard inclination vectors and the corresponding decision vectors as relational column vectors $i : N \leftrightarrow \mathbf{1}$ and $Bi : N \leftrightarrow \mathbf{1}$, respectively.

We develop a column-wise enumeration of the set $B(I)$ of decision vectors with relation-algebraic means. The influence function B is given by the rule 'following only unanimous trend-setters', which means that an agent follows his trend setters only if they all have the same inclination. In [7] we prove that:

Theorem 6.1. *For each inclination vector $i : N \leftrightarrow \mathbf{1}$, the decision vector $Bi : N \leftrightarrow \mathbf{1}$ under the rule 'following only unanimous trend-setters' is given by*

$$Bi = (i \cap (\overline{d} \cup (d \cap D^{\mathsf{T}}i \cap D^{\mathsf{T}}\overline{i}))) \cup (d \cap \overline{D^{\mathsf{T}}\overline{i}}),$$

where $d := depend(D)$.

The relation-algebraic expression $(i \cap (\overline{d} \cup (d \cap D^{\mathsf{T}}i \cap D^{\mathsf{T}}\overline{i}))) \cup (d \cap \overline{D^{\mathsf{T}}\overline{i}})$ is built from i using unions, intersections, complements and left-compositions with constants only. If we replace the column vector $i : N \leftrightarrow \mathbf{1}$ by the membership relation $\mathsf{E} : N \leftrightarrow 2^N$ that column-wisely enumerates all inclination vectors and adapt simultaneously the type $[N \leftrightarrow \mathbf{1}]$ of d to the type $[N \leftrightarrow 2^N]$ of E by a right-composition with the universal row vector $\mathsf{L} : \mathbf{1} \leftrightarrow 2^N$, we get the relation

$$Dvec(D) := (\mathsf{E} \cap (\overline{d\mathsf{L}} \cup (d\mathsf{L} \cap D^{\mathsf{T}}\mathsf{E} \cap D^{\mathsf{T}}\overline{\mathsf{E}}))) \cup (d\mathsf{L} \cap \overline{D^{\mathsf{T}}\overline{\mathsf{E}}}) \tag{24}$$

of type $[N \leftrightarrow 2^N]$ that column-wisely enumerates the set $B(I)$ of decision vectors.

6.2   Computing the Group Decisions

Next, we deliver a relation-algebraic specification of the group decisions under majority as decision rule via a row vector.

We assume that a row vector $m : \mathbf{1} \leftrightarrow 2^N$ is available such that for all $X \in 2^N$ we have $m_{\perp,X}$ iff $|X| \geq \lfloor \frac{|N|}{2} \rfloor + 1$. In RELVIEW such a vector can be easily obtained with the help of the base operation *cardfilter* as

$$m := \overline{\overline{cardfilter(\mathsf{L}, w)}}^{\mathsf{T}}, \tag{25}$$

where the first argument $L : 2^N \leftrightarrow 1$ describes the entire powerset 2^N, and the second argument $w : W \leftrightarrow 1$ determines the threshold for majority by its length, i.e., fulfills $|W| = [\frac{|N|}{2}] + 1$. In [7] we show the following result:

Theorem 6.2. *Let, based on the specifications (24) and (25), the row vector $gdv(D)$ of type $[1 \leftrightarrow 2^N]$ be defined by*

$$gdv(D) := m\, syq(\mathsf{E}, Dvec(D)),$$

where $\mathsf{E} : N \leftrightarrow 2^N$ is the membership relation. Then we have for all $X \in 2^N$: If the decision vector $Bi : N \leftrightarrow 1$ equals the X-column of $Dvec(D)$, then $gdv(D)_{\perp,X}$ holds iff the number of 1-entries in Bi is at least $[\frac{|N|}{2}] + 1$.

6.3 Computing the Hoede-Bakker Index

We assume that the player $k \in N$, on which the sets I_k^{++}, I_k^{+-}, I_k^{-+} and I_k^{--} depend, is described by a column point $p : N \leftrightarrow 1$ in the relational sense. As the definitions of the sets use the values $gd(Bi)$ for $i \in I$, we assume that the group decision row vector $g := gdv(D)$ is at hand. In [7] we prove the following:

Theorem 6.3. *Let, depending on the column point $p : N \leftrightarrow 1$ and the row vector $g : 1 \leftrightarrow 2^N$, the four vectors $ipp(p,g)$, $ipm(p,g)$, $imp(p,g)$ and $imm(p,g)$ of type $[1 \leftrightarrow 2^N]$ be defined as follows, where $\mathsf{E} : N \leftrightarrow 2^N$ is the membership relation:*

$$ipp(p,g) := p^\mathsf{T}\mathsf{E} \cap g \qquad ipm(p,g) := p^\mathsf{T}\mathsf{E} \cap \overline{g}$$
$$imp(p,g) := p^\mathsf{T}\overline{\mathsf{E}} \cap g \qquad imm(p,g) := p^\mathsf{T}\overline{\mathsf{E}} \cap \overline{g}$$

Then we have for all $X \in 2^N$: If the X-column of E equals the inclination vector $i : N \leftrightarrow 1$, then we have that $ipp(p,g)_{\perp,X}$ holds iff $i \in I_k^{++}$, $ipm(p,g)_{\perp,X}$ holds iff $i \in I_k^{+-}$, $imp(p,g)_{\perp,X}$ holds iff $i \in I_k^{-+}$, and $imm(p,g)_{\perp,X}$ holds iff $i \in I_k^{--}$.

In other words, the row vector $ipp(p,g)$ precisely designates those columns of the membership relation E which belong to the set I_k^{++}, and the remaining three row vectors do the same for the sets I_k^{+-}, I_k^{-+} and I_k^{--}, respectively.

6.4 Computing the Influence Indices

We assume a coalition S of agents to be described by a column vector $s : N \leftrightarrow 1$, and an agent $j \in N$ to be described by a column point $p : N \leftrightarrow 1$. We compute the possibility influence index of S on j. Since it is defined by means of the sizes of the sets $I_{S \to j}$ and $I_{S \to j}^*(B)$, we need to describe these sets within relation algebra. $I_{S \to j}$ and $I_{S \to j}^*(B)$ are subsets of I_S. In [7] the following is shown:

Theorem 6.4. *Assume $s : N \leftrightarrow 1$ to be a description of the coalition $S \subseteq N$ and the row vector $is(s)$ of type $[1 \leftrightarrow 2^N]$ to be defined as*

$$is(s) := [s^\mathsf{T}, s^\mathsf{T}]\, \overline{(\overline{\pi\mathsf{E}} \cup \rho\mathsf{E}) \cap (\overline{\rho\mathsf{E}} \cup \pi\mathsf{E})},$$

where $\mathsf{E} : N \leftrightarrow 2^N$ is the membership relation, and $\pi : N \times N \leftrightarrow N$ and $\rho : N \times N \leftrightarrow N$ are the projection relations. Then we have for all $X \in 2^N$: If the X-column of E equals the inclination vector $i : N \leftrightarrow \mathbf{1}$, then $is(s)_{\bot,X}$ holds iff $i \in I_S$.

Hence, the row vector $is(s)$ precisely designates those columns of the membership relation E which belong to the set I_S. Next, we deliver the relation-algebraic specification of the set $I_{S \to j}$, where $j \in N$ is described by the column point $p : N \leftrightarrow \mathbf{1}$. In [7] we prove the following theorem:

Theorem 6.5. *Assume* $s : N \leftrightarrow \mathbf{1}$ *describes the coalition* $S \subseteq N$, *the column point* $p : N \leftrightarrow \mathbf{1}$ *describes agent* $j \in N$, *the column point* $q \subseteq s$ *describes agent* $k \in S$, *and the row vector* $potinf(s,p)$ *of type* $[\mathbf{1} \leftrightarrow 2^N]$ *is defined by*

$$potinf(s,p) := ((r \cup r') \cap \overline{r \cap r'}) \, inj(is(s)^\mathsf{T}),$$

where $r := p^\mathsf{T}\mathsf{E}\, inj(is(s)^\mathsf{T})^\mathsf{T}$ *and* $r' := q^\mathsf{T}\mathsf{E}\, inj(is(s)^\mathsf{T})^\mathsf{T}$ *with* $\mathsf{E} : N \leftrightarrow 2^N$ *as membership relation. Then we have for all* $X \in 2^N$: *If the* X-column of E *equals the inclination vector* $i : N \leftrightarrow \mathbf{1}$, *then* $potinf(s,p)_{\bot,X}$ *holds iff* $i \in I_{S \to j}$.

Hence, we relation-algebraically specify a row vector that precisely designates those columns of E which are inclination vectors of potential influence of S on j.

To obtain a row vector $inf(s,p,D)$ of type $[\mathbf{1} \leftrightarrow 2^N]$ that precisely designates those columns of the membership relation $\mathsf{E} : N \leftrightarrow 2^N$ which are inclination vectors of influence of S on j, i.e., members of $I^*_{S \to j}(B)$, we use the equation

$$I^*_{S \to j}(B) = I_{S \to j} \cap \{i \in I_S \mid (Bi)_j = i_S\}.$$

The relation-algebraic specification of $I^*_{S \to j}(B)$ is given by the row vector

$$inf(s,p,D) := potinf(s,p) \cap \overline{(r \cup r') \cap \overline{r \cap r'}} \; inj(is(s)^\mathsf{T}) \qquad (26)$$

with r and r' given by $r := p^\mathsf{T} Dvec(D)\, inj(is(s)^\mathsf{T})^\mathsf{T}$ and $r' := q^\mathsf{T}\mathsf{E}\, inj(is(s)^\mathsf{T})^\mathsf{T}$.

6.5   Computing the Sets of Followers

For modeling sets of followers we use *column vectors*. The relations R and Q column-wisely enumerate I_S and $B(I_S)$, respectively, and the column point q is used for specifying for $i \in I_S$ the specific Boolean value i_S. In [7] we show that:

Theorem 6.6. *Assume* $s : N \leftrightarrow \mathbf{1}$ *to describe the coalition* $S \subseteq N$, *and the column point* $q \subseteq s$ *to describe some player* $k \in S$. *Furthermore, let* $\mathsf{E} : N \leftrightarrow 2^N$ *be the membership relation. If the column vector* $follow(D,s)$ *of type* $[N \leftrightarrow \mathbf{1}]$ *is defined as*

$$follow(D,s) := syq(Q^\mathsf{T}, R^\mathsf{T} q)$$

with relations $R := \mathsf{E}\, inj(is(s)^\mathsf{T})^\mathsf{T}$ *and* $Q := Dvec(D)\, inj(is(s)^\mathsf{T})^\mathsf{T}$, *then for all* $j \in N$ *we have* $follow(D,s)_{j,\bot}$ *iff* $j \in F_B(S)$.

7   Concluding Remarks

We have presented different measures of centrality that capture complementary aspects of a node's position in a network. As remarked in [19], the measures based on degree, closeness, and betweenness imply different "theories" of how centrality might affect group processes: centrality as activity, as independence, and as control. Despite this fact, all centrality measures should have some features in common, e.g., they should rank highest the most central node. As concluded in [19] all the three measures of network centrality agree in assigning the maximum centrality score to the star, and the minimum centrality score to a cycle and complete networks. Between these extremes, the three measures of network centrality may differ significantly in their rankings of networks. In a given application, one centrality measure or a combination of some measures might be more appropriate than another measure or a combination of measures.

Many centrality measures have not been discussed in this paper. A very interesting work is e.g., [1], where the intercentrality of a node in a network is investigated. Roughly speaking, it is the sum of the node's Bonacich centrality and its contribution to Bonacich centrality of other nodes. Apart from several sociological contributions to measuring centrality in social networks, also a game theoretic approach to centrality concepts is presented in the literature. For example, in [31] the authors propose a new definition of degree of centrality based on some extension of the Banzhaf index [2]. Also many works by Van den Brink and his co-authors deliver game theoretic measures of centrality in networks; see e.g. [32,41,42,43].

Despite the existence of numerous centrality measures, as remarked in [12] most of the sociologically interesting processes are not covered by the major measures. For instance, there are no measures appropriate for infection and gossip processes. It seems therefore important to investigate centrality measures that could fill that gap.

It has been proved by numerous works (see e.g. [5,6,7,8]) that the relation-algebraic approach to game theoretic problems is very appropriate and useful. There are still many more possibilities for combining relation algebra and REL-VIEW to investigate and solve problems from Game Theory and Social Choice Theory. One of them might be an application of the tools in question to some centrality measures.

References

1. Ballester, C., Calvo-Armengol, A., Zenou, Y.: Who's Who in Networks. Wanted: The Key Player. Econometrica 74, 1403–1417 (2006)
2. Banzhaf, J.: Weighted Voting Doesn't Work: A Mathematical Analysis. Rutgers Law Review 19, 317–343 (1965)
3. Behnke, R., Berghammer, R., Meyer, E., Schneider, P.: RELVIEW - A system for calculating with relations and relational programming. In: Astesiano, E. (ed.) ETAPS 1998 and FASE 1998. LNCS, vol. 1382, pp. 318–321. Springer, Heidelberg (1998)

4. Berghammer, R., Karger, B., von Ulke, C.: Relation-algebraic Analysis of Petri Nets with RELVIEW. In: Margaria, T., Steffen, B. (eds.) TACAS 1996. LNCS, vol. 1055, pp. 49–69. Springer, Heidelberg (1996)
5. Berghammer, R., Rusinowska, A., De Swart, H.: Applying Relational Algebra and RelView to Coalition Formation. EJOR 178/2, 530–542 (2007)
6. Berghammer, R., Rusinowska, A., De Swart, H.: An Interdisciplinary Approach to Coalition Formation. EJOR 195, 487–496 (2009)
7. Berghammer, R., Rusinowska, A., De Swart, H.: Applying Relational Algebra and RelView to Measures in a Social Network. EJOR 202, 182–195 (2010)
8. Berghammer, R., Bolus, S., Rusinowska, A., De Swart, H.: A Relation-algebraic Approach to Simple Games. EJOR 210, 68–80 (2011)
9. Berghammer, R., Schmidt, G., Winter, M.: RELVIEW and RATH – Two Systems for Dealing with Relations. In: De Swart et al. (eds.) pp. 1–16 (2003)
10. Bonacich, P.: Factoring and Weighting Approaches to Status Scores and Clique Identification. Journal of Mathematical Sociology 2, 113–120 (1972)
11. Bonacich, P.: Power and Centrality: A Family of Measures. American Journal of Sociology 92, 1170–1182 (1987)
12. Borgatti, S.P.: Centrality and Network Flow. Social Networks 27, 55–71 (2005)
13. Brink, C., Kahl, W., Schmidt, G. (eds.): Relational Methods in Computer Science. Advances in Computing Science. Springer, Heidelberg (1997)
14. DeGroot, M.H.: Reaching a Consensus. Journal of the American Statistical Association 69, 118–121 (1974)
15. DeMarzo, P., Vayanos, D., Zwiebel, J.: Persuasion Bias, Social Influence, and Unidimensional Opinions. Quarterly Journal of Economics 118, 909–968 (2003)
16. De Swart, H., Orłowska, E., Schmidt, G., Roubens, M. (eds.): Theory and Applications of Relational Structures as Knowledge Instruments. LNCS, vol. 2929. Springer, Heidelberg (2003)
17. De Swart, H., Orłowska, E., Schmidt, G., Roubens, M. (eds.): TARSKI 2006. LNCS (LNAI), vol. 4342. Springer, Heidelberg (2006)
18. Freeman, L.: A Set of Measures of Centrality Based on Betweenness. Sociometry 40, 35–41 (1977)
19. Freeman, L.: Centrality in Social Networks: Conceptual Clarification. Social Networks 1, 215–239 (1979)
20. Friedkin, N.E., Johnsen, E.C.: Social Influence and Opinions. Journal of Mathematical Sociology 15, 193–206 (1990)
21. Friedkin, N.E., Johnsen, E.C.: Social Positions in Influence Networks. Social Networks 19, 209–222 (1997)
22. Golub, B., Jackson, M.O.: Naïve Learning in Social Networks and the Wisdom of Crowds. American Economic Journal: Microeconomics 2(1), 112–149 (2010)
23. Goyal, S.: Connections: An Introduction to the Economics of Networks. Princeton University Press, Princeton (2007)
24. Grabisch, M., Rusinowska, A.: Measuring Influence in Command Games. Social Choice and Welfare 33, 177–209 (2009)
25. Grabisch, M., Rusinowska, A.: A Model of Influence in a Social Network. Theory and Decision 69(1), 69–96 (2010)
26. Grabisch, M., Rusinowska, A.: A Model of Influence with an Ordered Set of Possible Actions. Theory and Decision 69(4), 635–656 (2010)
27. Grabisch, M., Rusinowska, A.: Different Approaches to Influence Based on Social Networks and Simple Games. In: Van Deemen, A., Rusinowska, A. (eds.) Collective Decision Making: Views from Social Choice and Game Theory. Series Theory and Decision Library C, vol. 43, pp. 185–209. Springer, Heidelberg (2010)

28. Grabisch, M., Rusinowska, A.: A Model of Influence with a Continuum of Actions. GATE Working Paper, 2010-04 (2010)
29. Grabisch, M., Rusinowska, A.: Iterating Influence Between Players in a Social Network. CES Working Paper, 2010.89 (2010)
30. Grabisch, M., Rusinowska, A.: Influence Functions, Followers and Command Games. Games and Economic Behavior (forthcoming), doi: 10.1016/j.geb.2010.06.003
31. Grofman, B., Owen, G.: A Game Theoretic Approach to Measuring Degree of Centrality in Social Networks. Social Networks 4, 213–224 (1982)
32. Hendrickx, R., Borm, P., Van den Brink, R., Owen, G.: The VL Control Measure for Symmetric Networks. Social Networks 31, 85–91 (2009)
33. Hoede, C., Bakker, R.: A Theory of Decisional Power. Journal of Mathematical Sociology 8, 309–322 (1982)
34. Hu, X., Shapley, L.S.: On Authority Distributions in Organizations: Controls. Games and Economic Behavior 45, 153–170 (2003)
35. Hu, X., Shapley, L.S.: On Authority Distributions in Organizations: Equilibrium. Games and Economic Behavior 45, 132–152 (2003)
36. Jackson, M.O.: Social and Economic Networks. Princeton University Press, Princeton (2008)
37. Katz, L.: A New Status Index Derived from Sociometric Analysis. Psychometrika 18, 39–43 (1953)
38. Rusinowska, A.: Different Approaches to Influence in Social Networks. Invited tutorial at the COMSOC 2010 (Third International Workshop on Computational Social Choice, Düsseldorf) (2010), http://ccc.cs.uni-duesseldorf.de/ COMSOC-2010/slides/invitedrusinowska.pdf
39. Rusinowska, A., de Swart, H.: Generalizing and Modifying the Hoede-Bakker Index. In: de Swart, H., Orłowska, E., Schmidt, G., Roubens, M. (eds.) TARSKI 2006. LNCS (LNAI), vol. 4342, pp. 60–88. Springer, Heidelberg (2006)
40. Schmidt, G., Ströhlein, T.: Relations and Graphs, Discrete Mathematics for Computer Scientists. EATCS Monographs on Theoretical Computer Science. Springer, Heidelberg (1993)
41. Van den Brink, R., Gilles, R.: Measuring Domination in Directed Networks. Social Networks 22, 141–157 (2000)
42. Van den Brink, R.: The Apex Power Measure for Directed Networks. Social Choice and Welfare 19, 845–867 (2002)
43. Van den Brink, R., Borm, P., Hendrickx, R., Owen, G.: Characterization of the β-and the Degree Network Power Measure. Theory and Decision 64, 519–536 (2008)
44. Wasserman, S., Faust, K.: Social Network Analysis: Methods and Applications. Cambridge University Press, Cambridge (1994)

Synthesising Terminating Tableau Calculi for Relational Logics

(Invited Paper)

Renate A. Schmidt

School of Computer Science,
The University of Manchester, UK

Abstract. Tableau-based deduction is an active and well-studied area of several branches of logic and automated reasoning. In this paper we discuss the challenge of automatically generating tableau calculi from the semantic specification of logics, while guaranteeing soundness, completeness and termination, when possible.

1 Introduction

Tableau-based deduction has a long tradition in several branches of logic and automated reasoning, with the history going to the nineteen sixties, and very early work in the nineteen thirties. Tableau calculi are now available for all kinds of logics, from classical logic and first-order logic to non-classical logics including intuitionistic logic, modal logics and description logic, and second-order logic. Tableau calculi come in many flavours, from ground semantic Smullyan-type calculi, to hypertableau, free-variable tableau calculi and disconnection tableau calculi. They are being used in many applications, from multi-agent systems, ontology reasoning and the semantic web, to diagnosis, testing and non-monotonic reasoning. And many implemented tableau provers are available.

Despite the multitudes of ways tableau calculi can be defined, even for the same class of logics, the body of work in the literature suggests that it is possible to develop tableau calculi in a systematic way for large classes of logics. This has been shown by studies such as [16,12,20] for modal logics, and [2] for intuitionistic logics. From overview papers such as [15] for modal logics, and [3] for description logics, it is apparent that, using the same techniques, tableau calculi can be systematically developed for large classes of logics.

The question arises if these techniques can be used to develop tableau calculi automatically from the specification of logics. The problem of finding sound and complete axiomatisations for arbitrary logics is known to be undecidable; so is finding sound and complete tableau calculi, or other kinds of deduction calculi. The problem of determining the decidability of a logic is also inherently undecidable. Even developing terminating deduction calculi for logics known to be decidable is undecidable. This means there can be no general solution to the problem of automatically generating sound, complete and terminating tableau calculi, even where it is possible.

H. de Swart (Ed.): RAMICS 2011, LNCS 6663, pp. 40–49, 2011.

For certain classes of logics it is however easy to write down sets of sound and complete tableau rules. In [22] I have shown that it is possible to synthesise tableau calculi for modal logics by translation to first-order clauses and refinement of first-order resolution.

It is also possible to synthesise tableau calculi directly. In our most recent work we have introduced a general framework for synthesising tableau calculi directly from the specification of a logic [25]. The framework incorporates a powerful blocking mechanism recently introduced for deciding expressive description logics [23]. This mechanism can be used in conjunction with other logics, including full first-order logic [5,6], and can be shown to unify standard blocking approaches. Although not many case studies have been undertaken yet, we believe this framework can accommodate the generation of tableau calculi for most known modal logics, description logics and relational logics, and many non-classical logics. In contrast to [22], the generated calculi are based only on standard tableau-style elimination rules.

This paper is an overview of the tableau synthesis framework introduced in [25,26]. In brief, the tableau synthesis method works as follows. The user defines the formal semantics of the given logic in a many-sorted first-order language. The semantic specification is then transformed into tableau rules. The tableau synthesis process consists of three stages: (i) Automated synthesis of a tableau calculus from the specification of the semantics of a logic or logical theory (Section 2), (ii) refining the rules (Section 3), and (iii) adding blocking to ensure termination, for decidable logics (Section 4). These stages are described in the next three sections. Cases where the framework has been applied are discussed in Section 5.

2 Tableau Calculus Synthesis

The first stage of the tableau synthesis framework is the rule synthesis stage. In this stage an initial set of tableau rules is generated from the specification of the semantics of a logic that must be specified by the user.

There are two high-level specification languages: the object language for defining the syntax of the logic and the meta-language for specifying the semantics of the logic. The *object language* $\mathcal{L}$ is a many-sorted propositional language expressive enough to define the syntax of modal logics, description logics, relational logics and other non-classical logics.

The *meta-language* $FO(\mathcal{L})$ is an extension of the object language $\mathcal{L}$ in which formulae of the logic are represented as terms and connectives as functions. The meta-language is a many-sorted language with a designated domain sort and designated domain symbols for encoding the semantics of the connectives of the logic and encoding properties of characteristic interpretations. The meta-language has the full expressivity of many-sorted first-order logic with function symbols and equality.

Let E denote any formula of the object language $\mathcal{L}$, and let ϕ^+ and ϕ^- denote any formulae of the meta-language $FO(\mathcal{L})$. We say a semantic specification S is *normalised*, if it consists of three types of sentences:

(S^+) $\forall \overline{x}\, (\nu(E(\overline{p}), \overline{x}) \to \phi^+(\overline{p}, \overline{x}))$

(S^-) $\forall \overline{x}\, (\phi^-(\overline{p}, \overline{x}) \to \nu(E(\overline{p}), \overline{x}))$

(S^b) Any $FO(\mathcal{L})$-sentence without occurrences of non-atomic $\mathcal{L}$-formulae.

Here, ν denotes the 'holds' predicate. $\overline{x}$ denotes a sequence of n variables ranging over the domain sort, where E is a formula (with free variables $\overline{p}$) interpreted as an n-ary relation. $\overline{p}$ denotes a sequence of m propositional variables of the object language.

The idea is that S^+ and S^- sentences define the semantics of positive and negative occurrences of formulae E—typically connectives of $\mathcal{L}$. For example, for defining the semantics of the modal box operator, $E(\overline{p})$ would be $\Box p$, and $\phi^+(\overline{p}, \overline{x})$ and $\phi^-(\overline{p}, \overline{x})$ might be $\forall y\big(R(x, y) \to \nu(p, y)\big)$, where R denotes the accessibility relation in the underlying semantics. Intuitively, this specifies that for any modal formula p, $\Box p$ is true in a state x iff p is true in every R-successor of x. In general, $\phi^+(\overline{p}, \overline{x})$ and $\phi^-(\overline{p}, \overline{x})$ do not need to coincide.

Any additional properties, such as frame correspondence properties of the accessibility relation in modal logic, can be specified as type S^b sentences. For example, that R is a transitive relation can be specified by the formula

$$\forall x \forall y \forall z \big((R(x, y) \wedge R(y, z)) \to R(x, z)\big).$$

A specification is transformed into tableau rules by first transforming each sentence into Skolemised implicational form and then rewriting it as a rule. The three types of sentences are transformed respectively to:

(1) $$\nu(E(\overline{p}), \overline{x}) \to \bigvee_{j=1}^{J} \bigwedge_{k=1}^{K_j} \psi_{jk}$$

(2) $$\neg\nu(E(\overline{p}), \overline{x}) \to \bigvee_{j=1}^{J} \bigwedge_{k=1}^{K_j} \psi_{jk}$$

(3) $$\bigvee_{j=1}^{J} \bigwedge_{k=1}^{K_j} \psi_{jk}.$$

The ψ_{jk} denote literals (with free variables $\overline{p}$ and $\overline{x}$). These are then recast as these tableau rules:

(ρ^+) $$\dfrac{\nu(E(\overline{p}), \overline{x})}{\psi_{11}, \ldots, \psi_{1K_1} \mid \cdots \mid \psi_{J1}, \ldots, \psi_{JK_J}}$$

(ρ^-) $$\dfrac{\neg\nu(E(\overline{p}), \overline{x})}{\psi_{11}, \ldots, \psi_{1K_1} \mid \cdots \mid \psi_{J1}, \ldots, \psi_{JK_J}}$$

(ρ^b) $$\dfrac{}{\psi_{11}, \ldots, \psi_{1K_1} \mid \cdots \mid \psi_{J1}, \ldots, \psi_{JK_J}}.$$

The antecedents of the implications (1) and (2) have respectively become the premises of the ρ^+ and ρ^- rules. The succedents of the implications are respectively the disjunctive normal forms of the right hand and left hand sides of the S^+ and S^- sentences. These have become the conclusions of the ρ^+ and ρ^- rules. The ρ^+ rule is the *decomposition rule* for positive occurrences of formulae of the form $E(\bar{p})$ and the ρ^- rule is the decomposition rule for negative occurrences of formulae of the form $E(\bar{p})$.

The S^b sentences (3) are first transformed into Skolemised disjunctive normal form and then reformulated as ρ^b rules. The ρ^b rules are referred to as *theory rules*.

For example, the generated decomposition rules for the modal box operator are

$$
(4) \qquad \frac{\nu(\Box p, x)}{\neg R(x, y) \mid \nu(p, y)} \quad \text{and} \quad \frac{\neg\nu(\Box p, x)}{R(x, f(p, x)),\ \neg\nu(p, f(p, x))}.
$$

$f(p, x)$ in the right rule is the Skolem term introduced during the transformation to Skolemised implicational form for the quantifier $\exists y$ of the semantic definition of $\Box$. Because of the way that Skolemisation is defined, $f(p, x)$ is uniquely associated with $\Box$, p and x. The intuition of the right rule is that for each formula of the form $\neg\nu(\Box\psi, s)$ on the current branch a domain element $f(\psi, s)$ is created and the appropriate instantiations of the conclusions, namely $R(s, f(\psi, s))$ and $\neg\nu(\psi, f(\psi, s))$ are added to the current branch. An alternative to using Skolem terms is to introduce fresh constants, uniquely associated with the premise, in term-introducing rules.

The transitivity property of R is transformed to the rule

$$
\frac{}{\neg R(x, y) \mid \neg R(y, z) \mid R(x, z)}.
$$

In addition, the generated calculus includes default *closure rules*, one for each $\mathcal{L}$ sort and each interpreted predicate symbol. These are the closure rules added for modal logics.

$$
(5) \qquad \frac{\nu(p, x),\ \neg\nu(p, x)}{\bot} \qquad\qquad \frac{R(x, y),\ \neg R(x, y)}{\bot}
$$

Let S_L denote the semantic specification of a logic L. We denote the tableau calculus generated from S_L by T_L. The generated calculi are Smullyan type tableau calculi with either two premises (in the case of closure rules), one premise (in the case of positive and negative decomposition rules) or no premises (in the case of theory rules). The rules operate on ground formulae of the form $(\neg)\nu(\psi, s)$ or literals where the predicate symbol are interpreted symbols, such as $R(s, t)$ and $\neg R(s, t)$ in the case of modal logic.

A *tableau derivation*, or *tableau*, for a calculus T_L is a finitely branching, ordered tree whose nodes are sets of ground formulae in the tableau language. At the start, the tableau derivation is initialised with the set $\{\nu(\phi, a) \mid \phi \in \mathcal{S}\}$, where $\mathcal{S}$ is a set of $\mathcal{L}$-formulae to be tested for satisfiability and a denotes a fresh constant of the domain sort.

In the inference process we take for granted that any rule is applied only once to the same set of premises. This ensures that tableau derivations are strict. We also take for granted that any variables occurring in the conclusions of a rule, that do not occur in any of the premises of the rule, are instantiated only with (ground) Skolem terms occurring on the current branch. In the left rule of (4), y is an example of a variable that occurs in the conclusions but not in the premises of the rule. Further, we assume that the rules are applied non-deterministically in a tableau derivation.

Let $T_L(S)$ denote a complete tableau derivation built by applying the rules of the calculus T_L and starting with a set S of $\mathcal{L}$-formulae as input. That is, we assume that all branches in the tableau derivation are either closed or fully expanded (in the limit).

We say a calculus T_L is *sound* iff for every set S of formulae satisfiable in an S_L-model, any complete tableau derivation $T_L(S)$ is open. A calculus T_L is said to be *complete* iff for any S_L-unsatisfiable set S of formulae there is a tableau derivation $T_L(S)$ that is closed.

Theorem 1 ([25]). *If S_L is any normalised semantic specification of a logic L then T_L is sound.*

If a normalised semantic specification satisfies certain well-definedness conditions (for definitions see [25]) then the generated tableau calculus is also complete.

Theorem 2 ([25]). *If S_L is any well-defined semantic specification of a logic L then T_L is complete.*

Actually we prove that generated tableau calculi are constructively complete, which is a slightly stronger property than completeness. It means that for every open branch in a tableau there is a model, which reflects all the formulae occurring on the branch [25].

3   Tableau Calculus Refinement

The second stage of the tableau calculus synthesis process involves refinement of the generated rules, because the degree of branching in the generated rules is higher than is necessary. What is exploited is that under certain conditions it is possible to replace rules by rules with better properties.

One refinement modifies rules one-by-one by attempting to move formulae in conclusion positions to premise positions and appropriate sign switching. This replaces one premise or no premise rules by multiple premise rules, thereby constraining rule application. For example, rule refinement of the rules

$$\frac{\nu(\Box p, x)}{\neg R(x,y) \mid \nu(p,y)} \quad \text{and} \quad \frac{}{\neg R(x,y) \mid \neg R(y,z) \mid R(x,z)}$$

leads to the replacement of these with the rules

$$\frac{\nu(\Box p, x),\ R(x,y)}{\nu(p,y)} \quad \text{and} \quad \frac{R(x,y),\ R(y,z)}{R(x,z)}.$$

When a certain general admissibility condition is satisfied (for details, see [25]) this kind of rule refinement preserves soundness and constructive completeness. The condition is inductive and, hence, second-order. Automatic verification of this condition is therefore not possible in general.

The condition can be proved in the two cases above, thus justifying the refinement. Replacing the disjunction rule

$$\frac{\nu(\neg(p \wedge q), x)}{\neg\nu(p, x) \mid \neg\nu(q, x)} \qquad \text{by} \qquad \frac{\nu(\neg(p \wedge q), x), \; \nu(p, x)}{\neg\nu(q, x)}$$

is however not justified in the framework without other changes to the calculus.

When there is enough expressivity in the language of the logic for capturing its own semantics, the rules of the generated calculi can be reformulated without the use of the ν predicate. For hybrid logics the refined calculus is similar to semantic labelled tableau calculi with explicit accessibility relations.

If the last refinement is not possible, the calculi obtained can be viewed as semantic tableau calculi operating on signed formulae.

4   Adding Blocking

The third stage of the tableau calculus synthesis process is attempting to ensure termination. Obviously, such an attempt can only be successful if the logic of interest is decidable. For some logics, for example, propositional logic and modal logic K, the synthesised tableau calculi are already terminating, and no further effort is needed. This is the exception rather than the norm.

A tableau calculus T_L is *terminating* (for satisfiability) iff for every *finite* set of concepts S every closed tableau $T_L(S)$ is finite and every open tableau $T_L(S)$ has a finite open branch.

A standard way of turning ground semantic tableau methods into decision procedures is to add blocking. The idea of blocking is to modify the tableau inference process in such a way so that ideally finite models are found if they exist. This can be done by reusing or identifying domain terms. Loop checking mechanisms developed for modal logic, hybrid logic and description logic tableau algorithms are typically based on comparing sets of formulae that hold for the same domain terms [18,3,17,7,13]. They have been principally designed and used for logics with some form of tree model property. Approaches based on systematically reusing domain terms have been used for minimal model generation for classical logic [10,11]. This approach finds finite models when they exist. Another possibility is using conjectured equality constraints between domain terms and equality reasoning [19,5,23,8].

Any of these blocking approaches can be integrated into the generated tableau calculi. Here, we discuss the integration of the *unrestricted blocking* mechanism introduced in [23] into the framework. This approach is based on a blocking rule and equality reasoning. An advantage of this form of blocking is that

- it is generic and can be combined with most logics,
- it can be added to existing calculi with minimal intervention (and not only tableau calculi),
- it can be restricted for particular purposes [19,5,6], for example, to simulate the other blocking techniques [19],
- it can be used speculatively, and
- it can be used even for logics that are not decidable.

The idea of restricted and unrestricted blocking is that two terms on a branch are identified, possibly restricted by certain effectively testable side conditions being true. If this leads to a contradiction then this choice was not good and needs to be undone. This can be realised by adding the following cut rule, called *unrestricted blocking rule*, to a calculus.

$$\overline{x \approx y \mid x \not\approx y}$$

As this rule adds equalities and inequalities to branches, equality reasoning will need to be provided by the calculus. This means the calculus needs to be extended, unless the calculus already makes provision for appropriate equality reasoning. An easy solution is to extend the calculus with the standard equality rules. In our framework, putting the standard equality axioms into the semantic specification leads to the generation of standard equality rules.

Equality reasoning requires special care to keep the search space under control. First, it is essential to prevent the same inferences being performed repeatedly with equal terms. This can be prevented by restricting the application of the rules in the calculus to one representative term in each equivalence class of terms. Alternatively, it can be achieved by endowing the set of terms with a well-founded ordering and limiting the application of all rules just to the terms that are minimal under the ordering in any equivalence class. The ordering could be the subterm ordering, or the order in which terms are introduced by term-introducing rules (δ rules).

Because the rules operate on ground formulae, another possibility is to realise equality reasoning via ordered rewriting [21]. This is how equality reasoning for restricted and unrestricted blocking has been implemented in the METTEL prover [29,30] and the MSPASS prover [6].

Regardless of how equality reasoning is realised it is crucial that tableau derivations are constructed in a fair way. In general, this means that formulae are selected fairly for rule application and also branches are selected fairly during the derivation [26,21].

Let T_L be a sound and constructively complete tableau calculus for a semantic specification S_L of a logic L. We have shown [24,27] that T_L extended with the unrestricted blocking mechanism is sound and constructively complete for S_L. Furthermore, the extended calculus is terminating for L, if the following conditions both hold:

- There is a finite closure operator sub (defined on sets of formulae of the language of L) such that T_L is compatible with sub, that is, for every set S

of formulae, all $\mathcal{L}$-formulae occurring in a tableau derivation $T(\mathcal{S})$ belong to sub.
– L has the effective finite model property with respect to S_L.

This implies that the extensions of the generated and refined tableau calculi with unrestricted blocking are sound and (constructively) complete. Moreover, if it is known that the given logic has the effective finite model property with respect to a finite semantic specification then they are terminating as well.

5   Applying the Framework

Concrete applications of the tableau synthesis framework can be found in [25], where propositional intuitionistic logic is considered as an example, and the long version [26], where also the description logic $\mathcal{SO}$ is considered. The semantics of propositional intuitionistic logic is not Boolean and it is restricted by a background theory. $\mathcal{SO}$ is the extension of the description logic $\mathcal{ALC}$ with singleton concepts (nominals) and transitive roles, and is an analogue of the hybrid version of the multi-modal logic $K4$. The tableau calculi produced for these logics after refinement resemble those of existing calculi but the termination mechanism is different.

In [4] we have applied the framework to develop a tableau-based algorithm for testing the admissibility of modal rules. For this we needed to devise a tableau calculus for a modal logic that has not been considered before.

By design the method can generate terminating tableau calculi for expressive description logics such as $\mathcal{ALBO}$ [23] and $\mathcal{ALBO}^{\mathsf{id}}$ [26]. $\mathcal{ALBO}$ extends the description logic $\mathcal{ALC}$ with the Boolean role operators, role inverse, and singleton concepts (nominals). $\mathcal{ALBO}^{\mathsf{id}}$ includes also role identity and is expressively equivalent to the relational calculus [28] without relational composition or Peirce logic (the logical version of Peirce algebras [9]) without relational composition.

6   Concluding Remarks

The tableau synthesis framework ensures that the generated tableau calculi are sound and constructively complete for logics defined by well-defined semantic specifications. Adding the unrestricted blocking mechanism produces a terminating tableau calculus, whenever the logic is known to have the effective finite model property.

One of the difficulties has been to design the framework in such a way that the specification languages are not too complicated and over-burdening, while at the same time covering as many logics as is possible. In many cases semantic definitions found in the literature can be directly encoded in the specification languages of the framework. In some cases some reformulation of the syntactic and semantic definitions might be necessary, but because of the inherent undecidability of deduction calculus synthesis, there is no hope that adequate reformulations can always be found. Nevertheless, we believe that the framework is applicable to most known first-order definable modal, description and relational logics. Non first-order definable logics such as propositional dynamic logic are certainly beyond the scope of the framework as it is currently defined.

Our immediate goal is to implement the framework as a tableau calculus synthesis tool. Given the semantic specification of a logic the tool will automatically generate the corresponding rules and output them as a calculus. This tool could then be combined with prover engineering platforms such as LoTREC [14] and the Tableau Workbench [1] (possibly extended with unrestricted blocking) to obtain implemented tableau provers. Our ultimate goal is to develop a tool for automatically generating implemented tableau provers from logic specifications, which I think is within the realm of current possibilities.

Acknowledgements

I thank Dmitry Tishkovsky, Hilverd Reker, Mohammad Khodadadi and Ullrich Hustadt for useful discussions. Financial support from the UK Engineering and Physical Sciences Research Council is gratefully acknowledged.

References

1. Abate, P., Goré, R.: The Tableaux Work Bench. In: Cialdea Mayer, M., Pirri, F. (eds.) TABLEAUX 2003. LNCS, vol. 2796, pp. 230–236. Springer, Heidelberg (2003)
2. Avellone, A., Miglioli, P., Moscato, U., Ornaghi, M.: Generalized tableau systems for intermediate propositional logics. In: Galmiche, D. (ed.) TABLEAUX 1997. LNCS, vol. 1227, pp. 43–61. Springer, Heidelberg (1997)
3. Baader, F., Sattler, U.: An overview of tableau algorithms for description logics. Studia Logica 69, 5–40 (2001)
4. Babenyshev, S., Rybakov, V., Schmidt, R.A., Tishkovsky, D.: A tableau method for checking rule admissibility in S4. Electronic Notes in Theoretical Computer Science 262, 17–32 (2010)
5. Baumgartner, P., Schmidt, R.A.: Blocking and other enhancements for bottom-up model generation methods. In: Furbach, U., Shankar, N. (eds.) IJCAR 2006. LNCS (LNAI), vol. 4130, pp. 125–139. Springer, Heidelberg (2006)
6. Baumgartner, P., Schmidt, R.A.: Blocking and other enhancements for bottom-up model generation methods (2008) (manuscript)
7. Bolander, T., Bräuner, T.: Tableau-based decision procedures for hybrid logic. Journal of Logic and Computation 16(6), 737–763 (2006)
8. Bonacina, M.P., Lynch, C.A., de Moura, L.: On deciding satisfiability by theorem proving with speculative inferences. Journal of Automated Reasoning (2010) (published online)
9. Brink, C., Britz, K., Schmidt, R.A.: Peirce algebras. Formal Aspects of Computing 6(3), 339–358 (1994)
10. Bry, F., Manthey, R.: Proving finite satisfiability of deductive databases. In: Börger, E., Kleine Büning, H., Richter, M.M. (eds.) CSL 1987. LNCS, vol. 329, pp. 44–55. Springer, Heidelberg (1988)

11. Bry, F., Torge, S.: A deduction method complete for refutation and finite satisfiability. In: Dix, J., Fariñas del Cerro, L., Furbach, U. (eds.) JELIA 1998. LNCS (LNAI), vol. 1489, pp. 1–17. Springer, Heidelberg (1998)
12. Fariñas del Cerro, L., Gasquet, O.: A general framework for pattern-driven modal tableaux. Logic Journal of the IGPL 10(1), 51–83 (2002)
13. Cialdea Mayer, M., Cerrito, S.: Nominal substitution at work with the global and converse modalities. In: Beklemishev, L., Goranko, V., Shehtman, V. (eds.) Advances in Modal Logic, vol. 8, pp. 57–74. College Publications (2010)
14. Gasquet, O., Herzig, A., Longin, D., Sahade, M.: LoTREC: Logical tableaux research engineering companion. In: Beckert, B. (ed.) TABLEAUX 2005. LNCS (LNAI), vol. 3702, pp. 318–322. Springer, Heidelberg (2005)
15. Goré, R.: Tableau methods for modal and temporal logics. In: D'Agostino, M., Gabbay, D., Hähnle, R., Posegga, J. (eds.) Handbook of Tableau Methods, pp. 297–396. Kluwer, Dordrecht (1999)
16. Heuerding, A.: Sequent calculi for proof search in some modal logics. PhD thesis, Universität Bern (1998)
17. Horrocks, I., Sattler, U.: A tableau decision procedure for SHOIQ. Journal of Automated Reasoning 39(3), 249–276 (2007)
18. Hughes, G.E., Cresswell, M.J.: An Introduction to Modal Logic. Routledge, London (1968)
19. Hustadt, U., Schmidt, R.A.: On the relation of resolution and tableaux proof systems for description logics. In: Dean, T. (ed.) Proceedings IJCAI 1999, pp. 110–115. Morgan Kaufmann, San Francisco (1999)
20. Massacci, F.: Single step tableaux for modal logics: Computational properties, complexity and methodology. Journal of Automated Reasoning 24(3), 319–364 (2000)
21. Reker, H., Schmidt, R.A., Tishkovsky, D.: Clausal tableau decision procedures for the two-variable fragment with equality (2010) (manuscript)
22. Schmidt, R.A.: Developing modal tableaux and resolution methods via first-order resolution. In: Governatori, G., Hodkinson, I., Venema, Y. (eds.) Advances in Modal Logic, vol. 6, pp. 1–26. College Publications, London (2006)
23. Schmidt, R.A., Tishkovsky, D.: Using tableau to decide expressive description logics with role negation. In: Aberer, K., Choi, K.-S., Noy, N., Allemang, D., Lee, K.-I., Nixon, L.J.B., Golbeck, J., Mika, P., Maynard, D., Mizoguchi, R., Schreiber, G., Cudré-Mauroux, P. (eds.) ASWC 2007 and ISWC 2007. LNCS, vol. 4825, pp. 438–451. Springer, Heidelberg (2007)
24. Schmidt, R.A., Tishkovsky, D.: A general tableau method for deciding description logics, modal logics and related first-order fragments. In: Armando, A., Baumgartner, P., Dowek, G. (eds.) IJCAR 2008. LNCS (LNAI), vol. 5195, pp. 194–209. Springer, Heidelberg (2008)
25. Schmidt, R.A., Tishkovsky, D.: Automated synthesis of tableau calculi. In: Giese, M., Waaler, A. (eds.) TABLEAUX 2009. LNCS, vol. 5607, pp. 310–324. Springer, Heidelberg (2009)
26. Schmidt, R.A., Tishkovsky, D.: Automated synthesis of tableau calculi, (2010) (manuscript submitted for publication)
27. Schmidt, R.A., Tishkovsky, D.: Using tableau to decide description logics with full role negation and identity, (2010) (manuscript submitted for publication)
28. Tarski, A.: On the calculus of relations. J. of Symb. Logic 6(3), 73–89 (1941)
29. Tishkovsky, D.: METTEL, http://www.mettel-prover.org
30. Tishkovsky, D., Schmidt, R.A., Khodadadi, M.: METTEL: A generic tableau prover, (2011) (manuscript submitted for publication)

From Arrow's Impossibility to Schwartz's Tournament Equilibrium Set

(Invited Tutorial)

Felix Brandt

Institut für Informatik
Technische Universität München
brandtf@in.tum.de

Abstract. Perhaps the most influential result in social choice theory is Arrow's impossibility theorem, which states that a seemingly modest set of desiderata cannot be satisfied when aggregating preferences [1]. While Arrow's theorem might appear rather negative, it can also be interpreted in a positive way by identifying what *can* be achieved in preference aggregation.

In this talk, I present a number of variations of Arrow's theorem—such as those due to Mas-Colell and Sonnenschein [8] and Blau and Deb [2]—in their choice-theoretic version. The critical condition in all these theorems is the assumption of a rationalizing binary relation or equivalent notions of choice-consistency. The bulk of my presentation contains three escape routes from these results. The first one is to ignore consistency with respect to a variable set of alternatives altogether and require consistency with respect to a variable electorate instead. As Smith [12] and Young [14] have famously shown, this essentially characterizes the class of scoring rules, which contains plurality and Borda's rule. For the second escape route, we factorize choice-consistency into two parts, contraction-consistency and expansions-consistency [11]. While even the mildest dose of the former has severe consequences on the possibility of choice, varying degrees of the latter characterize a number of appealing social choice functions, namely the top cycle, the uncovered set, and the Banks set [3,9,4]. Finally, I suggest to redefine choice-consistency by making reference to the *set* of chosen alternatives rather than individual chosen alternatives [6]. It turns out that the resulting condition is a weakening of transitive rationalizability and can be used to characterize the minimal covering set and the bipartisan set. Based on a two decades-old conjecture due to Schwartz [10], the tournament equilibrium set can be characterized by the same condition or, alternatively, by a weak expansion-consistency condition from the second escape route. Whether Schwartz's conjecture actually holds remains a challenging combinatorial problem as well as one of the enigmatic open problems of social choice theory.

Throughout the presentation I will discuss the algorithmic aspects of all considered social choice functions. While some of the mentioned functions can be easily computed, other ones do not admit an efficient algorithm unless P equals NP [13,5,7].

H. de Swart (Ed.): RAMICS 2011, LNCS 6663, pp. 50–51, 2011.
© Springer-Verlag Berlin Heidelberg 2011

References

1. Arrow, K.J.: Social Choice and Individual Values, 2nd edn. Cowles Foundation, New Haven (1963)
2. Blau, J.H., Deb, R.: Social decision functions and the veto. Econometrica 45, 871–879 (1977)
3. Bordes, G.: Consistency, rationality and collective choice. Review of Economic Studies 43, 451–457 (1976)
4. Brandt, F.: Minimal stable sets in tournaments. Journal of Economic Theory (forthcoming, 2011)
5. Brandt, F., Fischer, F.: Computing the minimal covering set. Mathematical Social Sciences 56, 254–268 (2008)
6. Brandt, F., Harrenstein, P.: Set-rationalizable choice and self-stability. Journal of Economic Theory (forthcoming, 2011)
7. Brandt, F., Fischer, F., Harrenstein, P., Mair, M.: A computational analysis of the tournament equilibrium set. Social Choice and Welfare 34, 597–609 (2010)
8. Mas-Colell, A., Sonnenschein, H.: General possibility theorems for group decisions. Review of Economic Studies 39, 185–192 (1972)
9. Moulin, H.: Choosing from a tournament. Social Choice and Welfare 3, 271–291 (1986)
10. Schwartz, T.: Cyclic tournaments and cooperative majority voting: A solution. Social Choice and Welfare 7, 19–29 (1990)
11. Sen, A.K.: Choice functions and revealed preference. Review of Economic Studies 38, 307–317 (1971)
12. Smith, J.H.: Aggregation of preferences with variable electorate. Econometrica 41, 1027–1041 (1973)
13. Woeginger, G.J.: Banks winners in tournaments are difficult to recognize. Social Choice and Welfare 20, 523–528 (2003)
14. Young, H.P.: Social choice scoring functions. SIAM Journal on Applied Mathematics 28, 824–838 (1975)

Automated Engineering of Relational and Algebraic Methods in Isabelle/HOL
(Invited Tutorial)

Simon Foster[1], Georg Struth[1], and Tjark Weber[2]

[1] Department of Computer Science, The University of Sheffield
{s.foster,g.struth}@dcs.shef.ac.uk
[2] Computer Laboratory, University of Cambridge
tw333@cam.ac.uk

Abstract. We present a new integration of relational and algebraic methods in the Isabelle/HOL theorem proving environment. It consists of a fine grained hierarchy of algebraic structures based on Isabelle's type classes and locales, and a repository of more than 800 facts obtained by automated theorem proving. We demonstrate further benefits of Isabelle for hypothesis learning, duality reasoning, theorem instantiation, and reasoning across models and theories. Our work forms the basis for a reference repository and a program development environment based on algebraic methods. It can also be used by mathematicians for exploring and integrating new variants.

1 Introduction

Kleene and relation algebras provide semantics for programs and a basis for integrating logics for computing systems. Relational and algebraic reasoning is at the core of popular program development methods for imperative, concurrent and functional programs. Algebraic approaches seem particularly useful for modelling and analysing the information and control flow in computing systems. Specifications are often very compact, concise and generic in this setting; proofs are based on first-order equational reasoning, hence calculational, and often much shorter than set-theoretic ones. This makes such approaches very suitable for automation. Applications in program development, however, may require additional mechanisms for reasoning about data structures or data types and higher-order features for fixed points or (co)recursion.

Interactive theorem proving (ITP) systems such as Isabelle/HOL [29] and Coq [6], and special-purpose systems [1], have been used to implement Kleene and relation algebras [32,21,30]. It has also been shown that automated theorem proving (ATP) systems are quite successful at proving algebraic theorems at textbook level [16,17] and able to support simple program analysis tasks [4].

The strengths and weaknesses of ITP and ATP are almost orthogonal: Apart from supporting theory hierarchies and proof management, ITP systems offer the expressiveness of higher-order logic for modelling and reasoning. But they require considerable mathematical expertise and sophisticated tactics for feeding

H. de Swart (Ed.): RAMICS 2011, LNCS 6663, pp. 52–67, 2011.

manual proofs into the tool. ATP systems, in contrast, have traditionally been optimised towards proof search performance; even non-expert users can discharge many calculational proof obligations fully automatically. But these tools are typically limited to first-order logic, and proof management, hypothesis learning and modular reasoning are usually not supported.

Currently, however, there are strong trends to make these two worlds converge. Isabelle, in particular, is being transformed into a theorem proving environment that combines higher-order modelling and reasoning with ATP systems, satisfiability modulo theories (SMT) solvers, decision procedures and counterexample generators. So does this make the best of both worlds available for the automated engineering of relational and algebraic methods?

This paper comes to an overall very positive answer. It proposes Isabelle as a way forward in engineering relational and algebraic methods. Algebraic theorem proving is now perhaps just as easy with Isabelle as with ATP systems. Important additional benefits arise from Isabelle's higher-order features and SMT integration. Our main contributions are as follows.

We implement a theory hierarchy for relation and Kleene algebras using Isabelle's local specification facilities. This hierarchy includes many popular variants such as basic process algebras, probabilistic Kleene algebras, demonic refinement algebras, omega algebras, Kleene algebras with (anti)domain and modal Kleene algebras. Theorems are inherited across the hierarchy by subclass and sublocale proofs, and the difference between reasoning in Kleene algebras and relation algebras often vanishes.

We prove more than 800 facts in this setting, all of them by ATP using Isabelle's Sledgehammer tool [8]. It turns out that Sledgehammer is very good at learning hypotheses for proofs. Since Isabelle reconstructs all proofs produced by Sledgehammer's external ATP systems with an internally verified tool, the degree of automation is slightly lower than by standalone ATP, and a few difficult theorems do not succeed by ATP alone in one full sweep. Nevertheless, most of the basic facts attempted could be proved automatically in one single step.

We show that Isabelle's higher-order features are very valuable for automated theory engineering. Dualities can be formalised and used to obtain theorems for free. Higher-order variables can be used for instantiating theorems, for instance, from abstract Galois connections or conjugations. Set-theoretic specifications of algebras with explicit carrier sets support mechanised reasoning in universal algebra. By formally linking abstract algebras with concrete models, we obtain a seamless transition between pointwise and pointfree reasoning.

These results suggest that our Isabelle/HOL formalisation has considerable potential for turning relation and Kleene algebras into program development and verification tools that are relatively lightweight yet offer a high degree of automation. We propose it as the basis for a standardised repository for algebraic methods from which further variants, a more extensive library of theorems and proofs, and more concrete program analysis environments can be engineered.

Beyond these technical contributions, another main purpose of this paper is to serve as a tutorial introduction to the relational and algebraic methods community.

2   Isabelle/HOL

Isabelle/HOL [29] is a popular ITP system based on higher-order logic. It has recently mutated from a metalogical framework for specifying logics [31] into a theorem proving environment with integrated decision procedures, ATP systems, SMT solvers and counterexample generators. Intricate mathematical proofs and industrial verification tasks have been carried out with this tool (e.g. [25,22]). For our purposes, the following features of Isabelle are particularly interesting.

First, Isabelle offers a user interface with notation support [29, §4]. Standard notations for Kleene and relation algebras can therefore be defined and used. LATEX code can automatically be generated from proofs and specifications. In fact, this paper was generated from our Isabelle theory files, with the added benefit that its entire technical content has been formally verified.

Second, the proof language Isar [36] supports a proof style that is natural and easily readable for humans. Equational reasoning, in particular, can be translated almost unchanged into Isabelle [3], and Isar proofs could be published directly in mathematical texts.

Third, Isabelle offers facilities for theory hierarchies and modules through local specifications [15]. A local specification consist of parameters and assumptions: for instance, the operations of Boolean algebra, and the axioms satisfied by these operations. Hierarchies of local specifications can be established through extension or proof. Relation algebra, for instance, can be specified as an extension (*subclass* in Isabelle parlance) of Boolean algebra, and all theorems proved in the latter context become automatically available in the former.

Fourth, Isabelle's Sledgehammer tool [8] integrates various external ATP systems. To obtain trustworthy facts, all external ATP proofs are reconstructed either by proof search using the internal Isabelle-verified theorem prover Metis [18], or more directly by Isar. Sledgehammer selects hypotheses among local verified lemmas and calls the external ATP systems. Since Metis is less efficient, the external provers can be used iteratively to minimise hypothesis sets before Metis is called. Alternatively, an Isar proof can be generated that attempts a stepwise reconstruction of external proofs. SMT solvers, notably Z3, have recently been integrated [9] as an alternative to ATPs. Finally, counterexample generators such as Nitpick [7] are now part of Isabelle, too. They allow a game of proof and refutation when developing and prototyping theories.

In contrast to previous rather monolithic ATP proofs in Kleene and relation algebra, where theorems were often attempted directly from the axioms, the user now owns the means of production: proofs can be performed at any level of granularity, from fully automated proofs to textbook-style manual Isar proof scripts, in which the individual proof steps can be automated.

Examples of these features and the different styles of proof are given in the remainder of this paper. Isabelle is well documented; and more information can be found at the tool web site [19]. Our complete repository, including all facts used in this paper, can be found online [33]. All hypotheses used in the proofs presented can easily be found by searching their names in the repository.

3   Implementing a Kleene Algebra Hierarchy

We informally use the term "Kleene algebra" for a family of algebras based on variants of idempotent semirings—or dioids—which are extended with operations for finite or infinite iteration. Different variants correspond to different system semantics and different intended applications, including processes, probabilistic systems, program refinement, sequential or concurrent program analysis. Program semantics for partial or total correctness and formalisms such as dynamic, temporal or Hoare logics can be obtained by adding further axioms.

Isabelle's local specification facilities allow us to build a modular theory hierarchy for Kleene algebras and to inherit theorems across theories and models. We outline the approach in this section. To simplify the presentation, the code in this paper sometimes differs slightly from that in the repository.

Our hierarchy starts with the axiomatic class of join semilattices.

Class *join-semilattice* $=$ *plus-ord* $+$
assumes *add-assoc*: $(x+y)+z = x+(y+z)$
and *add-comm*: $x+y = y+x$
and *add-idem*: $x+x = x$

It expands a predefined class *plus-ord*, which provides notation for addition and order and connects them via $x \leq y \equiv x + y = y$ by the semilattice axioms.

Linking the equational view on semilattices with the order-based one requires showing that join semilattices are partial orders.

Sublass (in *join-semilattice*) *order*
Proof
  fix $x\ y\ z$
  show $x \leq x$ **by** (*metis add-idem leq-def*)
  show $x \leq y \implies y \leq x \implies x = y$ **by** (*metis add-comm leq-def*)
  show $x \leq y \implies y \leq z \implies x \leq z$ **by** (*metis add-assoc leq-def*)
  show $x < y \longleftrightarrow x \leq y \land \neg\, (y \leq x)$ **by** (*metis strict-leq-def*)
qed

The individual proof goals are prescribed by Isabelle, in particular that for $<$ which is usually a definition. Isabelle's built-in ATP system Metis is called on each goal with the hypotheses indicated. More information about such proofs can be found in the following sections. All Isabelle theorems for partial orders are now available for join semilattices.

The next level of our hierarchy implements variants of semirings and dioids, whose multiplication symbol is provided by a predefined class *mult*.

Class *near-semiring = plus + mult +*
  assumes *mult-assoc*: $(x \cdot y) \cdot z = x \cdot (y \cdot z)$
  and *add-assoc'*: $(x+y)+z = x+(y+z)$
  and *add-comm'*: $x+y = y+x$
  and *distr*: $(x+y) \cdot z = x \cdot z + y \cdot z$

Class *near-dioid = near-semiring + plus-ord +*
  assumes *idem*: $x+x = x$

Near-dioids form the basis of process algebras like CCS or ACP [5]. By definition, near-diods are near-semirings, and all near-semiring theorems are automatically inherited. But the link with semilattices requires proof.

Sublass (in *near-dioid***)** *join-semilattice* — automatic proof omitted

Other variants of semirings and dioids can be obtained by extension, e.g.,

Class *pre-dioid = near-dioid +*
  assumes *subdistl*: $z \cdot x \leq z \cdot (x+y)$

Class *semiring = near-semiring +*
  assumes *distl*: $x \cdot (y+z) = x \cdot y + x \cdot z$

Class *dioid = semiring + near-dioid*

Predioids form the basis for game algebras [14] and probabilistic Kleene algebras [27]. Semirings and dioids have various applications in different fields of mathematics and computing. We have extended these structures with additive and multiplicative units in the usual ways. Different variants are needed again for different applications.

At the next level of the hierarchy we add a Kleene star to our variants of semirings with 1. We only show the extremal points of this level.

Class *left-near-kleene-algebra = near-dioid-one + star +*
  assumes *star-unfoldl*: $1+x \cdot x^\star \leq x^\star$
  and *star-inductl*: $z+x \cdot y \leq y \longrightarrow x^\star \cdot z \leq y$

Class *kleene-algebra = left-kleene-algebra-zero +*
  assumes *star-inductr*: $z+y \cdot x \leq y \longrightarrow z \cdot x^\star \leq y$

Isolation of the left star unfold and left induction axiom is important for variants such as probabilistic Kleene algebras. A right unfold law can be proved from the three axioms for all variants except near Kleene algebras.

At the final level of the hierarchy we extend different algebraic variants with an omega operation for infinite iteration. Since omega algebras [10] will not appear any further in this paper, we refer to our repository for details.

Our repository also contains semigroups, semirings and Kleene algebras with domain and antidomain [11,13], including modal Kleene algebras [28] and demonic refinement algebras [37]. Domain semirings essentially subsume Kleene algebras with tests [23]. Most theories have been developed to the present state of knowledge. A proof environment for dynamic logics, temporal logics or Hoare logic can be obtained from them with minor effort. Structures such as concurrent Kleene algebras, action algebras and action lattices, and Kleene algebras with converse are under development.

4   Integrating Relation Algebras

Our implementation of relation algebras follows Roger Maddux's book [26], which itself is based on Alfred Tarski's original paper [35]. It uses Huntington's axioms for Boolean algebras. These are rather minimalist, and we have added definitions for the partial order, the maximal and minimal element, and meet.

Class *boolean-algebra* = *plus-ord* + *uminus* + *one* + *zero* + *mult* +
  assumes *join-assoc*: $(x+y)+z = x+(y+z)$
  and *join-comm*: $x+y = y+x$
  and *compl*: $x = -((-x)+(-y))+(-((-x)+y))$
  and *one-def*: $x+(-x) = 1$
  and *zero-def*: $(-\ 1) = 0$
  and *meet-def*: $x{\cdot}y = -((-x)+(-y))$

For applications in Section 6, we introduce the concept of *conjugate* functions over a Boolean algebra [20], which gives rise to Boolean algebras with operators and Galois connections. In particular, they yield theorems for free.

Definition (**in** *boolean-algebra*)
  conjugation-p f g $\equiv \forall\, x\ y.\ (f(x){\cdot}y = 0 \longleftrightarrow x{\cdot}g(y) = 0)$

   Next we obtain relation algebras from Boolean algebras and show that every relation algebra is a dioid.

Class *relation-algebra* = *boolean-algebra* + *composition* + *unit* + *converse* +
  assumes *comp-assoc*: $(x;y);z = x;(y;z)$
  and *comp-unitr*: $x;e = x$
  and *comp-distr*: $(x+y);z = x;z + y;z$
  and *conv-invol*: $(x^{\smile})^{\smile} = x$
  and *conv-add*: $(x+y)^{\smile} = x^{\smile}+y^{\smile}$
  and *conv-contrav*: $(x;y)^{\smile} = y^{\smile};x^{\smile}$
  and *comp-res*: $x^{\smile};(-(x;y)) \leq -y$

Sublocale *relation-algebra* $\subseteq$ *dioid-one-zero* (*op* +) (*op* ;) (*op* $\leq$) (*op* <) *0* *e*
— automatic proof omitted

In this case, we establish a sublocale (instead of a subclass) relationship because different signatures need to be matched.
   To link relation algebras with Kleene algebras, we add a reflexive-transitive closure operation.

Class *relation-algebra-rtc* = *relation-algebra* + *star* +
  assumes *rtc-unfoldl*: $e+x;x^{\star} \leq x^{\star}$
  and *rtc-inductl*: $z+x;y \leq y \longrightarrow x^{\star};z \leq y$
  and *rtc-inductr*: $z+y;x \leq y \longrightarrow z;x^{\star} \leq y$

Sublocale *relation-algebra-rtc* ⊆
   kleene-algebra (*op* +) (*op* ;) (*op* ≤) (*op* <) *0 e* (*op* ⋆)
  — automatic proof omitted

All facts for Kleene algebra are now available for relation algebras; the difference between reasoning in Kleene algebra and relational reasoning becomes often invisible for users.

Our implementation of relation algebra includes the most important textbook concepts and theorems about functions, subidentities or tests, domain and range elements, vectors and points.

Relation algebras have previously been implemented in Isabelle/HOL by Gritzner and von Oheimb [30]. At that time, however, type classes, locales and ATP integration were not available, and proving relational theorems required strong user interaction. Their system therefore uses a translation to relational atom structures and complex algebras to increase the degree of automation, which was certainly impressive in 1997. A comparison nicely illustrates the technological advances in automated and interactive theorem proving over the last decade.

5   Automated Algebraic Proofs with Isabelle

Our development of a repository of verified theorems for relational and algebraic methods in Isabelle is largely based on ATP via Sledgehammer. To our knowledge, it is the first larger case study on automated algebraic reasoning within Isabelle.

In this paper, we can only present a few examples out of the more than 800 proofs in our repository. Our main intention is to illustrate the range of possibilities available and to compare this new approach to our previous work with standalone ATP systems.

We show that, in Kleene algebras, the Kleene star indeed axiomatises the reflexive transitive closure. (In our repository, these proofs have been obtained in slightly weaker algebras.) The first proof goal, namely that $x^\star$ is reflexive, transitive, and extends x, can be stated as follows.

Lemma (in *kleene-algebra***)** *star-rtc1*: $1+x+x^\star{\cdot}x^\star \leq x^\star$
  — proof below

To prove this lemma we called Sledgehammer, which itself called a number of external ATP systems. In this case the external ATPs returned with proofs within seconds. In a second round, the ATP systems were rerun to minimise the hypothesis set needed. Finally, Isabelle's built-in ATP system Metis performed proof search on a selected hypothesis set to reconstruct the external proof within Isabelle. In our example, this yields

  by (*metis add-lub star-ext star-ref star-trans*)

This list shows that Sledgehammer has "learned" appropriate hypotheses for proving this fact. Instead of the star axioms, for instance, two lemmas are used that have previously been verified. This sort of proof management is not supported by standalone ATP systems.

Alternatively, the SMT solver Z3 can be used, whose proof is again reconstructed internally:

by (*smt add-lub star-ext star-ref star-trans*)

On our repository, the overall performance of Z3 is comparable to Metis, but both are less successful than the external ATP systems. This explains why hypothesis minimisation is an important feature of Sledgehammer.

A third variant is asking Sledgehammer for an Isar proof. In our example, the following Isar proof script is generated for lemma *star-rtc1*.

Proof –
  have $1 \leq x^\star$ **by** (*metis star-ref*)
  hence $1 \leq x^\star \wedge x \leq x^\star$ **by** (*metis star-ext*)
  hence $1 + x \leq x^\star$ **by** (*metis add-lub*)
  hence $1 + x \leq x^\star \wedge x^\star \cdot x^\star \leq x^\star$ **by** (*metis star-trans*)
  thus $1 + x + x^\star \cdot x^\star \leq x^\star$ **by** (*metis add-lub*)
qed

Individual proof steps are again verified by Metis, but, due to the change of granularity, Isar proofs can often be checked more quickly. The Isar proof feature of Sledgehammer is, however, still experimental and often rather unstable.

It now remains to show that $x^\star$ is the *least* element above x that is reflexive and transitive. The proof, which can be found in our repository, can again be fully automated, but Metis takes several minutes.

A particular benefit of the ITP approach is that users own the means of production. Instead of fully automated proofs, textbook-style proofs can be given as Isar scripts whenever Sledgehammer fails or the user chooses so. The following example shows such a proof in which, of course, the user's expertise is needed.

Lemma (**in** *kleene-algebra*) *star-slide*: $(x \cdot y)^\star \cdot x = x \cdot (y \cdot x)^\star$
Proof –
  have $x \cdot (y \cdot x)^\star \leq x \cdot (y \cdot (x \cdot y)^\star \cdot x + 1)$
    by (*metis add-comm mult-assoc star-unfoldr-eq star-slide1*
           mult-isor add-iso mult-isol)
  hence $x \cdot (y \cdot x)^\star \leq (x \cdot y \cdot (x \cdot y)^\star + 1) \cdot x$
    by (*metis distl mult-assoc mult-oner distr mult-onel*)
  hence $x \cdot (y \cdot x)^\star \leq (x \cdot y)^\star \cdot x$
    by (*metis add-comm star-unfoldl-eq*)
  thus *?thesis* **by** (*metis antisym-conv star-slide1*)
qed

In the first step, the well-known star unfold law $1 + x^\star \cdot x = x^\star$, which has previously been verified, is used. The second and third step use essentially distributivity and star unfold. The slide law $(x \cdot y)^\star \cdot x \leq x \cdot (y \cdot x)^\star$, which again has been verified before, is used in the final step.

Our experiments suggest that handwritten proofs in relation algebra and Kleene algebra can usually be translated directly into readable Isar scripts.

The theory hierarchy, combined with ATP systems and counterexample generators, helps finding the weakest structure in which theorems hold. We were able, for instance, to prove all identities that were known to hold in Kleene algebras and all formulas in omega algebras—including the above slide rule—without right star induction. These empirical observations suggest that weaker variants of Kleene (and omega) algebras may already be complete for the algebra of (omega-)regular events.

6   Higher-Order Features

Apart from managing ATP proofs, Isabelle's higher-order features are very useful for theory engineering. Here we give only two examples: the exploitation of duality for domain and range semirings, and the instantiation of theorems of Boolean algebras with operators in relation algebras.

The domain of a relation is the set of all states on which a relation is enabled. In the semiring of relations, $domain(x) \equiv \{(p,p) \mid \exists q.\,(p,q) \in x\}$. More abstractly, we use a domain operation d and the following axioms [13].

Class *domain-semiring = semiring-one-zero + plus-ord + domain-op +*
  assumes *d1*: $x+(d(x)\cdot x) = d(x)\cdot x$
  and *d2*: $d(x\cdot y) = d(x\cdot d(y))$
  and *d3*: $d(x)+1 = 1$
  and *d4*: $d(0) = 0$
  and *d5*: $d(x+y) = d(x)+d(y)$

We can prove that domain elements are precisely the fixpoints of the domain operation. This domain characterisation is not merely an equivalence between domain semiring identities; it involves an existential quantifier.

Lemma (in *domain-semiring***)** *d-fixpoint*: $(\exists y.\ x = d(y)) \longleftrightarrow d(x) = x$
  — automatic proof omitted

Semiring duality is duality with respect to opposition, that is, the order of multiplication is swapped. The notion extends to semirings with one and zero, and preservation of theorems can be expressed by the following lemma, which states that the opposite contravariant multiplication induces again a semiring[1].

Definition (in *mult***)** $x \odot y \equiv y \cdot x$

Lemma (in *semiring-one-zero***)** *dual-semiring-one-zero*:
  class.semiring-one-zero 0 1 (op +) (op ⊙)   — automatic proof omitted

In the context of domain semirings, the dual of domain is range:

Class *range-semiring = semiring-one-zero + plus-ord + range-op +*
  assumes *r1*: $x+(x\cdot r(x)) = x\cdot r(x)$
  and *r2*: $r(x\cdot y) = r(r(x)\cdot y)$

---
[1] The logically equivalent **Sublocale** *semiring-one-zero* $\subseteq$ *semiring-one-zero* 0 1 (op +) (op ⊙) is not accepted by Isabelle for technical reasons.

and *r3*: $r(x)+1 = 1$
and *r4*: $r(0) = 0$
and *r5*: $r(x+y) = r(x)+r(y)$

Sublocale *range-semiring* $\subseteq$ *domain-semiring r* (*op* +) (*op* $\leq$) (*op* <) *0 1* (*op* $\odot$)
 — automatic proof omitted

This sublocale expression states that range semirings are duals of domain semirings with respect to opposition. It allows us to obtain statements about range directly by dualising domain statements. In fact, all statements about range in our repository have been obtained this way. The following range export law, for instance, is automatically derived from its dual domain export law, which has been proved by other means.

Lemma (**in** *range-semiring*) *range-export*: $r(x \cdot r(y)) = r(x) \cdot r(y)$
 by (*metis dual.domain-export opp-mult-def*)

In the proof, the domain export law $d(d(x) \cdot y) = d(x) \cdot d(y)$, which has previously been proved, is dualised by the sublocale statement above. The above definition of $\odot$ is also needed in the proof. The hypotheses have been found automatically by Sledgehammer.

Next we show how abstract theorems about conjugate functions in Boolean algebras with operators can automatically be instantiated to concrete theorems about relation algebras. In our hierarchy, relation algebras form a subclass of Boolean algebras. Hence conjugation is available for relation algebras as well.

The following lemma shows that the functions $\lambda y.\, x; y$ and $\lambda y.\, x^{\smile}; y$ are conjugate in relation algebras. This is, in fact, one of the famous Schröder rules.

Lemma (**in** *relation-algebra*) *schroeder-1*: $(x;y) \cdot z = 0 \longleftrightarrow y \cdot (x^{\smile};z) = 0$
Proof –
 have $(x;y) \cdot z = 0 \longleftrightarrow (z^{\smile};x) \cdot y^{\smile} = 0$ **by** (*metis conv-invol peirce*)
 thus *?thesis* **by** (*metis conv-invol conv-zero conv-contrav conv-times meet-comm*)
qed

Lemma (**in** *relation-algebra*) *schroeder-1-var*:
 conjugation-p ($\lambda\ y\ .\ x\ ;\ y$) ($\lambda\ y\ .\ x^{\smile}\ ;\ y$)
 by (*metis conjugation-p-def schroeder-1*)

Lemma *schroeder-1* proves the Schröder law explicitly from Peirce's formula $(x;y) \cdot z^{\smile} = 0 \longleftrightarrow (y;z) \cdot x^{\smile} = 0$, which can be found in our repository. It is used in the proof of Lemma *schroeder-1-var* to express the conjugation property. The following modular law of relation algebra, for instance, can then be obtained automatically by instantiating an abstract modular law of Boolean algebras with operators that holds for conjugate functions.

Lemma (**in** *boolean-algebra*) *modular-1*:
 assumes *conjugation-p f g* **shows** $f(x) \cdot y \leq f(x \cdot g(y)) \cdot y$
 — proof omitted

Corollary (**in** *relation-algebra*) *modular-1'*: $(x;y) \cdot z \leq (x;(y \cdot (x^{\smile};z))) \cdot z$
 by (*metis schroeder-1-var modular-1*)

Such higher-order features are particularly useful for modal semirings and modal Kleene algebras, where dualities, conjugations and Galois connections relate the forward and backward box and diamond operators. Dualities can then be used as theorem transformers and conjugations as theorem generators. Theorems for free can thus be effectively realised by theory engineering in Isabelle.

7   Abstract versus Set-Theoretic Classes

This section discusses an alternative set-theoretic specification of algebraic structures that uses explicit carrier sets. The abstract type-based approach described in the previous sections is usually sufficient to reason *in* algebraic structures, e.g., to prove identities in Kleene algebra. However, limitations show up when reasoning *about* these structures, for instance about subalgebras. Explicit carrier sets overcome these limitations in Isabelle/HOL. They allow specifications that are more appropriate for mechanising model theory or universal algebra.

To keep our example simple, we show a carrier-based specification for domain semigroups [11] instead of domain semirings.

Class *carrier-semigroup* $=$ *mult* $+$
  fixes *carrier* :: $'a$ *set*
  assumes *m-closed*: $[\![x{\in}carrier;\ y{\in}carrier]\!] \Longrightarrow x{\cdot}y \in carrier$
  and *m-assoc*: $[\![x{\in}carrier;\ y{\in}carrier;\ z{\in}carrier]\!] \Longrightarrow (x{\cdot}y){\cdot}z = x{\cdot}(y{\cdot}z)$

Class *carrier-domain-semigroup* $=$ *carrier-semigroup* $+$ *domain-op* $+$
  assumes *d-closed*: $x{\in}carrier \Longrightarrow d(x) \in carrier$
  and *d1*: $x{\in}carrier \Longrightarrow d(x){\cdot}x = x$
  and *d2*: $[\![x{\in}carrier;\ y{\in}carrier]\!] \Longrightarrow d(x{\cdot}d(y)) = d(x{\cdot}y)$
  and *d3*: $[\![x{\in}carrier;\ y{\in}carrier]\!] \Longrightarrow d(d(x){\cdot}y) = d(x){\cdot}d(y)$
  and *d4*: $[\![x{\in}carrier;\ y{\in}carrier]\!] \Longrightarrow d(x){\cdot}d(y) = d(y){\cdot}d(x)$

In contrast to our previous abstract specifications, each assumption is now relativised to the carrier set and closure conditions for the operations have been added, as usual in algebra.

By using carrier sets, we can now prove automatically that the set of domain elements in a domain semigroup forms a domain subsemigroup. Because Isabelle/HOL does not offer dependent types, this would be difficult, if not impossible, to express without explicit carrier sets.

Lemma (in *carrier-domain-semigroup*) *domain-subsemigroup*:
  class.carrier-domain-semigroup (*op* $\cdot$) $\{x{\in}carrier.\ d(x){=}x\}$ *d*
  — automatic proof omitted

The lemma states that the set of all elements x in the carrier that satisfy $d(x) = x$ endowed with the operations $\cdot$ and d forms a domain semigroup with carrier.

This example suggests that metalogical statements could still be proved by ATP when using classes with explicit carrier sets, although set theory is now involved. Additional experiments suggest that carrier sets may cause substantial overhead and more fragile proof automation, but further evidence is needed.

The abstract and the set-theoretic level can be linked, and theorems can be transferred between them. Given an abstract algebraic structure, the universal set over its type, i.e., $\{x.\ True\}$, constitutes a suitable carrier set for the corresponding set-theoretic structure. Conversely, given a structure with explicit carrier set C, the subtype of all elements in C constitutes a suitable base type for the corresponding abstract structure. However, due to Isabelle/HOL's lack of dependent types, this subtype can be defined only when C does not depend on local parameters.

8   Integrated Point-Wise and Point-Free Reasoning

We have discussed how a hierarchy of algebraic structures can be defined in Isabelle, and we explained how this hierarchy is useful for organising theory engineering. However, in intended models, the theorems thus obtained are conditional: that is, $0^* = 1$, *provided* 0, 1 and $\cdot^*$ denote the respective operations of, for instance, a Kleene algebra. We cannot apply these theorems in concrete models unless we know that the operations in these models satisfy the axioms of (in this case) Kleene algebra. Three important models of Kleene algebras are sets of traces, formal languages, and binary relations [12]. In this section we sketch how abstract "point-free" reasoning in relation and Kleene algebra can be formally linked with "point-wise" reasoning in concrete models. A full account can again be found in our repository.

A *trace* is given by a list of odd length whose first element is a state, and in which states and actions alternate. Isabelle provides pair types $'\alpha \times '\beta$ and a polymorphic list type $'\alpha$ *list*, but has limited support for predicative subtypes. Therefore, the following (equivalent) characterisation of traces is easier to work with formally: a trace is a pair consisting of an initial state and a list of transitions, where each transition is a pair of action and successor state.

Types $('\sigma, '\alpha)$ *trace* $= '\sigma \times ('\alpha \times '\sigma)$ *list*

We define functions *first* and *last* that extract the first and last state of a trace, respectively. Multiplication of traces t and u is a partial operation that is defined when the last state of t is equal to the first state of u.

Definition $t \cdot u \equiv$ *if* $last(t) = first(u)$ *then* $(\pi_1(t), \pi_2(t)\ @\ \pi_2(u))$ *else undefined*

Here π_1, π_2 are the projections for pairs, and @ denotes list concatenation.

Multiplication can be lifted to a complex product on sets of traces in the usual way. HOL is a logic of total functions; *undefined* above—contrary to common mathematical usage—is a constant of the logic that merely denotes some completely unspecified value. In the complex product, we only consider pairs of traces whose product is defined.

Definition $T \cdot U \equiv \bigcup t \in T.\ \bigcup u \in \{u \in U.\ last(t) = first(u)\}.\ \{t \cdot u\}$

The empty set is the multiplicative zero, and the set $\bigcup_p \{(p, [])\}$ of all single-state traces is the multiplicative unit. In fact, sets of traces form a Kleene algebra (where addition is given by set union, the order coincides with the subset relation, and the star operation is given by arbitrary iterations of multiplication, as in language theory). Isabelle provides the **Interpretation** command to formally establish this relationship.

Interpretation *kleene-algebra* $(op \cup)$ $(op \cdot)$ $(op \subseteq)$ $(op \subset)$ $\emptyset$ $(\bigcup p. \{(p, [])\})$ $(\cdot^\star)$
  — proof omitted

Isabelle sets up a proof obligation that requires the user to show that the operations listed indeed satisfy the axioms of Kleene algebras. Simple axioms can be verified automatically, while harder ones—in particular verification of the star axioms, which now require induction—need user interaction. The hierarchical approach leads to well-structured interpretation proofs: the fact that sets of traces form left Kleene algebras (with 0) and Kleene algebras (with 0) in which also the right star induction axiom holds can be verified incrementally. This is useful, for instance, when attempting completeness proofs for the weakest possible axiomatisation.

In addition to the trace model of Kleene algebras, we have formalised the models of formal languages (i.e., sets of words, where words are implemented as lists, and word multiplication is given by list concatenation) and binary relations (i.e., sets of ordered pairs, with multiplication given by relative product).

Having established, for instance, that binary relations form a Kleene algebra, Isabelle immediately makes all abstract theorems of Kleene algebra proved in the system available for binary relations as well. For instance,

Lemma $y^\star \circ x^\star \leq x^\star \circ y^\star \implies (x + y)^\star \leq x^\star \circ y^\star$
  — automatic proof omitted

is the instance of the abstract Church-Rosser theorem of Kleene algebra for binary relations (with $\circ$ denoting relative product and $\cdot^\star$ the reflexive-transitive closure of a relation). We can therefore seamlessly switch back and forth between point-free abstract reasoning (at the algebraic level) and point-wise concrete reasoning (in the model of binary relations). Krauss and Nipkow [24] artfully explore this connection to decide (in)equations of binary relations using an equivalence checker for regular expressions. Making their integrated decision procedure for regular expressions available for deciding identities in Kleene algebra could increase the proof automation for this class significantly.

Models such as traces, languages and binary relations carry, of course, a richer structure than what has been captured by Kleene or relation algebras. This can to a certain extent be captured abstractly by defining Kleene algebras over quantales or relation algebras over complete Boolean algebras. While a specification of these structures in Isabelle is straightforward, the suitability of Sledgehammer in this higher-order context remains another interesting open question.

9   Future Directions

The work in this paper documents only the initial steps towards an Isabelle repository for algebraic methods. The directions for future work that arise from this work are perhaps more important than the results obtained so far. We envisage three main directions:

(a) The creation of a standardised repository for relational and algebraic methods that includes the most important variants, models and theorems, and reflects the state of the art in the field.
(b) The integration of this repository into a development and verification environment for programs and computing systems that combines lightweight relational specification languages with heavyweight automation.
(c) The exploration of more advanced mathematics, e.g., the model theory and universal algebra of Kleene algebras and relation algebras within Isabelle.

These directions can best be addressed through a joint effort within the RAMiCS community, and the repository, its notation, conceptualisation, structure and design is therefore open to additions and debate. A possible way forward is the creation of a Wiki to which any researcher in the area will be able to contribute through a moderated process. A minimal requirement would be that all documents checked in must compile with Isabelle.

Using standalone ATP systems for reasoning automatically about Kleene and relation algebras showed that proofs of calculational statements at textbook level can usually be automated. But there are several limitations, in particular the fact that reasoning about data structures and data types such as numbers, arrays, lists is not sufficiently supported, and that proofs by induction are not possible. The presence of both SMT solvers and higher-order features in Isabelle/HOL seems of great benefit here and certainly deserves further exploration. Automating large parts of program analyses in this new setting seems possible, although we cannot provide any empirical evidence yet. The development and integration of decision procedures for fragments of Kleene algebras and relation algebras seems also very beneficial in this respect.

Finally, we are not aware that a systematic formalisation of model theory or universal algebra in Isabelle/HOL has so far been attempted. Our proof experiments show that some simple metamathematical concepts and proofs can efficiently be handled by ATP, which might facilitate this endeavour. Completeness proofs for variants of Kleene algebras are particularly important for integrating decision procedures and further enhancing automation.

10   Conclusion

We have shown that Isabelle/HOL is a highly useful environment for automated theorem proving in relation algebras and variants of Kleene algebras that overcomes previous limitations of standalone ATP proofs with these structures. Main advantages include theory hierarchies, proof management, hypothesis learning, cross-theory reasoning, automatisation of duality and abstraction/instantiation, integration of abstract and model-based reasoning and other higher-order features that enable metatheory reasoning. These suitably complement the sheer proof power of ATP systems on calculational proofs. The integration of SMT solvers into Isabelle promises additional benefits for reasoning about data structures and data types.

These results provide further evidence that algebraic and relational methods are very suitable as lightweight formal methods with heavyweight automation; and that our Isabelle repository is a significant step in that direction.

But there is still scope for improvement from the tool side, too. The existing automation gap between the external ATP systems and the internal proof reconstruction needs to be closed, and a wider range of ATP systems should be integrated. For this purpose, ATP output should be further standardised (cf. TSTP [34]). Other valuable features would be better ATP support for order-based reasoning (ordered chaining calculi [2]), better sort or type support, and enhanced techniques for hypothesis learning and, more generally, large hypothesis sets.

Acknowledgments. The authors would like to thank Andy Gordon, Tony Hoare, Peter Jipsen and Makarius Wenzel for fruitful discussions. We acknowledge funding from EPSRC grants EP/F067909/1 and EP/G031711/1.

References

1. Aboul-Hosn, K., Kozen, D.: KAT-ML: an interactive theorem prover for Kleene algebra with tests. J. Applied Non-Classical Logics 16(1-2), 9–34 (2006)
2. Bachmair, L., Ganzinger, H.: Ordered chaining calculi for first-order theories of transitive relations. J. ACM 45(6), 1007–1049 (1998)
3. Bauer, G., Wenzel, M.: Calculational reasoning revisited (an Isabelle/Isar experience). In: Boulton, R.J., Jackson, P.B. (eds.) TPHOLs 2001. LNCS, vol. 2152, pp. 75–90. Springer (2001)
4. Berghammer, R., Struth, G.: On automated program construction and verification. In: Bolduc, C., Desharnais, J., Ktari, B. (eds.) MPC 2010. LNCS, vol. 6120, pp. 22–41. Springer, Heidelberg (2010)
5. Bergstra, J.A., Fokkink, W.J., Ponse, A.: Process algebra with recursive operations. In: Bergstra, J.A., Ponse, A., Smolka, S.A. (eds.) Handbook of Process Algebra, pp. 333–389. Elsevier, Amsterdam (2001)
6. Bertot, Y., Castéran, P.: Interactive Theorem Proving and Program Development, Coq'Art: the Calculus of Inductive Constructions. Springer, Heidelberg (2004)
7. Blanchette, J.C., Nipkow, T.: Nitpick: A counterexample generator for higher-order logic based on a relational model finder. In: Kaufmann, M., Paulson, L.C. (eds.) ITP 2010. LNCS, vol. 6172, pp. 131–146. Springer, Heidelberg (2010)
8. Böhme, S., Nipkow, T.: Sledgehammer: Judgement day. In: Giesl, J., Hähnle, R. (eds.) IJCAR 2010. LNCS, vol. 6173, pp. 107–121. Springer, Heidelberg (2010)
9. Böhme, S., Weber, T.: Fast LCF-style proof reconstruction for Z3. In: Kaufmann, M., Paulson, L.C. (eds.) ITP 2010. LNCS, vol. 6172, pp. 179–194. Springer, Heidelberg (2010)
10. Cohen, E.: Separation and reduction. In: Backhouse, R., Oliveira, J.N. (eds.) MPC 2000. LNCS, vol. 1837, pp. 45–59. Springer, Heidelberg (2000)
11. Desharnais, J., Jipsen, P., Struth, G.: Domain and antidomain semigroups. In: Berghammer, R., Jaoua, A.M., Möller, B. (eds.) RelMiCS 2009. LNCS, vol. 5827, pp. 73–87. Springer, Heidelberg (2009)
12. Desharnais, J., Möller, B., Struth, G.: Kleene algebra with domain. ACM TOCL 7(4), 798–833 (2006)
13. Desharnais, J., Struth, G.: Internal axioms for domain semirings. Science of Computer Programming 76(3), 181–203 (2011)

14. Goranko, V.: The basic algebra of game equivalence. Studia Logica 75, 221–238 (2003)
15. Haftmann, F., Wenzel, M.: Local theory specifications in isabelle/Isar. In: Berardi, S., Damiani, F., de'Liguoro, U. (eds.) TYPES 2008. LNCS, vol. 5497, pp. 153–168. Springer, Heidelberg (2009)
16. Höfner, P., Struth, G.: Automated reasoning in kleene algebra. In: Pfenning, F. (ed.) CADE 2007. LNCS (LNAI), vol. 4603, pp. 279–294. Springer, Heidelberg (2007)
17. Höfner, P., Struth, G.: On automating the calculus of relations. In: Armando, A., Baumgartner, P., Dowek, G. (eds.) IJCAR 2008. LNCS (LNAI), vol. 5195, pp. 50–66. Springer, Heidelberg (2008)
18. Hurd, J.: System description: The Metis proof tactic. In: Benzmueller, C., Harrison, J., Schuermann, C. (eds.) ESHOL 2005, pp. 103–104. arXiv.org (2005)
19. Isabelle website, http://isabelle.in.tum.de/ (accessed February 20, 2011)
20. Jónsson, B., Tarski, A.: Boolean algebras with operators, Part I. American Journal of Mathematics 73, 891–939 (1951)
21. Kahl, W.: Calculational relation-algebraic proofs in Isabelle/Isar. In: Berghammer, R., Möller, B., Struth, G. (eds.) RelMiCS 2003. LNCS, vol. 3051, pp. 178–190. Springer, Heidelberg (2004)
22. Klein, G., et al.: seL4: Formal verification of an OS kernel. Comm. ACM 53(6), 107–115 (2010)
23. Kozen, D.: Kleene algebra with tests. ACM TOPLAS 19(3), 427–443 (1997)
24. Krauss, A., Nipkow, T.: Proof pearl: Regular expression equivalence and relation algebra. Journal of Automated Reasoning (to appear, 2011)
25. Mackenzie, D.: What in the name of Euclid is going on here? Science 307(5714), 1402–1403 (2005)
26. Maddux, R.D.: Relation Algebras. Elsevier, Amsterdam (2006)
27. McIver, A., Weber, T.: Towards automated proof support for probabilistic distributed systems. In: Sutcliffe, G., Voronkov, A. (eds.) LPAR 2005. LNCS (LNAI), vol. 3835, pp. 534–548. Springer, Heidelberg (2005)
28. Möller, B., Struth, G.: Algebras of modal operators and partial correctness. Theoretical Computer Science 351(2), 221–239 (2006)
29. Nipkow, T., Paulson, L.C., Wenzel, M.T.: Isabelle/HOL A Proof Assistant for Higher-Order Logic. LNCS, vol. 2283. Springer, Heidelberg (2002)
30. von Oheimb, D., Gritzner, T.F.: RALL: Machine-supported proofs for relation algebra. In: McCune, W. (ed.) CADE 1997. LNCS, vol. 1249, pp. 380–394. Springer, Heidelberg (1997)
31. Paulson, L.C.: Isabelle: The next seven hundred theorem provers. In: Lusk, E.L., Overbeek, R.A. (eds.) CADE 1988. LNCS, vol. 310, pp. 772–773. Springer, Heidelberg (1988)
32. Struth, G.: Abstract abstract reduction. J. Logic and Algebraic Programming 66(2), 239–270 (2006)
33. Struth, G.: et al.: Isabelle algebraic methods repository (2011), http://www.dcs.shef.ac.uk/~georg/isa (accessed February 20, 2011)
34. Sutcliffe, G., Suttner, C.: The TPTP problem library for automated theorem proving, http://www.tptp.org (accessed February 20, 2011)
35. Tarski, A.: On the calculus of relations. J. Symbolic Logic 6(3), 73–89 (1941)
36. Wenzel, M.: Isabelle/Isar— a versatile environment for human-readable formal proof documents. Ph.D. thesis, Institut für Informatik, Technische Universität München, Germany (2002)
37. von Wright, J.: Towards a refinement algebra. Science of Computer Programming 51(1-2), 23–45 (2004)

Explaining Voting Paradoxes; Including Arrow's and Sen's Theorems
(Invited Tutorial)

Donald G. Saari

Institute for Mathematical Behavioral Sciences
University of California, Irvine CA 92697-5100
dsaari@uci.edu

Abstract. What makes the field of social choice so fascinating is that it is full of complexities, assertions about the impossibility of doing what seems quite natural to do, and many mysteries. Adding to its allure is the importance of the topic; these mysteries can indicate and predict sources of societal complications. While some of these difficulties will be described in this tutorial, be assured that this will not be still another hand-wringing, 'what can go wrong' session. Instead, a main theme will be to provide hope coupled with positive assertions and new research directions. To accomplish this objective, it is necessary to explain why many of these difficulties arise, and that will be done with an emphasis on 'intuition'. In fact, expect a 'hands-on', interactive session.

As examples, the year 2011 is the sixtieth anniversary of Arrow's seminal Impossibility Theorem and the forty-first anniversary of Sen's 'Impossibility of a Paretian Liberal'. The starkly negative assertions of both results have imposed roadblocks to progress by indicating that what we want to do cannot be done. In simplified terms, Arrow's result implies that once there are more than two candidates, 'no voting rule is fair'. Indeed, it is not uncommon for a new voting method to be challenged in terms of Arrow's result. Sen's assertion, on the other hand, suggests that individual rights are impossible in that even a surprisingly minimal level of liberalism cannot coexist with Pareto's requirement of societal acceptance when there is universal agreement.

Although both assertions have proved to be strongly influential (in part because they impose barriers for progress), the issue raised in the tutorial is whether they really mean what we have thought they have for these many decades. As it will be shown, they do not. Instead, as it will be developed, both results admit new and radically different conclusions, where some of them replace discouragement with hope!

An important gain offered by this analysis is that it opens new directions of research. As hints of what they may be, a new interpretation of Arrow's result immediately suggests a wide array of previously unrecognized concerns that arise in almost all disciplines from social choice to computing to even engineering and nanotechnology. A similar added value comes from knowing what really causes

H. de Swart (Ed.): RAMICS 2011, LNCS 6663, pp. 68–69, 2011.

Sen's result; it opens the door to provide new ways to examine philosophical issues such as the dysfunctional behavior that can accompany society when it is evolving from one social norm to another.

Similarly, for centuries, even millennia, knowledgeable people have been deeply concerned about the paradoxes of voting. They should be; rather than amusing mathematics oddities, these paradoxes indicate how an election 'winner' need not be the one whom the voters really wanted. What adds significance to the concern is that 'winners' affect our future. This uncomfortable reality is what motivates 'electoral reform' movements, such as what is going on in the US and UK.

Complexities associated with voting paradoxes, however, have made voting systems difficult to analyze. Because of this, it is fair to say that 'reform movements' tend to fight the 'last war'; the proposed 'reform' may correct a problem that emerged in a previous election, but it need not prevent other, new kinds of difficulties from occurring.

Fortunately, a new approach has been developed to analyze voting rules. This methodology makes it much easier to address many of these 'voting theory' concerns. The fundamentals of this procedure will be introduced in an intuitive manner. Armed with the basics of this technique, participants will be able to construct any number of new paradoxical behaviors, and to analyze all three-candidate settings.

References

1. Saari, D.G.: Disposing Dictators, Demystifying Voting Paradoxes. Cambridge University Press, Cambridge (2008)
2. Saari, D.G.: Geometry of Voting. In: Arrow, K., Sen, A., Suzumura, K. (eds.) Handbook of Social Choice & Welfare, Ch. 27. Elsevier, Amsterdam (2010)

Relation Algebraic Approaches to Fuzzy Relations

Relations

(Invited Tutorial)

Michael Winter*

Department of Computer Science, Brock University,
St. Catharines, Ontario, Canada, L2S 3A1
mwinter@brocku.ca

Most information we encounter on a daily basis is flooded with imprecision and uncertainty. For example, we might say that Peter is tall without having a precise definition of *tallness*. Peter might be 2.00m high and is, therefore, considered tall. On the other hand, there might be Jon, measuring 1.60m, which we probably do not consider to be tall. It is not clear whether Charles with a height of 1.80m is tall or not. Most likely we would say that Charles is *somewhat* tall, meaning that 'tall' better describes Charles than it does Jon, but not as good as it describes Peter. To handle such kind of information, Zadeh [24] introduced the concept of fuzzy sets. A fuzzy set is a mapping d from a set X, the so-called universe, to the unit interval $[0,1]$. The value $d(x)$ is called the degree of membership of x in the fuzzy set d. For example, the fuzzy set t of tall persons will probably assign the degree 1 to Peter, the degree 0 to Jon, and some suitable value in between such as 0.5 to Charles. Using the minimum and the maximum operation on $[0,1]$ one can easily define intersection and union of fuzzy sets. This has been generalized by using so-called triangular norm/conorms instead of minimum and maximum. Two years after Zadeh's initial paper, Goguen [5] generalized the concept of fuzzyness even further to L-fuzzy sets. He replaced the unit interval $[0,1]$ by an arbitrary complete distributive lattice L to represent degrees of membership. One application of L-fuzziness is given when multiple criteria contribute towards the degree of membership. For example, we might consider a car *hot* if it is fast and if it has a powerful AV system, i.e., the speed as well as the AV system contribute towards the degree of membership of any car in the set of hot cars. Consequently, a car that has both will probably have degree 1 (largest element of the lattice), and a car that does not have either will probably have degree 0 (smallest element of the lattice). However, there might be a car that is fast without having an AV system and another car that is slow but does have a powerful AV system. Both cars should get a degree of membership somewhere between 0 and 1. Using the unit interval would require to rank the cars with respect to 'hotness' since every pair of values in $[0,1]$ can be compared. Using distributive lattices allows to model this situation without an artificial ranking. Similar to fuzzy sets, one can define intersection and union of L-fuzzy sets based on the lattice meet and join or, alternatively, using semigroup

* The author gratefully acknowledge support from the Natural Sciences and Engineering Research Council of Canada.

H. de Swart (Ed.): RAMICS 2011, LNCS 6663, pp. 70–73, 2011.

operations on L. Based on either theory, fuzzy sets or L-fuzzy sets, one may also define fuzzy relations as fuzzy sets of pairs. All standard operations on relations such as converse and composition can be generalized using arbitrary joins, binary meet (or minimum), or suitable semigroup operations.

A fuzzy controller is a controller based on fuzzy methods. The so-called Mamdani approach for constructing such a controller is based on several components, the rule base, a decision module, a fuzzification, and a defuzzyfication. This structure is summarized in the figure below.

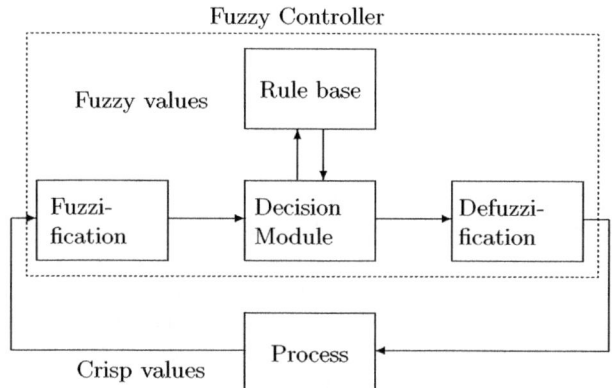

A fuzzy controller is usually formulated using linguistic entities, i.e., abstract notions represented by common words from every day language such as 'extremely high speed', 'hot water', 'very heavy rain' and so on. Variables ranging over those abstract entities are called linguistic variables. They are understood as variables over suitable L-fuzzy sets. As indicated in the example above, linguistic entities are often built up from two components. First of all, there is a basic notion of an abstract entity as 'high speed'. This basic notion may be modified by an adverb as 'very', 'roughly', 'more or less' or 'extremely'. These adverbs can be seen as linguistic modifiers, i.e., functions mapping L-fuzzy sets to L-fuzzy sets. The rule base is given by a set of simple rules using linguistic variables. In most cases those rules are formulated as conditional expressions of the form: If x is in M, then $y = N$. In such a rule x and y are linguistic in/output variables and M and N are fuzzy sets. The decision module describes to which degree a rule of the rule base is applied to a given input. Finally, the fuzzyfication and the defuzzyfication convert crisp input values into fuzzy inputs and fuzzy results into crisp results, respectively.

The calculus of binary relation in its different forms [2,8,9,10,13,16] has been used intensively in applications of mathematics and computer science. For some examples we refer to [1,11,12], the COST Action 274 TARSKI [14,15], and the conference series on Relational Methods in Computer Science (RelMiCS), of course. All of the approaches above are based on algebraic properties of relations, i.e., they handle relations as first-order entities instead of defining them as sets of pairs of certain elements. As a consequence (semi)automatic proof systems become easily available and can be used in applications. On the other hand, relation algebras may not be representable as binary relations on a set (or between multiple sets). In addition, the calculus of relations also includes fuzzy and L-fuzzy relations as models. For example, it is well-known that a reasonable category of L-relations constitutes a Dedekind category as introduced in [9]. There have been multiple attempts to capture fuzzyness and fuzzy concepts within these theories. The papers [3,6,7] introduce and investigate the concept of fuzzy relation algebras as an algebraic formalization of fuzzy relations. Those algebras are equipped with a semi-scalar multiplication, i.e., an operation mapping an element from $[0,1]$ and a fuzzy relation to a fuzzy relation. Using this operation it was even possible to characterize when a fuzzy relation is crisp, i.e., all degrees of membership are either 0 or 1. Unfortunately, this approach cannot be generalized to arbitrary lattices L because such a semi-scalar multiplication might not exist. Furthermore, it was shown in [17] that there is no formula in the theory of Dedekind categories expressing the fact that a given L-fuzzy relation is crisp. Therefore, the concept of Goguen categories was introduced and studied in a series of papers [17,18,19,20,21] and summarized in a book [22]. Goguen categories are based on a Dedekind category with two extra operators $R^{\uparrow}$, the support of R, and $R^{\downarrow}$, the kernel of R, mapping a relation R to two crisp relations. In fact, $R^{\uparrow}$ is the greatest crisp relation contained in R, and $R^{\downarrow}$ is the smallest crisp relation containing R. Recently, weaker theories based on fewer axioms and/or only one of the two operators have been studied [4,23].

In this tutorial we want to provide an overview of the theory of Goguen categories and related structures. In particular, we want to investigate their relationship, opportunities to define relational operators based on semigroup operations, the representation theory of Goguen categories, and the uniqueness of the kernel operation with respect to a given Dedekind category. A major part of the tutorial will focus on applying this theory to the development of an L-fuzzy controller. We will describe how to model every component of a controller including linguistic variables, rules, linguistic modifiers, and the fuzzyfication and defuzzyfication procedure as elements, operations, and construction in a Goguen category. The controller itself is described by a single term in that language. This allows mathematical reasoning about the controller using the relational calculus. In particular, properties such as totality of the controller, i.e., the property that the controller produces an output for every possible input, can be proven. Alternatively one could even use a converse approach by taking a desired property such as totality and consider it as invariant in a stepwise refinement process leading to the actual controller. We will illustrate the whole approach by constructing a concrete controller regulating the temperature of a chamber growing bacteria colonies. During the development process we will treat the totality of the controller as an invariant property. In particular, we will make the controller parametric in certain semigroup operations and compute simple conditions for them guaranteeing the totality of the controller. This example will also demonstrate the advantage of using lattices instead of the unit interval $[0,1]$ in the development of certain controllers where the notion of an optimal solution is based on multiple criteria.

References

1. Bird, R., de Moor, O.: Algebra of Programming. Prentice-Hall, Englewood Cliffs (1997)
2. Freyd, P., Scedrov, A.: Categories, Allegories. North-Holland, Amsterdam (1990)
3. Furusawa, H.: Algebraic Formalisations of Fuzzy Relations and their Representation Theorems. PhD-Thesis, Kyushu University (1998)
4. Furusawa, H., Kawahara, Y., Winter, M.: Dedekind Categories with Cutoff Operators. Accepted by Fuzzy Sets and Systems (2010)
5. Goguen, J.A.: L-fuzzy sets. J. Math. Anal. Appl. 18, 145–157 (1967)
6. Kawahara, Y., Furusawa, H.: Crispness and Representation Theorems in Dedekind Categories. Kyushu University (1997), doi:TR 143
7. Kawahara, Y., Furusawa, H.: An Algebraic Formalisation of Fuzzy Relations. Fuzzy Sets and Systems 101, 125–135 (1999)
8. Maddux, R.D.: Relation Algebras. Studies in Logic and the Foundations of Mathematics, vol. 150. Elsevier Science, Amsterdam (2006)
9. Olivier, J.P., Serrato, D.: Catégories de Dedekind. Morphismes dans les Catégories de Schröder. C.R. Acad. Sci. Paris 290, 939–941 (1980)
10. Olivier, J.P., Serrato, D.: Squares and Rectangles in Relational Categories - Three Cases: Semilattice, Distributive lattice and Boolean Non-unitary. Fuzzy Sets and Systems 72, 167–178 (1995)
11. Schmidt G., Ströhlein T.: Relationen und Graphen. Springer, Heidelberg (1989); English version: Relations and Graphs. Discrete Mathematics for Computer Scientists. EATCS Monographs on Theoret. Comput. Sci. Springer, Heidelberg (1993)
12. Schmidt, G.: Relational Mathematics. Encyclopedia of Mathematics and Its Application 132 (2011)
13. Schmidt, G., Hattensperger, C., Winter, M.: Heterogeneous Relation Algebras. In: Brink, C., Kahl, W., Schmidt, G. (eds.) Relational Methods in Computer Science, Advances in Computer Science, Springer, Vienna (1997)
14. de Swart, H., Orłowska, E., Schmidt, G., Roubens, M. (eds.): Theory and Applications of Relational Structures as Knowledge Instruments. LNCS, vol. 2929. Springer, Heidelberg (2003)
15. de Swart, H., Orłowska, E., Schmidt, G., Roubens, M. (eds.): TARSKI 2006. LNCS (LNAI), vol. 4342. Springer, Heidelberg (2006)
16. Tarski, A.: On the calculus of relations. Journal of Symbolic Logic 6(3), 73–89 (1941)
17. Winter, M.: A new Algebraic Approach to L-Fuzzy Relations convenient to study Crispness. INS Information Science 139(3-4), 233–252 (2001)
18. Winter, M.: Relational Constructions in Goguen Categories. In: de Swart, H. (ed.) RelMiCS 2001. LNCS, vol. 2561, pp. 212–227. Springer, Heidelberg (2002)
19. Winter, M.: Derived Operations in Goguen Categories. TAC Theory and Applications of Categories 10(11), 220–247 (2002)
20. Winter, M.: Representation Theory of Goguen Categories. Fuzzy Sets and Systems 138, 85–126 (2003)
21. Winter, M.: Goguen Categories. JoRMiCS 1, 339–357 (2005)
22. Winter, M.: Goguen Categories - A Categorical Approach to L-Fuzzy Relations. Trends in Logic 25 (2007)
23. Winter, M.: Arrow Categories. Fuzzy Sets and Systems 160, 2893–2909 (2009)
24. Zadeh, L.A.: Fuzzy sets. Information and Control 8, 338–353 (1965)

A First-Order Calculus for Allegories

Bahar Aameri[1] and Michael Winter[2,*]

[1] Department of Computer Science,
University of Toronto,
Toronto, Ontario, Canada, M5S 3G8
bahar@cs.toronto.edu
[2] Department of Computer Science,
Brock University,
St. Catharines, Ontario, Canada, L2S 3A1
mwinter@brocku.ca

Abstract. In this paper we present a language and first-order calculus for formal reasoning about relations based on the theory of allegories. Since allegories are categories, the language is typed in Church-style. We show soundness and completeness of the calculus and demonstrate its usability by presenting the RelAPS system; a proof assistant for relational categories based on the calculus presented here.

1 Introduction

Binary relations and categories of relations in particular have been widely used in mathematics and computer science for various purposes. For some very interesting applications we refer to [4,5,10,16,17,18]. The most general and probably most influential theory is given by the notion of an allegory and its extensions due to P. Freyd and A. Scedrov [5,10]. Further extensions of this theory including fuzzy relations have been proposed throughout the literature, e.g., [15,16,17,20].

Many systems supporting theorem proving in general and for relations in particular have been developed during the past years. However, most systems show serious disadvantages if the theory of allegories is considered. In the following we want to explore some of those systems.

A typical example for a fully automated system is Prover9 [13]. Prover9 is an automated theorem prover for first order and equational logic. Automatic features are very helpful in any theorem prover, whether fully automatic or interactive. However, system like Prover9 normally share two problems. First of all, the language of the system is not typed. The typing contained in the theory of allegories could be modeled by partial operations within an untyped language. However, this would produce additional proof obligations verifying that all entities are well defined. This is a serious disadvantage of the system in this context since the type information could be part of the language and checked in advance. Secondly, the calculus used and the proofs generated are

---

* The author gratefully acknowledges support from the Natural Sciences and Engineering Research Council of Canada.

H. de Swart (Ed.): RAMICS 2011, LNCS 6663, pp. 74–91, 2011.
© Springer-Verlag Berlin Heidelberg 2011

tailored for automatic proving. As a consequence they are usually very hard to read for a human being.

In another project, a semi-automated proof system has been developed for basic category-theoretic reasoning [12]. It is based on a first order sequent calculus that captures the basic properties of categories, functors and natural transformations as well as a small set of proof tactics that automate proof search in this calculus. Since it is a automated system, it has similar problems as Prover9. Typing is not part of its languages, hence, like Prover9, this would produce additional proof obligations. The system is also based on fixed theory and no additional operations or theories can be added to it. Therefore, it is not possible to work within the theory of allegories since the system only supports basic category theory.

RALF [3] was designed as a special purpose proof assistant for heterogeneous relation algebras with the goal of supporting proofs in a calculational style. RALF has a graphical user interface which represents theorems as trees, i.e., every term is displayed as a tree where the leaves are constants and variables and the nodes are the relational operations. This makes some terms hard to read. The system is based on a fixed axiomatization; the axioms of a heterogeneous relation algebra. It is not possible to work with weaker and/or stronger theories within the system. Furthermore, the system is no longer supported and there is no working version any longer available.

RALL [14] is another theorem proving system for heterogeneous relation algebra which has the ability of automatic proving for small theorems. It uses the Isabelle/HOL type system to support reasoning within abstract heterogeneous relation algebras with minimal effort. However, RALL limits itself to reasoning within representable relation algebras. The system works by translating relation-algebraic formulas into higher-order logic. As a consequence the system is incomplete. Moreover, this method cannot be applied to weaker structures like allegories. A further consequence of this method is that the proofs generated are proofs of the translated formulas within a fully automated system, which explains why they are very hard to read. Nevertheless, the idea to utilize Isabelle/HOL and similar systems like PVS in order to implement the theory of allegories is an interesting idea that should be investigated in future work. The first-order theory presented in this paper can be used to achieve a sound and complete implementation. Even a combination of both systems, the RelAPS system of the current paper and the implementation in HOL, is of interest. For example, certain automation could be done via one of those systems.

In this paper we present a language and first-order calculus for formal reasoning about relations. We use the theory of allegories as an underlying axiom system, i.e., a theory that is based on categories. Therefore, relations are typed and the regular operations on relations are partial, i.e., can only be applied if the relations involved have suitable type. The goal of designing the language and the calculus was to mimic human reasoning as closely as possible. The basic language was already presented and used in [19,20] in the context of fuzzy relations.

Instead of using the theory of allegories as a basis it would be possible to have a more general system and to define allegories and its operations as a theory. Notice that this could not be done with equality because equality is not finitely axiomatizable in first-order logic. However, we did not take this route because of the following reason. The equational theory of allegories as well as the equational theory of representable allegories are decidable [5,9]. Furthermore, even though the latter theory cannot finitely be axiomatized, a hierarchy of theories starting with allegories leading to representable allegories can be defined. The decision procedure indicates which theory is needed to proof the equation in question. The decision algorithm as well as a proof generation procedure have been implemented and integrated into the system [8]. Unfortunately, this feature is currently disabled because an error in the proof presented in [9] ([5] does not provide a detailed proof) was identified. Once this problem is resolved the feature will be available again.

The rules of natural deduction mimic human reasoning very closely. However, the explicit tree structure together with the practice of making assumptions and discarding them later by applying certain rules seems not to be very suitable for computer support. In particular, discarding assumptions is not a local operation; it affects whole subtrees. On the other hand, the sequent calculus keeps track of assumptions on the left-hand side of a sequent which makes every rule a local rule. The fact that the right-hand side of a sequent, i.e., the assertion or the goal of the proof, is also a list of formulas (including the empty list) does not really model human reasoning. There are versions of the sequent calculus that have a single conclusion on the right hand side, called single-conclusion calculi or intuitionistic sequent calculi. However, these calculi only cover intuitionistic logic, and, hence, they are weaker in their expressive power. Therefore, we developed a calculus which is mixture of both calculi. It is based on a sequent, i.e., keeps rules local, but has exactly one formula on the right-hand side of that sequent, i.e., there is exactly one assertion or goal at any time of the proof.

The language and the calculus have been implemented in the RelAPS system. This system is an interactive proof assistant that was designed in order to work with allegories and any possible extension thereof avoiding the problems mentioned in the systems above.

The rest of the paper is organized as follows. In Section 2 we present the basics of the theory of allegories. The Sections 3 and 4 are dedicated to the formal language and the new calculus. Finally, we present a short introduction to the RelAPS system in Section 5.

2   Relational Preliminaries

Throughout this paper, we use the following notation. To indicate that a morphism R of a category $\mathcal{R}$ has source A and target B we write $R : A \to B$. The collection of all morphisms $R : A \to B$ is denoted by $\mathcal{R}[A, B]$ and the composition of a morphism $R : A \to B$ followed by a morphism $S : B \to C$ by $R; S$. Last but not least, the identity morphism on A is denoted by $\mathbb{I}_A$.

We recall briefly some fundamentals on allegories [5] and relational constructions within them. For further details we refer to [5,16,20]. Furthermore, we assume that the reader is familiar with the basic notions from category theory. For unexplained material we refer to [2].

Definition 1. *An allegory $\mathcal{R}$ is a category satisfying the following:*

1. *For all objects A and B the collection $\mathcal{R}[A, B]$ is a meet semi-lattice, whose elements are called relations. Meet and the induced ordering are denoted by $\sqcap$ and $\sqsubseteq$, respectively.*
2. *There is a monotone operation $\breve{\ }$ (called converse) such that for all relations $Q : A \to B$ and $R : B \to C$ the following holds: $(Q; R)^{\breve{\ }} = R^{\breve{\ }}; Q^{\breve{\ }}$ and $(Q^{\breve{\ }})^{\breve{\ }} = Q$.*
3. *For all relations $Q : A \to B$, $R, S : B \to C$ we have $Q; (R \sqcap S) \sqsubseteq Q; R \sqcap Q; S$.*
4. *For all relations $Q : A \to B, R : B \to C$ and $S : A \to C$ the modular law $Q; R \sqcap S \sqsubseteq Q; (R \sqcap Q^{\breve{\ }}; S)$ holds.*

In this paper we only need some basic properties of relations in some examples. We have listed those properties in the following lemma:

Lemma 1. *Let $Q; Q' : A \to B$ and $R, R' : B \to C$ be relations. Then we have the following:*

1. *Composition is monotonic, i.e., $Q \sqsubseteq Q'$ and $R \sqsubseteq R'$ implies $Q; R \sqsubseteq Q'; R'$.*
2. *$Q \sqsubseteq Q; Q^{\breve{\ }}; Q$.*
3. *$\mathbb{I}_A^{\breve{\ }} = \mathbb{I}_A$.*

A proof of the previous lemma can be found in any of [5,10,16,17,18,20].

Besides the basic theory of allegories, the following two extension are of particular interest.

Definition 2. *An allegory $\mathcal{R}$ is called distributive if:*

1. *For every pair of objects A and B the class $\mathcal{R}[A, B]$ is a distributive lattice with smallest element $\perp\!\!\!\perp_{AB}$. Union is denoted by $\sqcup$.*
2. *For all relations $Q : A \to B$, $R, S : B \to C$ we have $Q; \perp\!\!\!\perp_{BC} = \perp\!\!\!\perp_{AC}$ and $Q; (R \sqcup S) = Q; R \sqcup Q; S)$.*

A distributive allegory is called a division allegory if for all relations $R : B \to C$ and $S : A \to C$ there is a relation $S/R : A \to B$ (called the left residual of S and R) so that for all $Q : A \to B$ we have $Q; R \sqsubseteq S$ iff $Q \sqsubseteq S/R$.

3   A First-Order Language for Allegories

The language of allegories is two-sorted. One kind of terms will denote objects, and the other kind denotes relations. In addition, the language is typed, i.e., every relational term has a source and a target. We have chosen a Church-style typing system, i.e., every relational variable, constant symbol, and function symbol has a fixed (and known) type.

Because of its differences to a regular language for first-order logic we describe the syntax of our language in the following section in detail. In particular the dependency of relational variables on object terms, and, hence, the notion of free and bound variables, is unusual and needs a proper definition.

3.1   Syntax

In order to provide a proper language for allegories, we require a countable set of object variables V_{obj} and a countable set of object constant symbols C_{obj}. The two sets V_{obj} and C_{obj} as well as similar sets introduced later are supposed to be disjoint, i.e., $V_{obj} \cap C_{obj} = \emptyset$. Notice that the current version does not allow function or predicate symbols in object terms. As indicated in Section 6 this will be added in a future version allowing several extension to user defined operations.

Definition 3. *The set of object terms consists of object variables and object constant symbols.*

We also require the following components:

- V_{rel} is a countable set of relational variables. Each variable r has a type $t_1 \to t_2$ where t_1 and t_2 are object terms. To indicate that the variable r has type $t_1 \to t_2$ we write $r : t_1 \to t_2$,
- C_{rel} is a countable set of relational constant symbols. Each constant symbol c has a type $t_1 \to t_2$ where t_1 and t_2 are object terms. To indicate that the constant symbol c has type $t_1 \to t_2$ we write $c : t_1 \to t_2$,
- F is a countable set of typed function symbols. Each function symbol f has a type $\{(t_1 \to s_1), ..., (t_n \to s_n)\} \to (t \to s)$ where $t_1, s_1, ..., t_n, s_n, t, s$ are object terms. To indicate that the variable f has type $\{(t_1 \to s_1), ..., (t_n \to s_n)\} \to (t \to s)$ we write $f : \{(t_1 \to s_1), ..., (t_n \to s_n)\} \to (t \to s)$.

Notice that the type information of all three entities are given by object terms. This allows to specify dependency and relationship between the type of parameters. For example, distributive allegories add a smallest relation $\perp\!\!\!\perp$ and a union operation $\sqcup$ to the theory of allegories. The smallest relation exists between every pair of morphisms, and union can be applied to relations between the same objects. This can be specified by defining $\perp\!\!\!\perp : a \to b$ and $\sqcup : \{(a \to b), (a \to b)\} \to (a \to b)$ where a and b are object variables.

Based on the components above we define relational terms as follows:

Definition 4. *The set of relational terms of type $s_1 \to s_2$, where s_1 and s_2 are object terms is defined recursively as follows:*

1. *If $r : s_1 \to s_2$ is a relational variable, then r is a relational term of type $s_1 \to s_2$.*

2. If $c : s_1 \to s_2$ is a relational constant symbol, then c is a relational term of type $s_1 \to s_2$.

3. If s is an object term, then $\mathbb{I}_s$ is a relational term of type $s \to s$.

4. If t is a relational term of type $s_1 \to s_2$, then $t^{\smile}$ is a relational term of type $s_2 \to s_1$.

5. If t_1 and t_2 are relational terms of type $s_1 \to s_2$, then $t_1 \sqcap t_2$ is a relational term of type $s_1 \to s_2$.

6. If t_1 and t_2 are relational terms of type $s_1 \to s_2$ resp. $s_2 \to s_3$, then $t_1 ; t_2$ is a relational term of type $s_1 \to s_3$.

7. If $t_1, ..., t_n$ are relational terms of type $s_1 \to s_1', ..., s_n \to s_n'$ and f is a n-ary function symbol with type $f : \{(s_1 \to s_1'), ..., (s_n \to s_n')\} \to (s \to s')$, then $f(t_1, ..., t_n)$ is a relational term of type $s \to s'$.

In order to define formulas we need an additional component, a countable set P of typed predicate symbols. Each predicate symbol p has a type $\{(t_1 \to s_1), ..., (t_n \to s_n)\}$ where $t_1, s_1, ..., t_n, s_n$, are object terms. To indicate that the predicate symbol p has type $\{(t_1 \to s_1), ..., (t_n \to s_n)\}$ we write $p : \{(t_1 \to s_1), ..., (t_n \to s_n)\}$. Finally we can define the set of formulas.

Definition 5. *The set of formulas is defined recursively as follows:*

1. $\perp$ is a formula.

2. If t_1 and t_2 are relational terms of type $s_1 \to s_2$, then $t_1 = t_2$ is a formula.

3. If $t_1, \ldots, t_n$ are relational terms of type $s_1 \to s_1', ..., s_n \to s_n'$ and p is a n-ary predicate symbol with type $\{(s_1 \to s_1'), ..., (s_n \to s_n')\}$, then $p(t_1, \ldots, t_n)$ is a formula.

4. If φ_1 and φ_2 are formulas, then $\varphi_1 \wedge \varphi_2$ is a formula.

5. If φ_1 and φ_2 are formulas, then $\varphi_1 \vee \varphi_2$ is a formula.

6. If φ_1 and φ_2 are formulas, then $\varphi_1 \to \varphi_2$ is a formula.

7. If φ is a formula, then $\neg\varphi$ is a formula.

8. If φ is a formula and $r : s_1 \to s_2$ is a relation variable, then $(\forall r : s_1 \to s_2)\varphi$ is a formula.

9. If φ is a formula and a is an object variable, then $(\forall a)\varphi$ is a formula.

10. If φ is a formula and $r : s_1 \to s_2$ is a relation variable, then $(\exists r : s_1 \to s_2)\varphi$ is a formula.

11. If φ is a formula and a is an object variable, then $(\exists a)\varphi$ is a formula.

Notice that there is no equality on object terms. This predicate is not invariant under isomorphisms, i.e., constitute a property that is of no/minor interest in category theory since almost all concepts are defined 'up to isomorphism'. A property not invariant under isomorphisms is sometimes called 'evil'.

In the next step, we want to introduce the concept of free variables in a formula. Obviously, there are two kinds of variables: object variables and relational variables. The set of object variables in an object term is defined as usual. Since relational variables and constants are typed with object terms, the notion of an object variable in a relational term is not standard.

Definition 6. *The set of object variables $OV(t)$ and the set of relational variables $RV(t)$ of a relational term t is defined recursively as follows:*

1. $OV(r) = OV(s_1) \cup OV(s_2)$ *for a relational variable* $r : s_1 \to s_2$,
2. $OV(c) = OV(s_1) \cup OV(s_2)$ *for a relational constant symbol* $c : s_1 \to s_2$ *in* C_{rel},
3. $OV(t^{\smile}) = OV(t)$,
4. $OV(t_1; t_2) = OV(t_1 \sqcap t_2) = OV(t_1) \cup OV(t_2)$,
5. $OV(f(t_1, ..., t_n)) = OV(t_1) \cup ... \cup OV(t_n)$ *for every function symbol* $f : \{(s_1 \to s_1'), ..., (s_n \to s_n')\} \to (s \to s')$.
6. $RV(r) = \{r\}$ *for every relational variable* r,
7. $RV(c) = \emptyset$ *for every relational constant symbol* c *in* C_{rel},
8. $RV(t^{\smile}) = RV(t)$,
9. $RV(t_1; t_2) = RV(t_1 \sqcap t_2) = RV(t_1) \cup RV(t_2)$,
10. $RV(f(t_1, ..., t_n)) = RV(t_1) \cup ... \cup RV(t_n)$ *for every function symbol* f.

Now we can define free object and relational variables in a formula.

Definition 7. *The set of free object variables $OFV(\varphi)$ and the set of free relational variables $RFV(\varphi)$ of a formula φ is defined as follows:*

1. $OFV(\bot) = \emptyset$,
2. $OFV(t_1 = t_2) = OV(t_1) \cup OV(t_2)$,
3. $OFV(p(t_1, ..., t_n)) = OV(t_1) \cup ... \cup OV(t_n)$,
4. $OFV(\varphi_1 \otimes \varphi_2) = OFV(\varphi_1) \cup OFV(\varphi_2)$ *where* $\otimes \in \{\wedge, \vee, \to\}$,
5. $OFV(\neg\varphi) = OFV(\varphi)$,
6. $OFV((Qa)\varphi) = OFV(\varphi) \setminus \{a\}$ *where* $Q \in \{\forall, \exists\}$,
7. $OFV((Qr : s_1 \to s_2)\varphi) = OFV(\varphi) \cup OV(r)$ *where* $Q \in \{\forall, \exists\}$,
8. $RFV(\bot) = \emptyset$,
9. $RFV(t_1 = t_2) = RV(t_1) \cup RV(t_2)$,
10. $RFV(p(t_1, ..., t_n)) = RV(t_1) \cup ... \cup RV(t_n)$,
11. $RFV(\varphi_1 \otimes \varphi_2) = RFV(\varphi_1) \cup RFV(\varphi_2)$ *where* $\otimes \in \{\wedge, \vee, \to\}$,
12. $RFV(\neg\varphi) = RFV(\varphi)$,
13. $RFV((Qa)\varphi) = RFV(\varphi)$ *where* $Q \in \{\forall, \exists\}$,
14. $RFV((Qr : s_1 \to s_2)\varphi) = RFV(\varphi) \setminus \{r\}$ *where* $Q \in \{\forall, \exists\}$,

3.2   Semantics

As for the syntax, the dependency of relational variables on object terms requires a careful definition of the semantics of the language. In particular, the notion of an environment, i.e., a function mapping variables to values, is not standard since the collection of possible values for a relational variable may depend on the values of certain object variables. For example, assume a is an object variable, $r : a \to a$ is a relational variable, and that the environment maps a to the object A and r to a relation $R : A \to A$. An update of a to the object B now requires that r is mapped to a relation with source and target B since $r : a \to a$.

As a first step to a proper definition of the semantics we need a universe where all syntactic entities can be interpreted by suitable values.

Definition 8. *A pre-model $\mathcal{P}$ consists of the following data:*

1. *$|\mathcal{P}|$ a non-empty allegory,*
2. *For each constant symbol $c \in C_{obj}$ a constant $c^{\mathcal{P}} \in \mathrm{Obj}_{|\mathcal{P}|}$.*

In order to define the semantics of terms and formulas we have to replace the free variables of the formula by actual values. Those values are stored in so called environments.

Definition 9. *An object environment σ_o over a pre-model $\mathcal{P}$ is a function from the set of object variables to the objects of $|\mathcal{P}|$.*

We are now ready to define the value of an object term in a pre-model.

Definition 10. *The value $\mathcal{V}_{\mathcal{P}}$ of object terms under the environment σ_o is defined by:*

- *$\mathcal{V}_{\mathcal{P}}(a)(\sigma_o) = \sigma_o(a)$ for every object variable a,*
- *$\mathcal{V}_{\mathcal{P}}(c)(\sigma_o) = c^{\mathcal{P}}$ for every constant symbol $c \in C_{obj}$.*

In the next definition we define an environment for both relational and object variables.

Definition 11. *An environment $\sigma = (\sigma_o, \sigma_r)$ over a pre-model $\mathcal{P}$ is a pair of functions so that σ_o is an object environment over $\mathcal{P}$ and σ_r maps each relational variable $r : s_1 \to s_2$ to a relation $\sigma_r(r) : \mathcal{V}_{\mathcal{P}}(s_1)(\sigma_o) \to \mathcal{V}_{\mathcal{P}}(s_2)(\sigma_o)$.*

In the following σ_o and σ_r will always refer to the object and relational part of an environment σ, respectively. Similarly, we will write $\sigma(a)$ instead of $\sigma_o(a)$ for object variables a, and $\sigma(r : s_1 \to s_2)$ instead of $\sigma_r(r : s_1 \to s_2)$ for relational variables $r : s_1 \to s_2$.

Storing a new value for a variable in an environment is called update. Such an update of an environment yields again an environment. Recall that updating an object variable may change the collection of relations that a relational variable might be mapped to.

Definition 12. *The update $\sigma[A/a]$ of σ at the object variable a with the object A is defined by:*

$$\sigma[A/a](b) = \begin{cases} \sigma(b) & \text{iff } a \neq b, \\ A & \text{iff } a = b, \end{cases}$$

$$\sigma[A/a](r : s_1 \to s_2) = \begin{cases} \sigma(r : s_1 \to s_2) & \text{iff } s_1 \neq a \text{ and } s_2 \neq a, \\ R : \sigma[A/a](s_1) \to \sigma[A/a](s_2) & \text{iff } s_1 = a \text{ or } s_2 = a \end{cases}$$

for an arbitrary relation $R : \sigma[A/a](s_1) \to \sigma[A/a](s_2)$.
The update $\sigma[R/r : s_1 \to s_2]$ at the relation variable $r : s_1 \to s_2$ with the relation $R : \sigma(s_1) \to \sigma(s_2)$ is defined by:

$$\sigma[R/r : s_1 \to s_2](a) = \sigma(a),$$

$$\sigma[R/r : s_1 \to s_2](q : s_1 \to s_2) = \begin{cases} \sigma(q : s_1 \to s_2) & \text{iff } r : s_1' \to s_2' \neq s_1 \to s_2, \\ R & \text{iff } r : s_1' \to s_2' = q : s_1 \to s_2. \end{cases}$$

To ascribe meaning to all formulas, we need, besides a non empty allegory, an appropriate interpretation of each of the constant, function and predicate symbols.

Definition 13. *A relational model $\mathcal{M}$ is a pre-model with the following data:*

1. *For each $c : s_1 \to s_2$ in C_{rel} and environment σ a constant $c_\sigma^{\mathcal{M}} : \sigma(s_1) \to \sigma(s_2)$ so that $\sigma(s_1) = \sigma'(s_1)$ and $\sigma(s_2) = \sigma'(s_2)$ implies $c_\sigma^{\mathcal{M}} = c_{\sigma'}^{\mathcal{M}}$,*
2. *For each function symbol $f : \{(t_1 \to s_1), ..., (t_n \to s_n)\} \to (t \to s)$ in F and environment σ, a n-ary function $f_\sigma^{\mathcal{M}}$ which is mapping $|\mathcal{M}|[\sigma(t_1), \sigma(s_1)] \times ... \times |\mathcal{M}|[\sigma(t_n), \sigma(s_n)]$ to $|\mathcal{M}|[\sigma(t), \sigma(s)]$ so that $\sigma(t_1) = \sigma'(t_1), ..., \sigma(t_n) = \sigma'(t_n), \sigma(t) = \sigma'(t)$ and $\sigma(s_1) = \sigma'(s_1), ..., \sigma(s_n) = \sigma'(s_n), \sigma(s) = \sigma'(s)$ implies $f_\sigma^{\mathcal{M}} = f_{\sigma'}^{\mathcal{M}}$,*
3. *For each predicate symbol p in P with type $(t_1 \to s_1), ..., (t_n \to s_n)$ and environment σ, a subset $p_\sigma^{\mathcal{M}} \subseteq \{|\mathcal{M}|[\sigma(t_1), \sigma(s_1)] \times ... \times |\mathcal{M}|[\sigma(t_n), \sigma(s_n)]\}$ so that $\sigma(t_1) = \sigma'(t_1), ..., \sigma(t_n) = \sigma'(t_n), \sigma(t) = \sigma'(t)$ and $\sigma(s_1) = \sigma'(s_1), ..., \sigma(s_n) = \sigma'(s_n), \sigma(s) = \sigma'(s)$ implies $p_\sigma^{\mathcal{M}} = p_{\sigma'}^{\mathcal{M}}$.*

Notice that the interpretation of constant, function, and predicate symbols depend on (object) environments. Simply indexing these entities with objects of the allegory does not work because the type information is given by object terms which contain variables, constants, and dependencies between parameters.

In the previous definition, an object environment would be sufficient because it is just needed to get the value of an object term. Note that the value of an object term in a model $\mathcal{M}$ is the same as that in the pre-model $\mathcal{P}$ it contains, i.e., $\mathcal{V}_{\mathcal{M}}(s)(\sigma) = \mathcal{V}_{\mathcal{P}}(s)(\sigma_o)$.

Now we are ready to define the value of relational terms and the validity of formulas. Both definitions are done inductively on the structure of the language.

Definition 14. *Let $\mathcal{M}$ be a relational model and σ be an environment. The value $\mathcal{V}_{\mathcal{M}}$ of terms under the environment σ is defined by:*

1. *$\mathcal{V}_{\mathcal{M}}(r : s_1 \to s_2)(\sigma) = \sigma(r : s_1 \to s_2)$ for every relational variable $r : s_1 \to s_2$,*
2. *$\mathcal{V}_{\mathcal{M}}(c : s_1 \to s_2)(\sigma) = c_\sigma^{\mathcal{M}}$ for every constant $c : s_1 \to s_2$ in C_{rel},*
3. *$\mathcal{V}_{\mathcal{M}}(f(t_1, ..., t_n))(\sigma) = f_\sigma^{\mathcal{M}}(\mathcal{V}_{\mathcal{M}}(t_1)(\sigma), ..., \mathcal{V}_{\mathcal{M}}(t_n)(\sigma))$*
4. *$\mathcal{V}_{\mathcal{M}}(\mathbb{I}_a)(\sigma) = \mathbb{I}_{\sigma(a)}$,*
5. *$\mathcal{V}_{\mathcal{M}}(t^\smile)(\sigma) = (\mathcal{V}_{\mathcal{M}}(t)(\sigma))^\smile$,*
6. *$\mathcal{V}_{\mathcal{M}}(t_1 \sqcap t_2)(\sigma) = \mathcal{V}_{\mathcal{M}}(t_1)(\sigma) \sqcap \mathcal{V}_{\mathcal{M}}(t_2)(\sigma)$,*
7. *$\mathcal{V}_{\mathcal{M}}(t_1; t_2)(\sigma) = \mathcal{V}_{\mathcal{M}}(t_1)(\sigma); \mathcal{V}_{\mathcal{M}}(t_2)(\sigma)$.*

The next step is to define the validity of formulas.

Definition 15. *Let $\mathcal{M}$ be a relational model, and σ be an environment. The validity of a formula in $\mathcal{M}$ under σ is defined inductively as follows:*

1. *$\mathcal{M} \models_\sigma t_1 = t_2$ iff $\mathcal{V}_{\mathcal{M}}(t_1)(\sigma) = \mathcal{V}_{\mathcal{M}}(t_2)(\sigma)$,*
2. *$\mathcal{M} \models_\sigma p(t_1, ..., t_n)$ iff $(\mathcal{V}_{\mathcal{M}}(t_1)(\sigma), ..., \mathcal{V}_{\mathcal{M}}(t_n)(\sigma)) \in p_\sigma^{\mathcal{M}}$,*

3. $\mathcal{M} \models_\sigma \varphi_1 \wedge \varphi_2$ iff $\mathcal{M} \models_\sigma \varphi_1$ and $\mathcal{M} \models_\sigma \varphi_2$,

4. $\mathcal{M} \models_\sigma \varphi_1 \vee \varphi_2$ iff $\mathcal{M} \models_\sigma \varphi_1$ or $\mathcal{M} \models_\sigma \varphi_2$,

5. $\mathcal{M} \models_\sigma \varphi_1 \rightarrow \varphi_2$ iff $\mathcal{M} \models_\sigma \neg\varphi_1$ or $\mathcal{M} \models_\sigma \varphi_2$,

6. $\mathcal{M} \models_\sigma \neg\varphi$ iff $\mathcal{M} \not\models_\sigma \varphi$

7. $\mathcal{M} \models_\sigma (\forall r : s_1 \rightarrow s_2)\varphi$ iff $\mathcal{M} \models_{\sigma[R/r:s_1 \rightarrow s_2]} \varphi$ for all relations $R : \sigma(s_1) \rightarrow \sigma(s_2)$,

8. $\mathcal{M} \models_\sigma (\forall a)\varphi$ iff $\mathcal{M} \models_{\sigma[A/a]} \varphi$ for all objects A,

9. $\mathcal{M} \models_\sigma (\exists r : s_1 \rightarrow s_2)\varphi$ iff $\mathcal{M} \models_{\sigma[R/r:s_1 \rightarrow s_2]} \varphi$ for some relation $R : \sigma(s_1) \rightarrow \sigma(s_2)$,

10. $\mathcal{M} \models_\sigma (\exists a)\varphi$ iff $\mathcal{M} \models_{\sigma[A/a]} \varphi$ for some object A.

As usual we write $\mathcal{M} \models \varphi$ if $\mathcal{M} \models_\sigma \varphi$ holds for all environment σ and $\models \varphi$ if $\mathcal{M} \models \varphi$ holds for all relational models $\mathcal{M}$.

For our language versions of a coincidence and a substitution lemma hold.

Lemma 2. *Let a be an object variable, $r : s_1 \rightarrow s_2$ be a relational variable, s, s' be object terms, t, t' be relational terms, φ be a formula, and $\mathcal{M}$ be a relational model. Furthermore, suppose σ_1 and σ_2 are environments over $\mathcal{M}$ so that $\sigma_1(a) = \sigma_2(a)$ for all free object variables in s, t or φ and $\sigma_1(r : s_1 \rightarrow s_2) = \sigma_2(r : s_1 \rightarrow s_2)$ for all free relational variables r in t or φ, respectively. Then we have the following:*

1. $\mathcal{V}_\mathcal{M}(s)(\sigma_1) = \mathcal{V}_\mathcal{M}(s)(\sigma_2)$

2. $\mathcal{V}_\mathcal{M}(t)(\sigma_1) = \mathcal{V}_\mathcal{M}(t)(\sigma_2)$.

3. $\mathcal{M} \models_{\sigma_1} \psi$ iff $\mathcal{M} \models_{\sigma_2} \psi$.

4. $\mathcal{V}_\mathcal{M}(s'[s/a])(\sigma) = \mathcal{V}_\mathcal{M}(s')(\sigma[\mathcal{V}_\mathcal{M}(s)(\sigma)/a])$.

5. $\mathcal{V}_\mathcal{M}(t'[t/r])(\sigma) = \mathcal{V}_\mathcal{M}(t')(\sigma[\mathcal{V}_\mathcal{M}(t)(\sigma)/r])$.

6. $\mathcal{V}_\mathcal{M}(t'[s/a])(\sigma) = \mathcal{V}_\mathcal{M}(t')(\sigma[\mathcal{V}_\mathcal{M}(s)(\sigma)/a])$.

7. $\mathcal{M} \models_\sigma \varphi[t/r]$ iff $\mathcal{M} \models_{\sigma[\mathcal{V}_\mathcal{M}(t)(\sigma)/r]} \varphi$.

8. $\mathcal{M} \models_\sigma \varphi[s/a]$ iff $\mathcal{M} \models_{\sigma[\mathcal{V}_\mathcal{M}(s)(\sigma)/a]} \varphi$.

For a proof of the lemma above we refer to [1].

4   A First-Order Calculus

In this section we introduce our first-order logic calculus for relational categories. The calculus is formulated in a sequent style [6] but with exactly one formula on the right-hand side. It has three different types of rules; axioms, which represent the basic tautologies of logic and the axioms of the theory of allegories, structural rules, which operate on the sequent of formula in a judgment, and logical rules, which are concerned with the logical operations.

Definition 16. *The axioms in Figures 1 and the rules of Figure 2 and 3 constitute the formal calculus of allegories. If Γ is a sequence of formulas and φ is a formula, then we write $\Gamma \vdash \varphi$ to indicate that there is a derivation in the calculus ending in that sequence.*

$$\varphi \vdash \varphi \quad (\text{Axiom})$$

$$\vdash (\forall a)(\forall b)(\forall r : a \rightarrow b)\mathbb{I}_a; r = r$$

$$\vdash (\forall a)(\forall b)(\forall r : a \rightarrow b)r; \mathbb{I}_b = r$$

$$\vdash (\forall a_1)(\forall a_2)(\forall a_3)(\forall a_4)(\forall r : a_1 \rightarrow a_2)(\forall q : a_2 \rightarrow a_3)(\forall u : a_3 \rightarrow a_4)(r; q); u = r; (q; u)$$

$$\vdash (\forall a)(\forall b)(\forall r : a \rightarrow b)r \sqcap r = r$$

$$\vdash (\forall a)(\forall b)(\forall r : a \rightarrow b)(\forall q : a \rightarrow b)(\forall u : a \rightarrow b)(r \sqcap q) \sqcap u = r \sqcap (q \sqcap u)$$

$$\vdash (\forall a)(\forall b)(\forall r : a \rightarrow b)(r^{\smile})^{\smile} = r$$

$$\vdash (\forall a)(\forall b)(\forall r : a \rightarrow b)(\forall q : a \rightarrow b)(r \sqcap q)^{\smile} = r^{\smile} \sqcap q^{\smile}$$

$$\vdash (\forall a)(\forall b)(\forall r : a \rightarrow b)(\forall q : b \rightarrow c)(r; q)^{\smile} = q^{\smile}; r^{\smile}$$

$$\vdash (\forall a)(\forall b)(\forall r : a \rightarrow b)(\forall q : b \rightarrow c)(\forall u : b \rightarrow c)r; (q \sqcap u) = r; (q \sqcap u) \sqcap r; q \sqcap r; u$$

$$\vdash (\forall a)(\forall b)(\forall r : a \rightarrow b)(\forall q : b \rightarrow c)(\forall u : a \rightarrow c)r; q \sqcap u = r; (q \sqcap r^{\smile}; u) \sqcap r; q \sqcap u$$

Fig. 1. Axioms

Weakening rule $\quad \dfrac{\Gamma \vdash \psi}{\Gamma, \varphi \vdash \psi} \text{ Weak}$

Contraction rule $\quad \dfrac{\Gamma, \varphi, \varphi \vdash \psi}{\Gamma, \varphi \vdash \psi} \text{ Cont.}$

Permutation rule $\dfrac{\Gamma, \varphi_2, \varphi_1 \vdash \psi}{\Gamma, \varphi_1, \varphi_2 \vdash \psi} \text{ Perm.}$

Cut rule $\quad \dfrac{\Gamma \vdash \varphi \quad \Gamma, \varphi \vdash \psi}{\Gamma \vdash \psi} \text{ Cut}$

Fig. 2. Structural rules

Note that the =L rule in the form presented here (and implemented in the system) is not really a left rule since the equation appears on the right-hand side of $\vdash$. The rule can alternatively be formulated as

$$\dfrac{\Gamma, t_1 = t_2 \vdash \psi[t_1/r]}{\Gamma, t_1 = t_2 \vdash \psi[t_2/r]} =\text{L}'$$

left logical rules	right logical rules
$$\frac{\Gamma \vdash t_1 = t_2 \quad \Gamma \vdash \psi[t_1/r]}{\Gamma \vdash \psi[t_2/r]} \ =\text{L}$$	$$\frac{}{\vdash t = t} \ =\text{R}$$
$$\frac{\Gamma, \varphi_1, \varphi_2 \vdash \psi}{\Gamma, \varphi_1 \wedge \varphi_2 \vdash \psi} \ \wedge\text{L}$$	$$\frac{\Gamma \vdash \varphi_1 \quad \Gamma \vdash \varphi_2}{\Gamma \vdash \varphi_1 \wedge \varphi_2} \ \wedge\text{R}$$
$$\frac{\Gamma, \varphi_1 \vdash \psi \quad \Gamma, \varphi_2 \vdash \psi}{\Gamma, \varphi_1 \vee \varphi_2 \vdash \psi} \ \vee\text{L}$$	$$\frac{\Gamma \vdash \varphi_1}{\Gamma \vdash \varphi_1 \vee \varphi_2} \ \vee\text{R} \qquad \frac{\Gamma \vdash \varphi_2}{\Gamma \vdash \varphi_1 \vee \varphi_2} \ \vee\text{R}$$
$$\frac{\Gamma \vdash \varphi_1 \quad \Gamma, \varphi_2 \vdash \psi}{\Gamma, \varphi_1 \rightarrow \varphi_2 \vdash \psi} \ \rightarrow\text{L}$$	$$\frac{\Gamma, \varphi_1 \vdash \varphi_2}{\Gamma \vdash \varphi_1 \rightarrow \varphi_2} \ \rightarrow\text{R}$$
$$\frac{\Gamma \vdash \varphi}{\Gamma, \neg\varphi \vdash \psi} \ \neg\text{L}$$	$$\frac{\Gamma, \varphi \vdash \bot}{\Gamma \vdash \neg\varphi} \ \neg\text{R}$$
$$\frac{\Gamma, \varphi[t/r] \vdash \psi}{\Gamma, (\forall r : s_1 \rightarrow s_2)\varphi \vdash \psi} \ \forall\text{L (rel)}$$	$$\frac{\Gamma \vdash \varphi}{\Gamma \vdash (\forall r : s_1 \rightarrow s_2)\varphi} \ \forall\text{R (rel)}$$ If r does not occur free in any formula of Γ.
$$\frac{\Gamma, \varphi[s/a] \vdash \psi}{\Gamma, (\forall a)\varphi \vdash \psi} \ \forall\text{L (obj)}$$	$$\frac{\Gamma \vdash \varphi}{\Gamma \vdash (\forall a)\varphi} \ \forall\text{R (obj)}$$ If a does not occur free in any formula of Γ.
$$\frac{\Gamma, \varphi \vdash \psi}{\Gamma, (\exists r : s_1 \rightarrow s_2)\varphi \vdash \psi} \ \exists\text{L (rel)}$$ If r does not occur free in any formula of Γ and in ψ.	$$\frac{\Gamma \vdash \varphi[t/r]}{\Gamma \vdash (\exists r : s_1 \rightarrow s_2)\varphi} \ \exists\text{R (rel)}$$
$$\frac{\Gamma, \varphi \vdash \psi}{\Gamma, (\exists a)\varphi \vdash \psi} \ \exists\text{L (obj)}$$ If a does not occur free in any formula of Γ and in ψ.	$$\frac{\Gamma \vdash \varphi[s/a]}{\Gamma \vdash (\exists a)\varphi} \ \exists\text{R (obj)}$$

$$\frac{\Gamma, \neg\varphi \vdash \bot}{\Gamma \vdash \varphi} \ \text{PBC}$$

Fig. 3. Logical rules

having the equation on the left-hand side. We have chosen the version presented in Figure 3 because that version seems more convenient to use. In particular this rule models the way equational reasoning is implemented using the 'Working Area' in RelAPS (see Section 5).

Theorem 1 (Soundness and Completeness). *The calculus is sound and complete, i.e., $\vdash \varphi$ iff $\models \varphi$ for all formulas φ.*

The previous theorem was shown in [1]. The soundness proof uses a straight-forward induction on the structure of the derivation. The completeness result was obtained following Henkin's method, i.e., it was actually shown that every consistent theory in our language has a model in the sense of Definition 13. Significant modifications to the original proof had to be made in order to cope with the categorical structure of the models.

5   The System RelAPS

The RelAPS system is an interactive theorem prover that can be downloaded via the following web page: `http://www.cosc.brocku.ca/~mwinter/RelAPS/`. The purpose of the RelAPS system is to provide an environment in which a user may construct a proof of certain theorems as close to a hand-written proof as possible. This provides the benefits of having a system ensuring that an individual proof-step is executed properly, while it remains the responsibility of the user to complete the proof. It should be mentioned that it is not the aim of the system to provide automated deduction.

   In order to achieve the goal mentioned above the following design decisions have been made:

1. Allegories are only the beginning of a whole hierarchy of relational categories. For an overview we refer to [11]. Therefore, the system was designed to handle extensions of the theory of allegories such as distributive allegories, division allegories and so on.
2. A lot of proofs in the theory of relations are either based on equational and/or inclusion based reasoning or on a chain of equivalences. For example, in order to proof that a partial identity is idempotent, i.e., $Q \sqsubseteq \mathbb{I}$ implies $Q = Q;Q$, one could argue as follows:

$$
\begin{aligned}
Q &\sqsubseteq Q;Q^{\smile};Q & &\text{Lemma 1(2)} \\
&\sqsubseteq Q;\mathbb{I}^{\smile};Q & &\text{assumption} \\
&= Q;\mathbb{I};Q & &\text{Lemma 1(3)} \\
&= Q;Q, & &\text{identity axiom} \\
\text{and } Q;Q &\sqsubseteq Q;\mathbb{I} & &\text{assumption} \\
&= Q. & &\text{identity axiom}
\end{aligned}
$$

Another example is the following proof of $(Q/R)/S = Q/(S;R)$ for suitable relations in a division allegory:

$$X \sqsubseteq (Q/R)/S \iff X;S \sqsubseteq Q/R \qquad \text{residual axiom}$$
$$\iff X;S;R \sqsubseteq Q \qquad \text{residual axiom}$$
$$\iff X \sqsubseteq Q/(S;R). \qquad \text{residual axiom}$$

In both examples we use a chain of inclusion or equivalences, and in each step we apply an axiom, an assumption or a lemma. The RelAPS system has a special 'working area' to perform proofs in that style. On a first look it seems that there is a mismatch between the calculational style and the calculus presented. Actually, the calculational style can be seen as a different presentation of a proof in natural deduction using some derived rules such as transitivity and symmetry of equality and equivalence and previously shown properties of the operations such as monotonicity of ';'. In order to use this style the user has to specify and to proof the required properties of the operations in advance. After that converting the proof above into a natural deduction version and vice versa is straight forward. A more detailed discussion on this topic was presented in [7].

3. The previous examples also show that humans usually use certain properties of operations and predicates without mentioning. Composition is associative which is used in the second proof. Both, composition and converse, are monotonic which is used in the first proof. Similar examples can be constructed in which the symmetry or an anti-monotonic behavior is used implicitly. The RelAPS system is capable of handling this kind of reasoning.

4. As already mentioned in the introduction we have chosen a logic calculus that mimics human reasoning closely. The proofs is constructed bottom-up and it is complete when all its subtrees have been ended by the application of axioms.

5.1 A Short Tour through the System

Upon starting RelAPS a dialog requires the user to specify which theory should be loaded. When the program is started for the first time, the only option presented is the (default) theory of allegories. Later on, user defined theories together with their set of operations and constants will also be available. After selecting an appropriate theory the user may access the RelAPS interface, which is shown in Figure 4.

The 'Assertions' (or goal) window displays the assertion of the current proof corresponding to the right-hand side of $\vdash$ in a derivation. The text area of the 'Assertions' window simply displays the current state of the assertion being worked with. The user may only work with one assertion at a time.

The 'Assumptions' window displays the assumptions that are associated with the current proof. This corresponds with the sequence Γ on the left-hand side of $\vdash$. The text area of the 'Assumptions' window allows to select any assumption in order to apply logical rules. Therefore, the order of the formulas is not important,

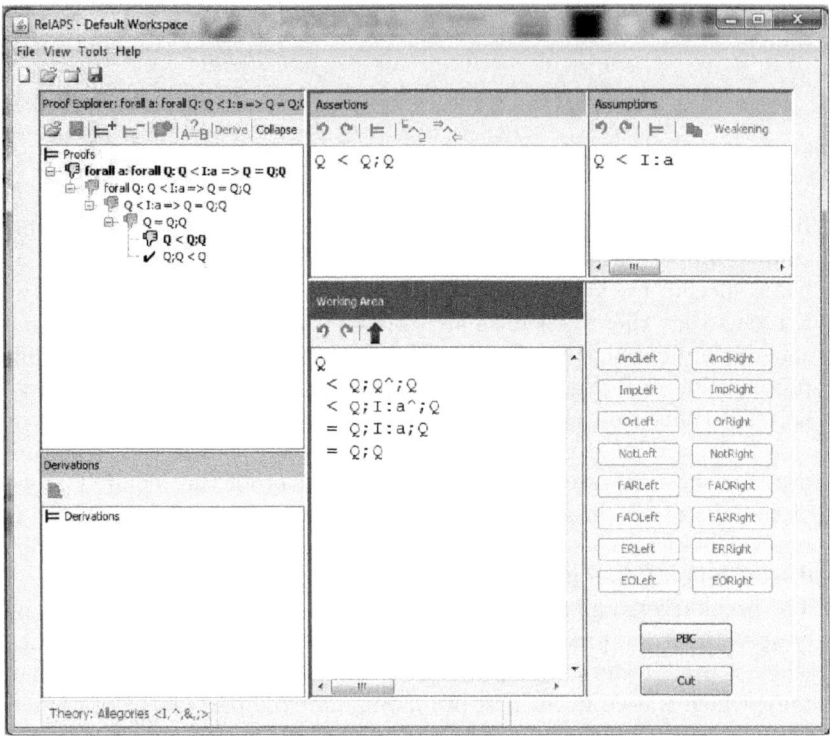

Fig. 4. The RelAPS Interface

i.e., the Permutation rule is implemented implicitly. The 'Weakening' button corresponds to the Weakening rule and removes a selected assumption from the current proof. The 'Duplicate' button implements the Contraction rule by duplicating a selected assumption.

The buttons on the right side of 'Working Area' are used to apply the logical rules, introduced in Section 4. All derivation buttons are disabled by default except PBC and Cut since the corresponding rules can always be applied. The right-hand rule buttons are enabled based on current formula in the 'Assumption' window. An appropriate left-hand rule button will be enabled when the user selects an assumption in the 'Assumptions' window. There are no button for the (=L) and (=R) rules. These rules can be applied by using the 'Working Area' window described below.

When the user selects a term of either an equation or inclusion in the 'Assertions' or 'Assumption' window, the corresponding 'Derive' button ($\models$) will become enabled. This allows the user to move the selection to the 'Working Area' in order to use equation or inclusion based reasoning. A proof in the 'Working Area' is based on the equational rules (=L) and (=R). If a subterm of the current term is selected, a menu immediately pops up allowing the user to apply an axiom, an assumption, or a previously proved theorem to the current selection. By pressing the Apply button (⇑) the original term in the 'Assertions' or 'Assumption' window will be replaced by latest term in the 'Working Area'. Instead of selecting a term, the user can also select an equation or inclusion and move the whole formula to the 'Working Area' in order to start a chain of equivalences. Notice that in the 'Working Area' certain properties of operations such as monotonicity, associativity, and symmetry are applied automatically. The system keeps track of those properties for each of the operations defined.

Each time a rule is applied the Axiom rule is automatically checked by the system. In that step the system actually checks whether the assertion is among the formulas in the assumption window, i.e., the Weakening rule is implicitly used in this process.

There are two additional buttons in the 'Assertion' window allowing the user to split an equation into two inclusions, and to split an equivalence into two implications.

In the 'Proof Explorer' window the tree of the current proof is displayed. The user can switch between the different subtrees by clicking the appropriate assertion in that tree view. Subtrees that have already successfully been proved are tagged by 'thumb up', and all other subtrees are tagged by 'thumb down'. This window also contains buttons to save and load proofs and to enter a new theorem.

A user may define a new theory. In order to do so the user has to define new operations and axioms, which can be done by selecting the appropriate function of the 'Tools' menu. During this process the user may also specify certain properties of operations such as monotonicity, associativity, and symmetry. Once specified the system requires the property of the operation either as an axiom or that it is shown as a theorem. Afterwards the property can be used in the 'Working Area' automatically as described above. A new theory can be saved and used later. In particular, it will be available as a selection in the staring dialog.

6   Conclusion and Future Work

There is plenty of further work that could improve the RelAPS system. In this section we want to sketch four of those projects.

The main focus in the future should be on automating some proof steps, particularly very basic steps. Certain sub-theories of allegories, such as the equational theory, are known to be decidable. Once it has been implemented, the system could suggest to the user that the current proof obligation is in a certain sub-theory that can be decided. If the user chooses to let the system finish the proof, the corresponding algorithm is used to find that proof.

More flexible user-defined operations is another way to extend the system. Currently it is not possible to define functions that take objects as parameters and return objects. The user only can define functions that return relations. In addition, function symbols could take a mixture of relational and object terms as parameters which would be useful for relational constructions such as splittings. Such an extension would also require a modification of the language studied in this paper.

Even though user defined predicate symbols are already part of the language, they have not been implemented in the RelAPS system, i.e., the only defined predicate symbols are '<', '>', and '='. Here the same flexibility as outlined in the previous paragraph for function symbols might be useful.

Producing LaTeX output of the proofs is another possible project. A researcher could use the system to prove theorems which are then automatically checked for correctness and use the LaTeX output for publications. For any textual representation of a RelAPS proof it would be very helpful if the user could specify the level of detail in which the proof should be presented. Depending on that level each step in the textual representation could summarize several steps in the actual derivation.

References

1. Aameri, B.: Extending RelAPS to First-Order Logic. MSc. Thesis, Brock University (2010)
2. Asperti, A., Longo, G.: Categories, Types, and Structures. Foundation of Computing Series. MIT Press, Cambridge (1991)
3. Berghammer, R., Hattensperger, C., Schmidt, G.: RALF:A Relation Algebraic Formula manipulation system and proof checker. In: Nivat, M., Rattray, C., Rus, T., Scollo, G. (eds.) Proc. 3rd Conference on Algebraic Methodology and Software Technology, Workshops in Computing, pp. 407–408. Springer, Heidelberg (1994)
4. Bird, R., de Moor, O.: Algebra of Programming. Prentice-Hall, Englewood Cliffs (1997)
5. Freyd, P., Scedrov, A.: Categories, Allegories. North-Holland, Amsterdam (1990)
6. Gentzen, G.: Untersuchungen über das logische Schließen I. Mathematische Zeitschrift 39(2), 176–210 (1934)
7. Glanfield J.: RelAPS: A Proof System for Relational Categories. Presented at RelMICS/AKA: Relations and Kleene Algebra in Computer Science. Manchester, UK. August (2006).
8. Glanfield, J.: Towards Automated Derivation in the Theory of Allegories. MSc. Thesis, Brock University (2008)
9. Gutiérrez, C.: The Arithmetic and Geometry of Allegories - normal forms and complexity of a fragment of the theory of relations. PhD thesis, Wesleyan University (1999)
10. Johnstone, P.T.: Sketches of an Elephant, A Topos Theory Compendium, vol. 1, p. 42. Oxford Logic Guides (2002)
11. Kahl, W.: Refactoring Heterogeneous Relation Algebras around Ordered Categories and Converse. J. Rel. Meth. Comp. Sci. (JoRMiCS) 1, 277–313 (2004)
12. Kozen, D., Kreitz, C., Richter, E.: Automating Proofs in Category Theory. In: Furbach, U., Shankar, N. (eds.) IJCAR 2006. LNCS (LNAI), vol. 4130, pp. 392–407. Springer, Heidelberg (2006)
13. McCune, W.: Prover9: Automated Theorem Prover for First Order and Equational Logic, http://www.cs.unm.edu/~mccune/prover9
14. Oheim, D.V., Gritzner, T.F.: RALL: Machine Supported Proofs for Relation Algebra. In: McCune, W. (ed.) CADE 1997. LNCS, vol. 1249, pp. 380–394. Springer, Heidelberg (1997)
15. Olivier, J.P., Sarrato, D.: Catégories de Dedekind. Morphismes Dans Les Categories de Schröder. C. R. Acad. Sci. Paris 290, 939–941 (1980)

16. Schmidt, G., Ströhlein, T.: Relationen und Graphen. Springer, Heidelberg (1989); English version: Relations and Graphs. Discrete Mathematics for Computer Scientists. EATCS Monographs on Theoret. Comput. Sci. Springer, Heidelberg (1989)
17. Schmidt, G., Hattensperger, C., Winter, M.: Heterogeneous Relation Algebras. In: Brink, C., Kahl, W., Schmidt, G. (eds.) Relational Methods in Computer Science, Advances in Computing Science, pp. 39–53. Springer, Vienna (1997)
18. Schmidt, G.: Relational Mathematics. Cambridge University Press, Cambridge (2010)
19. Winter, M.: A new algebraic approach to L-fuzzy relations convenient to study crispness. Information Science 139, 233–252 (2001)
20. Winter, M.: Goguen categories. A categorical approach to L-fuzzy relations. Trend in Logic 25 (2007)

Relational Modelling and Solution of Chessboard Problems

Rudolf Berghammer

Institut für Informatik, Christian-Albrechts-Universität Kiel
Olshausenstraße 40, D-24098 Kiel, Germany
rub@informatik.uni-kiel.de

Abstract. We describe a simple computing technique for solving independence and domination problems on rectangular chessboards. It rests upon relational modelling and uses the BDD-based tool RELVIEW for the evaluation of the relation-algebraic expressions that specify the problems' solutions and the visualization of the computed results. The technique described in the paper is very flexible and especially appropriate for experimentation. It can easily be applied to other chessboard problems.

1 Introduction

Problems concerning chessboards have been of interest for puzzle solvers as well as mathematicians and computer scientists for a very long time. A classical problem is the 8-queens problem. It asks for the number of possibilities to place 8 non-attacking queens on the classical chessboard with 64 squares. In the year 1850 the 92 possibilities have been found by F. Nauck [13], and E. Pauls [14] showed about 20 years later that 92 is indeed the total number of possibilities. Later on the 8-queens problem was modified and generalized in manifold ways by considering other chessboard topologies (rectangles, toroidal boards, staircase boundaries, boards with holes) and chess pieces (kings, rooks, bishops and knights). See, for example, [1,17,18] for more details.

The 8-queens problem is related to the independence (that is, mutually exclusive attacks) of chess pieces. Another interesting class of chessboard problems concerns domination. In the case of queens and the classical chessboard the queens domination problem asks for the minimum number of queens needed to dominate (i.e., attack) all squares of the chessboard and the number of possibilities to achieve such minimal dominations. It is known that 5 queens suffice and there are 4860 possibilities. Also chessboard domination has been studied already in the 19th century. According to [9], the first explicit statement of this problem was due to Abbe Durand in the year 1861, followed one year later by C.F. de Jaenisch [10]. Domination in view of queens and bishops is e.g., investigated in [8] and the rook domination problem is e.g., studied in [7]. As in the case of chessboard independence problems also the monographs [17,18] have to be mentioned because of their fundamental results.

Backtracking is one of the oldest techniques for solving chessboard problems. In this paper we propose another technique. It is based on relation algebra [15] as

H. de Swart (Ed.): RAMICS 2011, LNCS 6663, pp. 92–108, 2011.

methodological modelling means and consists essentially in the relation-algebraic specification of extremal independent and dominating sets of vertices of a given graph and the enumeration of these sets, as well as the specification of the "attack graphs" of the chess pieces in the same way. To evaluate the resulting relation-algebraic expressions and to visualize the solutions, the tool RELVIEW [2] is used. Integral parts of our technique are membership relations and size-comparison relations on powersets. If these specific relations are implemented via simple Boolean arrays, then exponential space is required and classical back-tracking is head and shoulders above the relation-algebraic approach. However, based on [11,12], in RELVIEW relations and a lot of operations on them are implemented very efficiently via binary decision diagrams (BDDs) [6] and, therefore, our method is no longer considerably slower than backtracking. We believe that it possesses some advantages. It is simple. The correctness proofs of the algorithms are very formal and they use only elementary properties of relations. All this drastically reduces the danger of making errors. Furthermore, our method is flexible and can easily be adapted to other board topologies and kinds of moves. In combination with the concise RELVIEW programs and the tool's visualization facilities this is ideal for experimenting while avoiding unnecessary overhead.

2   Relation Algebra and RELVIEW

If X and Y are sets, then a subset R of the direct product $X \times Y$ is a relation with source X and target Y. We denote the set (type) of all relations with source X and target Y (i.e., the powerset $2^{X \times Y}$ of $X \times Y$) by $[X \leftrightarrow Y]$ and write $R : X \leftrightarrow Y$ instead of $R \in [X \leftrightarrow Y]$. A (typed) relation $R : X \leftrightarrow Y$ corresponds to a predicate on $X \times Y$. If X and Y are finite, then we may consider R also as a Boolean matrix. These interpretations are well suited for many purposes and Boolean matrices are also used as one of the graphical representations of relations within RELVIEW. Therefore, in this paper we often use predicate and Boolean matrix terminology and notation. In particular, we speak of rows, columns and components of relations and write $R(x, y)$ instead of $\langle x, y \rangle \in R$ or $x \, R \, y$. We will use the following basic operations on relations: $\overline{R}$ (complement), $R \cup S$ (union), $R \cap S$ (intersection), R^{T} (transposition) and $R; S$ (composition). Furthermore, we will use the special relations O (empty relation), L (universal relation), and I (identity relation). Here we overload the symbols, i.e., avoid the binding of types to them. Finally, if $R : X \leftrightarrow Y$ is included in $S : X \leftrightarrow Y$ we write $R \subseteq S$ and equality of R and S is denoted as $R = S$.

A vector is a relation v with the specific set $\mathbf{1} := \{\perp\}$ as target. Since in $v(x, \perp)$ the argument $\perp$ is irrelevant, we write in the following $v(x)$ instead of $v(x, \perp)$. Vectors correspond to predicates on their sources and in the Boolean matrix model they are Boolean column vectors. We say that $v : X \leftrightarrow \mathbf{1}$ describes the subset Y of X if for all $x \in X$ we have $x \in Y$ iff $v(x)$. In such a case $inj(v) : Y \leftrightarrow X$ denotes the embedding relation of Y into X. This means that for all $y \in Y$ and $x \in X$ we have $inj(v)(y, x)$ iff $y = x$. To model sets we also will use the relation-level equivalents of the set-theoretic symbol "$\in$", i.e., membership *relations* $M : X \leftrightarrow 2^X$. These specific relations are defined by $M(x, Y)$ iff $x \in Y$, for all $x \in X$ and $Y \in 2^X$. A Boolean matrix representation of M requires exponential space. However, in [11] a BDD-implementation is presented, the number of vertices of which is linear in the size of X. A combination of embedding and membership relations allows a *column-wise enumeration* of a subset of a powerset. More specifically, if $v : 2^X \leftrightarrow \mathbf{1}$ describes a subset $\mathfrak{S}$ of 2^X in the sense defined above, then for all $x \in X$ and $Y \in \mathfrak{S}$ we have $(M; inj(v)^{\mathsf{T}})(x, Y)$ iff $x \in Y$. Using Boolean matrix terminology this means that the elements of $\mathfrak{S}$ are described precisely by the columns of the relation $M; inj(v)^{\mathsf{T}} : Y \leftrightarrow X$.

Given a direct product $X \times Y$, there are the projections which decompose a pair $u = \langle u_1, u_2 \rangle$ into its first component[1] u_1 and its second component u_2. For a relation-algebraic approach, it is very useful to consider instead of these functions the corresponding *projection relations* $\pi : X \times Y \leftrightarrow X$ and $\rho : X \times Y \leftrightarrow Y$ such that, given any $u \in X \times Y$, it holds that $\pi(u, x)$ iff $u_1 = x$ and that $\rho(u, y)$ iff $u_2 = y$. Projection relations algebraically allow us to specify the *pairing* $R \| S : X \times X' \leftrightarrow Y \times Y'$ of relations $R : X \leftrightarrow Y$ and $S : X' \leftrightarrow Y'$ in such a way that $(R \| S)(u, v)$ is equivalent to $R(u_1, v_1)$ and $S(u_2, v_2)$, for all $u \in X \times X'$ and $v \in Y \times Y'$. We get this property via the definition $R \| S = \pi; R; \sigma^{\mathsf{T}} \cap \rho; S; \tau^{\mathsf{T}}$, where $\pi : X \times X' \leftrightarrow X$ and $\rho : X \times X' \leftrightarrow X'$ are the two projection relations of $X \times X'$ and $\sigma : Y \times Y' \leftrightarrow Y$ and $\tau : Y \times Y' \leftrightarrow Y'$ are those of $Y \times Y'$.

As already mentioned, we use RELVIEW to evaluate the relation-algebraic expressions we will develop in this paper. RELVIEW is a specific purpose computer algebra system for the visualization and manipulation of relations and for relational prototyping and programming. It is written in C and makes full use of the X-windows graphical user interface. The underlying technique is based on a very efficient BDD-implementation of relations. Details and applications can be found, for example, in [2,3,4,5,11,12].

3   Chessboard Independence and Domination Problems

Given a classical chessboard and a chess piece P, an undirected graph may be formed with the 64 squares of the board as vertices and with two vertices being adjacent if they are different and P situated at one is able to move by one step to the other. This directly generalizes to boards with $m > 0$ rows and $n > 0$ columns. E.g., if the mn squares of the $m \times n$ board correspond to the elements of the direct product $V := X \times Y$ of the sets $X := \{1, \ldots, m\}$ and $Y := \{1, \ldots, n\}$, then the pairs $u, v \in V$ form an edge $\{u, v\}$ in the (undirected) *rooks graph* iff they are different and, furthermore, $u_1 = v_1$ or $u_2 = v_2$, i.e., if the corresponding rooks are arranged on different squares and the squares are in the same row or the same column. In a similar way the *kings graph*, the *bishops graph*, the *queens graph* and the *knights graph* are defined by means of the chess pieces' moves.

With regard to independence, for the chess piece P and an $m \times n$ chessboard then the following two questions are equivalent:

---

[1] Throughout this paper, pairs u are assumed to be of the form $u = \langle u_1, u_2 \rangle$, i.e., the first component of u is denoted by u_1 and the second component by u_2.

a) What is the largest number of non-attacking copies of P that can be placed on the board?
b) What is the *independence number* $\alpha(G_P)$ of the chessboard graph G_P for P?

The independence number $\alpha(G)$ of an undirected graph $G = (V, E)$ is the size of a maximum independent set, where an *independent* (or stable) set is a subset of the set V of vertices in which no pair of different vertices is adjacent. Furthermore, the number of possibilities to arrange a largest number of non-attacking copies of P on the chessboard equals the size of the set of all maximum independent sets of the undirected graph G_P,

In the same way the chess domination problem for the chess piece P can be reduced to a classical graph-theoretic problem. Namely, if a set of vertices of an undirected graph $G = (V, E)$ is called *dominating* (or absorbing) if for all vertices outside of it there exists at least one adjacent vertex inside of it, then the following questions are equivalent:

a) What is the least number of copies of P that have to stand on the board to ensure that each empty square is attacked by at least one copy of P?
b) What is the *domination number* $\gamma(G_P)$ of the chessboard graph G_P for P?

The domination number $\gamma(G)$ of an undirected graph $G = (V, E)$ is the size of a minimum dominating set of vertices. Again the number of possibilities to arrange a least number of copies of P such that all squares, where no piece stands, are attacked equals the size of the set of all minimum dominating sets of G_P.

The *upper domination number* $\Gamma(G)$ of an undirected graph G is the largest size of a minimal (wrt. set inclusion) dominating set. In the literature independence and domination are also combined, leading to the problem of determining the smallest size of a subset that is at the same time independent and dominating, i.e., a *kernel* of the given graph. The corresponding graph parameter is called *independent domination number* and denoted by $i(G)$.

4   Computation of Independent and Dominating Sets

Having reduced the chessboard problems we consider in this paper to classical graph-theoretic problems, in this section we show how to solve the latter ones using relation algebra. The remaining task of relation-algebraically specifying the undirected graphs for given row and column numbers and chess pieces is postponed to the next section.

Assume $G = (V, E)$ to be an undirected graph. Then we can construct from G a directed graph $G_* = (V, R)$ with relation $R : V \leftrightarrow V$ by defining $R(x, y)$ iff $\{x, y\} \in E$, i.e., iff x and y are adjacent in G, for all $x, y \in V$. The relation R is symmetric, that is, we have $R = R^\mathsf{T}$. As edges of undirected graphs are 2-element sets of vertices, R is also irreflexive, i.e., $R \subseteq \bar{\mathsf{I}}$. Obviously, there is a 1-1-correspondence between undirected graphs on V and directed graphs on V with symmetric and irreflexive edges relations. Since directed graphs are nothing else than relations on sets of vertices, in the following we identify $G_* = (V, R)$

with $R : V \leftrightarrow V$ and investigate independence and domination in the context of symmetric and irreflexive relations only. We, furthermore, assume the carrier sets to be finite. This assumption is needed when we ask for extremal sets w.r.t. size.

Let $R : V \leftrightarrow V$ be a symmetric and irreflexive relation. If we specify independence and domination within predicate logic, then $Y \in 2^V$ is independent iff the following formula (I) holds, and dominating iff the following formula (D) holds.

(I) $\forall x, y : x \in Y \wedge y \in Y \rightarrow \neg R(x, y)$ (D) $\forall x : x \notin Y \rightarrow \exists y : y \in Y \wedge R(x, y)$

In both formulae (I) and (D) the quantifiers range over the set V. Starting with (I), we can calculate as given below, where $\mathsf{M} : V \leftrightarrow 2^V$ is a membership relation and the relation L has type $[\mathbf{1} \leftrightarrow V]$, i.e., is a transposed universal vector:

$$
\begin{aligned}
\forall x, y : x \in Y \wedge y \in Y \rightarrow \neg R(x, y) &\Leftrightarrow \forall x : x \in Y \rightarrow \forall y : y \in Y \rightarrow \neg R(x, y) \\
&\Leftrightarrow \forall x : \mathsf{M}(x, Y) \rightarrow \neg \exists y : \mathsf{M}(y, Y) \wedge R(x, y) \\
&\Leftrightarrow \forall x : \mathsf{M}(x, Y) \rightarrow \neg (R; \mathsf{M})(x, Y) \\
&\Leftrightarrow \neg \exists x : \mathsf{M}(x, Y) \wedge (R; \mathsf{M})(x, Y) \\
&\Leftrightarrow \neg \exists x : \mathsf{L}(\bot, x) \wedge (\mathsf{M} \cap R; \mathsf{M})(x, Y) \\
&\Leftrightarrow \neg(\mathsf{L}; (\mathsf{M} \cap R; \mathsf{M}))(\bot, Y) \\
&\Leftrightarrow \overline{\mathsf{L}; (\mathsf{M} \cap R; \mathsf{M})}^{\mathsf{T}}(Y)
\end{aligned}
$$

As a consequence, the set Y is independent iff the Y-component of the vector $\overline{\mathsf{L}; (\mathsf{M} \cap R; \mathsf{M})}^{\mathsf{T}}$ is true so that, by the definition given in Section 2, the vector

$$
indset(R) := \overline{\mathsf{L}; (\mathsf{M} \cap R; \mathsf{M})}^{\mathsf{T}} \tag{1}
$$

of type $[2^V \leftrightarrow \mathbf{1}]$ describes the set $\mathfrak{I}$ of all independent sets of V as a subset of 2^V in the sense of Section 2. To get from (D) a relation-algebraic specification of the set $\mathfrak{D}$ of all dominating sets, we calculate for $Y \in 2^V$ as follows:

$$
\begin{aligned}
\forall x : x \notin Y \rightarrow \exists y : y \in Y \wedge R(x, y) &\Leftrightarrow \forall x : \neg \mathsf{M}(x, Y) \rightarrow \exists y : \mathsf{M}(y, Y) \wedge R(x, y) \\
&\Leftrightarrow \forall x : \overline{\mathsf{M}}(x, Y) \rightarrow (R; \mathsf{M})(x, Y) \\
&\Leftrightarrow \neg \exists x : \overline{\mathsf{M}}(x, Y) \wedge \neg (R; \mathsf{M})(x, Y) \\
&\Leftrightarrow \neg \exists x : \mathsf{L}(\bot, x) \wedge (\overline{\mathsf{M}} \cap \overline{R; \mathsf{M}})(x, Y) \\
&\Leftrightarrow \neg(\mathsf{L}; (\overline{\mathsf{M}} \cap \overline{R; \mathsf{M}}))(\bot, Y) \\
&\Leftrightarrow \overline{\mathsf{L}; (\overline{\mathsf{M}} \cap \overline{R; \mathsf{M}})}^{\mathsf{T}}(Y)
\end{aligned}
$$

This leads to the following vector $domset(R) : 2^V \leftrightarrow \mathbf{1}$ that describes $\mathfrak{D}$ as a subset of the powerset 2^V, where the relations M and L in (2) are as (1):

$$
domset(R) := \overline{\mathsf{L}; (\overline{\mathsf{M}} \cap \overline{R; \mathsf{M}})}^{\mathsf{T}} \tag{2}
$$

To obtain from (1) and (2) vectors that describe the set $\mathfrak{I}_m$ of all maximum independent sets and the set $\mathfrak{D}_m$ of all minimum dominating sets, resp., there are two possibilities.

The first one is to use relation-algebraic specifications of greatest and least elements in combination with the *size-comparison relation* $\mathsf{C} : 2^V \leftrightarrow 2^V$ that relates $X, Y \in 2^V$ iff $|X| \leq |Y|$. Then we get for the set of all maximum independent sets the vector description $maxindset(R) : 2^V \leftrightarrow \mathbf{1}$, where

$$maxindset(R) := max(\mathsf{C}, indset(R)). \tag{3}$$

In (3) the vector $max(S, v) = v \cap \overline{\overline{S}^{\mathsf{T}}; v} : X \leftrightarrow \mathbf{1}$ describes for a pre-order relation $S : X \leftrightarrow X$ and a vector $v : X \leftrightarrow \mathbf{1}$ the set of all greatest elements; cf. [15]. Analogously, for the set of all minimum dominating sets we obtain

$$mindomset(R) := min(\mathsf{C}, domset(R)) \tag{4}$$

of type $[2^V \leftrightarrow \mathbf{1}]$, where now $min(S, v) = max(S^{\mathsf{T}}, v) = v \cap \overline{\overline{S}; v} : X \leftrightarrow \mathbf{1}$ describes, for S and v as above, the set of all least elements.

In [12] it is shown that a size-comparison relation $\mathsf{C} : 2^V \leftrightarrow 2^V$ can be implemented by a BDD with $\mathcal{O}(|V|^2)$ vertices. By a modification of this implementation in the same thesis also an operation *filter* is developed such that for all $k > 0$ the vector $filter(k) : 2^V \leftrightarrow \mathbf{1}$ describes the subset $\{Y \in 2^V \mid |Y| < k\}$ of the powerset 2^V. This offers an alternative method for obtaining the sets of extremal sets of $\mathfrak{I}$ and $\mathfrak{D}$. Consider the vector $card(k) := filter(k+1) \cap \overline{filter(k)}$. As it describes the set of all subsets of V with size k, by the vector

$$indset(R, k) := indset(R) \cap card(k) \tag{5}$$

of type $[2^V \leftrightarrow \mathbf{1}]$ the set $\mathfrak{I}_k$ of all sets of $\mathfrak{I}$ with size k is described, and by

$$domset(R, k) := domset(R) \cap card(k), \tag{6}$$

a vector of the same type, the set $\mathfrak{D}_k$ of all sets of $\mathfrak{D}$ with size k is described, both as subsets of 2^V. Practical experiments with RELVIEW have shown that in the case of larger boards the specifications (5) and (6) are much more efficient than the specifications (3) and (4) and lead, even when applied iteratively, much faster to the solutions of our chessboard problems. The efficiency of (5) and (6) even can be increased if the filter-process via the vector $card(k)$ and the descriptions of the sets $\mathfrak{I}$ and $\mathfrak{D}$ are intertwined. E.g., in the case of independent sets we worked with the variant

$$indset'(R, k) := \mathsf{L}; \overline{\overline{\mathsf{M} \cap R; \mathsf{M}} \cap (card(k); \mathsf{L})^{\mathsf{T}}}^{\mathsf{T}} \tag{7}$$

that follows from (5) by simple relation-algebraic reasoning. Let c abbreviate the expression $card(k)$. Then we obtain the equivalence of (5) and (7) by

$$\overline{\overline{\mathsf{L}; (\mathsf{M} \cap R; \mathsf{M})}^{\mathsf{T}} \cap c} = \overline{\overline{\mathsf{L}; (\mathsf{M} \cap R; \mathsf{M})}^{\mathsf{T}} \cap \mathsf{L}; \overline{c; \mathsf{L}}^{\mathsf{T}}}^{\mathsf{T}}$$

$$= \overline{\mathsf{L}; (\mathsf{M} \cap R; \mathsf{M}) \cup \mathsf{L}; \overline{c; \mathsf{L}}^{\mathsf{T}}}^{\mathsf{T}}$$

$$= \overline{\mathsf{L}; ((\mathsf{M} \cap R; \mathsf{M}) \cup \overline{c; \mathsf{L}}^{\mathsf{T}})}^{\mathsf{T}}$$

$$= \overline{\mathsf{L}; (\overline{\overline{\mathsf{M} \cap R; \mathsf{M}}} \cup \overline{c; \mathsf{L}}^{\mathsf{T}})}^{\mathsf{T}}$$

$$= \mathsf{L}; \overline{\overline{\mathsf{M} \cap R; \mathsf{M}} \cap (c; \mathsf{L})^{\mathsf{T}}}^{\mathsf{T}},$$

where $c = \overline{\overline{c; \mathsf{L}}; \mathsf{L}} = \mathsf{L}; \overline{\overline{c; \mathsf{L}}}^{\mathsf{T}}$ is used in the first step. The position of the filter expression $(card(k); \mathsf{L})^{\mathsf{T}}$ in (7) was found with the help of RELVIEW, since during the evaluation of (1) the explosion of the number of BDD-vertices was caused by the composition of L and $\mathsf{M} \cap R; \mathsf{M}$. But we won't go into details here.

It is obvious that the specifications of this section also can be used to compute largest minimal dominating sets, i.e., $\Gamma(G)$, and minimum kernels, i.e., $i(G)$, if the undirected graph G is described by a symmetric and irreflexive edges relation. Furthermore, all sets of sets we have specified in this section can be enumerated as columns of relations using the technique described in Section 2. For instance, $\mathsf{M}; inj(indset(R))^{\mathsf{T}} : V \leftrightarrow \mathfrak{I}$ column-wisely enumerates the set $\mathfrak{I}$ of all independent sets and $\mathsf{M}; inj(maxindset(R))^{\mathsf{T}} : V \leftrightarrow \mathfrak{I}_m$ the set $\mathfrak{I}_m$ of all maximum ones. Executing such column-wise enumerations with RELVIEW not only immediately leads to the graph parameters and solutions of chess problems we are interested in. The tool also allows us to mark vertices of graphs with vectors, i.e., columns of relations. This is very useful for the visualization of results.

5   Computation of Chessboard Relations

To model an $m \times n$ chessboard, we assume sets $X := \{1, \ldots, m\}$ and $Y := \{1, \ldots, n\}$ for the rows and columns, resp., and represent the squares by the elements of $V := X \times Y$. In Section 3 we have already mentioned the notion of an (undirected) chessboard graph for a chess piece P and in Section 4 that it suffices to consider instead of the graphs the corresponding symmetric and irreflexive relations. As the specifications of (extremal) independent and dominating sets in Section 4, also the specifications of the *chessboard relations* in this section are based on relation algebra, supported by one additional fact. As already mentioned, all carrier sets are assumed to be finite. Now, we suppose them to be equipped with a linear strict-order and the corresponding partial successor function to be available as a relation. To be more precise, if $z_1 < z_2 < \ldots < z_n$ is the ordering of a given finite set Z, then we suppose a relation $\mathsf{S}_Z : Z \leftrightarrow Z$ to be available such that for all $x, y \in Z$ it holds $\mathsf{S}_Z(x, y)$ iff there exists $i \in \mathbb{N}$ such that $x = z_i$ and $y = z_{i+1}$. In terms of order theory, S_Z is the *cover relation* (or the Hasse diagram) of the strict-order $<$ and, thus, the latter relation equals the transitive closure of the former. In RELVIEW, the strict-order is implicitly given via the internal enumeration of the set Z within the tool, and the relation S_Z may be computed by the pre-defined operation succ.

For given $m, n > 0$, we assume $1 < 2 < \ldots < m$ to be the ordering of the $m \times n$ chessboard's row set X and $1 < 2 < \ldots < n$ to be the ordering of its column set Y. Graphically, we suppose the rows to be numbered from bottom to top and the columns from left to right, both in ascending order so that $\langle 1, 1 \rangle$ is the lowermost-leftmost square and $\langle m, n \rangle$ is the uppermost-rightmost one. Using the pairing construction of Section 2, the building blocks for the construction of the chessboard relations (except the knights relation) can be specified, viz. by

$$up(\mathsf{S}_X) := \mathsf{S}_X \,\|\, \mathsf{I} \qquad right(\mathsf{S}_Y) := \mathsf{I} \,\|\, \mathsf{S}_Y \qquad (8)$$

for the unidirectional vertical and horizontal one-square moves, resp., and by

$$pdiag(\mathsf{S}_X, \mathsf{S}_Y) := \mathsf{S}_X \parallel \mathsf{S}_Y \qquad\qquad ndiag(\mathsf{S}_X, \mathsf{S}_Y) := \mathsf{S}_X^{\mathsf{T}} \parallel \mathsf{S}_Y \qquad (9)$$

for the unidirectional one-square moves in the positive diagonal direction ("$\nearrow$") and the negative diagonal direction ("$\searrow$") of the chessboard, resp. That these four relations, each of type $[V \leftrightarrow V]$, in fact specify the claimed moves can be seen as follows. Assume $u, v \in V$ to be squares of the chessboard. With the help of the point-wise description of pairing, we can calculate as follows:

$$up(\mathsf{S}_X)(u, v) \Leftrightarrow (\mathsf{S}_X \parallel \mathsf{I})(u, v)$$
$$\Leftrightarrow \mathsf{S}_X(u_1, v_1) \wedge \mathsf{I}(u_2, v_2) \Leftrightarrow u_1 + 1 = v_1 \wedge u_2 = v_2$$

This equivalence shows that $up(\mathsf{S}_X)$ specifies for all squares $u \in V$ the move from u to the square $\langle u_1 + 1, u_2 \rangle$, i.e., to the vertical upper neighbour square, if it exists. Furthermore, for all $u, v \in V$ the subsequent equivalence holds:

$$pdiag(\mathsf{S}_X, \mathsf{S}_Y)(u, v) \Leftrightarrow (\mathsf{S}_X \parallel \mathsf{S}_Y)(u, v)$$
$$\Leftrightarrow \mathsf{S}_X(u_1, v_1) \wedge \mathsf{S}_Y(u_2, v_2) \Leftrightarrow u_1 + 1 = v_1 \wedge u_2 + 1 = v_2$$

So, $pdiag(\mathsf{S}_X, \mathsf{S}_Y)$ specifies for all squares $u \in V$ the move to its neighbour square $\langle u_1 + 1, u_2 + 1 \rangle$ w.r.t. the positive diagonal direction, again only if it exists. In the same way it can be shown that $right(\mathsf{S}_Y)$ specifies the moves to neighbour squares on the right, i.e., it holds $right(\mathsf{S}_Y)(u, v)$ iff $u_1 = v_1$ and $u_2 + 1 = v_2$ for all $u, v \in V$, and $ndiag(\mathsf{S}_X, \mathsf{S}_Y)$ specifies the moves to the neighbour squares w.r.t. the negative diagonal direction, i.e., it holds $ndiag(\mathsf{S}_X, \mathsf{S}_Y)(u, v)$ iff $u_1 - 1 = v_1$ and $u_2 + 1 = v_2$ for all $u, v \in V$.

Having specified the basic moves relation-algebraically, it is rather simple to specify the rooks and the bishops chessboard relations. A rook precisely attacks all squares of its row and its column. Using the building blocks (8) for the vertical and horizontal one-square moves, this leads to the rooks chessboard relation (or rooks "attack relation") $rook(\mathsf{S}_X, \mathsf{S}_Y) : V \leftrightarrow V$ as

$$rook(\mathsf{S}_X, \mathsf{S}_Y) := up(\mathsf{S}_X)^{\Diamond} \cup right(\mathsf{S}_Y)^{\Diamond}, \qquad (10)$$

where $R^{\Diamond} := R^+ \cup (R^+)^{\mathsf{T}}$ denotes the *symmetric closure* of the *transitive closure* $R^+ := \bigcup_{i>0} R^i$ of a relation R (the powers of R are inductively defined by $R^1 := R$ and $R^{i+1} := R; R^i$). In (10) the expression $up(\mathsf{S}_X)^{\Diamond}$ specifies the moves on the columns and the expression $right(\mathsf{S}_Y)^{\Diamond}$ those on the rows. The rule that a bishop precisely attacks the squares of the two diagonals he stands on, can be expressed with the help of the building blocks of (9) for diagonal moves as

$$bishop(\mathsf{S}_X, \mathsf{S}_Y) := pdiag(\mathsf{S}_X, \mathsf{S}_Y)^{\Diamond} \cup ndiag(\mathsf{S}_X, \mathsf{S}_Y)^{\Diamond}; \qquad (11)$$

this specifies the bishops chessboard relation $bishop(\mathsf{S}_X, \mathsf{S}_Y) : V \leftrightarrow V$. A queen can move as a rook and as a bishop. As a consequence, the queens chessboard relation $queen(\mathsf{S}_X, \mathsf{S}_Y) : V \leftrightarrow V$ is the union of the rooks chessboard relation and the bishops chessboard relation, i.e., we have

$$queen(\mathsf{S}_X, \mathsf{S}_Y) := rook(\mathsf{S}_X, \mathsf{S}_Y) \cup bishop(\mathsf{S}_X, \mathsf{S}_Y). \qquad (12)$$

A king precisely attacks all squares adjacent to the square he stands on. If we translate this rule into the language of relation algebra, use $R^{\text{M}} := R \cup R^{\text{T}}$ as symmetric closure of a relation R and $(R \cup S)^{\text{M}} = R^{\text{M}} \cup S^{\text{M}}$, then we get

$$king(S_X, S_Y) := (up(S_X) \cup right(S_Y) \cup pdiag(S_X, S_Y) \cup ndiag(S_X, S_Y))^{\text{M}} \quad (13)$$

as specification of the kings chessboard relation $king(S_X, S_Y) : V \leftrightarrow V$.

What remains is the knights chessboard relation $knight(S_X, S_Y) : V \leftrightarrow V$. A knight precisely attacks those squares which can be reached by moving two squares horizontally and then one square vertically or by moving two squares vertically and then one square horizontally. Obviously, the property $(S_X \| S_Y^2)(u, v)$ holds iff the square $v \in V$ is reached from the square $u \in V$ by vertically moving one square upwards and then horizontally moving two squares to the right. In the same way all other possibilities for a knight's move can be specified. If we use again the symmetric closure notation and the law $(R \cup S)^{\text{M}} = R^{\text{M}} \cup S^{\text{M}}$ for symmetric closures, then we get the knights chessboard relation as

$$knight(S_X, S_Y) := ((S_X^2 \| S_Y) \cup (S_X \| S_Y^2) \cup ((S_X^2)^{\text{T}} \| S_Y) \cup (S_X^{\text{T}} \| S_Y^2))^{\text{M}}. \quad (14)$$

The informal arguments we have used to obtain the relation-algebraic specifications (10) to (14) also can be replaced by formal reasoning. We will demonstrate this for the rooks chessboard relation only. The pairing of relations fulfills $(R_1 \| R_2); (S_1 \| S_2) = (R_1; S_1) \| (R_2; S_2)$. Consequently it holds that $(R \| S)^n = R^n \| S^n$ for all $n > 0$. Using this fact, the equation $\text{I} = \text{I}^n$, and that the strict-orders on X and Y are the transitive closures of their covering relations S_X and S_Y, resp., we can calculate for all $u, v \in V$ as follows (n ranges over $\mathbb{N}$)

$$
\begin{aligned}
(S_X \| \text{I})^+(u, v) &\Leftrightarrow \exists n : n > 0 \wedge (S_X \| \text{I})^n(u, v) \\
&\Leftrightarrow \exists n : n > 0 \wedge S_X^n(u_1, v_1) \wedge \text{I}^n(u_2, v_2) \\
&\Leftrightarrow S_X^+(u_1, v_1) \wedge u_2 = v_2 \\
&\Leftrightarrow u_1 < v_1 \wedge u_2 = v_2
\end{aligned}
$$

and, applying this and the corresponding results for the remaining three expressions $(S_X \| \text{I})^+(v, u), (\text{I} \| S_Y)^+(u, v)$ and $(\text{I} \| S_Y)^+(v, u)$, we get

$$
\begin{aligned}
&rook(S_X, S_Y)(u, v) \\
&\Leftrightarrow up(S_X)^{\Diamond}(u, v) \vee right(S_Y)^{\Diamond}(u, v) \\
&\Leftrightarrow (S_X \| \text{I})^+(u, v) \vee (S_X \| \text{I})^+(v, u) \vee (\text{I} \| S_Y)^+(u, v) \vee (\text{I} \| S_Y)^+(v, u) \\
&\Leftrightarrow (u_1 < v_1 \wedge u_2 = v_2) \vee (v_1 < u_1 \wedge v_2 = u_2) \vee \\
&\quad\quad (u_1 = v_1 \wedge u_2 < v_2) \vee (v_1 = u_1 \wedge v_2 < u_2) \\
&\Leftrightarrow (u_1 \neq v_1 \wedge u_2 = v_2) \vee (u_1 = v_1 \wedge u_2 \neq v_2) \\
&\Leftrightarrow u \neq v \wedge (u_1 = v_1 \vee u_2 = v_2).
\end{aligned}
$$

The last line of this equivalence is the formalization of the fact that a rook on square u attacks the square v,

In RELVIEW relations and graphs interactively can be manipulated on the screen. This allows to play and experiment with further board topologies without large effort. Very regular non-standard chessboards, like cylindric and toroidal ones, even automatically can be constructed using variants of the relation-algebraic specifications developed in this section.

6   Experimental Results Concerning Independence

The specifications of the last sections can be transformed immediately into code for RELVIEW. In the following we only present some of the results on chess-board independence we have obtained with the tool and do not mention results concerning dominance. For symmetry reasons we may assume $m \leq n$.

It is rather obvious that for such a chessboard the largest number of non-attacking rooks equals m and that there are exactly $n(n-1)\cdots(n-m+1)$, that is, $\frac{n!}{(n-m)!}$ possibilities to arrange m rooks on the board. This result was confirmed by our RELVIEW experiments, as shown by the left-hand table of Figure 1 for the values $3 \leq m \leq n \leq 8$.

In the other table of Figure 1 the largest numbers of non-attacking kings and, separated by a slash, the numbers of possibilities to arrange them, are shown. The largest number of non-attacking kings is given by $\lfloor \frac{m+1}{2} \rfloor \lfloor \frac{n+1}{2} \rfloor$, where the floor expression $\lfloor r \rfloor$ specifies the integer part of the real number r. This generalizes the result of [18] for quadratic chessboards. Our result can be proved by a simple generalization of the proof of [18,17], since also an $m \times n$ chessboard can be partitioned into $\lfloor \frac{m}{2} \rfloor \lfloor \frac{n}{2} \rfloor$ 2×2 parts and, if m and/or n are odd, additional 1×2 parts for the bottom row, 2×1 parts for the rightmost column, and a 1×1 part for the lower right-hand corner. For the four board sizes 2×8, 2×9, 3×8, and 3×9, resp., this partitioning is illustrated in Figure 2 (again produced with the help of RELVIEW), where the black squares denote the positions of the 4, 5, 8, and 10 kings, resp. Very little is known about the number of possibilities to arrange a maximum number of non-attacking kings. The only result we are aware of is from [17] and says that for m and n being odd numbers there is only one arrangement. This is due to the fact that the placement of the king in the 1×1 part uniquely determines the arrangement of all other kings. See again Figure 2 and compare also with the entries of the right-hand table of Figure 1.

The two tables of Figure 3 show some of the experimental resulta we have obtained for the arrangement of non-attacking knights (left-hand table) and

	3	4	5	6	7	8
3	6	24	60	120	210	336
4		24	120	360	840	1680
5			120	720	2520	6720
6				720	5040	20160
7					5040	40320
8						40320

	3	4	5	6	7	8
3	4/1	4/9	6/1	6/16	8/1	8/25
4		4/79	6/27	6/408	8/81	8/1847
5			9/1	9/64	12/1	12/125
6				9/3600	12/256	12/26040
7					16/1	16/625
8						16/281571

Fig. 1. Independence for rooks and kings

Fig. 2. Partitioning of rectangular chessboards

	3	4	5	6	7	8
3	5/2	6/3	8/2	9/4	11/1	12/2
4		8/6	10/3	12/3	14/3	16/3
5			13/1	15/2	18/1	20/2
6				18/2	21/2	24/2
7					25/1	28/2
8						32/2

	3	4	5	6	7	8
3	4/8	6/1	7/3	8/4	9/5	10/9
4		6/16	8/1	8/81	10/9	10/400
5			8/32	10/1	11/9	12/25
6				10/64	12/1	12/729
7					12/128	14/1
8						14/256

Fig. 3. Independence for knights and bishops

bishops (right-hand table). In [17] it is shown that for $3 \leq m = n$ the largest number of non-attacking knights equals $\lfloor \frac{m^2+1}{2} \rfloor$. The used arguments can be modified in such a way that they also prove the equality of $\lfloor \frac{mn+1}{2} \rfloor$ and the largest number of non-attacking knights under the assumption $m \leq n$. The number $\lfloor \frac{mn+1}{2} \rfloor$ is the maximum of the number w of white squares and the number b of black squares, resp. If both m and n are odd and, as usual, the lowermost-leftmost square $\langle 1, 1 \rangle$ is black, then it holds $w = b - 1$ and there exists for $m, n \geq 5$ exactly one maximum arrangement: Put the knights on the black squares. Otherwise, we have $w = b$. If additionally $m > 4$, the arrangement of all knights on the black squares or, alternatively, on the white squares, are the only maximum independent arrangements. In the case $m = 4$ and $n \geq 5$ besides these two possibilities there is a third one: Place n knights on row 1 and n knights on row 4 (note, that $\lfloor \frac{4n+1}{2} \rfloor = \frac{4n}{2} = 2n$). For $m = n = 4$ even three further possibilities exist, as demonstrated by the RELVIEW-pictures of Figure 4. Each of them shows the 4×4 knights chessboard graph and three of the six maximum independent sets; the latter indicated by black vertices.

The largest numbers of non-attacking bishops for $m = n$ is $2m - 2$ and there are 2^m possibilities. This result of [18] is indicated by the diagonal of the right-hand table of Figure 3. Experimental results for the arrangement of bishops on rectangular boards also can be found in [16]. In almost all cases they coincide

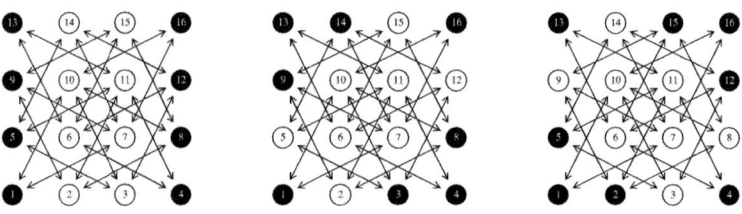

Fig. 4. Three of the 6 arrangements of 8 non-attacking knights on the 4×4 chessboard

with our results; we only have corrected the case $m = 4$ and $n = 8$ (in [16] $m = 8$ and $n = 4$) from $10/144$ to $10/400$ and the case $m = 6$ and $n = 8$ (in [16] $m = 8$ and $n = 6$) from $12/324$ to $12/729$. Our results also indicate that for $m < n$ it is possible to arrange at most $m + n - 2$ non-attacking bishops if m and n are even, and $m + n - 1$ non-attacking bishops otherwise. This corresponds with the results given in [16]. Furthermore, our RELVIEW experiments have shown that on non-quadratic boards maximum arrangements of non-attacking bishops are possible such that some of the bishops are placed in the interior of the board. In the case of quadratic boards all bishops of a maximum independent arrangement have to be placed on the outer ring of squares; cf. [17,18].

A bishop always stays on squares of a single colour. Hence the bishops graph possesses two connected components. This fact can be used to reduce the costs for solving the bishops independence problem. Suppose $B : V \leftrightarrow V$ to be the bishops chessboard relation. If we use the vector $p : V \leftrightarrow \mathbf{1}$ to describe the set $\{\langle 1, 1\rangle\}$ (i.e., the black lowermost-leftmost square), then the vector $b := p \cup B^+; p : V \leftrightarrow \mathbf{1}$ describes the subset V_b of V consisting of the black squares and, hence, its complement $\bar{b}$ describes the subset V_w of the white squares. The restriction of B to V_b is obtained via $B_b := inj(b); B; inj(b)^\mathsf{T} : V_b \leftrightarrow V_b$ and its restriction to V_w via $B_w := inj(\bar{b}); B; inj(\bar{b})^\mathsf{T} : V_w \leftrightarrow V_w$. If we identify undirected graphs and their edges relations, then from the independence numbers of B_b and B_w we get that of B as their sum. Furthermore, the number of maximum independent sets of B is the product of the numbers of maximum independent sets of B_b and B_w. We have computed via this approach e.g., the numbers for the relation B of the 13×16 chessboard. RELVIEW delivered for B_b as well as for B_w the bishops independence number 14 and exactly 233 maximum independent sets. Hence, for the 13×16 board the largest number of non-attacking bishops is $14 + 14 = 28$, as expected, and the number of possibilites to place them is $233 \cdot 233 = 54289$. These results also can be found in [16].

7   Proof of the Bishops Independence Number

Without giving a proof, in [16] it is mentioned that for $m < n$ it is possible to put at most $m + n - 1$ non-attacking (independent) bishops on an $m \times n$ chessboard if m is odd or n is odd, and $m + n - 2$ non-attacking bishops otherwise. In this section we present a proof of this result. In contrast to rooks, kings and knights where, as mentioned in the last section, proofs of the corresponding independence numbers can be obtained by modifications of the proofs of [17,18], for bishops new ideas and a more complex construction are necessary. Only the basic idea is the same: To prove that α is the independence number of a chess piece P, give an arrangement of α copies of P on the board and then show that an arrangement of more than α copies is impossible.

In the case of bishops the first step is constructive and has been found with the help of RELVIEW experiments by considering the cases $n = m+1$, $n = m+2$ and so forth. To enhance its presentation, we divide it into two parts and start with the following fact.

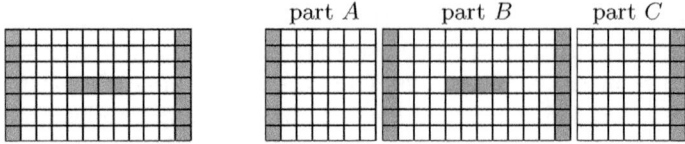

Fig. 5. Maximum arrangements of non-attacking bishops

Proposition 7.1. *Assume an $m \times n$ chessboard to be given, where $0 < m < n \leq 2m$. Then it is possible to arrange on it $m + n - 1$ non-attacking bishops if m is odd or n is odd, and $m + n - 2$ non-attacking bishops otherwise.*

Proof. First, let m be odd and $d := \frac{m+1}{2}$. Then $m + n - 1$ non-attacking bishops can be arranged in form of three groups as follows:

a) Arrange m bishops along the board's first column, that is, on the squares $\langle 1, y \rangle$, where $1 \leq y \leq m$.

b) Arrange $n - m - 1$ bishops along row d of the board starting with square $\langle d, d + 1 \rangle$ and ending with square $\langle d, n - d \rangle$.

c) Arrange m bishops along the board's last column, that is, on the squares $\langle y, n \rangle$, where $1 \leq y \leq m$.

For $m = 7$ and $n = 12$ this arrangement is visualized in the left-hand picture of Figure 5, where the black squares denote the positions of the 18 bishops, and for $n = m + 1$ the second group of bishops becomes empty since here $n - d = m + 1 - \frac{m+1}{2} = d$. That on row d exactly $n - m - 1$ bishops stand as second group follows from $n - d - (d + 1) + 1 = n - 2d = n - 2\frac{m+1}{2} = n - m - 1$. By a case analysis it can be checked that all $m + (n - m - 1) + m = m + n - 1$ bishops of the entire board are in fact non-attacking: Since the three groups of bishops are placed on the same column and row, resp., there are no attacks possible within the groups. Attacks between the first and the third group are impossible, since the range of coverage to the right of a bishop on column 1 ends with column m and $m < n$, and the range of coverage to the left of a bishop on column n ends with column $n - m + 1$ and $1 < n - m + 1$. Let p_d be the positive diagonal through $\langle d, d \rangle$ and n_d be the negative diagonal through $\langle d, d \rangle$. Then attacks between the first and the second group are impossible, since bishops of the first group only can attack pieces standing on or above p_d or on or below n_d and, conversely, bishops of the second group only can attack pieces standing below p_d or above n_d. Similar considerations show that attacks between the third and the second group are impossible.

Next, we assume that m is even and n is odd. We define $d := \frac{n+1}{2}$ and proceed as in the first case. The only difference is that we now use the middle column of the board for the arrangement of the second group of bishops, i.e., replace b) as follows:

d) Arrange $n - m - 1$ bishops along column d of the board starting with square $\langle m - d + 2, d \rangle$ and ending with square $\langle d - 1, d \rangle$.

That $1 \leq m - d + 2$ and $d - 1 \leq m$ follows from the assumption $n \leq 2m$. In the case $n = m + 1$ we get $m - d + 2 = \frac{m}{2} + 1$ and $d - 1 = \frac{m}{2}$ and the second group becomes empty. Again a simple case analysis shows that all bishops on the entire board are non-attacking. Since the calculation $d - 1 - (m - d + 2) + 1 = 2d - m - 2 = 2\frac{n+1}{2} - m - 2 = n - m - 1$ proves that $n - m - 1$ is in fact the number of bishops of the second group that, now vertically, is arranged on the board's column d, we get the total number of non-attacking bishops on the entire board as $m + (n - m - 1) + m = m + n - 1$.

What remains is the case that m and n are even. Here we can arrange the second group of bishops either horizontally or vertically. If we use the first possibility, this means that we replace b) as follows, using $d := \frac{m}{2}$ as number of the row the bishops are placed on.

e) Arrange $n - m - 2$ bishops along row d of the board starting with square $\langle d, d + 2 \rangle$ and ending with square $\langle d, n - d - 1 \rangle$.

By the calculation $n - d - 1 - (d + 2) + 1 = n - 2d - 2 = n - 2\frac{m}{2} - 2 = n - m - 2$ we prove that there stand in fact $n - m - 2$ bishops along row d. Hence on the entire board there are $m + (n - m - 2) + m = m + n - 2$ bishops. Again their independence can be proved by a simple case analysis. □

Next, we show that the restriction $n \leq 2m$ in Proposition 7.1 is unnecessary. The proof of the proposition can be seen as a recursive arrangement algorithm, with the arrangement of the proof of Proposition 7.1 as termination case.

Proposition 7.2. *Given an $m \times n$ chessboard with $0 < m < n$, it is possible to arrange on it $m + n - 1$ non-attacking bishops if m is odd or n is odd, and $m + n - 2$ non-attacking bishops otherwise.*

Proof. We use induction on $d := n - 2m$. The induction base is $d \leq 1$. In this case we have $n \leq 2m$ and the claim follows from Proposition 7.1.

For the induction step, assume $d > 1$. Then it holds that $2m < n$. We divide the given board into three parts. Part A consists of the m columns from 1 until m, part B of the $n - 2m$ columns from $m + 1$ until $n - m$, and part C of the m columns from $n - m + 1$ until n. Since $m > 0$ implies $n - 2m - 2m < n - 2m = d$ and $n - 2m$ is odd iff n is odd, from the induction hypothesis we get for the $m \times (n - 2m)$ board given by part B that it is possible to arrange on it $m + n - 2m - 1$ non-attacking bishops if m is odd or n is odd, and $m + n - 2m - 2$ non-attacking bishops if m is even and n is even, resp.

We complete these two arrangements in each case by m non-attacking bishops along the leftmost column of part A, i.e., the first column of the entire board, and m non-attacking bishops along the rightmost column of part C, i.e., the last column of the entire board. For $m = 7$ and $n = 26$ the result is visualized in the right-hand picture of Figure 5, where part B equals the chessboard of the figure's left-hand picture. By a case analysis it is again easy to verify that all positioned bishops are non-attacking. Their number is $m + (m + n - 2m - 1) + m = m + n - 1$ if m is odd or n is odd, and $m + (m + n - 2m - 2) + m = m + m - 2$ if m is even and and n is even. This concludes the induction step. □

And here is the claimed result with regard to independence of bishops on proper rectangular chessboards. Note that $m \neq n$ is necessary since for quadratic boards the largest number of non-attacking bishops is always $m + n - 2$ – and does not depend on the fact that m and/or n are even or odd, resp.

Proposition 7.3. *The largest number of non-attacking bishops that can be placed on an $m \times n$ chessboard, where $0 < m < n$, is $m + n - 1$ if m is odd or n is odd, and $m + n - 2$ if m is even and n is even.*

Proof. Let α denote the largest number of non-attacking bishops standing on the board. Since the board has $m + n - 1$ positive diagonals and there can stand at most one non-attacking bishop on each, it holds that $\alpha \leq m + n - 1$.

If m is odd or n is odd, then Proposition 7.2 shows $m + n - 1 \leq \alpha$. In combination with $\alpha \leq m + n - 1$ this yields $m + n - 1 = \alpha$. As we will show in a moment, if m and n are even it holds that $\alpha \leq m + n - 2$. Hence, in this case Proposition 7.2 yields $\alpha = m + n - 2$.

We prove $\alpha \leq m + n - 2$ for even m and n by contradiction. So, assume an arrangement of $m + n - 1$ non-attacking bishops on the board. Then there is exactly one bishop on each of the $m + n - 1$ positive diagonals. As already mentioned, we assume the lowermost-leftmost square $\langle 1, 1 \rangle$ to be black. Since m and n are even, we have $\frac{m+n}{2} - 1$ black positive diagonals and $\frac{m+n}{2}$ white positive diagonals. Hence, $\frac{m+n}{2} - 1$ of the $m + n - 1$ bishops are on black squares and the remaining $\frac{m+n}{2}$ ones are on white squares. But there is also exactly one of the $m + n - 1$ non-attacking bishops on each of the $m + n - 1$ negative diagonals. In the given case we have $\frac{m+n}{2}$ black negative diagonals and $\frac{m+n}{2} - 1$ white negative diagonals. This leads to the contradiction that now $\frac{m+n}{2}$ of the $m + n - 1$ bishops are on black squares and $\frac{m+n}{2} - 1$ are on white squares. □

In [17,18] it is shown that for $m = n$ the arrangement of the bishops on the top row of the chessboard determines the arrangement of all $m + n - 2$ non-attacking bishops because all bishops are placed on the outer ring of squares. Hence, there are 2^m possible arrangements (cf. again the diagonal of the right-hand table of Figure 3). If $m < n$, then some of the bishops of a maximum independent arrangement may occur in the interior of the board and the simple argument of [17,18] does not work anymore. We have not found a general formula for the number of maximum independent arrangements of bishops on proper rectangular boards. Only the case $m + 1 = n$ is simple. Here there exists exactly one arrangement for the $m + m + 1 - 1 = 2m$ non-attacking bishops: Place m of them on column 1 and the remaining m on column n.

8   Conclusion

We have described a simple computing technique for solving independence and domination problems on rectangular chessboards. It uses relation algebra as a methodological means and consists of the development of relation-algebraic specifications for independent and dominating sets of graphs and relations, resp., and a representation of the chess pieces' graphs as relations. To evaluate the specifications and to visualize the computed results we have used the tool RELVIEW. We have provided some of our experimental results concerning independence. Based on them, we have given a proof of the bishops independence number for proper rectangular boards.

Because of the global approach that RELVIEW takes, of course, it cannot compete for all intents and purposes with specifically tailored algorithms for chess problems. An example for the latter is the FPGA-based approach to solve the n-queens problem for $n = 26$, yielding 22.317.699.616.364.044 possibilities (TU Dresden, 2009, for details see URL http://queens.inf.tu-dresden.de).

Nowadays, systematic experiments are accepted as a way for obtaining new mathematical insights and, hence, tools for symbolic manipulation, prototypic computations, animation and visualization become increasingly important as one proceeds in investigations. Our opinion is that the attraction and general usefulness of our approach in this area lies in its flexibility, its large application area, the formal precision of the calculations and the concise form of the developed algorithms and, particular in view of RELVIEW, the computational power of the tool as a result of the use of BDDs and the manifold animation and visualization possibilities. We hope that the readers agree after reading the present paper.

References

1. Bell, J., Stevens, B.: A survey of known results and research areas for n queens. Discr. Math. 309, 1–31 (2009)
2. Berghammer, R., Neumann, F.: RELVIEW – An OBDD-Based Computer Algebra System for Relations. In: Ganzha, V.G., Mayr, E.W., Vorozhtsov, E.V. (eds.) CASC 2005. LNCS, vol. 3718, pp. 40–51. Springer, Heidelberg (2005)
3. Berghammer, R.: Applying relation algebra and RELVIEW to solve problems on orders and lattices. Acta. Infor. 45, 211–236 (2008)
4. Berghammer, R., Rusinowska, A., de Swart, H.: An interdisciplinary approach to coalition formation. Europ. J. Operat. Res. 195, 487–496 (2009)
5. Berghammer, R., Rusinowska, A., de Swart, H.: Applying relation algebra and RELVIEW to measures in a social network. EJOR 202, 182–198 (2010)
6. Bryant, R.E.: Symbolic Boolean manipulation with ordered binary decision diagrams. ACM Comput. Surv. 24, 293–318 (1992)
7. Chen, H.-C., Ho, T.-Y.: The rook problem on saw-toothed chessboards. Appl. Math. Lett. 21, 1234–1237 (2008)
8. Cockayne, E.J.: Chessboard domination problems. Discr. Math. 86, 13–20 (1990)
9. Gibbons, P.B., Webb, J.A.: Some new results for the queens domination problem. Austral. J. of Combinat. 15, 145–160 (1997)
10. de Jaenisch, C.F.: Applications de l'analyse mathematiques au jeu des echecs. Petrograd (1882)
11. Leoniuk, B.: ROBDD-basierte Implementierung von Relationen und relationalen Operationen mit Anwendungen. Dissertation, Universität Kiel (2001)

12. Milanese, U.: Zur Implementierung eines ROBDD-basierten Systems für die Manipulation und Visualisierung von Relationen. Dissertation, Universität Kiel (2003)
13. Nauck, F.: Briefwechsel mit allen für alle. Illustrierte Zeitung 15, 182 (1850)
14. Pauls, E.: Das Maximalproblem der Damen auf dem Schachbrett. Deutsche Schachzeitung 29, 129–134 (1874)
15. Schmidt, G., Ströhlein, T.: Relations and graphs. Discrete Mathematics for Computer Scientists. Springer, Heidelberg (1993)
16. Steinbach, B., Posthoff, C.: New results based on Boolean models. In: Steinbach, B. (ed.) Proc. 9th Int. Workshop on Boolean Problems, TU Freiberg, pp. 29–36. (2010)
17. Watkins, J.J.: Across the board: The mathematics of chessboard problems. Princeton University Press, Princeton (2004)
18. Yaglom, A.M., Yaglom, I.M.: Challenging mathematical problems with elementary solutions, vol. I. Holden-Day Inc. (1964)

A Functional, Successor List Based Version of Warshall's Algorithm with Applications

Rudolf Berghammer

Institut für Informatik, Christian-Albrechts-Universität Kiel
Olshausenstraße 40, D-24098 Kiel, Germany
`rub@informatik.uni-kiel.de`

Abstract. We show how formally and systematically to develop a purely functional version of Warshall's algorithm for computing transitive closures by combining the unfold-fold technique, relation-algebra and data refinement. It is based on an implementation of relations by lists of successor lists. The final version can immediately be implemented in HASKELL. This resulting HASKELL program has the same runtime complexity as the traditional imperative array-based implementation of Warshall's algorithm. We also demonstrate how it can be re-used as component in other functional algorithms.

1 Introduction

The computation of the transitive closure R^+ of a (binary) relation R on a set X has many practical applications. This is mainly due to the fact that, if R specifies the set of edges of a directed graph $G = (X, R)$ with X as set of vertices, then the relation R^+ relates two vertices $x, y \in X$ of G if and only if y is reachable from x via a nonempty path. Usually the task of computing the transitive closure R^+ is solved by Warshall's algorithm (published in [26]). Its traditional implementation in an imperative programming language is based on a representation of the relation R by a 2-dimensional Boolean array. This leads to a simple and efficient in-situ program with three nested loops that needs $O(m^3)$ steps, where m is the cardinality of the carrier set X of R.

In the last years also some array-based algorithms for the computation of reflexive-transitive closures have been published, which have sub-cubic runtime. Most of them are based upon sub-cubic algorithms for matrix multiplication, that is, on Strassen's seminal method (see [24]) and its refinements. There is also a refinement of Warshall's algorithm (more precisely, Floyd's extension for the all-pairs shortest paths problem [14]) that computes the transitive closure of a relation in time $O(n^{2.5})$; cf. [15]. But all these algorithms pay for their exponents by an intricacy that, particularly with regard to practice, makes it difficult to implement them correctly (w.r.t. the input/output behaviour as well as the theoretical runtime bound) in a conventional programming language. The complexity of the algorithms also leads in the O-estimation to such large constant coefficients that, again concerning practice, results are computed faster normally

H. de Swart (Ed.): RAMICS 2011, LNCS 6663, pp. 109–124, 2011.

only in the case of very dense relations or very large carrier sets. However, both cases hardly appear in practical applications.

In a lot of practical applications arrays are unfit for representing relations and graphs. In particular this holds if both R and R^+ are of "medium density" or even sparse. Such relations/graphs appear, for instance, in computational linguistics when computing the so-called subtype relation in HPSG-type signatures (see, for example, [21]), in XML-query processing since XML structures are (directed) trees plus few reference links (see, e.g., [11]), and in the context of ordered sets if a partial order is computed from the cover relation (the Hasse diagram). In all these applications a representation of relations by, say, successor lists (also known as adjacency lists), lists of pairs or even look-up tables (i.e., characteristic functions) is much more economic. But such a representation sacrifices the simplicity and efficiency of the above mentioned imperative implementation of Warshall's algorithm. Moreover, the traditional method of imperatively updating an array representing the relation/graph is alien to the purely functional programming paradigm which restricts the use of side effects.

In the present paper we show how systematically and by applying formal methods a purely functional version of Warshall's algorithm can be developed that uses an implementation of relations by lists of successor lists and also has a cubic runtime. In the first step we develop from a relation-algebraic problem specification by a combination of the unfold-fold-technique and relation-algebraic reasoning a functional algorithm for computing R^+ that solely is based on relation algebra and the generation of (relational) vectors via disjoint unions of (relational) points. To obtain from it a version that works on lists of successor sets we then use data refinement. In the first refinement step we represent relations on a set X by functions from X to its powerset 2^X and vectors on X by elements of the powerset 2^X. Next, we refine the functions to lists over 2^X and the subsets of X to lists over X. Going at last from the lists over 2^X to lists of lists over X, we obtain a version in the functional programming language HASKELL. Finally, we demonstrate how this HASKELL program can be re-used as component to construct further functional algorithms that solve well-known graph-theoretic problems.

The computation of transitive closures of relations and many of the applications can be seen as graph-theoretic problems. Related to our work are, therefore, all the approaches to program graph algorithms in a functional programming language. In the meantime functional graph algorithms have a tradition of more than two decades. Here we only want to mention some papers that deal with different aspects; for more references see the second section of [13]. In [4] transformational programming is applied to develop some simple functional reachability algorithms. The papers [18,19] deal with the specification and functional computation of the depth-first search forest and presents some classical applications (like topological sorting, testing for cycles, strongly connected components) in the functional style. To achieve a linear runtime as in the imperative case, monads are used to mimic the imperative marking technique. How relation algebra and the features of the functional-logic extension Curry (see [3]) of

HASKELL can be employed to solve some problems on relations and graphs in a very high-level declarative style is shown in [7]. All the papers just mentioned regard graphs as monolithic data. In contrast with this, in [12,13] graphs are inductively generated. This approach allows to write many graph algorithms in the typical functional style using pattern matching. Also [16] discusses an inductive definition of graphs, but restricted to directed cycle-free graphs and without presenting an implementation.

2   Relation-Algebraic Preliminaries

We denote the set (or type) of all (binary) relations with source X and target Y by $[X \leftrightarrow Y]$ and write $R : X \leftrightarrow Y$ instead of $R \in [X \leftrightarrow Y]$. If the sets X and Y are finite, we may consider R as a Boolean matrix with $|X|$ rows and $|Y|$ columns. We assume the reader to be familiar with the basic operations on relations, viz. R^{T} (transposition), $\overline{R}$ (complement), $R \cup S$ (join), $R \cap S$ (meet), $R; S$ (composition), the predicate indicating $R \subseteq S$ (inclusion) and the special relations O (empty relation), L (universal relation) and I (identity relation). Furthermore, we assume the reader to know the most fundamental laws of relation algebra like $\mathsf{I}; R = R$, $R^{\mathsf{T}^{\mathsf{T}}} = R$, $(R; S)^{\mathsf{T}} = S^{\mathsf{T}}; R^{\mathsf{T}}$, $R; (S \cup T) = R; S \cup R; T$, and the following one, in [22] called *Schröder equivalences*.

$$Q; R \subseteq S \iff Q^{\mathsf{T}}; \overline{S} \subseteq \overline{R} \iff \overline{S}; R^{\mathsf{T}} \subseteq \overline{Q} \tag{1}$$

We also will use the relation-algebraic specifications of the following properties: *reflexivity* $\mathsf{I} \subseteq R$, *irreflexivity* $R \subseteq \overline{\mathsf{I}}$, *transitivity* $R; R \subseteq R$, *symmetry* $R = R^{\mathsf{T}}$, *injectivity* $R; R^{\mathsf{T}} \subseteq \mathsf{I}$ and *surjectivity* $\mathsf{L}; R = \mathsf{L}$. For more details concerning the algebraic treatment of relations and its manifold connections to graph theory we refer to the textbooks [22,8].

A (relational) *vector* is a relation v which satisfies $v = v; \mathsf{L}$ and a (relational) *point* is an injective and surjective vector. For vectors the targets are irrelevant. We, therefore, consider in the following mostly vectors $v : X \leftrightarrow \mathbf{1}$ with a specific singleton set $\mathbf{1} = \{\bot\}$ as target and omit in such cases the second component $\bot$ in a pair, i.e., write $x \in v$ instead of $(x, \bot) \in v$. Then v *describes* the subset $\{x \in X \mid x \in v\}$ of X. If X is finite, a vector of $[X \leftrightarrow \mathbf{1}]$ can be considered as a Boolean matrix with $|X|$ rows and exactly one column, i.e., as a Boolean column vector in the usual sense, and the set it describes is given by the components with entry 1. In the Boolean matrix model a point of $[X \leftrightarrow \mathbf{1}]$ is a Boolean column vector in which exactly one entry is 1. This means that a point $p : X \leftrightarrow \mathbf{1}$ describes a singleton subset of X, or an element of X if we identify a singleton set $\{x\} \subseteq X$ with the only element $x \in X$ it contains. Later we will use that if p describes $x \in X$, then $(y, z) \in p; p^{\mathsf{T}}$ is equivalent to $y = x$ and $z = x$.

3   Transitive Closures and Warshall's Algorithm

Given $R : X \leftrightarrow X$, its *reflexive-transitive closure* $R^* : X \leftrightarrow X$ is defined as the least reflexive and transitive relation that contains R and its *transitive closure*

$R^+ : X \leftrightarrow X$ is defined as the least transitive relation that contains R. It is well-known (cf. [22,1,8]) that R^* and R^+ can also be specified via least fixed point constructions. The least fixed point of $\tau_R : [X \leftrightarrow X] \rightarrow [X \leftrightarrow X]$, that is defined by $\tau_R(Q) = I \cup R; Q$, is R^*, and the least fixed point of $\sigma_R : [X \leftrightarrow X] \rightarrow [X \leftrightarrow X]$, where $\sigma_R(Q) = R \cup R; Q$, is R^+. From these specifications we obtain by fixed point considerations (see e.g., again [22,1]) the following equations.

$$\mathsf{O}^* = \mathsf{I} \qquad R^* = \mathsf{I} \cup R^+ \qquad R^+ = R; R^* \qquad (R \cup S)^* = R^*; (S; R^*)^* \qquad (2)$$

A simple fixed point argument also shows that R is a transitive relation if and only if $R = R^+$. In [1] the rightmost equation of (2) is called the *star-decomposition* rule. This rule has a nice graph-theoretic interpretation. If in a directed graph $G = (X, R \cup S)$ the edges are coloured with two colours, say r and s, then each path in G can be decomposed into a (possibly empty) initial part p_0 with r-edges only, followed by a list of paths $p_1, \ldots, p_m$, where $m = 0$ is possible and each path p_i from the list starts with an s-edge and then consists of $n \geq 0$ subsequent r-edges only.

Now, let $R : X \leftrightarrow X$ be the relation of a finite directed graph $G = (X, R)$ with vertex set $X = \{v_1, \ldots, v_m\}$. The idea behind Warshall's algorithm is to consider for i, with $0 \leq i \leq m$, the subset $X_i := \{v_1, \ldots, v_i\}$ of X and the relation $R_i : X \leftrightarrow X$ such that for all $x, y \in X$ it holds

$$(x, y) \in R_i \iff \begin{cases} \text{there exists a path } (z_1, \ldots, z_k) \text{ from } x \text{ to} \\ y \text{ such that } k > 1 \text{ and } z_2, \ldots, z_{k-1} \in X_i. \end{cases} \qquad (3)$$

Using graph-theoretic reasoning, for all $x, y \in X$ and i such that $1 \leq i \leq m$ it can be shown that $(x, y) \in R_i$ if and only if $(x, y) \in R_{i-1}$ or $(x, v_i) \in R_{i-1}$ and $(v_i, y) \in R_{i-1}$. Since, furthermore, for all $x, y \in X$ it holds that $(x, y) \in R_0$ if and only if $(x, y) \in R$ and $(x, y) \in R_m$ if and only if $(x, y) \in R^+$, the transitive closure R^+ is the limit of the finite chain $R = R_0 \subseteq R_1 \subseteq \ldots \subseteq R_{m-1} \subseteq R_m = R^+$. The generation of this chain is exactly what the traditional implementation of Warshall's algorithm realizes in-situ by means of a Boolean array as follows.

$$\begin{aligned}
&\textbf{for } i = 1 \textbf{ to } m \textbf{ do} \\
&\quad \textbf{for } j = 1 \textbf{ to } m \textbf{ do} \\
&\quad\quad \textbf{for } k = 1 \textbf{ to } m \textbf{ do} \\
&\quad\quad\quad R[j, k] := R[j, k] \vee (R[j, i] \wedge R[i, k])
\end{aligned} \qquad (4)$$

If in the algorithm of (4) the Boolean array is assumed to be initialized with the relationships of the relation R, then after the i-th turn of the loop it holds for all j, k, with $1 \leq j, k \leq m$, the equivalence of $R[j, k]$ and $(v_j, v_k) \in R_i$. The two inner loops serve for the transformation of the R_i-realization via the array into the R_{i+1}-realization.

4   Computing Transitive Closures Using Relation Algebra

In the following we develop a purely functional algorithm for computing R^+ that solely is based on relation algebra. The main idea is to express (3) in terms of relation algebra. To this end we consider for a given relation $R : X \leftrightarrow X$ a vector $v : X \leftrightarrow \mathbf{1}$ and the partial identity relation $\mathsf{I}_v := \mathsf{I} \cap v; v^{\mathsf{T}}$ induced by v. If we assume that v describes the subset X_v of X, then a little reflection shows for all $x, y \in X$ that $(x, y) \in R; (\mathsf{I}_v; R)^*$ if and only if there exists a path $(z_1, \ldots, z_k)$ from x to y such that $k > 1$ and $z_2, \ldots, z_{k-1} \in X_v$. This motivates the following relation-algebraic specification as starting point.

$$\mathrm{warsh}(R, v) : X \leftrightarrow X \qquad \mathrm{warsh}(R, v) = R; (\mathsf{I}_v; R)^* \tag{5}$$

Using the first equation of (2), the definition of warsh in (5) implies for the empty vector $\mathsf{O} : X \leftrightarrow \mathbf{1}$ that

$$\mathrm{warsh}(R, \mathsf{O}) = R; (\mathsf{I}_{\mathsf{O}}; R)^* = R; (\mathsf{O}; R)^* = R; \mathsf{O}^* = R; \mathsf{I} = R,$$

which corresponds (since O describes $\emptyset$) to the equivalence of $(x, y) \in R_0$ and $(x, y) \in R$ for all $x, y \in X$. In the same manner we get from the third equation of (2) for the universal vector $\mathsf{L} : X \leftrightarrow \mathbf{1}$ the equation

$$\mathrm{warsh}(R, \mathsf{L}) = R; (\mathsf{I}_{\mathsf{L}}; R)^* = R; (\mathsf{I}; R)^* = R; R^* = R^+,$$

which corresponds (since L describes X) to the equivalence of $(x, y) \in R_m$ and $(x, y) \in R^+$ for all $x, y \in X$. Our goal is to obtain an inductive specification of (5) so that a later implementation in HASKELL can be based on pattern matching. In respect thereof, the two equations just shown correspond to the induction base and the termination case. They, therefore, motivate to consider a vector $v : X \leftrightarrow \mathbf{1}$ such that $v \neq \mathsf{L}$, to take an arbitrary point $p : X \leftrightarrow \mathbf{1}$ for that $p \subseteq \overline{v}$ holds, and to express $\mathrm{warsh}(R, v \cup p)$ in terms of $\mathrm{warsh}(R, v)$.

As a preparatory step we consider the corresponding partial identity relations. Because of the Schröder equivalences (1) the assumption $p \subseteq \overline{v}$ is equivalent to $v; p^{\mathsf{T}} \subseteq \overline{\mathsf{I}}$ and also to $p; v^{\mathsf{T}} \subseteq \overline{\mathsf{I}}$. Using these two properties we can calculate as follows.

$$
\begin{aligned}
\mathsf{I} \cap (v \cup p); (v \cup p)^{\mathsf{T}} &= \mathsf{I} \cap (v; v^{\mathsf{T}} \cup v; p^{\mathsf{T}} \cup p; v^{\mathsf{T}} \cup p; p^{\mathsf{T}}) \\
&= \mathsf{I} \cap (v; v^{\mathsf{T}} \cup p; p^{\mathsf{T}}) & v; p^{\mathsf{T}} \cup p; v^{\mathsf{T}} \subseteq \overline{\mathsf{I}} \\
&= (\mathsf{I} \cap v; v^{\mathsf{T}}) \cup p; p^{\mathsf{T}} & p; p^{\mathsf{T}} \subseteq \mathsf{I}
\end{aligned}
$$

Hence, we have $\mathsf{I}_{v \cup p} = \mathsf{I}_v \cup p; p^{\mathsf{T}}$. Now, the induction step of the inductive specification of the function warsh of (5) we are looking for follows from the subsequent calculation that uses the unfold-fold technique known from Burstall and Darlington's transformational programming system [10].

$$
\begin{aligned}
\mathrm{warsh}(R, v \cup p) &= R; (\mathsf{I}_{v \cup p}; R)^* & \text{unfold} \\
&= R; ((\mathsf{I}_v \cup p; p^{\mathsf{T}}); R)^* & \text{see above} \\
&= R; (\mathsf{I}_v; R \cup p; p^{\mathsf{T}}; R)^* \\
&= R; (\mathsf{I}_v; R)^*; (p; p^{\mathsf{T}}; R; (\mathsf{I}_v; R)^*)^* & \text{by (2)} \\
&= \mathrm{warsh}(R, v); (p; p^{\mathsf{T}}; \mathrm{warsh}(R, v))^* & \text{fold} \\
&= \mathrm{warsh}(R, v); (\mathsf{I} \cup (p; p^{\mathsf{T}}; \mathrm{warsh}(R, v))^+) & \text{by (2)} \\
&= \mathrm{warsh}(R, v); (\mathsf{I} \cup p; p^{\mathsf{T}}; \mathrm{warsh}(R, v)) & \text{see below} \\
&= \mathrm{warsh}(R, v) \cup \mathrm{warsh}(R, v); p; p^{\mathsf{T}}; \mathrm{warsh}(R, v)
\end{aligned}
$$

Here the correctness of the last but one step (hint: see below) follows from

$$p; p^{\mathsf{T}}; warsh(R, v); p; p^{\mathsf{T}}; warsh(R, v) \subseteq p; \mathsf{L}; p^{\mathsf{T}}; warsh(R, v)$$
$$= p; p^{\mathsf{T}}; warsh(R, v) \qquad p \text{ vector}$$

because this is the transitivity of $p; p^{\mathsf{T}}; warsh(R, v)$ and, as a consequence (see Section 3), the relation equals its transitive closure.

If we apply, as customary in functional programming, a let-clause to avoid the multiple calls of the function *warsh*, then the three equations just shown lead to the functional algorithm (6) for computing transitive closures. It only uses the constants and operations of relation algebra on the data structure side.

$$transcl : [X \leftrightarrow X] \to [X \leftrightarrow X]$$
$$transcl(R) = warsh(R, \mathsf{L})$$

$$warsh : [X \leftrightarrow X] \times [X \leftrightarrow \mathbf{1}] \to [X \leftrightarrow X] \qquad (6)$$
$$warsh(R, \mathsf{O}) \quad = R$$
$$warsh(R, v \cup p) = \texttt{let } S = warsh(R, v)$$
$$\texttt{in } S \cup S; p; p^{\mathsf{T}}; S$$

In the second equation of the auxiliary function *warsh* it is implicitly assumed that $v \neq \mathsf{L}$ and that p is a point of type $[X \leftrightarrow \mathbf{1}]$ with $p \subseteq \bar{v}$. If X is finite and of the form $X = \{v_1, \ldots, v_m\}$, then the universal vector $\mathsf{L} : X \leftrightarrow \mathbf{1}$ can be represented as union $p_1 \cup \ldots \cup p_m$ of m pairwise disjoint points $p_1, \ldots, p_m : X \leftrightarrow \mathbf{1}$, where p_i describes the element $v_i \in X$, with $1 \leq i \leq m$. In such a case the call $transcl(R)$ of the main function *transcl* leads to the total number of $m + 1$ calls of the function *warsh*.

It should be mentioned that the implicit assumption on v and p in (6) can be avoided if we suppose a choice function *point* to be at hand (as, for instance, in the relation-algebraic tool RELVIEW [6]) such that a call $point(v)$ yields a point that is contained in the nonempty vector v. For the point $p := point(v)$ then it holds that $v = (v \cap \bar{p}) \cup p$. Using this property in combination with a conditional, the inductive specification of the function *warsh* of (6) then can be reformulated as a recursive function (that now decreases the vector argument) as follows.

$$warsh : [X \leftrightarrow X] \times [X \leftrightarrow \mathbf{1}] \to [X \leftrightarrow X]$$
$$warsh(R, v) = \texttt{if } v = \mathsf{O} \texttt{ then } R$$
$$\texttt{else let } p = point(v) \qquad (7)$$
$$S = warsh(R, v \cap \bar{p})$$
$$\texttt{in } S \cup S; p; p^{\mathsf{T}}; S$$

But the version of the function *warsh* given in (7) does not directly lead to the final HASKELL program in the typical inductive (pattern matching based) functional style we are aiming at. Therefore, in the remainder of the paper we concentrate on the development of the function *warsh* of (6). It also should be mentioned that (6) and (7) can immediately be translated into HASKELL if a HASKELL library for relation algebra is at hand. The only such libraries we are aware of are described in [17] and [23]. But their use leads to a less efficient final program compared with the one we will develop in the next sections.

5 From Relation Algebra to Successor Functions

In the next two sections we show how, by the application of data refinement, the functional algorithm (6) can be transformed step-wise into a version that is based on a representation of relations by means of lists of successor sets and (as we will demonstrate in Section 7) can immediately be implemented in HASKELL using the pre-defined HASKELL datatype for lists only.

The first data refinement, which is treated in this section, represents relations $F : X \leftrightarrow X$ by functions $f : X \to 2^X$ such that for all $x, y \in X$ it holds that $(x, y) \in F$ if and only if $y \in f(x)$. In terms of graph theory, relations are represented by functions which map the vertices to their successor sets, i.e., by so-called successor functions. Furthermore, we represent vectors (and points) of type $[X \leftrightarrow \mathbf{1}]$ by the subsets (and elements, respectively) of X they describe.

Now, suppose that the input relation $R : X \leftrightarrow X$ of *transcl* is represented by the function $r : X \to 2^X$. Since the universal vector $\mathsf{L} : X \leftrightarrow \mathbf{1}$ describes (i.e., is represented by) the set X we obtain the following version of the main function *transcl* of (6) that now works on functions instead of relations,

$$
\begin{aligned}
&transcl : (X \to 2^X) \to (X \to 2^X) \\
&transcl(r) = warsh(r, X)
\end{aligned}
\tag{8}
$$

To obtain a corresponding new version of the auxiliary function *warsh* of (6) we assume again that $r : X \to 2^X$ represents $R : X \leftrightarrow X$. The induction base $warsh(r, \emptyset) = r$ is a consequence of the fact that the empty vector $\mathsf{O} : X \leftrightarrow \mathbf{1}$ is represented by the empty set $\emptyset$. For the remaining case, assume that the vector $v : X \leftrightarrow \mathbf{1}$ is represented by the subset V of X, the point $p : X \leftrightarrow \mathbf{1}$ is represented by the element $e \in X \setminus V$ and the relation $S : X \leftrightarrow X$ of the let-clause of (6) is represented by the function $s : X \to 2^X$. For all $x, y \in X$ we then can calculate as follows, where the fist steps use the definition of relational union and composition and the point-property mentioned at the end of Section 2. The conditional is introduced in the sixth step to enhance readability and to prepare a later translation into HASKELL.

$$
\begin{aligned}
&(x, y) \in S \cup S; p; p^\mathsf{T}; S \\
\Longleftrightarrow\ &(x, y) \in S \lor (x, y) \in S; p; p^\mathsf{T}; S \\
\Longleftrightarrow\ &(x, y) \in S \lor \exists i, j \in X : (x, i) \in S \land (i, j) \in p; p^\mathsf{T} \land (j, y) \in S \\
\Longleftrightarrow\ &(x, y) \in S \lor \exists i, j \in X : (x, i) \in S \land i = e \land j = e \land (j, y) \in S \\
\Longleftrightarrow\ &(x, y) \in S \lor ((x, e) \in S \land (e, y) \in S) \\
\Longleftrightarrow\ &y \in s(x) \lor (e \in s(x) \land y \in s(e)) \\
\Longleftrightarrow\ &y \in s(x) \lor \text{if } e \in s(x) \text{ then } y \in s(e) \text{ else } \textit{false} \\
\Longleftrightarrow\ &y \in s(x) \lor \text{if } e \in s(x) \text{ then } y \in s(e) \text{ else } y \in \emptyset \\
\Longleftrightarrow\ &y \in s(x) \lor y \in \text{if } e \in s(x) \text{ then } s(e) \text{ else } \emptyset \\
\Longleftrightarrow\ &y \in \text{if } e \in s(x) \text{ then } s(x) \cup s(e) \text{ else } s(x)
\end{aligned}
$$

If we apply the familiar λ-notation to denote anonymous functions, then the relationship just proved in combination with β-conversion shows that the anonymous function $\lambda x \bullet \texttt{if } e \in s(x) \texttt{ then } s(x) \cup s(e) \texttt{ else } s(x)$ (where x ranges over X) represents the relation $S \cup S; p; p^\mathsf{T}; S : X \leftrightarrow X$. Using additionally that the vector $v \cup p : X \leftrightarrow \mathbf{1}$ is represented by the subset $V \cup \{e\}$ of X we, finally, obtain the following new version of the auxiliary function $warsh$ of (6).

$$
\begin{aligned}
warsh &: (X \to 2^X) \times 2^X \to (X \to 2^X) \\
warsh(r, \emptyset) &= r \\
warsh(r, V \cup \{e\}) &= \texttt{let } s = warsh(r, V) \\
&\quad \texttt{in} \quad \lambda x \bullet \texttt{if } e \in s(x) \texttt{ then } s(x) \cup s(e) \texttt{ else } s(x)
\end{aligned}
\tag{9}
$$

In analogy to the version of the function $warsh$ of (6) in the second equation of its version of (9) it is implicitly assumed that $V \neq X$ and $e \in X \setminus V$.

6   From Successor Functions to Lists of Sets

Having replaced relations by functions, vectors by sets and points by elements, in the second refinement step we now represent the arguments and the results of the function $transcl$ of (8) and the function $warsh$ of (9) by lists. For the argument and the result of $transcl$ and, hence, also for the first argument and the result of $warsh$ we take lists over 2^X, i.e., elements of $(2^X)^*$, and for the second argument of $warsh$ we take lists over X, i.e., elements of X^*. To simplify the presentation *we assume for the following that the set X consists of the natural numbers $0, 1, \ldots, m$.* The additional number 0 (in Section 3 we assumed $X = \{v_1, \ldots, v_m\}$ as carrier set[1]) is motivated with a view to a later use of HASKELL lists. Here 0 is the index of the first list element.

The just made assumption on the carrier set X of all relations we consider allows to represent the function $r : X \to 2^X$ of (8) and (9) by the list $rs \in (2^X)^*$ of length $m + 1$ such that for all $x \in X$ the x-th component of rs — in the sequel we use the HASKELL notation $rs!!x$ for this construction — equals the set $r(x)$. In the same way the function $s : X \to 2^X$ used in the `let`-clause of $warsh$ is represented by a list $ss \in (2^X)^*$ of length $m + 1$ such that $ss!!x = s(x)$ for all $x \in X$. The second argument of $warsh$ is represented by the *increasingly sorted* list of the elements it contains, where we additionally do not allow *multiple occurrences of elements.*

Because the set X is represented by the increasingly sorted list of the natural numbers from 0 to m and m equals the length of the list rs minus 1, from the just introduced list representation we get the following new version of the main function $transcl$ of the algorithm. In this version we apply the HASKELL notation $[0..m]$ for the increasingly sorted list of the natural numbers from 0 to m to prepare the later translation into HASKELL.

$$
\begin{aligned}
transcl &: (2^X)^* \to (2^X)^* \\
transcl(rs) &= warsh(rs, [0..|rs| - 1])
\end{aligned}
\tag{10}
$$

---

[1] Using $X = \{v_0, v_1, \ldots, v_m\}$ in Section 3 would cause problems with the definition of the subsets $X_i := \{v_0, \ldots, v_i\}$, since $\emptyset = X_i$ then requires i to be -1.

To obtain a list-based version of the auxiliary function *warsh*, first, we represent the anonymous function of the `let`-clause of (9) by a list over 2^X. By assumption it holds that $ss!!x = s(x)$ for all $x \in X$, that is, all elements x of the list $[0..m]$. Using a notation similar to HASKELL's well-known list comprehension we, therefors, get for the function $\lambda x \bullet$ `if` $e \in s(x)$ `then` $s(x) \cup s(e)$ `else` $s(x)$ the list representation $[$`if` $e \in ss!!x$ `then` $ss!!x \cup ss!!e$ `else` $ss!!x \mid x \in [0..m]]$. It is obvious that this list comprehension coincides with the list comprehension $[$`if` $e \in ms$ `then` $ms \cup ss!!e$ `else` $ms \mid ms \in ss]$. If we suppose that the list $vs \in X^*$ represents the set V of the function of (9), then the list $e : vs$ with the additional first element $e \in X \setminus V$ represents the set $V \cup \{e\}$. This property in connection with the equality of the above list comprehensions and the fact that the empty list $[]$ represents the empty set $\emptyset$ shows that the function *warsh* of (9) is correctly implemented by the subsequent version on lists of sets.

$$
\begin{aligned}
&warsh : (2^X)^* \times X^* \to (2^X)^* \\
&warsh(rs, []) \quad = rs \\
&warsh(rs, e : vs) = \texttt{let } ss = warsh(rs, vs) \\
&\qquad\qquad\qquad\quad \texttt{in } \; [\texttt{if } e \in ms \texttt{ then } ms \cup ss!!e \texttt{ else } ms \mid ms \in ss]
\end{aligned}
\tag{11}
$$

By (11) we have reached the desired inductive functional style for the function *warsh*. Note that in the algorithm no longer an implicit assumption on the list *vs* and the element *e* in the second pattern is required.

The use of lists to represent the arguments and results of the function *transcl* of (8) and the function *warsh* of (9) is an important design decision of the data refinement process. Since the sizes of the lists are not changed by the algorithm, functional arrays would constitute an alternative manner of representation. Our decision for lists is motivated by the anonymous function appearing in (9) since such constructions frequently immediately and elegantly can be implemented in HASKELL by list comprehensions. For arrays HASKELL does not provide a corresponding language construct.

7   From Lists of Sets to List of Lists and HASKELL

Now, we are in the position to translate the functions of (10) and (11) into the functional programming language HASKELL and to show that the resulting HASKELL program runs in cubic time, i.e., has the same runtime complexity as the traditional imperative array-based implementation of Warshall's algorithm. We assume the reader to be familiar with HASKELL. Otherwise, he may consult one of the well-known textbooks about it, for example [9,25].

To obtain an implementation of the functions *transcl* and *warsh* of (10) and (11), respectively, in HASKELL we apply again data refinement and represent lists over 2^X, i.e., lists of sets of integers, by lists of lists of integers. For the appearing lists of integers we require the same properties as in Section 5. That is, these lists (and, as a consequence, all successor lists of the input of the main function) have to be increasingly sorted and without multiple occurrences of elements. The reason for this design decision — a precondition on the input — will become clear later when we consider the program's runtime complexity.

If we go from lists of sets of integers to list of lists of integers and formulate the result in HASKELL using the type-declarations

$$
\begin{aligned}
&\texttt{type Vertex = Int} \\
&\texttt{type Vertexset = [Vertex]} \\
&\texttt{type Relation = [Vertexset]}
\end{aligned}
\tag{12}
$$

for the universe containing the relation's carrier sets, its subsets and the relations, and the HASKELL function

$$
\begin{aligned}
&\texttt{vertices :: Relation -> Vertexset} \\
&\texttt{vertices rs = [0..length rs - 1]}
\end{aligned}
\tag{13}
$$

that yields the carrier set (vertices) of a relation (graph), then version (10) of the main function of the algorithm becomes the following HASKELL function.

$$
\begin{aligned}
&\texttt{transcl :: Relation -> Relation} \\
&\texttt{transcl rs = warsh rs (vertices rs)}
\end{aligned}
\tag{14}
$$

Subsets of X are implemented by HASKELL lists. Hence, set-membership $e \in ms$ can directly be implemented by the pre-defined HASKELL function `elem`. Assuming additionally a HASKELL function `cup` to be at hand that implements set union on the list implementations of sets, a straightforward translation of the function of (11) into HASKELL code looks as follows.

```
warsh :: Relation -> Vertexset -> Relation
warsh rs [] = rs
warsh rs (e:vs) =                                    (15)
   let ss = warsh rs vs
   in  [if elem e ms then cup ms (ss!!e) else ms | ms <- ss]
```

From Section 4 we know already that a call `transcl rs` leads to the total number of $|X| + 1$, that is, $m + 2$ calls of the HASKELL function `warsh`. The list specified by the list comprehension of `warsh` consists of $|X| = m + 1$ successor lists. Since each of these $|X|$ successor lists possesses at most $|X|$ elements and the `elem` test of HASKELL requires linear time in the length of the list argument, the entire list comprehension of the HASKELL function of (15) can be evaluated in time $O(|X|^2)$ if the call `cup ms (ss!!e)` only requires time $O(|X|)$. The list component `ss!!e` of `ss` can be computed in time $O(|X|)$.

A straightforward implementation of set union on a list implementation of sets (like HASKELL's pre-defined `union` function) requires quadratic runtime. But we can do better using that, because of the precondition, the two lists `ms` and `ss!!e` appearing in the list comprehension of `warsh` are increasingly sorted and without multiple occurrences of elements. On such specific lists an obvious linear implementation of set union is given by the following HASKELL function `cup` that merges two sorted lists into a sorted one and removes at the same time all multiple occurrences of elements. The declarations (12) until (16) constitute a HASKELL program for computing transitive closures that is based on a representation of relations via lists of successor lists and runs in cubic time like the traditional imperative version of Warshall's algorithm.

```
cup :: Vertexset -> Vertexset -> Vertexset
cup [] ys = ys
cup xs [] = xs
cup (x:xs) (y:ys) =                                        (16)
    case compare x y of EQ -> x : cup xs ys
                        LT -> x : cup xs (y:ys)
                        GT -> y : cup (x:xs) ys
```

We conclude this section with two modifications of the HASKELL program we have developed so far.

The first modification concerns the precondition on the input of the whole program. The cubic runtime remains preserved if in the HASKELL function of (14) the call warsh rs (vertices rs) is replaced by the following call:

$$\text{warsh (map (nub.sort) rs) (vertices rs)} \qquad (17)$$

In the HASKELL expression (17) the pre-defined HASKELL function nub removes duplicate elements from a list, sort is the pre-defined HASKELL sorting function on lists, the dot "." denotes HASKELL's function composition operation and map is HASKELL's pre-defined higher-order function that applies a function to each component of a list. The advantage of this modification is that its correctness does no longer depend on the fact that the successor lists of the input are strictly increasing, without sacrificing the runtime complexity.

The second modification concerns the element test in the list comprehension of (15). Again the cubic runtime of the whole HASKELL program remains preserved if in the HASKELL function warsh of (15) the call elem e ms is replaced by the call iselem e ms of the following HASKELL function.

```
iselem :: Vertex -> Vertexset -> Bool
iselem x [] = False
iselem x (y:ys) =
    case compare x y of EQ -> True                         (18)
                        GT -> iselem x ys
                        LT -> False
```

Practical experiments have shown that by this modification the runtime — depending on the input — to a greater or lesser extent is improved. This is because (18) takes advantage of the fact that the successor lists are increasingly sorted.

8   Some Graph-Theoretic Applications

As shown e.g., in [22,8], a lot of problems are closely related to transitive closures. In this section we present some simple functional graph algorithms which are formulated in HASKELL and are based on the HASKELL program of Section 7.

Suppose that $C : X \leftrightarrow X$ denotes the reflexive-transitive closure R^* of a relation $R : X \leftrightarrow X$. Due to the second equation of (2), the list representation cs of C in the sense of Section 6 is obtained from the list representation rs of R by inserting each $x \in X$ into the x-component of cs if not yet contained. In HASKELL the latter modification can easily be realized as follows.

$$
\begin{aligned}
&\texttt{rtc :: Relation -> Relation} \\
&\texttt{rtc rs =} \\
&\quad \texttt{let insert x xs = cup xs [x]} \\
&\quad \texttt{in zipWith insert (vertices rs) (transcl rs)}
\end{aligned}
\tag{19}
$$

Here we use an auxiliary function `insert` for list insertion and the pre-defined HASKELL function `zipWith` that takes a binary function and two lists and returns the list of corresponding pairs via zipping with the function. From (19) we immediately get `(rtc rs)!!x` as HASKELL expression for the graph-theoretic *descendants* (reachable vertices), since the set of descendants of $x \in X$ in the directed graph $G = (X, R)$ is given by the x-component of the list cs.

As next application we consider cycles. Given a directed graph $G = (X, R)$, a vertex $x \in X$ *lies on a cycle* if and only if $(x, x) \in R^+$. The latter is equivalent to $x \in cs!!x$, where now cs is the list representation of the transitive closure R^+. We, therefore, obtain the set of the vertices lying on a cycle by the elements of X that satisfy the predicate $\lambda x \bullet x \in cs!!x$. For the latter, HASKELL provides a pre-defined function `filter`. Altogether, we obtain the following result.

$$
\begin{aligned}
&\texttt{oncycle :: Relation -> Vertexset} \\
&\texttt{oncycle rs =} \\
&\quad \texttt{let cs = transcl rs} \\
&\quad \texttt{in filter (\ x -> elem x (cs!!x)) (vertices rs)}
\end{aligned}
\tag{20}
$$

Using (20) and the pre-defined emptiness test `null` on lists testing for *cycle-freeness* is now possible via `null (oncycle rs)`.

Our third application deals with sources and the testing of strong connectedness. A vertex $x \in X$ is called a *source* or an *initial vertex* of the directed graph $G = (X, R)$ if the set of its descendants equals X, and G is *strongly connected* if each vertex is a source. From the latter description we get at once the following HASKELL test function.

$$
\begin{aligned}
&\texttt{connected :: Relation -> Bool} \\
&\texttt{connected rs = (sources rs) == (vertices rs)}
\end{aligned}
\tag{21}
$$

What remains is the task to formulate the HASKELL function sources that computes the set of sources. Here we follow exactly the pattern of (20).

$$
\begin{aligned}
&\texttt{sources :: Relation -> Vertexset} \\
&\texttt{sources rs =} \\
&\quad \texttt{let cs = rtc rs} \\
&\quad \texttt{in filter (\ x -> cs!!x == (vertices rs)) (vertices rs)}
\end{aligned}
\tag{22}
$$

For the fourth application, we assume $R : X \leftrightarrow X$ as relation of a cycle-free directed graph $G = (X, R)$. Then the relation $R^- := R \cap \overline{R; R^+}$ is called the *transitive reduction* of R. It is obtained from R by removing all edges which can be bypassed by a path with at least two edges and constitutes the least subrelation S of R such that $S^+ = R^+$. (In case of a partial order P the transitive reduction $(P \cap \overline{\mathsf{I}})^-$ coincides with the cover relation — the Hasse diagram — of P.) From the definition of R^- as (set-theoretic) difference of R and $R; R^+$ we immediately obtain that if $R : X \leftrightarrow X$ is represented (in the sense of Section 5) by $r : X \to 2^X$ and $R; R^+$ is represented by $s : X \to 2^X$, then R^- is represented by $\lambda x \bullet r(x) \setminus s(x)$. Going from successor functions to lists of sets and afterwards to lists of lists and HASKELL leads to the following program for transitive reductions, where $\setminus\setminus$ is the pre-defined HASKELL operation for list difference[2] and comp implements the composition of relations.

$$
\begin{aligned}
&\texttt{transred :: Relation -> Relation}\\
&\texttt{transred rs = zipWith (\textbackslash\textbackslash) rs (comp rs (transcl rs))}
\end{aligned}
\tag{23}
$$

What remains is the development of the HASKELL function comp. Here we follow exactly the method applied in the Sections 5 to 7. Let $R, S : X \leftrightarrow X$ and suppose R and S to be be represented by the functions $r : X \to 2^X$ and $s : X \to 2^X$, respectively. Then we have for all $x, y \in X$ that

$$
\begin{aligned}
(x, y) \in R; S &\iff \exists i \in X : (x, i) \in R \wedge (i, y) \in S\\
&\iff \exists i \in X : i \in r(x) \wedge y \in s(i)\\
&\iff y \in \bigcup\{s(i) \mid i \in r(x)\}.
\end{aligned}
$$

Hence, the relation $R; S$ is represented by the function $\lambda x \bullet \bigcup\{s(i) \mid i \in r(x)\}$. If we now represent the functions r and s by lists rs and ss, respectively, i.e., as in Section 6, then the function $\lambda x \bullet \bigcup\{s(i) \mid i \in r(x)\}$ is represented by the list comprehension $[\bigcup\{ss!!i \mid i \in rs!!x\} \mid x \in X]$. For the translation into HASKELL we assume rs and ss to be the HASKELL counterparts of rs and ss, respectively. Then the HASKELL list for the set $\bigcup\{ss!!i \mid i \in rs!!x\}$ is obtained by the union (via the function cup) of all sorted lists ss!!i, where i ranges over the elements of rs!!x. Such a repeated application of cup corresponds to a fold over a list that exactly consists of all lists ss!!i, with the empty list [] as initial value. If we use right-fold, in HASKELL realized by the pre-defined function foldr, we get foldr cup [] [ss!!i | i <- rs!!x] as implementation of $\bigcup\{ss!!i \mid i \in rs!!x\}$ and the above list comprehension, that represents the composition $R; S$ we want to implement, immediately yields the following HASKELL function.

$$
\begin{aligned}
&\texttt{comp :: Relation -> Relation -> Relation}\\
&\texttt{comp rs ss =}\\
&\quad\texttt{[foldr cup [] [ss!!i | i <- rs!!x] | x <- (vertices rs)]}
\end{aligned}
\tag{24}
$$

---

[2] As in the case of the functional version of Warshall's algorithm the practical runtime of (23) is improved if a user-defined HASKELL function for list difference is used that takes advantage of the fact that all successor lists are increasingly sorted.

Having a HASKELL program for relational composition at hand, it is very simple to compute for a directed graph $G = (X, R)$ the set of vertices which lie on an odd cycle. A little reflection shows that $x \in X$ lies on an odd cycle if and only if $(x, x) \in R; (R; R)^*$, Assuming ss as list representation of the relation $R; (R; R)^*$, the latter property holds if and only if $x \in ss!!x$. From this an application of `filter` with the HASKELL version of the predicate $\lambda x \bullet x \in ss!!x$ immediately leads to the following result.

```
onoddcycle :: Relation -> Vertexset
onoddcycle rs =
    let ss = comp rs (rtc (comp rs rs))
    in  filter (\ x -> elem x (ss!!x)) (vertices rs)
```
(25)

By means of (25) it is very easy to test an undirected graph (i.e., a graph with an irreflexive and symmetric relation) to be bipartite. This is the case if and only if the expression `onoddcycle rs` evaluates to the empty list.

For our final application we need a HASKELL function for the transposition of relations $R : X \leftrightarrow X$. From the equivalence of $(y, x) \in R$ and $(x, y) \in R^\mathsf{T}$ for all $x, y \in X$ we get that if R is represented by $r : X \to 2^X$, then R^T is represented by $\lambda x \bullet \{y \in X \mid x \in r(y)\}$. As a consequence, the list representation rs of R leads to the list representation $[\{y \in X \mid x \in rs!!y\} \mid x \in X]$ of R^T. A translation of the latter into HASKELL leads to the following function.

```
transp :: Relation -> Relation
transp rs =
    let ve = vertices rs
    in  [filter (\ y -> elem x (rs!!y)) ve | x <- ve]
```
(26)

The application treats confluence. A directed graph $G = (X, R)$ is *confluent* if any two vertices with common ancestors have common descendants. Relation-algebraically this can be described by the inclusion $R^{*\mathsf{T}}; R^* \subseteq R^*; R^{*\mathsf{T}}$ or, equivalently, the equation $R^{*\mathsf{T}}; R^* \cap \overline{R^*; R^{*\mathsf{T}}} = O$. Let the HASKELL lists `ss` and `ts` represent the relations R^* and $R^{*\mathsf{T}}$, respectively. Then the list `comp ts ss` represents $R^{*\mathsf{T}}; R^*$, the list `comp ss ts` represents $R^*; R^{*\mathsf{T}}$ and, along the lines of (23), the list `zipWith (\\) (comp ts ss) (comp ss ts)` represents their difference. It remains to test whether the difference is the empty relation. Since this means to check whether each of the successor lists of its list representation is empty, it can be done by concatenating all successor lists and then testing the result to be empty. Altogether we arrive at the following solution, where the pre-defined HASKELL function `concat` concatenates a list of lists to a single list.

```
confluent :: Relation -> Bool
confluent rs
    let ss = rtc rs
        ts = transp ss
    in null (concat (zipWith (\\) (comp ts ss) (comp ss ts)))
```
(27)

In [22] it is shown that the inclusions $R^{*\mathsf{T}}; R^* \subseteq R^*; R^{*\mathsf{T}}$ and $R^\mathsf{T}; R^* \subseteq R^*; R^{*\mathsf{T}}$ are equivalent. Hence, in (27) the expression `comp ts ss` can be replaced by comp (`transp rs`) `ss`. Experiments have shown that by this modification the practical runtimes are slightly improved.

9   Concluding Remarks

A detailed analysis shows that the HASKELL function `comp` of (24) has a cubic runtime (like classical matrix multiplication) and the same holds for the function `transred` for computing transitive reductions. In [2] it is shown that algorithms for transitive reductions in general have the same runtime complexity as algorithms for (reflexive-)transitive closures. The complexity of testing confluence usually is studied for rewriting systems (see e.g., [20]) and not for general relations / graphs as we do. Our test function `confluence` requires cubic runtime and we are not aware of a faster algorithm for testing confluence in our general setting.

In view of runtime complexity the HASKELL functions `oncycle`, `connected`, `sources` and `onoddcycle` we have presented in Section 8 cannot compete with the well-known quadratic algorithms (in the number of vertices) or linear algorithms (in the number of edges) specifically tailored for the given problems, since their runtime complexities are dominated by the cubic costs for computing transitive closures. Despite of this disadvantage we believe that they (and a lot of others which can be obtained from purely relation-algebraic problem specifications in a similar way) have their benefits and usefulness. Such programs are easy to construct, lead to high-level and succinct solutions and the correctness proofs are usually relatively simple. For many practical problems even their performance is satisfactory. Due to these properties they are very suitable for prototyping purposes and as oracles for algorithm testing.

Another aspect is teaching. At present there exist only a few textbooks on algorithmics that base on the functional paradigm. All of them treat graphs and other relational structures sparsely, in contrast with books that base on imperative programming. Graphs are nothing else as relations on vertices and a lot of questions of graph theory are closely related to relational properties and problems. We think that — as extension of e.g., [22,5,8], where relation-algebra and imperative programming is combined for problem soving — also a combination of relation algebra and functional programming is an excellent means to teach the development and programming of graph algorithms and, in excess thereof, also of algorithms on other relation-based discrete structures like orders, lattices, Petri nets and games.

Acknowledgement. I want to thank B. Braßel, J. Christiansen and F. Huch for valuable discussions and the unknown referees for their helpful remarks.

References

1. Aarts, A., et al.: Fixed point calculus. Inform. Proc. Lett. 53, 131–136 (1996)
2. Aho, A., Garey, M., Ullman, J.: The transitive reduction of a directed graph. SIAM J. of Comput. 1, 131–137 (1972)

3. Antoy, S., Hanus, M.: Functional logic programming. Comm. of the ACM 53, 74–85 (2010)
4. Berghammer, R., Ehler, H., Zierer, H.: Development of graph algorithms by program transformation. In: Göttler, H., Schneider, H.-J. (eds.) WG 1987. LNCS, vol. 314, pp. 206–218. Springer, Heidelberg (1988)
5. Berghammer, R., von Karger, B.: Algorithms from relational specification. In: Brink, C., Kahl, W., Schmidt, G. (eds.) Relational Methods in Computer Science, pp. 131–149. Springer, Heidelberg (1997)
6. Berghammer, R., Neumann, F.: RELVIEW – An OBDD-Based Computer Algebra System for Relations. In: Ganzha, V.G., Mayr, E.W., Vorozhtsov, E.V. (eds.) CASC 2005. LNCS, vol. 3718, pp. 40–51. Springer, Heidelberg (2005)
7. Berghammer, R., Fischer, S.: Implementing relational specifications in a constraint functional language. Electr. Notes on Theor. Comput. Sci. 177, 169–183 (2007)
8. Berghammer, R.: Ordnungen, Verbände und Relationen mit Anwendungen (Orders, lattices and relations with applications). Vieweg-Teubner, Stuttgart (2008)
9. Bird, R.: Introduction to functional programming using HASKELL, 2nd edn. Prentice-Hall, Englewood Cliffs (1998)
10. Burstall, R.M., Darlington, J.: A transformation system for developing recursive programs. J. of the ACM 24, 44–67 (1977)
11. Chen, Y.: On the evaluation of large and sparse graph reachability queries. In: Bhowmick, S.S., Küng, J., Wagner, R. (eds.) DEXA 2008. LNCS, vol. 5181, pp. 97–105. Springer, Heidelberg (2008)
12. Erwig, M.: Functional programming with graphs. ACM SIGPLAN Notices 32, 52–65 (1997)
13. Erwig, M.: Inductive graphs and functional graph algorithms. J. of Funct. Progr. 11, 467–492 (2001)
14. Floyd, R.W.: Algorithm 97 (Shortest path). Comm. of the ACM 5, 345 (1962)
15. Fredmann, M.L.: New bounds on the complexity of the shortest path problem. SIAM J. on Comput. 5, 83–89 (1976)
16. Gibbons, J.: An initial algebra approach to directed graphs. In: Möller, B. (ed.) MPC 1995. LNCS, vol. 947, pp. 282–303. Springer, Heidelberg (1995)
17. Kahl, W.: Semigroupoid interfaces for relation-algebraic programming in HASKELL. In: Schmidt, R.A. (ed.) RelMiCS/AKA 2006. LNCS, vol. 4136, pp. 235–250. Springer, Heidelberg (2006)
18. King, D.J., Launchbury, J.: Structuring depth-first search algorithms in HASKELL. In: ACM Symposium on Principles of Programming, pp. 344–356. ACM, New York (1995)
19. Launchbury, J.: Graph algorithms with a functional flavour. In: Jeuring, J., Meijer, E. (eds.) AFP 1995. LNCS, vol. 925, pp. 308–331 (2005)
20. Lohrey, M.: Complexity results for confluence problems. In: Kutyłowski, M., Wierzbicki, T., Pacholski, L. (eds.) MFCS 1999. LNCS, vol. 1672, pp. 114–124. Springer, Heidelberg (1999)
21. Penn, G.: The algebraic structure of transitive closure and its application to attributed type signatures. Grammars 3, 295–312 (2000)
22. Schmidt, G., Ströhlein, T.: Relations and graphs. Discrete Mathematics for Computer Scientists. Springer, Heidelberg (1993)
23. Schmidt, G.: A proposal for a multilevel relational reference language. J. of Relat. Meth. in Comput. Sci. 1, 314–338 (2004)
24. Strassen, V.: Gaussian elimination is not optimal. Num. Math. 13, 354–356 (1969)
25. Thompson, S.: HASKELL – The craft of functional programming. Addison-Wesley, Reading (1999)
26. Warshall, S.: A theorem on Boolean matrices. J. of the ACM 9, 11–12 (1962)

Variable Side Conditions and Greatest Relations in Algebraic Separation Logic

Han-Hing Dang and Peter Höfner

Institut für Informatik, Universität Augsburg, D-86159 Augsburg, Germany
{h.dang,hoefner}@informatik.uni-augsburg.de

Abstract. When reasoning within separation logic, it is often necessary to provide side conditions for inference rules. These side conditions usually contain information about variables and their use, and are given within a meta-language, i.e., the side conditions cannot be encoded in separation logic itself. In this paper we discuss different possibilities how side conditions of variables—occurring e.g. in the ordinary or the hypothetical frame rule—can be characterised using algebraic separation logic. We also study greatest relations; a concept used in the soundness proof of the hypothetical frame rule. We provide one and only one level of abstraction for the logic, the side conditions and the greatest relations.

1 Introduction

Over the last years, separation logic (SL) (e.g. [11]) has been established as a formal system that allows reasoning and verification of imperative programs *including* shared mutable data structures. It is a proper extension of Hoare logic and has been used (a) to split data structures into logically connected regions which can then be analysed and reasoned about separately; (b) for the analysis of pointer variables and their update; and (c) for the dynamic assignment of "owners" of data regions under concurrent access to them.

One major instrument of SL is the frame rule. It allows adding arbitrary disjoint storage to the resources which are actually used by a command. Proof rules like the frame rule are often constrained by side conditions on the variables involved. Usually, they are formulated within a meta-language. This complicates reasoning in general and, more particularly, when building tools for automated reasoning (e.g. Smallfoot [1]) two layers have to be considered.

In a companion paper [5], an algebra for separation logic based on a relational semantics of commands has been presented. On this basis a restricted version of the frame rule has been proved in an abstract way. The proof itself is based on three assumptions: safety monotonicity, a frame property and preservation. The first two were also used by Reynolds [11], the third one was intended as an algebraic counterpart of the side condition of the frame rule.

In this paper we show that the concept of preservation as presented in [5] is too strict. Motivated by this observation, we discuss to which extent variable side conditions can be embedded into the algebraic framework. As a result we

H. de Swart (Ed.): RAMICS 2011, LNCS 6663, pp. 125–140, 2011.

characterise various side conditions at an algebraic level and provide one and only one level of abstraction for both the logic and the side conditions. Moreover we bring the hypothetical frame rule into our setting. This rule allows more general reasoning than the original one. As a further application for algebraic separation logic we give pointfree characterisations for greatest relations which play an important role in proving soundness of this particular frame rule.

2   The Frame Rule and the Set of Modified Variables

The frame rule [8] describes that a command can also be executed using a larger storage—as long as the command does not influence the additional storage:

$$\frac{\{p\}C\{q\}}{\{p*r\}C\{q*r\}} \quad MV(C) \cap FV(r) = \emptyset \;. \tag{1}$$

The expression $\{p\}\,C\,\{q\}$ denotes a slightly modified Hoare triple in partial correctness semantics where p and q are predicates about states and C is a command: as usual, the command C establishes the postcondition if the precondition is met. Additionally, the command C can always be executed whenever p is satisfied. The disjoint storage part, characterised by r, will remain unchanged as long as no free variable of r will be touched by any execution of C. This restriction on the usage of variables is described by the formula $MV(C) \cap FV(r) = \emptyset$.

Next to the side condition, two further assumptions on commands C are needed to prove the frame rule in SL: *safety monotonicity* and the *frame property*. The former guarantees that if C is executable from a state, it can also run on a state with a larger heap; the latter states that every execution of C can be tracked back to an execution of C running on states with a possible smaller heap.

Let us now take a closer look at the side condition of the frame rule and on the set $MV(C)$ of variables modified by a command C. Formally, the syntax of a command is given by[1]

$$exp ::= var \mid seq.var \mid \mathsf{tail}(seq) \mid \mathsf{head}(seq) \mid ...$$
$$comm ::= var := exp \mid \mathsf{dispose}\,exp$$
$$\mid \mathsf{skip} \mid comm\,;\,comm \mid \mathsf{if}\ bexp\ \mathsf{then}\ comm\ \mathsf{else}\ comm$$
$$\mid var := \mathsf{cons}\,(exp, \ldots, exp) \mid var := [exp] \mid [exp] := exp \;,$$

where *var* denotes variables, *exp* expressions and *bexp* boolean expressions. Moreover *seq* stands for sequences of values, . denotes concatenation and head, tail return the head or tail of a sequence resp. The command $v := \mathsf{cons}\,(e_1, ..., e_n)$ allocates n contiguous fresh cells and places the values of the expressions e_i in the current store as the contents of the i-th cell. The address of the first cell is stored in v while the following cells can be accessed via address arithmetic. The assignment $v := [e]$ dereferences the heap cell at the address given by the value of e and stores its value in v. An execution of $[e_1] := e_2$ assigns the value of e_2

---

[1] We provide more details on the commands used later on.

Table 1. $MV(C)$-set for commands C

C	$MV(C)$
$[e_1] := e_2$	$\emptyset$
$v := [e]$	$\{v\}$
dispose e	$\emptyset$
$v := \mathsf{cons}\,(e_1, ..., e_n)$	$\{v\}$
$C_1 \,;\, C_2$	$MV(C_1) \cup MV(C_2)$
if (b) then C_1 else C_2	$MV(C_1) \cup MV(C_2)$

to a cell located at the value of e_1. Finally, the command dispose e deallocates the heap cell located at the address which is the value of e.

Based on that, the set $MV(C)$ of modified variables (not heap cells!) of a command C can be determined inductively by the rules given in Table 1.

These definitions seem straightforward; however, they depend in an essential way on the syntax and structure of a command and not on its semantics.

For the following commands we assume that two variables x, y are available.

$$C_1 =_{df} (x := y) \qquad \text{and} \qquad C_2 =_{df} (x := y \,;\, y := 3 \,;\, y := x)\,.$$

After execution, both commands C_1 and C_2 have set the variable x to y and the value of y is the same as it was before the execution. However, C_2 modifies y during the execution. Hence $MV(C_1) = \{x\}$ and $MV(C_2) = \{x, y\}$. To connect commands with relations, the algebraic approach of [6] describes each command by an input-output relation between states, or, in other words, by a state transformer. (The details will be explained in the following sections). These relations only reflect the overall behaviour, hence cannot look at the syntactic structure of a given command. In particular, the commands C_1 and C_2 are indistinguishable for the algebraic approach. Usually, the set MV of modified variables lists all variables to which values are assigned. However, it would be interesting to determine the set of all variables which are "really changed" by a command as a relation. For the commands C_1 and C_2 this would be the set $\{x\}$.

3   Algebraic Separation Logic

Before looking at side conditions algebraically, we have to recapitulate the foundations of algebraic separation logic and its relational semantics.

A system's *state* is a pair consisting of a store and a heap; stores and heaps are partial functions from variables or addresses to values. To simplify the formal treatment, values and addresses are assumed to be integers.

$$Values = \mathbb{Z}\,, \qquad Stores = Vars \rightsquigarrow Values\,,$$
$$\{\mathrm{nil}\} \uplus Addresses \subseteq Values\,, \qquad Heaps = Addresses \rightsquigarrow Values\,,$$
$$States = Stores \times Heaps\,,$$

where *Vars* is the set of program variables, $\uplus$ denotes the disjoint union of sets and $M \rightsquigarrow N$ denotes the set of partial functions between M and N. Stores and heaps will be denoted by s and h, resp., while σ and τ stand for states. The store and heap part of a state σ are denoted by s_σ and h_σ, resp. As usual we define a domain operator dom and a range operator cod on relations and partial functions. For example, $\mathrm{dom}(s)$ for a store s returns the set of all variables with defined values; for a relation R a partial identity relation is returned. The constant nil denotes an improper reference and therefore heaps must not map nil to any value.

We follow the idea of [4,6] and define based on states, a *command* as a relation $C \in \mathit{Cmds} =_{df} \mathcal{P}(\mathit{States} \times \mathit{States})$. Relations offer a number of operations, including sequential composition $;$, choice $\cup$, converse $\check{}$ and complementation $\bar{}$. In general, relations and the structure $(\mathit{Cmds}, \subseteq, ;, I)$ in particular form Boolean quantales [2], where I is the identity relation. For the purpose of the paper, we restrict ourselves to relations; although most of the results could be also achieved in the more general quantale setting.

Next we recapitulate the relational semantics for the above mentioned concrete commands. For that we define $FV(e)$ as the set of all free variables occurring in an expression e. The value e^S of an expression is defined for an arbitrary store s only if $FV(e) \subseteq \mathrm{dom}(s)$. Moreover, we define an update operator on partial functions by $f \mid f' =_{df} f \cup \{(v,c) : (v,c) \in f' \wedge v \notin \mathrm{dom}(f)\}$ and abbreviate $\{(v,c)\} \mid f$ to $(v,c) \mid f$. We characterise the commands linking input states (s,h) and output states (s',h') by $R \mathrel{\widehat{=}} P$ to abbreviate the clause $(s,h)\,R\,(s',h') \Leftrightarrow_{df} P$. We require for each of the following commands C and expressions e occurring in C that $FV(e) \subseteq \mathrm{dom}(s)$.

$$[\![e_1] := e_2]\!]_c \mathrel{\widehat{=}} s' = s \wedge e_1^S \in \mathrm{dom}(h) \wedge h' = (e_1^S, e_2^S) \mid h \, ,$$

$$[\![v := [e]]\!]_c \mathrel{\widehat{=}} s' = (v, h(e^S)) \mid s \;\wedge\; e^S \in \mathrm{dom}(h) \wedge h' = h \, ,$$

$$[\![\mathsf{dispose}\ e]\!]_c \mathrel{\widehat{=}} s' = s \wedge e^S \in \mathrm{dom}(h) \wedge h' = h - \{(e^S, h(e^S))\} \, ,$$

$$[\![v := \mathsf{cons}\,(e_1, ..., e_n)]\!]_c \mathrel{\widehat{=}} \exists a \in \mathit{Addresses}\,.\; s' = (v, a) \mid s \;\wedge$$
$$a, \ldots, a + n - 1 \notin \mathrm{dom}(h) \wedge$$
$$h' = \{(a, e_1^S), \ldots, (a + n - 1, e_n^S)\} \mid h \, .$$

If a command is executable on a state, we assume that it can also run on larger states containing more variable declarations. Moreover, we assume that the command does not change the set of defined program variables. We assume $C \in \mathit{Cmds}$. $\sigma\,C\,\tau \Rightarrow \mathrm{dom}(s_\sigma) = \mathrm{dom}(s_\tau)$ and for all $X \subseteq \mathit{Vars}$. $\sigma \in \mathrm{dom}(C) \Rightarrow \exists \tau.\ \tau \in \mathrm{dom}(C) \wedge \mathrm{dom}(s_\tau) = \mathrm{dom}(s_\sigma) \cup X$. The semantics of an if–statement is defined as usual. Due to readability we will omit brackets $[\![\ ]\!]_c$ and have each statement stand for its semantics.

This forms the basis of algebraic separation logic—except for separating conjunction $*$ which is described in the next section.

4   States: Compatibility and Splitting

The separating conjunction $*$ of SL unites disjoint heap regions and allows reasoning about separate storage. The algebraic approach is more general and lifts this operation to general commands: a command is split into executions running on disjoint heap parts. By splitting we describe a generalised version of the separating conjunction $*$.

- Two stores s and s' are *compatible* iff $s = s' \vee \mathrm{dom}(s) \cap \mathrm{dom}(s') = \emptyset$.
- Two states $\sigma_1 = (s_1, h_1)$ and $\sigma_2 = (s_2, h_2)$ are *combinable* iff
$$(s_1, h_1) \mathbin{\#} (s_2, h_2) \Leftrightarrow_{df} s_1, s_2 \text{ are compatible} \wedge \mathrm{dom}(h_1) \cap \mathrm{dom}(h_2) = \emptyset .$$
- The *split* relation $\lhd$ is defined for states σ, σ_1 and σ_2 as
$$\sigma \lhd (\sigma_1, \sigma_2) \Leftrightarrow_{df} \sigma_1 \mathbin{\#} \sigma_2 \wedge \sigma = \sigma_1 * \sigma_2 ,$$
where $\sigma_1 * \sigma_2 = (s_1 \cup s_2, h_1 \cup h_2)$ if $\sigma_1 = (s_1, h_1)$ and $\sigma_2 = (s_2, h_2)$.
- The *join* relation $\rhd$ is the converse of $\lhd$, i.e., $\rhd = \lhd^{\smallsmile}$.
- The *Cartesian product* $C_1 \times C_2$ of two commands C_1, C_2 is given, as usual, by
$$(\sigma_1, \sigma_2)\,(C_1 \times C_2)\,(\tau_1, \tau_2) \Leftrightarrow_{df} \sigma_1\,C_1\,\tau_1 \wedge \sigma_2\,C_2\,\tau_2 .$$
It is well known that $\times$ and $;$ satisfy the exchange property

$$(R_1 \times R_2)\,;\,(S_1 \times S_2) = (R_1\,;\,S_1) \times (R_2\,;\,S_2) . \tag{2}$$

We assume for the rest of this paper that $;$ binds tighter than $\times$.

- The $*$ *composition* is defined by $C_1 * C_2 =_{df} \lhd\,;\,(C_1 \times C_2)\,;\,\rhd$.

By store compatibility it is required that both involved stores are either equal or both map from a disjoint set of variables. Therefore when joining and splitting states we are more liberal as in standard separation logic. The standard separating conjunction requires the stores involved to be equal.

For later properties we need to define a relation w.r.t. a command C that only allows store changes of C and excludes heap alteration. For states σ, σ', the *store-change* relation S_C for a command C is defined by
$$\sigma\,S_C\,\sigma' \Leftrightarrow_{df} h_\sigma = h_{\sigma'} \wedge changed(C) \subseteq \mathrm{dom}(s_\sigma) \wedge$$
$$(\exists\,\sigma_c, \sigma_c' \,.\, \sigma_c\,C\,\sigma_c' \wedge$$
$$s_{\sigma_c} \subseteq s_\sigma \wedge s_{\sigma_c'} \subseteq s_{\sigma'} \wedge s_\sigma - s_{\sigma_c} = s_{\sigma'} - s_{\sigma_c'}) ,$$
where $changed(C) =_{df} \bigcup_{(\tau_1, \tau_2) \in C} \mathrm{dom}(s_{\tau_1} - s_{\tau_2})$. This latter definition is motivated by the fact that $x \in \mathrm{dom}(s_{\tau_1} - s_{\tau_2}) \Leftrightarrow x \in \mathrm{dom}(s_{\tau_1}) \wedge s_{\tau_1}(x) \neq s_{\tau_2}(x)$. Given a command C the relation S_C changes each store variable of an input state as C would do. The first line of the definition ensures that all stores involved mention at least all variables that are changed by an arbitrary execution of C. This is necessary to ensure certain preservation properties given later. Next, we briefly sum up some results for the relation S_C needed later on.

Lemma 4.1. *For an arbitrary test r and commands C, D we have*

1. $S_r \subseteq I$,
2. $C * S_C \subseteq C * I$. *In particular,* $C * (\mathrm{emp}\,;\,S_C) \subseteq C$ *where* σ emp $\sigma' \Leftrightarrow_{df}$ $s_\sigma = s_\sigma' \wedge h_\sigma = \emptyset = h_{\sigma'}$,
3. *If* $C \subseteq D$ *then* $C * (r\,;\,S_D) \subseteq C * (r\,;\,S_C)$.
4. *If* $r\,;\,S_{C;D} \subseteq r\,;\,S_C\,;\,r\,;\,S_D$ *then* $(C;D) * (r\,;\,S_{C;D}) \subseteq (C * (r\,;\,S_C))\,;\,(D * (r\,;\,S_D))$.

Proof. (1) follows immediate from the definition. For (2) consider $\sigma \ (C * S_C) \ \sigma'$. Then by definition there exist states $\sigma_c, \sigma_c', \sigma_S, \sigma_S'$ with $\sigma = \sigma_c * \sigma_S \wedge \sigma_c \#$ $\sigma_S \wedge \sigma' = \sigma_c' * \sigma_S' \wedge \sigma_c' \# \sigma_S' \wedge \sigma_c \ C \ \sigma_c' \wedge \sigma_S \ S_C \ \sigma_S'$. By the definition of S_C this implies $s_{\sigma_c} = s_{\sigma_S}$ and $s_{\sigma_c'} = s_{\sigma_S'}$. Now set $\sigma_I =_{df} (\emptyset, h_{\sigma_S})$, then $\sigma = \sigma_c * \sigma_I \wedge \sigma_c \# \sigma_I \wedge \sigma_c \ C \ \sigma_c' \wedge \sigma_I \in I \wedge \sigma_c' \# \sigma_I \wedge \sigma' = \sigma_c' * \sigma_I$ holds.

Furthermore assume $\sigma \ C * (r \ ; \ S_D) \ \sigma'$. Again there exist $\sigma_c, \sigma_c', \sigma_r, \sigma_S$ with $\sigma = \sigma_c * \sigma_r \wedge \sigma_c \# \sigma_r \wedge \sigma_r \in r \wedge \sigma_r \ S_D \ \sigma_S \wedge \sigma_c \ C \ \sigma_c' \wedge \sigma' = \sigma_c' * \sigma_S \wedge \sigma_c' \# \sigma_S$ which implies $s_{\sigma_c} = s_{\sigma_r}$ and $s_{\sigma_c'} = s_{\sigma_S}$. Therefore, by $C \subseteq D$ and the definition of S_C also $\sigma_r \ S_C \ \sigma_S$ holds. Finally to prove (4), we calculate

$$
\begin{aligned}
(C \ ; \ D) * (r \ ; \ S_{C;D}) &= \vartriangleleft ; ((C \ ; \ D) \times (r \ ; \ S_{C;D})) ; \vartriangleright \\
&\subseteq \vartriangleleft ; ((C \ ; \ D) \times (r \ ; \ S_C \ ; \ r \ ; \ S_D)) ; \vartriangleright \\
&= \vartriangleleft ; ((C \times (r \ ; \ S_C)) ; (D \times (r \ ; \ S_D))) ; \vartriangleright \\
&\subseteq \vartriangleleft ; ((C \times (r \ ; \ S_C)) ; \vartriangleright ; \vartriangleleft ; (D \times (r \ ; \ S_D)) ; \vartriangleright \\
&= (C * (r \ ; \ S_C)) ; (D * (r \ ; \ S_D)) .
\end{aligned}
$$

This uses the definition of $*$, assumption, associativity, Exchange (2), neutrality of $(I \times I)$ and $(I \times I) \subseteq \vartriangleright ; \vartriangleleft$.   □

Informally, considering any execution of C then by $C * S_C$ only disjoint heap cells are added to the states involve, i.e., S_C changes exactly the same variables as C does. In particular in $C * (r; S_D)$, S_D alters the same variables of r as C does if $C \subseteq D$. Finally, the assumption $r \ ; \ S_{C;D} \subseteq r \ ; \ S_C \ ; \ r \ ; \ S_D$ states that r is not changed after all variable assignments of C.

5   The Frame Rule Algebraically

Besides commands the frame rule uses slightly modified Hoare triples $\{p\} \ C \ \{q\}$ (see Section 2). In the algebraic setting predicates (pre- and postconditions) can be modelled by tests as e.g. in [7]. In $Cmds$, tests are given by partial identity relations of the form $\{(\sigma, \sigma) \mid \sigma \in p\}$ for some set $p \in States$. We further abbreviate $(\sigma, \sigma) \in p$ to $\sigma \in p$. Using tests, an if–statement if p then C else C' is described by $p; C \cup \neg p; C'$, where $\neg p =_{df} \bar{p} \cap I$. It has further been shown that a (standard) Hoare triple $\{p\} \ C \ \{q\}$ is equivalent to $p; C \subseteq C; q$, where p, q are test elements. It expresses that q is reached under all C-transitions from p, i.e., the precondition p guarantees the postcondition q. The modified Hoare triples are characterised by

$$\{p\}C\{q\} \Leftrightarrow_{df} p \subseteq \mathrm{dom}(C) \wedge p \ ; C \subseteq C \ ; q . \tag{3}$$

Note that $\mathrm{dom}(C)$ for a relation C denotes the corresponding partial identity relation. This means that $\mathrm{dom}(C)$ is a subidentity of states. Informally, $p \subseteq \mathrm{dom}(C)$ states that C can be executed in all states satisfying p. By these definitions, the frame rule turns into the implication

$$p \ ; C \subseteq C \ ; q \Rightarrow (p * r) \ ; C \subseteq C \ ; (q * r) . \tag{4}$$

The frame rule as well as its algebraic counterpart (4) do not hold in general. As mentioned before, three assumptions are made to prove the frame rule: safety monotonicity, the frame property and the side condition.

Formally, a command C is *safety monotone* iff for all $\sigma, \sigma' \in States$. $\sigma \# \sigma' \wedge \sigma \in \mathrm{dom}(C) \Rightarrow \sigma * \sigma' \in \mathrm{dom}(C)$; a command C satisfies the *frame property* iff for all $\sigma, \sigma_{S_C}, \sigma_1 \in States$. $\sigma \in \mathrm{dom}(C) \wedge \sigma_{S_C} \in \mathrm{dom}(S_C) \wedge \sigma \# \sigma_{S_C} \wedge (\sigma * \sigma_1) \, C \, \sigma_1 \Rightarrow \exists \sigma_2, \sigma_S. \, \sigma \, C \, \sigma_2 \wedge \sigma_{S_C} \, S_C \, \sigma_S \wedge \sigma_2 \# \sigma_S \wedge \sigma_1 = \sigma_2 * \sigma_S$ [6]. These definitions can be given pointfree and purely algebraically:

- C is *safety-monotonic* iff $\mathrm{dom}(C) * I \subseteq \mathrm{dom}(C)$;
- C has the *frame property* iff $(\mathrm{dom}(C) \times \mathrm{dom}(S_C)) \, ; \triangleright \, ; C \subseteq (C \times S_C) \, ; \triangleright$.

Equivalence proofs[2] can be found in [6]. In the next section, we will have a closer look on the third assumption, namely the side condition.

6   Variable Preservation and Variable Side Conditions

Side conditions are often used to restrict the behaviour of commands. In the frame rule, $MV(C) \cap FV(r) = \emptyset$ guarantees that r still holds in the postcondition. In other words, C preserves r.

A first attempt to tackle this variable side conditions of the frame rule algebraically is given in [5]:

$$\triangleleft \, ; (C \times r) \subseteq C \, ; \triangleleft \, . \tag{5}$$

This equation states that whenever a state σ can be split into two states σ_c and σ_r such that C can be executed on σ_c, and σ_r satisfies r then each execution of C ends up in a state where σ_r can be retained completely unchanged. Using this, it is possible to prove the frame rule (4).

Unfortunately, Equation (5) is very restrictive and makes the algebraic approach incomplete. This means that not all instances satisfying the original frame rule satisfy the additional assumption (5). For example taking $C = (x := 1)$ and $r = (\mathrm{true})$ yields $\triangleleft \, ; (C \times r)$, which is not contained in $C \, ; \triangleleft$. In particular, consider a pair $(\sigma, (\tau_1, \tau_2)) \in \triangleleft \, ; (C \times r)$. We can assume that $s_{\tau_2}(x) = 2$. But there exists no such splitting in $C; \triangleleft$ since each τ_i has to satisfy $s_{\tau_i}(x) = 1$. This means the domain as well as the image of C need to be checked for combinability.

This yields another version of preservation. For arbitrary states $\sigma_1, \sigma_2, \tau_1$ and τ_2, the *combinability (joinability) relation* $\mathbb{X}$ on pairs of states is defined by

$$(\sigma_1, \sigma_2) \, \mathbb{X} \, (\tau_1, \tau_2) \Leftrightarrow_{df} \sigma_1 \# \sigma_2 \wedge \sigma_1 = \tau_1 \wedge \sigma_2 = \tau_2 \, .$$

The relation $\mathbb{X}$ is a test (a partial identity relation) that characterises those pairs of states that are combinable w.r.t. $\#$. We will use this relation to obtain the subcommand of a command that maintains combinability with a test r.

Using combinability, we can now define another version of side conditions. A command C *weakly preserves* r iff

$$\mathbb{X} \, ; (C \times r) \, ; \mathbb{X} \neq \emptyset \, . \tag{6}$$

---

[2] In this paper we deviate slightly from [6]. But the proofs can be adapted.

Pointwise this means that $\exists\,\sigma_1,\sigma_2,\sigma_r.\ \sigma_1 \# \sigma_r \wedge \sigma_1\,C\,\sigma_2 \wedge \sigma_r \in r \wedge \sigma_2 \# \sigma_r$. Informally, there is at least one execution of C such that an input state σ_1 as well as an output state σ_2 are combinable with a state σ_r. More precisely the relation $\mathbb{X}\,;(C \times r)\,;\mathbb{X}$ is removing all executions in C that do not maintain r. If the set is empty then r will definitely be changed by C. Simple consequences of the definition are that if C_1 and C_2 preserve r then $C_1 \cup C_2$ preserves r as well. Moreover, I and emp preserve r provided $r \neq \emptyset$.

In some sense, Equation (6) behaves really angelically: for an assertion r it searches for one "execution-path" in C that preserves r. Let us give an example: with $C =_{df}$ if $(x = 0)$ then skip else $x := 2$ and $r =_{df}\ (x \neq 2)$ we have $\mathbb{X}\,;(C \times r)\,;\mathbb{X} = \mathbb{X}\,;((x = 0) \times r)\,;\mathbb{X} \neq \emptyset$ since $\mathbb{X}\,;((x \neq 0\,;x := 2) \times r)\,;\mathbb{X} = \emptyset$.

Therefore this approach is not strong enough to capture the side conditions of the frame rule. Another disadvantage of Equation (6) is that algebraically inequalities cannot easily be used for equational (automated) reasoning.

Moreover, weak preservation is not closed under composition. Consider commands $C_1 =_{df}\ (x := 1), C_2 =_{df}\ (y := 2)$ and an assertion $r =_{df}\ (x = 3 \vee y = 4)$. Then $\mathbb{X}\,;(C_i \times r)\,;\mathbb{X} \neq \emptyset$ but $\mathbb{X}\,;(C_1\,;C_2 \times r)\,;\mathbb{X} = \emptyset$. It is not possible to force weak preservation to be closed under composition. This is based on the fact that relations cannot distinguish commands like $y := x$ from $y := x\,;x := 0\,;x := y$ (cf. Section 2). By equivalence of these commands this would imply that $y := x$ does not preserve $r =_{df}\ (x = 1)$.

To give a more appropriate definition we look at downward closure: a command C *downward-preserves* a test r iff $r \neq \emptyset$ and

$$\forall\,C' \subseteq C\ :\ \mathbb{X}\,;(C' \times r)\,;\mathbb{X} = \emptyset \Rightarrow C' = \emptyset\ . \tag{7}$$

Obviously, every command $C \neq \emptyset$ that downward-preserves r also weakly preserves r. But this definition is much more restrictive. Informally, the definition states that C preserves r only if all possible "executions of C" already preserve r. By this side condition, the above problem of the if-statement can be avoided: the problem of weak preservation is that if there is a choice between two execution paths, it suffices if one of these preserves r. If there is a choice of execution paths "inside" a command C, it can be split into two subcommands C_1 and C_2 with $C = C_1 \cup C_2$. Both commands have now to preserve r, i.e., $\mathbb{X}\,;(C_i \times r)\,;\mathbb{X} \neq \emptyset$ $(i \in \{1,2\})$. Moreover the definition is closed under $\cup$, but still not under $;$.

Now, we present a fourth possibility for algebraic side conditions. It is discussed in more detail since we will apply this condition in a small case study.

A command C *preserves* a test r iff

$$C * (r\,;S_C) \subseteq C * r\ . \tag{8}$$

Pointwise this spells out for all $\sigma, \sigma' \in \textit{States}$:

$$\exists\,\sigma_c,\sigma_r,\sigma'_c,\sigma_S.\ \sigma = \sigma_c * \sigma_r \wedge \sigma_c\,C\,\sigma'_c \wedge \sigma_r \in r \wedge \sigma_r\,S_C\,\sigma_S \wedge \sigma' = \sigma'_c * \sigma_S$$
$$\Rightarrow \exists\,\sigma_c,\sigma_r,\sigma'_c.\ \sigma = \sigma_c * \sigma_r \wedge \sigma_c\,C\,\sigma'_c \wedge \sigma_r \in r \wedge \sigma' = \sigma'_c * \sigma_r$$

assuming $\sigma_c \# \sigma_r$, $\sigma'_c \# \sigma_S$ and $\sigma'_c \# \sigma_r$. Informally, this inequation characterises commands that only modify variables which do not influence r.

Lemma 6.1. *For arbitrary commands C_1, C_2 and test r:*

- *If C_1, C_2 preserve r then $C_1 \cup C_2$ preserves r.*
- *Assume C, D preserves r. If $r \,;\, S_{C;D} \subseteq r \,;\, S_C \,;r\,;\, S_D$ and $(C * r) ; (D * r) \subseteq (C \,;\, D) * r$ then $C \,;\, D$ preserves r.*
- *I and emp preserve r, i.e., $I * (r \,;\, S_I) \subseteq I * r$ and $\mathsf{emp} * (r \,;\, S_{\mathsf{emp}}) \subseteq r$.*
- *C preserves I and emp.*

The proof is by straightforward calculations and by Lemma 4.1.

Using this definition of preservation (together with safety monotonicity and the frame property), it is again possible to prove the frame rule purely algebraically. Therefore the algebraic approach is still sound, but maybe still not complete. However, these new definitions make the algebraic frame rule more widely applicable and the restrictions are far smaller than before.

We use these results to model more complex side conditions of SL in the following section. In particular, we look at a variant of the frame rule and describe to which extent its side conditions can be included in the relational framework.

7   Variable Conditions in Information Hiding

In this section we present an approach to include more complex variable preservation conditions into the relational setting. For such side conditions we consider the *hypothetical frame rule* introduced in [9]. It uses the concept of information hiding. We give only some key concepts of that rule and present how reasoning with this proof rule can be captured by our relational approach. For more details concerning the frame rule we refer to [9].

We only treat a special case of the hypothetical frame rule. The ideas given can be easily generalised. The inference rule reads

$$\frac{\{p_1\}k_1\{q_1\}[X_1], \ \{p_2\}k_2\{q_2\}[X_2] \ \vdash \ \{p\}C\{q\}}{\{p_1 * r\}k_1\{q_1 * r\}[X_1, Y], \ \{p_2 * r\}k_2\{q_2 * r\}[X_2, Y] \ \vdash \ \{p * r\}C\{q * r\}} \ ,$$

where the side conditions are skipped for the moment. They will be given below. The semantics of $\vdash$ is as follows: if the triples on the left hand side of $\vdash$ hold, then C satisfies $\{p\}C\{q\}$. To explain the new type of triples above we consider $\{p_i\}k_i\{q_i\}[X_i]$. k_i denotes an identifier, i.e., a placeholder for a *local* command C_i, that is, a command that satisfies safety monotonicity and the frame property. Such commands are determined by environments η, i.e., mappings from identifiers to local commands. In particular the premise is quantified over all environments η that make $\vdash$ holds. The sets X_i in the triples list the variables which each k_i is allowed to change. Replacing k_i in the triple $\{p_i\}k_i\{q_i\}$ with a concrete $C_i = \eta(k_i)$, the triple can be interpreted with usual semantics.

The general hypothetical frame rule only considers an arbitrary number of $\{p_i\}k_i\{q_i\}[X_i]$ triples and therefore does not introduce any new concepts. To get an idea for the usage of this proof rule we consider again its premise. The command C denotes a command that uses during its execution the local commands $\eta(k_i)$ for an actual considered environment η. Now the hypothetical frame rule allows us to infer triples with more information. The pre- and postconditions of k_i now come with additional disjoint heap cells satisfying the predicate r. Moreover the sets of variables that the k_i may modify are extended by a common set Y. Intuitively this means that all k_i together can be seen as a module or package providing some functionality through its public methods, namely the k_i. Usually the concrete implementation remains hidden by an import of such module.

The premise of the proof rule expresses the described situation. The consequent of the rule gives a view of the module from its inside. It reveals all internally used variables and heap cells used by k_i. All k_i share some private variables and storage. In particular, to be able to work correctly on those resources, each k_i has to maintain a resource invariant r. Due to this behaviour the inference rule comes with more complex variable conditions than the ordinary frame rule. However, it is much more flexible than the ordinary frame rule and allows reasoning in a more realistic setting. The following side conditions come with the hypothetical frame rule to restrict the behaviour for k_i and C

(a) C does not modify any free variables of r, except through k_1 and k_2
(b) Y is disjoint from X_1, X_2, $FV(p_i)$, $FV(q_i)$, $FV(p)$, $FV(q)$ and $MV(C)$.

By these conditions module variables can only be modified within a module.

Before tackling these side conditions we first show that the hypothetical frame rule can be included in our relational approach. The first construct we consider are the triples of the form $\{p_i\}k_i\{q_i\}[X_i]$. For such triples we first define for an arbitrary set X of variables a command C_X by

$$C_X =_{df} \{(\sigma, \sigma') : X \subseteq \operatorname{dom}(s_\sigma) \cap \operatorname{dom}(s_{\sigma'}),\ s_\sigma|_{\overline{X}} = s_{\sigma'}|_{\overline{X}},\ h_\sigma = h_{\sigma'}\}$$

where $\overline{X} = Vars - X$. This command is used to ensure that all variables which are not in a set X have to preserve their value from the input to the output states. Each variable in X can be changed to an arbitrary value. In analogy to Equation (8) we say that a command k *preserves* C_X if $k * (C_X\,;\,S_k) \subseteq k * C_X$ By this we further define the new triples by

$$\{p_i\}k_i\{q_i\}[X_i] \Leftrightarrow_{df} \{p_i\}k_i\{q_i\} \wedge k_i \text{ preserves } C_{X_i} .$$

Note again that we cannot restrict variable modifications "inside" a command (cf. the example given at the end of Section 2). This is a major restriction of our relational approach. To verify the side condition (a) within *Cmds* we assume a syntactically given command C by the grammar of Section 2. The idea is to split the command C into subsequences C_k of C where no k_i occurs and verify preservation of $[\![C_k]\!]_c$ relationally. To ensure that each C_k of C before and after a k_i preserves r, we apply the following routine to C. We start by a set $Z =_{df} \{C\}$.

1. Repeat until no $C_j \in Z$ contains if then else : if there exists C_i with $C \equiv C_1$; (if p then C_2 else C_3) ; C_4 then $Z := Z - \{C\} \cup \{C_1\,;p\,;C_2\,;C_4,\ C_1\,;\neg p\,;C_3\,;C_4\}$
2. Repeat until no $C_j \in Z$ contains a k_i : if $\exists C_i .\ C \equiv C_1\,;\,k_i\,;\,C_2$ then $Z := Z - \{C\} \cup \{C_1, \operatorname{cod}(C_1\,;\,k_i)\,;\,C_2\}$

$\equiv$ denotes syntactical equivalence. Since C_2 is reached after $C_1\,;\,k_i$ it remains to consider $\operatorname{cod}(C_1\,;\,k_i)\,;\,C_2$. To consider the right command relation in C_j the routine appends tests to commands e.g. in $p\,;\,C_j$ although p is syntactically not a command. A concrete example is given in the next section.

Now each of these subcommands has to preserve r, i.e., it has to maintain the values of all free variables of r. By knowing the concrete structure of a command, Assumption (a) can be checked completely at the relational level.

Next we introduce an approach to characterise Assumption (b) relationally. We constrain the use of the internal variables Y of a module by the following inequations which are to be add to the premises of the proof rule.

$$k_i \text{ preserves } v_Y, \quad i \in \{1, 2\} \tag{9}$$

$$C \text{ preserves } v_Y, \tag{10}$$

where $v_Y =_{df} \{(\sigma, \sigma) : \sigma = (s, h), Y \subseteq \mathsf{dom}(s), h \in Heaps\}$.

By Assumption (9) no variable in Y is changed by any execution of k_i. Therefore also each execution of k_i starting from p_i and finishing in q_i preserves Y which the intention of requiring Y to be disjoint from $FV(p_i)$ and $FV(q_i)$. The same argumentation holds for requiring each X_i to be is disjoint from Y. By the second assumption also C does not modify any variables of Y.

In summary, the intended restrictions by all meta-level variable conditions of the hypothetical frame rule can be checked pointfree and purely relationally when knowing the concrete structure and syntax of commands. To demonstrate and clarify these conditions we present a short example in the next section.

8   A Relational Treatment of a Queue Module

In this section, we exemplarily replay the queue module example given in [9]. This module offers two methods to enqueue and to dequeue elements from a list which represents the shared and hidden storage of the module procedures. From the outside of the module the list cannot be seen, i.e., from that point of view only values will be cut off from or appended to an abstract sequence α stored in a variable Q. The following two interface specifications[3] are given

$$\{Q = \alpha \wedge z = n \wedge \mathsf{emp}\} \, \mathsf{enq} \, \{Q = \alpha \cdot n \wedge \mathsf{emp}\}[\{Q\}] \,,$$
$$\{Q = n \cdot \alpha \wedge \mathsf{emp}\} \, \mathsf{deq} \, \{Q = \alpha \wedge z = n \wedge \mathsf{emp}\}[\{Q\}]$$

and will be part of the antecedent of the hypothetical frame rule. The precondition for enq ensures that Q stores the sequence α of the hidden list while z stores an arbitrary value n to be appended at the end of the sequence. An environment η_1 could e.g. return $\eta_1(\mathsf{enq}) = (Q := Q \cdot z)$. In deq, the head of the sequence α is assigned to z and then deleted from α. A possible local command for that could be $\eta_1(\mathsf{deq}) = (z := \mathsf{head}(\alpha) \,;\, Q := \mathsf{tail}(\alpha))$.

These specifications can be embedded into the relational framework by simply requiring $\eta_1(\mathsf{enq}) * (C_{\{Q\}} \,;\, S_{\eta_1(\mathsf{enq})}) \subseteq \eta_1(\mathsf{enq}) * C_{\{Q\}}$ and $(Q = \alpha \,;\, z = n \,;\, \mathsf{emp}) \,;\, \eta_1(\mathsf{enq}) \subseteq \eta_1(\mathsf{enq}) \,;\, (Q = \alpha \cdot n \,;\, \mathsf{emp})$. The same can be done for deq.

The conclusion of the hypothetical frame rule reveals the resource invariant of the module. Concretely the resource invariant r for this module is $\mathsf{listseg}(Q, x, y) * (y \mapsto -, -)$ which ensures a list segment from x to y representing a sequence

---

[3] In [9] the specifications are used parametrical; for simplicity reasons we use values.

stored in Q. The predicate $\mathsf{listseg}(Q, x, y)$ is defined by $(x = y \wedge \alpha = \varepsilon \wedge \mathsf{emp}) \vee$ $(x \neq y \wedge \exists v, z.\ x \mapsto v, z * \mathsf{listseg}(\mathsf{tail}(Q), z, y))$. The last two cells starting from y reserve storage for a value that an execution of enq will append. By this an internal implementation mapped by η_2, i.e., local commands for enq and deq might look as follows

$$
\begin{array}{ll}
Q := Q \cdot z; & Q := \mathsf{tail}(Q); \\
t := \mathsf{cons}\,(-, -); & z := [x]\,;\, t := x\,;\, x := [x + 1]; \\
[y] := z\,;\, [y + 1] := t\,;\, y := t & \mathsf{dispose}\ t
\end{array}
$$

In particular the following triples are inferred by the hypothetical frame rule.

$$
\{(Q = \alpha \wedge z = n \wedge \mathsf{emp}) * r\}\ \mathsf{enq}\ \{(Q = \alpha \cdot n \wedge \mathsf{emp}) * r\}[\{Q\}, \{t, x, y\}],
$$
$$
\{(Q = n \cdot \alpha \wedge \mathsf{emp}) * r\}\ \mathsf{deq}\ \{(Q = \alpha \wedge z = n \wedge \mathsf{emp}) * r\}[\{Q\}, \{t, x, y\}].
$$

Again we can formulate these triples relationally with the variable condition modelled by $\eta_2(\mathsf{enq}) * (C_{\{Q\} \cup \{t,x,y\}}\ ;\ S_{\eta_2(\mathsf{enq})}) \subseteq \eta_2(\mathsf{enq}) * C_{\{Q\} \cup \{t,x,y\}}$. Next we consider for the hypothetical frame rule the command

$$
C = \mathsf{if}\ b\ \mathsf{then}\ (\eta_2(\mathsf{deq})\,;\, z = 1\,;\, \eta_2(\mathsf{enq}))\ \mathsf{else}\ (\eta_2(\mathsf{deq})\,;\, z = 0\,;\, \eta_2(\mathsf{enq}))\ .
$$

It is split into $C_1 = (b = \mathsf{true})\,;\, (\eta_2(\mathsf{deq})\,;\, z := 1\,;\, \eta_2(\mathsf{enq}))$ and $C_2 = (b = \mathsf{false})\,;\, (\eta_2(\mathsf{deq})\,;\, z := 0\,;\, \eta_2(\mathsf{enq}))$. Consequently we split both commands recursively at $\eta_2(\mathsf{deq})$ and $\eta_2(\mathsf{enq})$. This results in commands $C_{1_1} = (b = \mathsf{true})$, $C_{1_2} = (\mathsf{cod}(b = \mathsf{true}\,;\, \eta_2(\mathsf{deq}))\,;\, z := 1)$, $C_{1_3} = \mathsf{cod}(b = \mathsf{true}\,;\, \eta_2(\mathsf{deq})\,;\, z := 1)$, $C_{2_1} = (b = \mathsf{false})$, $C_{2_2} = (\mathsf{cod}(b = \mathsf{false}\,;\, \eta_2(\mathsf{deq}))\,;\, z := 0)$ and $C_{2_3} = \mathsf{cod}(b = \mathsf{false}\,;\, \eta_2(\mathsf{deq})\,;\, z := 0)$. All commands C_{i_j} preserve r. For example it can be shown for C_{1_2} that $C_{1_2} * (r\,;\, S_{C_{1_2}}) \subseteq C_{1_2} * r$.

This shows that restrictions by side conditions of the hypothetical frame rule and the rule itself can be expressed in the presented relational framework. Only the subcommands as given by the structure of a command is needed for a relation based argumentation which facilitates algebraic reasoning with this rule.

9   Greatest Local Relations

To conclude this work, we introduce the concept of *greatest local relations*[4]. As discussed before, the premise and the conclusion of the hypothetical frame rule are quantified over all possible environments η. According to [9], proving soundness of the inference rule can be simplified. It suffices to consider so-called greatest environments since those environments already capture any other environment. A greatest environments maps an identifier k_i of $\{p_i\}k_i\{q_i\}[X_i]$ to the greatest local relation satisfying it. For further details see [9].

Our basic intention in this work is to include greatest local relations and their variable conditions into our setting. This underpins that the relational approach is also able to capture concepts used in the hypothetical frame rule. We give a pointfree characterisation for these commands in the relational setting and explain how their variable conditions can be included.

---

[4] A local relation of [9] is a local command in our approach.

Definition 9.1. [[9]] Consider a triple $\{p\}k_i\{q\}[X]$. We define the greatest local relation $\mathsf{great}(p, q, X)$ satisfying that triple by the following conditions assuming $\sigma = (s, h)$ and $\sigma' = (s', h')$

1. σ is safe for $\mathsf{great}(p, q, X) \iff \sigma \in p * \mathsf{true}$
2. $\sigma \, \mathsf{great}(p, q, X) \, \sigma' \iff$
   (a) $\forall x \notin X. \ s(x) = s'(x)$
   (b) $\forall h_p, h_I. \ (h = h_p \cup h_I \wedge h_p \cap h_I = \emptyset \wedge (s, h_p) \in p)$
       $\Rightarrow \exists h_q. \ h_q \cap h_I = \emptyset \wedge h' = h_q \cup h_I \wedge (s', h_q) \in q$

By this definition every state $\sigma = (s, h)$ which $\mathsf{great}(p, q, X)$ is executed at can be split into two disjoint states $(s, h_p) \in p$ and an arbitrary state (s, h_I). Every variable of $\mathsf{dom}(s) - X$ cannot be modified by $\mathsf{great}(p, q, X)$ and a resulting state σ' can be split again into states (s', h_I) and $(s', h_q) \in q$.

We implicitly treat the variable condition (2a) by an extra assumption. Hence we omit X in $\mathsf{great}(p, q)$ and include the variable condition (2a) by requiring for the rest of this section $\mathsf{great}(p, q)$ preserves C_X with C_X as in Section 4.

For a relational characterisation of the remaining properties of Definition 9.1 we use the concept of residuals. In REL, the *right residual* $R \backslash S$ is defined as $\overline{R^\smile \, ; \overline{S}}$ (e.g., [12]). Residuals can equally be defined by the Galois connection $x \leq a \backslash b \Leftrightarrow_{df} a \cdot x \leq b$ in the general setting of quantales [2]. Therefore the presented theory lifts to a more abstract setting. The definition entails $a \cdot (a \backslash b) \leq b$.

Moreover, for this section we define $S =_{df} S_{C_X}$ and let $\top$ denote the universal relation. By these definitions, we characterise in our relational terms $\mathsf{great}(p, q)$ satisfying the properties (1) and (2b) as follows:

Lemma 9.2. *For tests p, q the greatest local relation $\mathsf{great}(p, q)$ can be characterised by*

$$\mathsf{great}(p, q) = (p * I) \, ; \mathsf{res}(p, q),$$

where $\mathsf{res}(p, q) = ((p \times \mathsf{dom}(S)) \, ; \triangleright) \backslash ((\top \, ; q \times S) \, ; \triangleright).$

Proof. We show that (2b) is satisfied by $\mathsf{res}(p, q)$. We have for arbitrary σ and τ

$$\sigma \ \overline{((p \times \mathsf{dom}(S)) \, ; \triangleright) \backslash ((\top \, ; q \times S) \, ; \triangleright)} \ \tau$$

$\Leftrightarrow \sigma \ \overline{\triangleleft \, ; (p \times \mathsf{dom}(S)) \, ; \overline{(\top \, ; q \times S) \, ; \triangleright}} \ \tau$

$\Leftrightarrow \neg(\exists \sigma_p, \sigma_S. \ \sigma = \sigma_p * \sigma_S \wedge \sigma_p \, \# \, \sigma_S \wedge \sigma_p \in p \wedge \sigma_S \in \mathsf{dom}(S)$
$\qquad \wedge \neg((\sigma_p, \sigma_S) \, (\top \, ; q \times S) \, ; \triangleright \, \tau))$

$\Leftrightarrow \neg(\exists \sigma_p, \sigma_S. \ \sigma = \sigma_p * \sigma_S \wedge \sigma_p \, \# \, \sigma_S \wedge \sigma_p \in p \wedge \sigma_S \in \mathsf{dom}(S)$
$\qquad \wedge \neg(\exists \sigma_q, \sigma'_S. \ \sigma_p \, \top \, \sigma_q \wedge \sigma_q \in q \wedge \sigma_S \, S \, \sigma'_S \wedge \tau = \sigma_q * \sigma'_S \wedge \sigma_q \, \# \, \sigma'_S))$

$\Leftrightarrow \forall \sigma_p, \sigma_S. \ \sigma = \sigma_p * \sigma_S \wedge \sigma_p \, \# \, \sigma_S \wedge \sigma_p \in p$
$\qquad \Rightarrow \exists \sigma_q, \sigma'_S. \ \sigma_q \in q \wedge \sigma_S \, S \, \sigma'_S \wedge \tau = \sigma_q * \sigma'_S \wedge \sigma_q \, \# \, \sigma'_S$

Moreover, by (1), in Definition 9.1 we know $\mathsf{dom}(\mathsf{great}(p, q)) \subseteq p * I$ and by pointwise calculations above we have $p * I \subseteq \mathsf{dom}(\mathsf{res}(p, q))$. Then $p * I = (p * I) * (p * I) \subseteq (p * I) \, ; \mathsf{dom}(\mathsf{res}(p, q)) = \mathsf{dom}((p * I) \, ; \mathsf{res}(p, q)) = \mathsf{dom}(\mathsf{great}(p, q))$. $\square$

The assumption $\sigma_S \, S \, \sigma'_S$ is used to change all variables of X in s_{σ_S}. Notice also that $\mathsf{res}(p, q)$ does not work on pairs of states, i.e., $\mathsf{res}(p, q) \subseteq States \times States$. With this characterisation, we can give relational proofs for the following results.

Lemma 9.3. *For all p, q, we have that $\mathsf{great}(p,q)$ satisfies $\{p\}\mathsf{great}(p,q)\{q\}$.*

Proof. For the proof we need a standard result from relation algebra: if R_1, R_2 and S_1 are subidentities with $R_1 \subseteq S_1$ then $(R_1 \times R_2)\,;(S_1 \times I) = R_1 \times R_2$.

To show $p\,;\mathsf{great}(p,q) \subseteq \mathsf{great}(p,q)\,;q$ it remains to prove $p\,;\mathsf{great}(p,q) \subseteq \top\,;q$. Both inequations are equivalent by standard semiring theory. We calculate assuming $p = p * \mathsf{emp} = p * (\mathsf{emp}; \mathsf{dom}(S))$

$$
\begin{aligned}
& p\,;\mathsf{great}(p,q) \\
={} & \vartriangleleft\,;(p \times \mathsf{emp})\,;\vartriangleright\,;\mathsf{res}(p,q) \\
={} & \vartriangleleft\,;(p \times \mathsf{emp})\,;(p \times \mathsf{dom}(S))\,;\vartriangleright\,;\mathsf{res}(p,q) \\
\subseteq{} & \vartriangleleft\,;(p \times \mathsf{emp})\,;(\top\,;q \times S)\,;\vartriangleright \\
={} & \vartriangleleft\,;(p\,;\top\,;q) \times (\mathsf{emp}\,;S)\,;\vartriangleright \\
\subseteq{} & p\,;\top\,;q \subseteq \top\,;q
\end{aligned}
$$

Moreover, $p \subseteq \mathsf{dom}(\mathsf{great}(p,q))$ follows from $p \subseteq p * I = \mathsf{dom}(\mathsf{great}(p,q))$. $\quad\square$

Next we characterise $\mathsf{great}(p,q)$ also as a local command. Informally, the inequation below is similar to the frame property and also characterises arbitrary storage not needed for $\mathsf{great}(p,q)$ to remain untouched.

Lemma 9.4. *For arbitrary tests p and q we have*

$$((p * I) \times \mathsf{dom}(S))\,;\vartriangleright\,;\mathsf{great}(p,q) \subseteq ((p * I)\,;\mathsf{great}(p,q) \times S)\,;\vartriangleright$$

Proof. We calculate for all $\sigma_p \in p, \sigma_2 \in \mathsf{dom}(S)$ and arbitrary σ_1, σ'

$$
\begin{aligned}
& \sigma_p \,\#\, \sigma_1 \wedge (\sigma_p * \sigma_1, \sigma_2)\,((p * I) \times \mathsf{dom}(S))\,;\vartriangleright\,;\mathsf{great}(p,q)\,\sigma' \\
\Leftrightarrow{} & \sigma_p \,\#\, \sigma_1 \wedge \sigma_p * \sigma_1 \in p*I \wedge (\sigma_p * \sigma_1) \,\#\, \sigma_2 \wedge (\sigma_p * \sigma_1) * \sigma_2\,\mathsf{great}(p,q)\,\sigma' \\
\Leftrightarrow{} & \sigma_p \,\#\, \sigma_1 \wedge \sigma_p \in p \wedge \sigma_p \,\#\, (\sigma_1 * \sigma_2) \wedge \sigma_p * (\sigma_1 * \sigma_2)\,\mathsf{great}(p,q)\,\sigma' \\
\Rightarrow{} & \sigma_p \,\#\, \sigma_1 \wedge \sigma_p \in p \wedge (\sigma_p * \sigma_1) \,\#\, \sigma_2 \wedge (\sigma_p * \sigma_1)\,\mathsf{great}(p,q) * S\,\sigma' \\
\Rightarrow{} & \exists \sigma'_1, \sigma_S.\; \sigma_p \,\#\, \sigma_1 \wedge \sigma_p \in p \wedge (\sigma_p * \sigma_1) \,\#\, \sigma_2 \wedge (\sigma_p * \sigma_1)\,\mathsf{great}(p,q)\,\sigma'_1 \wedge \\
& \sigma_2 S \sigma_S \wedge \sigma'_1 \,\#\, \sigma_S \wedge \sigma'_1 * \sigma_S = \sigma' \\
\Leftrightarrow{} & \exists \sigma'_1.\; (\sigma_p * \sigma_1, \sigma_2)\,((p * I)\,;\mathsf{great}(p,q) \times S)\,;\vartriangleright\,\sigma'.
\end{aligned}
$$

$\quad\square$

Lemma 9.5. *The relation $\mathsf{great}(p,q)$ is local, i.e., it satisfies the frame property and safety monotonicity.*

Proof. To see that $\mathsf{great}(p,q)$ satisfies the frame property, we know by Lemma 9.4

$$((p * I) \times \mathsf{dom}(S))\,;\vartriangleright\,;\mathsf{great}(p,q) \subseteq ((p * I)\,;\mathsf{great}(p,q) \times S)\,;\vartriangleright.$$

By $\mathsf{dom}(\mathsf{great}(p,q)) = p * I$ and isotony the claim follows. Finally for safety monotonicity: $\mathsf{dom}(\mathsf{great}(p,q)) * I = (p * I) * I = p * I = \mathsf{dom}(\mathsf{great}(p,q))$. $\quad\square$

Lemma 9.6. *Consider an arbitrary local command C that satisfies the triple $\{p\}C\{q\}$ and $S_C \subseteq S$, i.e., C modifies variables in X. Then $p\,;C \subseteq \mathsf{great}(p,q)$.*

Proof. First we show $C \subseteq ((p \times \mathsf{dom}(S)); \rhd) \setminus ((\top; q \times S); \rhd)$ which is equivalent to $(p \times \mathsf{dom}(S)); \rhd; C \subseteq (\top; q \times S); \rhd$. We calculate

$$
\begin{aligned}
& (p \times \mathsf{dom}(S)); \rhd; C \\
\subseteq\ & (p \times \mathsf{dom}(S)); (\mathsf{dom}(C) \times \mathsf{dom}(S)); \rhd; C \\
\subseteq\ & (p \times \mathsf{dom}(S)); (C \times S_C); \rhd \\
\subseteq\ & (p; C \times S); \rhd \\
\subseteq\ & (C; q \times S); \rhd \subseteq (T; q \times S); \rhd.
\end{aligned}
$$

Then $p; C \subseteq (p * I); C \subseteq (p * I); \mathsf{res}(p, q) = \mathsf{great}(p, q)$. $\qquad\square$

It can be seen that if a command C satisfies $\{p\}\, C\, \{q\}[X]$ then $p; C$ is a subset of $\mathsf{great}(p, q)$, i.e., $\mathsf{great}(p, q)$ captures every such local command C.

10   Related Work

There exist different approaches to the issue of variable preservation in proof rules. We briefly sum up some proposals of [3,10]. One idea is to omit the store and to put variable declarations on the heap, i.e., treat variables the same way as heap cells. Although this would give a uniform treatment, the logic itself would become more complicated, especially Hoare's variable assignment axiom. For our purpose this seems inadequate since the presented case study requires variable declarations to be present for disjoint states; such a resource-based treatment of variables would exacerbate treating standard examples.

The main approach taken in [3,10] was to introduce ownership predicates for variables that remain unchanged by variable substitutions. It ensures the permission to write to certain variables. This approach tracks ownership rights of variables in states and treats them by a special $*$ operation. This also shifts the treatment of variable conditions to the logic layer. Therefore the conditions can be verified within the logic. Our goal is rather constraining our relational approach by equations that model the side conditions. This allows including plain the store-heap model into our abstract and pointfree treatment without adding additional information like ownership rights. In addition our treatment also enables reasoning purely at the relational level.

11   Conclusion and Outlook

We have studied to which extent side conditions that naturally appear in the frame rule of separation logic can be handled with the same mechanism as the logic itself. The approach presented expresses the side conditions in the same algebra as the logic. In classical logic, side conditions require the introduction of a meta-language; in our algebraic version this is not the case.

As a further application of our approach we have formulated the hypothetical frame rule and its more complex variable side conditions in our relational setting. The result of this formulation is that our abstract approach can also be used for information hiding proof rules. Especially, greatest local relations can be characterised in a pointfree way in our setting.

As further work we will try to overcome the deficit of the relational approach by working on strings to represent the concrete structure of commands. Moreover, an abstract treatment of the hypothetical frame rule seems to be possible since its meta-level conditions can be lifted to an abstract level. As a consequence a wider range of models can be considered.

Acknowledgements. We are grateful to Bernhard Möller for many valuable remarks and comments. We are also most grateful to the referees who pointed out several flaws and helped to significantly increase the quality.

References

1. Berdine, J., Calcagno, C., O'Hearn, P.W.: Smallfoot: Modular Automatic Assertion Checking with Separation Logic. In: de Boer, F.S., Bonsangue, M.M., Graf, S., de Roever, W.-P. (eds.) FMCO 2005. LNCS, vol. 4111, pp. 115–137. Springer, Heidelberg (2006)
2. Birkhoff, G.: Lattice Theory, Colloquium Publications, 3rd edn., vol. XXV. American Mathematical Society, Providence (1967)
3. Bornat, R., Calcagno, C., Yang, H.: Variables as resource in separation logic. Electronic Notes in Theoretical Computer Science 155, 247–276 (2006)
4. Dang, H.-H., Höfner, P., Möller, B.: Towards algebraic separation logic. In: Berghammer, R., Jaoua, A.M., Möller, B. (eds.) RelMiCS 2009. LNCS, vol. 5827, pp. 59–72. Springer, Heidelberg (2009)
5. Dang, H.-H., Höfner, P., Möller, B.: Algebraic Separation Logic. Tech. Rep. 2010-06, Institute of Computer Science, University of Augsburg (2010)
6. Dang, H.-H., Höfner, P., Möller, B.: Algebraic Separation Logic. J. Logic and Algebraic Programming (accepted, 2011)
7. Kozen, D.: On Hoare logic, Kleene algebra, and types. In: Gärdenfors, P., Woleński, J., Kijania-Placek, K. (eds.) The Scope of Logic, Methodology, and Philosophy of Science: Volume One of the 11th Int. Congress Logic, Methodology and Philosophy of Science, Studies in Epistemology, Logic, Methodology, and Philosophy of Science, vol. 315, pp. 119–133. Kluwer, Dordrecht (2002)
8. O'Hearn, P.W., Reynolds, J.C., Yang, H.: Local reasoning about programs that alter data structures. In: Fribourg, L. (ed.) CSL 2001 and EACSL 2001. LNCS, vol. 2142, pp. 1–19. Springer, Heidelberg (2001)
9. O'Hearn, P.W., Reynolds, J.C., Yang, H.: Separation and information hiding. ACM Trans. Program. Lang. Syst. 31(3), 1–50 (2009)
10. Parkinson, M., Bornat, R., Calcagno, C.: Variables as Resource in Hoare logics. In: Proceedings of the 21st Annual IEEE Symposium on Logic in Computer Science, pp. 137–146. IEEE Computer Society, Los Alamitos (2006)
11. Reynolds, J.C.: An introduction to separation logic. In: Broy, M. (ed.) Engineering Methods and Tools for Software Safety and Security, pp. 285–310. IOS Press, Amsterdam (2009)
12. Schmidt, G., Ströhlein, T.: Relations and Graphs: Discrete Mathematics for Computer Scientists. Springer, Heidelberg (1993)

An Algebraic Approach to Preference Relations

Ivo Düntsch[1] and Ewa Orłowska[2]

[1] Brock University
St. Catharines, Ontario, Canada, L2S 3A1
duentsch@brocku.ca
[2] National Institute of Telecommunications
Szachowa 1, 04–894 Warsaw, Poland
orlowska@itl.waw.pl

Abstract. We define a class of structures – *preference algebras* – such that properties of preference relations can be expressed with their operations. We prove a discrete duality between preference algebras and preference relational structures.

1 Introduction

The concept of a preference structure appears in a variety of fields such as social choice theory, economics, and game theory [14,8,11], fuzzy logic [3,12], among others. For a recent concise introduction to preference structures we refer the reader to [7], and for a general presentation of relations and their applications the reader is invited to consult [13]. Typically, a preference is viewed as a binary relation, say P on a set of alternatives. The statement xPy is intuitively interpreted as x *is preferred to* y. Together with a preference relation a binary relation of indifference, I, it is often assumed; here, xIy is interpreted as x *is similar to* y or *there is no preference for* x *or* y. In any particular theory there are various axioms assumed for preference and indifference relations. Reusch [9,10] proposes the following (minimal) set of requirements on preference and indifference: For all alternatives x and y,

1. If x is preferred to y, then it is not the case that y is preferred to x.
2. If x is preferred to y, then it is not the case that x is indifferent to y.
3. x is indifferent to x.
4. If x is indifferent to y, then y is indifferent to x.

Postulates 1 and 2 may be interpreted as saying that the underlying preference structure is conflict-free.

In this paper we define a class of preference algebras which will be shown to correspond to the class of preference structures in the precise sense of discrete duality.

A preference algebra is a join of two mixed algebras, see [1], such that the properties of the preference and indifference relations can be expressed with their operations. The method of discrete duality employed in the paper enables us to

H. de Swart (Ed.): RAMICS 2011, LNCS 6663, pp. 141–147, 2011.
© Springer-Verlag Berlin Heidelberg 2011

consider logics of preference either with an algebraic semantics determined by preference algebras or with relational semantics determined by preference structures. The representation theorems presented in the paper yield the equivalence of these two semantics as it is shown in [5].

Given a class of preference algebras, Alg, and a class of preference relational structures (preference frames), Frm, a discrete duality between Alg and Frm is a tuple ⟨Alg, Frm, 𝔠𝔣, 𝔠𝔪⟩ such that 𝔠𝔣 is a mapping from Alg to Frm, 𝔠𝔪 is a mapping from Frm to Alg every A in Alg is embeddable into 𝔠𝔪 𝔠𝔣(A), and every F in Frm is embeddable into 𝔠𝔣 𝔠𝔪(F).

2   Definitions and Notation

Suppose that ⟨$B, \wedge, \vee, \neg, 0, 1$⟩ is a Boolean algebra. With some abuse of language we will identify algebras with their base set. If $A \subseteq B$ and $h : B \to B$ a function, then $h[A] \overset{\mathrm{df}}{=} \{h(a) : a \in A\}$ is the complex image of A under h. A *modal operator* on B is a function $f : B \to B$ which satisfies $f(0) = 0$, and $f(a \vee b) = f(a) \vee f(b)$ for all $a, b \in B$. A *sufficiency operator* on B is a function $g : B \to B$ which satisfies $g(0) = 1$, and $g(a \vee b) = g(a) \wedge g(b)$ for all $a, b \in B$.

A *mixed algebra* (MIA) [1] is a structure ⟨B, f, g⟩ where B is a Boolean algebra and $f, g : B \to B$ are functions on B such that

(2.1) f is a modal operator on B,

(2.2) g is a sufficiency operator on B.

and, for all ultrafilters F, G of B,

(2.3) $F \cap g[G] \neq \emptyset \Longleftrightarrow f[G] \subseteq F.$

The condition 2.3 is easily seen to be equivalent to the one originally given in [1].

A *weak MIA* is complete and atomic Boolean algebra with a modal operator f and a sufficiency operator g such that $f(p) = g(p)$ for all atoms p of B. Each MIA is a weak MIA, and the converse does not necessarily hold.

Mixed algebras are an extension of Jónsson – Tarski Boolean algebras [4] which add additive and normal operators to the Boolean operations by sufficiency operators which are co–additive and co–normal. Not every modal algebra can be extended to a MIA: If, for example, f is the identity function in a Boolean algebra B, then there is no sufficiency operator such that ⟨B, f, g⟩ is a MIA [1].

In terms of canonical structures, a modal operator f talks about R, while its sufficiency operator talks about $-R$. Reflexivity, for example, can be expressed by a modal operator, but to express irreflexivity one needs a sufficiency operator. Also, antisymmetry can be expressed by a mixed modal–sufficiency expression, but not by a modal or sufficiency expression alone [2].

Observe that condition (2.3) is second order, and it is known that the class of mixed algebras is not first order axiomatizable [2].

If $\langle B, f, g \rangle$ is a MIA, define a binary relation R_B on $\mathrm{Ult}(B)$ by $\langle F, G \rangle \in R_B \overset{\mathrm{df}}{\Longleftrightarrow} f[G] \subseteq F$. The structure $\mathfrak{Cf}(B) \overset{\mathrm{df}}{=} \langle \mathrm{Ult}(B), R_B \rangle$ is called the *canonical structure of* B.

The set of all binary relations on a set U is denoted by $\mathrm{Rel}(U)$; if $x \in U$, we let $R(x) \overset{\mathrm{df}}{=} \{z \in U : xRz\}$; the relational converse is denoted by $R^{\vee}$. For $R \in \mathrm{Rel}(U)$, we define two operators on 2^U by

$$(2.4) \qquad \langle R \rangle(S) \overset{\mathrm{df}}{=} \{x : (\exists y)[xRy \text{ and } y \in S]\} = \{x : R(x) \cap S \neq \emptyset\}.$$

$$(2.5) \qquad [[R]](S) \overset{\mathrm{df}}{=} \{x : (\forall y)[y \in S \Rightarrow xRy]\} = \{x : S \subseteq R(x)\}.$$

It is well known that $\langle R \rangle$ is a complete modal operator on the power set algebra of U, and that $[[R]]$ is a complete sufficiency operator, see e.g. [1]. Furthermore, $\langle 2^U, \langle R \rangle, [[R]] \rangle$ is a MIA [1], called the *complex algebra of* $\langle U, R \rangle$.

The following correspondences are well known:

Lemma 1. *1. R is reflexive if and only if $X \subseteq \langle R \rangle(X)$ for all $X \subseteq U$.*
2. R is symmetric if and only if $X \subseteq [[R]]([[R]](X))$ for all $X \subseteq U$.

3   Preference Frames and Preference Algebras

In this section we introduce preference frames and preference algebras. Here, we use as a basis the "most traditional preference model" [7], also known as a $\langle P, I \rangle$ *– structure*: A *preference frame* is a structure $\langle X, P, I \rangle$, where X is a nonempty set and P, I, are binary relations on X which satisfy

F$_1$. $P \cap P^{\vee} = \emptyset$.
F$_2$. $P \cap I = \emptyset$.
F$_3$. I is reflexive.
F$_4$. I is symmetric.

These axioms reflect the postulates 1 - 4 from Section 1. We shall usually denote a preference frame $\langle X, P, I \rangle$ just by its universe X.

A *preference algebra*

$$\langle B, \vee, \wedge, \neg, 0, 1, f_1, g_1, f_2, g_2 \rangle$$

is a structure such that $\langle B, \vee, \wedge, \neg, 0, 1 \rangle$ is a Boolean algebra, $\langle B, f_1, g_1 \rangle$ and $\langle B, f_2, g_2 \rangle$ are MIAs, and, for all $a \in B$,

A$_1$. $a \wedge f_1(g_1(a)) = 0$.
A$_2$. $a \wedge f_1(g_2(a)) = 0$.
A$_3$. $a \leq f_2(a)$.
A$_4$. $a \leq g_2(g_2(a))$.

These axioms correspond to the frame axioms listed above in the precise sense of the modal correspondence theory.

In Section 4 (resp. Section 5) we present an example of a preference algebra (resp. a preference frame). While we are not aware of any universal-algebraic preference structures, a great variety of preference frames can be found in the literature devoted to applications of preference relations.

4   The Complex Algebra of a Preference Frame

Let X be a preference frame. The complex algebra B_X of X has as its universe the powerset algebra of X, and the following distinguished operators: If $A \subseteq X$, then

$$f_P(A) \stackrel{\mathrm{df}}{=} \langle P \rangle(A), \qquad\qquad g_P(A) \stackrel{\mathrm{df}}{=} [[P]](A),$$

$$f_I(A) \stackrel{\mathrm{df}}{=} \langle I \rangle(A), \qquad\qquad g_I(A) \stackrel{\mathrm{df}}{=} [[I]](A).$$

B_X is called the *complex algebra of the preference frame* X, denoted by $\mathfrak{Cm}(X)$.

Theorem 1. *The complex algebra of a preference frame is a weak MIA satisfying* A_1 – A_4.

Proof. We have shown in [1] that both $\langle 2^X, \langle P \rangle, [[P]] \rangle$ and $\langle 2^X, \langle I \rangle, [[I]] \rangle$ are weak MIAs.

A_1: Let $A \subseteq X$, $x \in A$, and assume that $x \in \langle P \rangle [[P]](A)$. Then, there is some y such that xPy and $y \in [[P]](A)$. From the latter we infer that for all s, $s \in A$ implies yPs. Since $x \in A$, by the hypothesis we obtain yPx, which, together with xPy, contradicts the asymmetry F_1 of P.

A_2: Let $A \subseteq X$, $x \in A$, and assume that $x \in \langle P \rangle [[I]](A)$. Similar to A_1 there is some y such that xPy and yIx. Since I is symmetric, we also have xIy, contradicting F_2.

A_3 and A_4 follow directly from Lemma 1.

5   The Canonical Frame of a Preference Algebra

Suppose that $\langle B, f_1, g_1, f_2, g_2 \rangle$ is a preference algebra, and let $\mathrm{Ult}(B)$ be the set of ultrafilters on the Boolean algebra B. Define binary relations P_B and I_B on $\mathrm{Ult}(B)$ by

(5.1) $$F P_B G \Longleftrightarrow f_1[G] \subseteq F,$$

(5.2) $$F I_B G \Longleftrightarrow f_2[G] \subseteq F.$$

The structure $\langle \mathrm{Ult}(B), P_B, I_B \rangle$ is called the *canonical frame of the preference algebra* B, denoted by $\mathfrak{Cf}(B)$.

Theorem 2. *The canonical frame of a preference algebra is a preference frame.*

Proof. It is well known that A_3 implies the reflexivity of I, and A_4 implies the symmetry of I, see e.g. [1].

F_1: Assume that P_B is not asymmetric. then, there are ultrafilters F, G such that $f_1[G] \subseteq F$ and $f_1[F] \subseteq G$. By the latter condition and (2.3), there is some

$a \in F$ such that $g_1(a) \in G$. From $f_1[G] \subseteq F$ we obtain that $f_1(g_1(a)) \in F$. Since $a \in F$ and F is a proper filter, we have $0 \neq a \wedge f_1(g_1(a))$, contradicting A_1.

F_2: Assume that there are $F, G \in \mathrm{Ult}(B)$ such that FP_BG and FI_BG. Since I_B is symmetric, we also obtain GI_BF, i.e. $f_1[G] \subseteq F$ and $f_2[F] \subseteq G$. Then, by (2.3), $G \cap g_2[F] \neq \emptyset$, so there is some $a \in B$ such that $a \in F$ and $g_2(a) \in G$. Using $f_1[G] \subseteq F$ we see that $f_1(g_2(a)) \in F$, hence, $a \wedge f_1(g_2(a)) \in F$. Since F is a proper filter, this contradicts A_2.

6   The Duality Result

Theorem 3. *Suppose that $\langle X, P, I \rangle$ is a preference frame, and $\langle B, f_1, g_1, f_2, g_2 \rangle$ a preference algebra.*

1. *The mapping $h : B \to \mathfrak{Cm}\,\mathfrak{Cf}(B)$ defined by $h(a) = \{F \in \mathrm{Ult}(B) : a \in F\}$ is an embedding of preference algebras.*
2. *The mapping $k : X \to \mathfrak{Cf}\,\mathfrak{Cm}(X)$ defined by $k(x) = \{A \in 2^X : x \in A\}$ is an embedding of preference frames.*

Proof. 1. We have shown in [1], Proposition 12, that the mapping h embeds a MIA $\langle B, f, g \rangle$ into the complex algebra of its canonical frame. Following a referee's request, we make this fact explicit.

Since h is a Stone mapping it, is a Boolean embedding, and it is sufficient to show that h preserves the modal and sufficiency operators of preference algebras. Following [6] we show preservation of the operator g_1, that is $h(g_1(a)) = g_{P_{(2^X)}}(h(a))$.

First, observe that

$$F \in g_{P_{(2^X)}}(h(a)) \iff F \in [[P_B]]h(a) \iff (\forall G \in \mathrm{Ult}(B))[G \in h(a) \Rightarrow FP_BG],$$

and therefore by (2.3),

$$f_1(G) \in F \iff (\forall G \in \mathrm{Ult}(B))[a \in G \Rightarrow F \cap g_1(G) \neq \emptyset].$$

"$\subseteq$": Take $G \in \mathrm{Ult}(B)$ such that $a \in G$. Since $g_1(a) \in F$, $F \cap g_1(G) \neq \emptyset$.

"$\supseteq$": Suppose $g_1(a) \notin F$. Consider set $Z_{g_1} = \{b \in B : g_1^d(b) \notin F\}$, where $g_1^d(b) = \neg g_1(\neg b)$ and $\neg$ is the Boolean complement. Let G' be a filter generated by $Z_{g_1} \cup \{a\}$. G' is a proper filter, for suppose otherwise, then there is $a' \in Z_{g_1}$ such that $a' \wedge a = 0$, which yields $a \leq \neg a'$. Since g_1 is antitone, $g_1(\neg a') \leq g_1(a)$. By definition of Z_{g_1}, $g^d(a') \notin F$ and since F is maximal, $g_1(\neg a') = \neg g^d(a') \in F$. Hence, $g_1(a) \in F$, a contradiction. Thus G' can be extended to a prime filter, say G, containing it. Since $a \in G'$, $a \in G$. Hence, by the assumption, $F \cap g_1(G) \neq \emptyset$. It follows that for some $b \in G$, $g_1(b) \in F$. Then $\neg g_1(b) = g_1^d(\neg b) \notin F$ and hence $\neg b \in Z_{g_1} \subseteq G$, which yields $b \notin G$, a contradiction.

The proof of preservation of modal operators can also be found in [6].

2. Since $k(x)$ is the principal ultrafilter of 2^X generated by $\{x\}$, the mapping k is well defined. Let $x, y \in X$; we need to show that $xPy \iff k(x)P_{(2^X)}k(y)$ and $xIy \iff k(x)I_{(2^X)}k(y)$. First, observe that

$$(6.1) \qquad k(x)P_{(2^X)}k(y) \iff \langle P \rangle [k(y)] \subset k(x)$$

$$(6.2) \qquad\qquad\qquad \iff (\forall Y \subseteq X)[y \in Y \Rightarrow x \in \langle P \rangle(Y)]$$

$$(6.3) \qquad\qquad\qquad \iff (\forall Y \subseteq X)[y \in Y \Rightarrow P(x) \cap Y \neq \emptyset].$$

"$\Rightarrow$": Let xPy and $y \in Y$. Then, $P(x) \cap Y \neq \emptyset$, and hence by (6.3) we have $k(x)P_{(2^X)}k(y)$.

"$\Leftarrow$": Suppose that $k(x)P_{(2^X)}k(y)$ for some $x, y \in X$. Setting $Y \stackrel{\text{df}}{=} \{y\}$ and using (6.3) we obtain xPy.

Corollary 1

1. *Any preference frame can be embedded into the canonical frame of its complex algebra.*
2. *Any preference algebra can be embedded into the complex algebra of its canonical frame.*

7 Conclusion and Outlook

In this paper we introduced a class of preference algebras corresponding to the class of traditional $\langle P, I \rangle$ preference frames. We proved representation theorems for these classes, thus obtaining a discrete duality between them.

The preference structures considered in the paper can be extended in a natural way to the structures with multiple pairs of preference and indifference relations associated with agents making choices on a set of alternatives. Algebraic counterparts to these structures will be preference algebras constructed from multiple mixed algebras. Then some axioms reflecting relationships among preferences of various agents may be added. The duality results for those structures will open the way to a study of aggregation of preferences of the agents both in an algebraic and a relational framework.

Other directions for future work may include studying restrictions of the preference and/or indifference relation. In particular, transitivity of preference and/or indifference is of importance. The corresponding algebraic axioms are well known from correspondence theory. In some approaches to preference modeling, a third relation of incomparability, J, is introduced and postulated to be irreflexive and symmetric. In such preference structures, it is usually assumed that the relations P, $P^{\smile}$, I, and J form a partition of the set of all the pairs of alternatives. Further work is planned on discrete dualities for such and similar structures as, for example, interval orders or semi–orders.

Acknowledgment. We should like to thank the referees for thoughtful comments and pointers to further literature.

References

1. Düntsch, I., Orłowska, E.: Beyond modalities: Sufficiency and mixed algebras. In: Orłowska, E., Szałas, A. (eds.) Relational Methods in Computer Science Applications, pp. 277–299. Physica Verlag, Heidelberg (2000)
2. Düntsch, I., Orłowska, E.: Boolean algebras arising from information systems. Annals of Pure and Applied Logic 127, 77–98 (2004)
3. Fodor, J., Roubens, M.: Fuzzy Preference Modelling and Multicriteria Decision Support. North Holland, Dordrecht (1994)
4. Jónsson, B., Tarski, A.: Boolean algebras with operators I. American Journal of Mathematics 73, 891–939 (1951)
5. Orłowska, E., Rewitzky, I.: Discrete duality and its applications to reasoning with incomplete information. In: Kryszkiewicz, M., Peters, J.F., Rybiński, H., Skowron, A. (eds.) RSEISP 2007. LNCS (LNAI), vol. 4585, pp. 51–56. Springer, Heidelberg (2007)
6. Orłowska, E., Rewitzky, I., Düntsch, I.: Relational semantics through duality. In: MacCaull, W., Winter, M., Düntsch, I. (eds.) RelMiCS 2005. LNCS, vol. 3929, pp. 17–32. Springer, Heidelberg (2006)
7. Öztürk, M., Tsoukiàs, A., Vincke, P.: Preference modelling. In: Ehrgott, M., Greco, S., Figueira, J. (eds.) State of the Art in Multiple Criteria Decision Analysis, pp. 27–72. Springer, Heidelberg (2005)
8. Pauly, M.: Formal methods and the theory of social choice. In: Berghammer, R., Möller, B., Struth, G. (eds.) RelMiCS/AKA 2008. LNCS, vol. 4988, pp. 1–2. Springer, Heidelberg (2008)
9. Reusch, B.: On the Axiomatics of Rational Preference Structures (2006), http://lrb.cs.uni-dortmund.de/pdf/On20the%20Axiomatics%20of%20Rational%20Preference%20Structures2.pdf (retrieved February 2, 2011)
10. Reusch, B.: An axiomatic theory of (rational) preference structures, Technische Universität Dortmund (2008) (manuscript)
11. Roubens, M., Vincke, P.: Preference Modelling, Lecture Notes in Economics and Mathematical Sciences, vol. 250. Springer, Heidelberg (1985)
12. Saminger-Platz, S.: Basics of preference and fuzzy preference modeling. In: Berghammer, R., Moeller, B., Struth, G. (eds.) Relations and Kleene Algebra in Computer Science, pp. 25–40. Universität Augsburg, PhD Programme at RelMiCS10/AKA5 (2008)
13. Schmidt, G.: Relational Mathematics, Encyclopedia of Mathematics and its Applications, vol. 132. Cambridge University Press, Cambridge (2010)
14. de Swart, H.: Logic, Game Theory and Social Choice. Tilburg University Press (1999)

Relational and Multirelational Representation Theorems for Complete Idempotent Left Semirings

Hitoshi Furusawa[1] and Koki Nishizawa[2]

[1] Department of Mathematics and Computer Science, Kagoshima University
furusawa@sci.kagoshima-u.ac.jp
[2] Department of Information Systems, Tottori University of Environmental Studies
koki@kankyo-u.ac.jp

Abstract. Brown and Gurr have shown a relational representation theorem for quantales. Complete idempotent left semirings are a relaxation of quantales by giving up strictness and distributivity of composition over arbitrary joins from the left. We show a relational representation theorem for them. Multirelations are generalisation of relations. We also show a multirelational representation theorem for complete idempotent left semirings.

1 Introduction

A quantale has been introduced by Mulvey [6] as a complete sup-lattice together with an associative composition satisfying the distributive laws. The same structure was also investigated by Conway [3] under the name of standard Kleene algebras (**S**-algebras). In [1], a relational quantale has been defined to be a quantale whose elements are relations on a set, ordered by inclusion and forming a monoid under relational composition. Then, Brown and Gurr have shown a relational representation theorem for quantales. The construction of a relational quantale from a quantale uses a subset of the given quantale's underlying set satisfying two conditions, which has been called a generating set for the quantale.

Complete idempotent left semirings are a relaxation of quantales by giving up strictness and distributivity of composition over arbitrary joins from the left. In this paper, we define a relational complete idempotent left semiring similarly to relational quantales. Then, following the idea of [1], we show a relational representation theorem for complete idempotent left semirings. The relational representation theorem in [1] is a special case of our relational representation theorem.

Multirelations are a generalisation of relations. These are studied as a semantic domain of programs [9] and game logic [8]. In the context of modal logics, multirelations appear as neighbourhood models or Scott-Montague models [2]. Up-closed multirelations are known as one of the typical models for complete idempotent left semirings [7]. We also define a multirelational complete idempotent left semiring to be a complete idempotent left semiring whose elements

H. de Swart (Ed.): RAMICS 2011, LNCS 6663, pp. 148–163, 2011.

are multirelations on a set, ordered by inclusion and forming a monoid under multirelational composition. Then, we show a multirelational representation theorem for complete idempotent left semirings. In the case of multirelational representation, one of the conditions for generating sets may be dropped.

Brown and Gurr has shown that every completely coprime algebraic quantale is isomorphic to a relational complete quantale in which all joins are given by unions in [1]. In this paper, we investigate if analogous results hold for complete idempotent left semirings. This investigation exhibits a difference between the relational representation and the multirelational representation.

In Section 2, we recall notions and some basic results about complete idempotent left semirings. In Section 3, we introduce relational complete idempotent left semirings and show that every complete idempotent left semiring is isomorphic to a relational complete idempotent left semiring. This result is extended to an equivalence of categories between the category of complete idempotent left semirings and the category of relational complete idempotent left semirings. In Section 4, we introduce multirelational complete idempotent left semirings and show that every complete idempotent left semiring is isomorphic to a multirelational complete idempotent left semiring. This result is extended to an equivalence of categories between the category of complete idempotent left semirings and the category of multirelational complete idempotent left semirings. In Section 5, we prove that every completely coprime algebraic complete idempotent left semiring is isomorphic to a multirelational complete idempotent left semiring in which all joins are given by unions. Section 6 summarises this paper.

2   Complete Idempotent Left Semirings

Idempotent left semirings [5] are defined as follows.

Definition 1. An *idempotent left semiring*, or briefly an *IL-semiring* is a tuple $(S, +, \cdot, 0, 1)$ with a set S, two binary operations $+$ and $\cdot$, and $0, 1 \in S$ satisfying the following properties:

- $(S, +, 0)$ is an idempotent commutative monoid.
- $(S, \cdot, 1)$ is a monoid.
- For all $a, b, c \in S$, $a \cdot c + b \cdot c = (a + b) \cdot c$, $a \cdot b + a \cdot c \le a \cdot (b + c)$, and $0 \cdot a = 0$,

where the *natural order* $\le$ is given by $a \le b$ iff $a + b = b$.

We often abbreviate $a \cdot b$ to ab.

Remark 1. An IL-semiring S satisfying $ab + ac = a(b + c)$ and $a0 = 0$ for all $a, b, c \in S$ is an idempotent semiring.

Definition 2. A *complete* IL-semiring S is an IL-semiring satisfying the following properties: For each $A \subseteq S$,

- the least upper bound $\bigvee A$ of A exists in S, and
- $(\bigvee A)a = \bigvee \{xa \mid x \in A\}$ for each $a \in S$.

A complete IL-semiring preserving *right directed joins* is a complete IL-semiring satisfying

$$a(\bigvee A) = \bigvee \{ax \mid x \in A\}$$

for each element a and each non-empty directed subset A. A complete IL-semiring preserving *the right* 0 is a complete IL-semiring satisfying $a0 = 0$ for each element a. A complete IL-semiring preserving *the right* $+$ is a complete IL-semiring satisfying $ab + ac = a(b + c)$ for any elements a, b, and c.

Remark 2 (Nishizawa et al. [7]). Let Q be a complete IL-semiring.

1. Q preserves the right $+$ and right directed joins iff $\cdot$ distributes over all non-empty joins even from the left hand side.
2. Q preserves the right 0, $+$ and right directed joins iff Q is a quantale.

The following example is found by using the model searcher Mace4 [4].

Example 1. Let $B = \{\bot, 1, a, \top\}$ with an ordering defined by

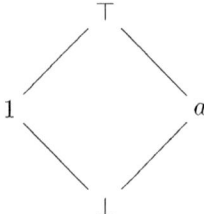

and binary operator $\cdot$ defined by

$\cdot$	$\bot$	1	a	$\top$
$\bot$	$\bot$	$\bot$	$\bot$	$\bot$
1	$\bot$	1	a	$\top$
a	$\bot$	a	a	$\top$
$\top$	$\bot$	$\top$	a	$\top$.

Then, $(B, +, \cdot, \bot, 1)$ is a complete IL-semiring. Observe that this complete IL-semiring is not a quantale since $a(1 + a) = \top \neq a = a1 + aa$.

We write CILS for the category whose objects are complete IL-semirings and whose arrows are completely join-preserving homomorphisms between them. We write CILS$_D$ for the full subcategory of CILS whose objects are complete IL-semirings preserving right directed joins. Similarly, we define CILS$_{0,+}$, CILS$_{0,+,D}$, and so on. The eight categories and forgetful functors between them form the cube of Fig. 1.

For a subset X of a complete IL-semiring S and for each $a \in S$, the set $\{x \in X \mid x \leq a\}$ is denoted by $d_X(a)$. The mapping $a \mapsto d_X(a)$ determines the function d_X from S to $\wp(S)$.

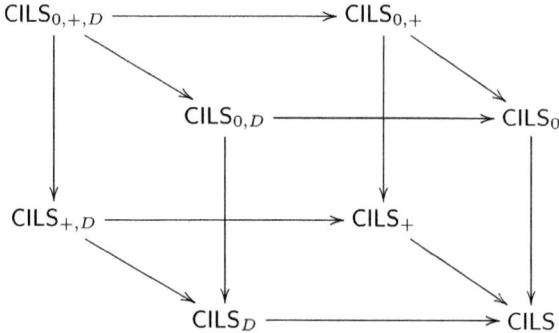

Fig. 1. The cube of complete IL-semirings

Definition 3. Let S be a complete IL-semiring. A subset $G \subseteq S$ is a *generating set* for S if $a \leq \bigvee d_G(a)$ for each $a \in S$. A generating set G is *decompositional* if for all $g \in G$ and $a, b \in S$, if $g \leq ab$ then there exists $h \in G$ such that $g \leq ah$ and $h \leq b$.

Let S be a quantale. A generating set for S in the sense of [1] is a decompositional generating set in this paper. As in the case of quantales [1], the decompositionality enables us to make multiplications in a complete IL-semiring correspond with relational compositions in the relational complete IL-semiring induced by it, which may be shown in Lemma 2 (via 4 of Remark 3) in this paper.

Example 2. Consider the complete IL-semiring B from Example 1. Though $\{1, a\}$ is a generating set for B, it is not decompositional since $1 \not\leq a = a1 = aa$ in spite of $1 \leq a\top$. Also, $\{\bot, 1, a\}$ is a non-decompositional generating set for B. On the other hand, B is the unique decompositional generating set for B.

Remark 3. Let S be a complete IL-semiring and G a generating set for S. Then, for any $g \in G$ and $a, b, c \in S$, the following holds.

1. $a = \bigvee d_G(a)$.
2. $a \leq b$ iff $d_G(a) \subseteq d_G(b)$.
3. $g \leq abc$ iff there exists $z \in S$ such that $g \leq az$ and for each $h \in d_G(z)$, $h \leq bc$.[1]
4. If G is decompositional, $g \leq abc$ iff there exists $h \in G$ such that $g \leq ah$ and $h \leq bc$.

Lemma 1. *Let S be a complete IL-semiring. Then S is a generating set for S itself. Moreover, S is decompositional.*

---

[1] The following similar (but slightly awkwarder) property holds.
  3'. For any $X \subseteq S$, $g \leq ab(\bigvee X)$ iff there exists $H \subseteq S$ such that $g \leq a(\bigvee H)$ and for each $h \in H$, $h \leq b(\bigvee X)$.

Proof. S is a generating set for S since $a \in d_S(a)$ for each $a \in S$. Suppose that $a \leq bc$. Putting $h = c$, we have $h \leq c$ and $a \leq bh$. ☐

Definition 4. An element a of a complete IL-semiring S is *completely coprime* (CCP) if, for each $X \subseteq S$, $a \leq \bigvee X$ implies that there exists $x \in X$ such that $a \leq x$.

The notion of completely coprime is synonymous with the notion of *completely join prime*. We write $\mathsf{CCP}(S)$ for the set of CCP elements of a complete IL-semiring S.

Definition 5. A complete IL-semiring S is *completely coprime algebraic* (CCPA) if $\mathsf{CCP}(S)$ is a generating set for S.

Example 3. CCPA quantales [1] are CCPA complete IL-semirings preserving the right 0, the right +, and right directed joins.

Example 4. Consider the complete IL-semiring B from Example 1. Then, $\{1, a\}$ is $\mathsf{CCP}(B)$. As we have seen in Example 2, $\mathsf{CCP}(B)$ is a generating set for B. Thus, B is CCPA.

In [1], it is shown that $\mathsf{CCP}(Q)$ of a CCPA quantale Q is a decompositional generating set for Q. However, $\mathsf{CCP}(S)$ of a CCPA complete IL-semiring S need not be decompositional.

Example 5. As we have seen in Example 4, the complete IL-semiring B from Example 1 is CCPA. But, as we have seen in Example 2, $\mathsf{CCP}(B)$ is not decompositional.

3   Relational Representation Theorem

Following the treatment in [1] quite closely, we provide a relational representation theorem for complete IL-semirings.

Definition 6. Let A be a set. A *relational complete IL-semiring on A* is a pair $(\mathcal{R}, I)$ of $\mathcal{R} \subseteq \wp(A \times A)$ and $I \in \mathcal{R}$ such that

- $(\mathcal{R}, \subseteq)$ is a complete join semilattice,
- $(\mathcal{R}, ;, I)$ is a monoid,
- for any $\chi \subseteq \mathcal{R}$ and $R \in \mathcal{R}$, $(\bigvee \chi) ; R = \bigvee \{P ; R \mid P \in \chi\}$,

where ; is relational composition, that is, $(x, y) \in P ; R$ iff there exists z such that $(x, z) \in P$ and $(z, y) \in R$.

Note that joins in a relational complete IL-semiring do not have to be given by unions.

Example 6. Relational quantales [1] are relational complete IL-semirings preserving the right 0, the right +, and right directed joins.

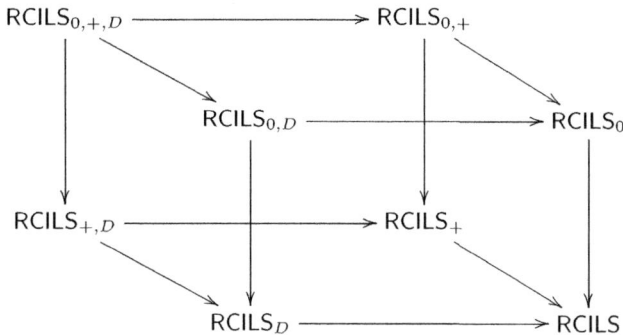

Fig. 2. The cube of relational complete IL-semirings

We write RCILS for the full subcategory of CILS whose objects are relational complete IL-semirings. As the case of CILS, we define $RCILS_D$, $RCILS_{0,+}$, $RCILS_{0,+,D}$, and so on. Then the eight categories and forgetful functors between them form the cube of Fig. 2.

Let S be a complete IL-semiring and G a generating set for S. For $a \in S$, we define the *relation* $\rho_G(a)$ *induced by* a and $\overline{\rho}_G(S) \subseteq \wp(S \times S)$ by

$$\rho_G(a) = \{(g,b) \in G \times S \mid g \leq ab\} \quad \text{and} \quad \overline{\rho}_G(S) = \{\rho_G(a) \mid a \in S\} \ ,$$

respectively. Then the mapping $a \mapsto \rho_G(a)$ determines the function ρ_G from S to $\overline{\rho}_G(S)$.

Lemma 2. *Let S be a complete IL-semiring and G a generating set for S. Then the following holds for any $a, b \in S$ and $X \subseteq S$.*

1. *$\rho_G(a) \subseteq \rho_G(b)$ iff $a \leq b$.*
2. *$(\overline{\rho}_G(S), \subseteq)$ is a complete join semilattice in which $\rho_G(\bigvee X)$ is the join of $\{\rho_G(x) \mid x \in X\}$.*
3. *$\rho_G(a) \, ; \rho_G(b) = \rho_G(ab)$ if G is decompositional.*

Proof. 1. ($\Rightarrow$) Note that $(g,1) \in \rho_G(a)$ for any $g \in d_G(a)$. Since $\rho_G(a) \subseteq \rho_G(b)$, $(g,1) \in \rho_G(b)$. Thus, $g \leq b$ for any $g \in d_G(a)$. Therefore $a = \bigvee d_G(a) \leq b$.
($\Leftarrow$) If $(g,c) \in \rho_G(a)$, $g \leq ac \leq bc$ since $a \leq b$ and $\cdot$ is monotone. So, $(g,c) \in \rho_G(b)$.
2. By 1, $\rho_G(x) \subseteq \rho_G(\bigvee X)$ for each $x \in X$. Thus, $\rho_G(\bigvee X)$ is an upper bound for $\{\rho_G(x) \mid x \in X\}$. Suppose that $\rho_G(c)$ is another upper bound. Then, by 1, $x \leq c$ for each $x \in X$, whence $\bigvee X \leq c$. So, by 1, $\rho_G(\bigvee X) \leq \rho_G(c)$. Therefore, $\rho_G(\bigvee X)$ is the least upper bound of $\{\rho_G(x) \mid x \in X\}$.

3.
$$(g,c) \in \rho_G(a) \, ; \rho_G(b)$$
$$\iff \exists h \in G. \, (g,h) \in \rho_G(a) \quad \text{and} \quad (h,c) \in \rho_G(b)$$
$$\iff \exists h \in G. \, g \le ah \quad \text{and} \quad h \le bc$$
$$\iff g \le abc \qquad\qquad\qquad\qquad \text{(by 4 of Remark 3)}$$
$$\iff (g,c) \in \rho_G(ab)$$
$\square$

Theorem 1. *For a complete IL-semiring S and a decompositional generating set G for S, the pair $(\overline{\rho}_G(S), \rho_G(1))$ is a relational complete IL-semiring on S. Moreover, S is isomorphic to $(\overline{\rho}_G(S), \rho_G(1))$.*

Proof. $(\overline{\rho}_G(S), \rho_G(1))$ is a relational complete IL-semiring on S and the function ρ_G from S to $\overline{\rho}_G(S)$ is a bijective and completely join-preserving homomorphism by Lemma 2.
$\square$

Example 7. Consider the set B from Example 1. B is the decompositional generating set for B and $\rho_B(\bot)$, $\rho_B(1)$, $\rho_B(a)$, and $\rho_B(\top)$ are relations on B, which are represented by the following Boolean matrices.

$$
\begin{array}{c c}
\begin{array}{c}
\quad \bot\ 1\ a\ \top \\
\begin{array}{c}
\bot \\ 1 \\ a \\ \top
\end{array}
\left[\begin{array}{cccc}
1 & 1 & 1 & 1 \\
0 & 0 & 0 & 0 \\
0 & 0 & 0 & 0 \\
0 & 0 & 0 & 0
\end{array}\right]
\end{array}
&
\begin{array}{c}
\quad \bot\ 1\ a\ \top \\
\begin{array}{c}
\bot \\ 1 \\ a \\ \top
\end{array}
\left[\begin{array}{cccc}
1 & 1 & 1 & 1 \\
0 & 1 & 0 & 1 \\
0 & 0 & 1 & 1 \\
0 & 0 & 0 & 1
\end{array}\right]
\end{array}
\\[4pt]
\rho_B(\bot) & \rho_B(1)
\end{array}
$$

$$
\begin{array}{c c}
\begin{array}{c}
\quad \bot\ 1\ a\ \top \\
\begin{array}{c}
\bot \\ 1 \\ a \\ \top
\end{array}
\left[\begin{array}{cccc}
1 & 1 & 1 & 1 \\
0 & 0 & 0 & 1 \\
0 & 1 & 1 & 1 \\
0 & 0 & 0 & 1
\end{array}\right]
\end{array}
&
\begin{array}{c}
\quad \bot\ 1\ a\ \top \\
\begin{array}{c}
\bot \\ 1 \\ a \\ \top
\end{array}
\left[\begin{array}{cccc}
1 & 1 & 1 & 1 \\
0 & 1 & 0 & 1 \\
0 & 1 & 1 & 1 \\
0 & 1 & 0 & 1
\end{array}\right]
\end{array}
\\[4pt]
\rho_B(a) & \rho_B(\top)
\end{array}
$$

The computation of relational composition on $\overline{\rho}_B(B)$ is as follows.

;	$\rho_B(\bot)$	$\rho_B(1)$	$\rho_B(a)$	$\rho_B(\top)$
$\rho_B(\bot)$	$\rho_B(\bot)$	$\rho_B(\bot)$	$\rho_B(\bot)$	$\rho_B(\bot)$
$\rho_B(1)$	$\rho_B(\bot)$	$\rho_B(1)$	$\rho_B(a)$	$\rho_B(\top)$
$\rho_B(a)$	$\rho_B(\bot)$	$\rho_B(a)$	$\rho_B(a)$	$\rho_B(\top)$
$\rho_B(\top)$	$\rho_B(\bot)$	$\rho_B(\top)$	$\rho_B(a)$	$\rho_B(\top)$

So, $(\overline{\rho}_B(B), \rho_B(1))$ is a relational complete IL-semiring. Moreover, it is isomorphic to B. Observe that joins are not given by unions since

$$\rho_B(a) + \rho_B(1) = \rho_B(\top) \ne \rho_B(a) \cup \rho_B(1) \ .$$

Also observe that $(\overline{\rho}_B(B), \rho_B(1))$ is not a relational quantale [1] since

$$\rho_B(a) \, ; (\rho_B(a) + \rho_B(1)) = \rho_B(\top) \ne \rho_B(a) = \rho_B(a) \, ; \rho_B(a) + \rho_B(a) \, ; \rho_B(1) \ .$$

Remark 4. The same construction taking a non-decompositional generating set need not provide a relational complete IL-semiring. For example, again, consider the complete IL-semiring B from Example 1. Taking $\mathsf{CCP}(B)$ from Example 4, we obtain the following four relations.

$$
\begin{array}{c}
\quad \perp\ 1\ a\ \top \\
\begin{array}{c} \perp \\ 1 \\ a \\ \top \end{array}
\begin{bmatrix} 0\ 0\ 0\ 0 \\ 0\ 0\ 0\ 0 \\ 0\ 0\ 0\ 0 \\ 0\ 0\ 0\ 0 \end{bmatrix}
\end{array}
\qquad
\begin{array}{c}
\quad \perp\ 1\ a\ \top \\
\begin{array}{c} \perp \\ 1 \\ a \\ \top \end{array}
\begin{bmatrix} 0\ 0\ 0\ 0 \\ 0\ 1\ 0\ 1 \\ 0\ 0\ 1\ 1 \\ 0\ 0\ 0\ 0 \end{bmatrix}
\end{array}
$$

$$\rho_{\mathsf{CCP}(B)}(\perp) \qquad\qquad \rho_{\mathsf{CCP}(B)}(1)$$

$$
\begin{array}{c}
\quad \perp\ 1\ a\ \top \\
\begin{array}{c} \perp \\ 1 \\ a \\ \top \end{array}
\begin{bmatrix} 0\ 0\ 0\ 0 \\ 0\ 0\ 0\ 1 \\ 0\ 1\ 1\ 1 \\ 0\ 0\ 0\ 0 \end{bmatrix}
\end{array}
\qquad
\begin{array}{c}
\quad \perp\ 1\ a\ \top \\
\begin{array}{c} \perp \\ 1 \\ a \\ \top \end{array}
\begin{bmatrix} 0\ 0\ 0\ 0 \\ 0\ 1\ 0\ 1 \\ 0\ 1\ 1\ 1 \\ 0\ 0\ 0\ 0 \end{bmatrix}
\end{array}
$$

$$\rho_{\mathsf{CCP}(B)}(a) \qquad\qquad \rho_{\mathsf{CCP}(B)}(\top)$$

It is obvious that the pair $(\overline{\rho}_{\mathsf{CCP}(B)}(B), \subseteq)$ is a complete join semilattice (in which all joins are given by unions). However, $\overline{\rho}_{\mathsf{CCP}(B)}(B)$ does not form a relational complete IL-semiring since it is not closed under composition. This fact can be checked by computing $\rho_{\mathsf{CCP}(B)}(a)\,;\,\rho_{\mathsf{CCP}(B)}(\top)$.

By Lemma 1, every complete IL-semiring has at least one decompositional generating set for itself. So, the following properties are immediate from the above theorem.

Corollary 1.   *1. Every complete IL-semiring is isomorphic to a relational complete IL-semiring on its underlying set.*
   2. Every complete IL-semiring preserving right directed joins is isomorphic to a relational complete IL-semiring preserving right directed joins on its underlying set.
   3. Every complete IL-semiring preserving the right 0 is isomorphic to a relational complete IL-semiring preserving the right 0 on its underlying set.
   4. Every complete IL-semiring preserving the right + is isomorphic to a relational complete IL-semiring preserving the right + on its underlying set.

The representation theorem [1, Corollary 3.13] for quantales by Brown and Gurr is induced from the above corollary.

   Let S and S' be complete IL-semirings. We write $\rho(a)$ and $\overline{\rho}(S)$ for $\rho_S(a)$ and $\overline{\rho}_S(S)$, respectively. For each completely join-preserving homomorphism f from S to S', we define $\overline{\rho}(f)\colon \overline{\rho}(S) \to \overline{\rho}(S')$ by $\overline{\rho}(f)(\rho(a)) = \rho(f(a))$ for each $a \in S$.

Lemma 3. *$\overline{\rho}$ is a functor from* CILS *to* RCILS.

Proof. By Theorem 1, $\overline{p}$ maps every object of CILS to an object of RCILS. By Lemma 2, $\overline{p}(f)$ is an arrow of RCILS for each arrow f of CILS. $\overline{p}(f \circ g) = \overline{p}(f) \circ \overline{p}(g)$ and $\overline{p}(\mathrm{id}_S) = \mathrm{id}_{\overline{p}(S)}$ hold by the definition of $\overline{p}$.   □

Corollary 2. *The restriction $\overline{p}_v$ of $\overline{p}$ to CILS_v is a functor from CILS_v to RCILS_v, where v is either $0, +, D, 0, +, 0, D, +, D,$ or $0, +, D$.*

We write ι for the inclusion functor from RCILS to CILS. The restriction ι_v of ι to RCILS_v is the inclusion functor from RCILS_v to CILS_v, where v is either $0, +, D, 0, +, 0, D, +, D,$ or $0, +, D$.
    These sixteen functors are visualised as follows.

$$\mathrm{CILS} \xrightarrow{\overline{p}} \mathrm{RCILS} \qquad \mathrm{CILS}_0 \xrightarrow{\overline{p}_0} \mathrm{RCILS}_0$$

$$\mathrm{CILS}_+ \xrightarrow{\overline{p}_+} \mathrm{RCILS}_+ \qquad \mathrm{CILS}_D \xrightarrow{\overline{p}_D} \mathrm{RCILS}_D$$

$$\mathrm{CILS}_{0,+} \xrightarrow{\overline{p}_{0,+}} \mathrm{RCILS}_{0,+} \qquad \mathrm{CILS}_{0,D} \xrightarrow{\overline{p}_{0,D}} \mathrm{RCILS}_{0,D}$$

$$\mathrm{CILS}_{+,D} \xrightarrow{\overline{p}_{+,D}} \mathrm{RCILS}_{+,D} \qquad \mathrm{CILS}_{0,+,D} \xrightarrow{\overline{p}_{0,+,D}} \mathrm{RCILS}_{0,+,D}$$

Proposition 1. *The functors ι and $\overline{p}$ form an equivalence of categories between* CILS *and* RCILS.

Proof. It holds that $\overline{p}(\iota((\mathcal{R}, I))) = \overline{p}(\mathcal{R})$ and $\iota(\overline{p}(S)) = \overline{p}(S)$. By Theorem 1, $\mathcal{R}$ is isomorphic to $\overline{p}(\mathcal{R})$ and so is S to $\overline{p}(S)$. The natural isomorphism is given by ρ.   □

Corollary 3. *The restriction ι_v of ι to RCILS_v and the restriction $\overline{p}_v$ of $\overline{p}$ to CILS_v form an equivalence of categories between CILS_v and RCILS_v, where v is either $0, +, D, 0, +, 0, D, +, D,$ or $0, +, D$.*

The equivalence [1, Proposition 4.2] between the category of quantales and the category of relational quantales is the case of $v = 0, +, D$.

4   Multirelational Representation Theorem

Definition 7. Let A be a set. A *multirelational complete IL-semiring on A* is a pair $(\mathcal{M}, I)$ of $\mathcal{M} \subseteq \wp(A \times \wp(A))$ and $I \in \mathcal{M}$ such that

- $(\mathcal{M}, \subseteq)$ is a complete join semilattice,
- $(\mathcal{M}, ;, I)$ is a monoid,

– for any $\chi \subseteq \mathcal{M}$ and $R \in \mathcal{M}$, $(\bigvee \chi) \,;R = \bigvee \{P \,;R \mid P \in \chi\}$,

where ; is multirelational composition, that is, $(x, Y) \in P \,;R$ iff there exists Z such that $(x, Z) \in P$ and for each $z \in Z$ $(z, Y) \in R$.

Note that joins in a multirelational complete IL-semiring do not have to be given by unions. A multirelational complete IL-semiring preserving right directed joins, the right 0, and the right + may be called multirelational quantale.

Example 8. Let A be a set and $R \subseteq A \times \wp(A)$. R is called up-closed if $(x, X) \in R$ and $X \subseteq Y$ implies $(x, Y) \in R$ for any $x \in A$ and $X, Y \subseteq A$. We write $\mathsf{UMRel}(A)$ for the set of up-closed multirelations on A. Also, we define $\mathbf{1} \subseteq A \times \wp(A)$ by $(x, X) \in \mathbf{1}$ iff $x \in X$. Then, $\mathbf{1} \in \mathsf{UMRel}(A)$ and the pair $(\mathsf{UMRel}(A), \mathbf{1})$ is a multirelational complete IL-semiring with $\bigvee \chi = \bigcup \chi$ for each $\chi \subseteq \mathsf{UMRel}(A)$.

Example 9. Let $(\mathcal{R}, I)$ be a relational complete IL-semiring. Defining, for each $R \in \mathcal{R}$, $R' = \{(x, \{y\}) \mid (x, y) \in R\}$ and $\mathcal{R}' = \{R' \mid R \in \mathcal{R}\}$, $(\mathcal{R}', I')$ is a multirelational complete IL-semiring. Considering the relational complete IL-semiring $(\overline{\rho}_B(B), \rho_B(1))$ from Example 7, $(\overline{\rho}_B(B)', \rho_B(1)')$ is a multirelational complete IL-semiring on B. In this case, joins are not given by unions.

We write MCILS for the full subcategory of CILS whose objects are multirelational complete IL-semirings. As the case of CILS, we define MCILS_D, $\mathsf{MCILS}_{0,+}$, $\mathsf{MCILS}_{0,+,D}$, and so on. Then the eight categories and forgetful functors between them form the cube Fig. 3.

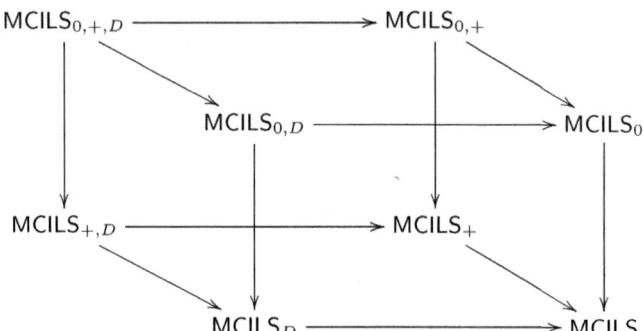

Fig. 3. The cube of multirelational complete IL-semirings

Let S be a complete IL-semiring and G a generating set for S. For $a \in S$, we define the *multirelation* $\mu_G(a)$ *induced by* a and $\overline{\mu}_G(S) \subseteq \wp(S \times \wp(S))$ by

$$\mu_G(a) = \{(g, d_G(x)) \in G \times \wp(S) \mid g \leq ax\} \text{ and } \overline{\mu}_G(S) = \{\mu_G(a) \mid a \in S\},$$

respectively.[2] Then the mapping $a \mapsto \mu_G(a)$ determines the function μ_G from S to $\overline{\mu}_G(S)$. Note that

$$(g, X) \in \mu_G(a) \iff \exists x \in S.\ (g, x) \in \rho_G(a) \text{ and } X = d_G(x) \ .$$

Also note that, for each $a \in S$, $\mu_G(a)$ is isomorphic to $\rho_G(a)$ by 2 of Remark 3.[3]

The following lemma is analogous to Lemma 2. However, generating sets in the following lemma need not be decompositional.

Lemma 4. *Let S be a complete IL-semiring and G a generating set for S. Then the following holds for any $a, b \in S$ and $X \subseteq S$.*

1. *$\mu_G(a) \subseteq \mu_G(b)$ iff $a \leq b$.*
2. *$(\overline{\mu}_G(S), \subseteq)$ is a complete join semilattice in which $\mu_G(\bigvee X)$ is the join of $\{\mu_G(x) \mid x \in X\}$.*
3. *$\mu_G(a)\,;\mu_G(b) = \mu_G(ab)$.*

Proof.   1. It follows from 1 of Lemma 2.
2. As 2 of Lemma 2, this holds by 1.
3. Applying 3 (instead of 4) of Remark 3 to the third equivalence, we have
$$(g, d_G(y)) \in \mu_G(a)\,;\mu_G(b)$$
$$\iff \exists z \in S.\,(g, d_G(z)) \in \mu_G(a) \text{ and } \forall h \in d_G(z).\,(h, d_G(y)) \in \mu_G(b)$$
$$\iff \exists z \in S.\,g \leq az \text{ and } \forall h \in d_G(z).\,h \leq by$$
$$\iff g \leq aby$$
$$\iff (g, d_G(y)) \in \mu_G(ab) \ .$$
$\square$

The following theorem is analogous to Theorem 1. However, again, generating sets in the following theorem need not be decompositional.

Theorem 2. *For a complete IL-semiring S and a generating set G for S, the pair $(\overline{\mu}_G(S), \mu_G(1))$ is a multirelational complete IL-semiring on S. Moreover, S is isomorphic to $(\overline{\mu}_G(S), \mu_G(1))$.*

Proof. Using Lemma 4 instead of Lemma 2, this is proved similarly to Theorem 1.
$\square$

---

[2] Replacing μ and $\overline{\mu}$ with μ' and $\overline{\mu}'$ defined by

$$\mu'_G(a) = \{(g, X) \in G \times \wp(S) \mid g \leq a(\bigvee X)\} \text{ and } \overline{\mu}'_G(S) = \{\mu'_G(a) \mid a \in S\} \ ,$$

we obtain the analogous results to ones in the rest of this paper. Indeed, we had adopted this construction till one of the reviewers pointed disadvantages of it. We could not have introduced μ without hints given by the reviewer. In the case, Lemma 4 might have been proved rather independently from Lemma 2 since the relationship between $\overline{\rho}_G(S)$ and $\overline{\mu}'_G(S)$ had not been so clear as the case of $\overline{\rho}_G(S)$ and $\overline{\mu}_G(S)$. Note that 3" from the footnote[1] had been prepared to prove $\mu'_G(a)\,;\mu'_G(b) = \mu'_G(ab)$. Also note that $(g, X) \in \mu'_G(a)$ iff there exists $x \in S$ such that $(g, d_G(x)) \in \mu_G(a)$ and $x = \bigvee X$.

[3] $\mu'_G(a)$ need not be isomorphic to $\rho_G(a)$. In fact, considering the complete IL-semiring B from Example 1, $\rho_{\mathsf{CCP}(B)}(a)$ is not isomorphic to $\mu'_{\mathsf{CCP}(B)}(a)$.

Example 10 (cf. Remark 4). Consider the complete IL-semiring B from Example 1. $CCP(B) = \{1, a\}$ is a generating set for B and $\mu_{CCP(B)}(\bot)$, $\mu_{CCP(B)}(1)$, $\mu_{CCP(B)}(a)$, and $\mu_{CCP(B)}(\top)$ are as follows.

$$\mu_{CCP(B)}(\bot) = \emptyset$$
$$\mu_{CCP(B)}(1) = \{(1, \{1\}), (1, \{1, a\}), (a, \{a\}), (a, \{1, a\})\}$$
$$\mu_{CCP(B)}(a) = \{(1, \{1, a\}), (a, \{1\}), (a, \{a\}), (a, \{1, a\})\}$$
$$\mu_{CCP(B)}(\top) = \{(1, \{1\}), (1, \{1, a\}), (a, \{1\}), (a, \{a\}), (a, \{1, a\})\}$$

The computation of multirelational composition on $\overline{\mu}_{CCP(B)}(B)$ is as follows.

;	$\mu_{CCP(B)}(\bot)$	$\mu_{CCP(B)}(1)$	$\mu_{CCP(B)}(a)$	$\mu_{CCP(B)}(\top)$
$\mu_{CCP(B)}(\bot)$	$\mu_{CCP(B)}(\bot)$	$\mu_{CCP(B)}(\bot)$	$\mu_{CCP(B)}(\bot)$	$\mu_{CCP(B)}(\bot)$
$\mu_{CCP(B)}(1)$	$\mu_{CCP(B)}(\bot)$	$\mu_{CCP(B)}(1)$	$\mu_{CCP(B)}(a)$	$\mu_{CCP(B)}(\top)$
$\mu_{CCP(B)}(a)$	$\mu_{CCP(B)}(\bot)$	$\mu_{CCP(B)}(a)$	$\mu_{CCP(B)}(a)$	$\mu_{CCP(B)}(\top)$
$\mu_{CCP(B)}(\top)$	$\mu_{CCP(B)}(\bot)$	$\mu_{CCP(B)}(\top)$	$\mu_{CCP(B)}(a)$	$\mu_{CCP(B)}(\top)$

So, $(\overline{\mu}_{CCP(B)}(B), \mu_{CCP(B)}(1))$ is a multirelational complete IL-semiring (in which all joins are given by unions). Moreover, it is isomorphic to B.

By Lemma 1, every complete IL-semiring has at least one generating set for itself. So, the following properties are immediate from the above theorem.

Corollary 4.   *1. Every complete IL-semiring is isomorphic to a multirelational complete IL-semiring on its underlying set.*
   2. Every complete IL-semiring preserving right directed joins is isomorphic to a multirelational complete IL-semiring preserving right directed joins on its underlying set.
   3. Every complete IL-semiring preserving the right 0 is isomorphic to a multirelational complete IL-semiring preserving the right 0 on its underlying set.
   4. Every complete IL-semiring preserving the right + is isomorphic to a multirelational complete IL-semiring preserving the right + on its underlying set.

This corollary induces a multirelational representation theorem for quantales.
    We write ι' for the inclusion functor from MCILS to CILS. The restriction ι'_v of ι' to MCILS$_v$ is the inclusion functor from MCILS$_v$ to CILS$_v$, where v is either $0, +, D, 0, +, 0, D, +, D,$ or $0, +, D$.
    Let S and S' be complete IL-semirings. We write $\mu(a)$ and $\overline{\mu}(S)$ for $\mu_S(a)$ and $\overline{\mu}_S(S)$, respectively. For each completely join-preserving homomorphism f from S to S', we define $\overline{\mu}(f) \colon \overline{\mu}(S) \to \overline{\mu}(S')$ by $\overline{\mu}(f)(\mu(a)) = \mu(f(a))$ for each $a \in S$.

Proposition 2. $\overline{\mu}$ *is a functor from* CILS *to* MCILS. *Moreover, the functors* ι' *and* $\overline{\mu}$ *form an equivalence of categories between* CILS *and* MCILS.

Corollary 5. *Let v be either $0, +, D, 0, +, 0, D, +, D,$ or $0, +, D$. The restriction $\overline{\mu}_v$ of $\overline{\mu}$ to* CILS$_v$ *is a functor from* CILS$_v$ *to* MCILS$_v$. *Moreover,* ι'_v *and* $\overline{\mu}_v$ *form an equivalence of categories between* CILS$_v$ *and* MCILS$_v$.

These sixteen functors are visualised as follows.

$$\text{CILS} \underset{\iota'}{\overset{\overline{\mu}}{\rightleftarrows}} \text{MCILS} \qquad\qquad \text{CILS}_0 \underset{\iota'_0}{\overset{\overline{\mu}_0}{\rightleftarrows}} \text{MCILS}_0$$

$$\text{CILS}_+ \underset{\iota'_+}{\overset{\overline{\mu}_+}{\rightleftarrows}} \text{MCILS}_+ \qquad\qquad \text{CILS}_D \underset{\iota'_D}{\overset{\overline{\mu}_D}{\rightleftarrows}} \text{MCILS}_D$$

$$\text{CILS}_{0,+} \underset{\iota'_{0,+}}{\overset{\overline{\mu}_{0,+}}{\rightleftarrows}} \text{MCILS}_{0,+} \qquad\qquad \text{CILS}_{0,D} \underset{\iota'_{0,D}}{\overset{\overline{\mu}_{0,D}}{\rightleftarrows}} \text{MCILS}_{0,D}$$

$$\text{CILS}_{+,D} \underset{\iota'_{+,D}}{\overset{\overline{\mu}_{+,D}}{\rightleftarrows}} \text{MCILS}_{+,D} \qquad\qquad \text{CILS}_{0,+,D} \underset{\iota'_{0,+,D}}{\overset{\overline{\mu}_{0,+,D}}{\rightleftarrows}} \text{MCILS}_{0,+,D}$$

5 CCPA Complete IL-semiring and Multirelational Complete IL-semiring

[1] has shown that, for a CCPA quantale Q, $\overline{\rho}_{\text{CCP}(Q)}(Q)$ is a relational quantale in which all joins are given by unions. The following proposition shows that, for each CCPA complete IL-semiring S, $(\overline{\rho}_{\text{CCP}(S)}(S), \subseteq)$ and $(\overline{\mu}_{\text{CCP}(S)}(S), \subseteq)$ are complete join semilattice in which all joins are given by unions.

Proposition 3. *Let S be a CCPA complete IL-semiring. Then, for $X \subseteq S$,*

1. $\bigvee\{\rho_{\text{CCP}(S)}(x) \mid x \in X\} = \bigcup\{\rho_{\text{CCP}(S)}(x) \mid x \in X\}$,
2. $\bigvee\{\mu_{\text{CCP}(S)}(x) \mid x \in X\} = \bigcup\{\mu_{\text{CCP}(S)}(x) \mid x \in X\}$.

Proof. 1 follows from

$$\begin{aligned}
&(g, y) \in \bigcup\{\rho_{\text{CCP}(S)}(x) \mid x \in X\} \\
\iff\ & \exists x \in X.\, (g, y) \in \rho_{\text{CCP}(S)}(x) \\
\iff\ & \exists x \in X.\, g \le xy \\
\iff\ & g \le \bigvee\{xy \mid x \in X\} \qquad\qquad \text{(since } g \text{ is CCP)} \\
\iff\ & g \le (\bigvee X)y \qquad\qquad\qquad \text{(since } \cdot \text{ preserves } \bigvee \text{ on the left)} \\
\iff\ & (g, y) \in \rho_{\text{CCP}(S)}(\bigvee X) \\
\iff\ & (g, y) \in \bigvee\{\rho_{\text{CCP}(S)}(x) \mid x \in X\} \quad \text{(by Lemma 2)} .
\end{aligned}$$

2 is proved similarly to 1.   □

However, for a CCPA complete IL-semiring S, $(\overline{\rho}_{\mathsf{CCP}(S)}(S), \rho_{\mathsf{CCP}(S)}(1))$ need not be a relational complete IL-semiring as we have seen in Remark 4. It is due to the fact that $\mathsf{CCP}(S)$ need not be decompositional, which demonstrated by Example 5, instead $\mathsf{CCP}(Q)$ is always decompositional for each CCPA quantale Q [1, Lemma 5.3].

On the other hand, the following proposition is immediate from 2 of Proposition 3 and Theorem 2.

Proposition 4. *For a CCPA complete IL-semiring S, $(\overline{\mu}_{\mathsf{CCP}(S)}(S), \mu_{\mathsf{CCP}(S)}(1))$ is a multirelational complete IL-semiring on S, in which all joins are given by unions. Moreover, S is isomorphic to $(\overline{\mu}_{\mathsf{CCP}(S)}(S), \mu_{\mathsf{CCP}(S)}(1))$.*

Corollary 6. 1. *Every CCPA complete IL-semiring is isomorphic to a multirelational complete IL-semiring on its underlying set, in which all joins are given by unions.*

2. *Every CCPA complete IL-semiring preserving right directed joins is isomorphic to a multirelational complete IL-semiring preserving right directed joins on its underlying set, in which all joins are given by unions.*

3. *Every CCPA complete IL-semiring preserving the right 0 is isomorphic to a multirelational complete IL-semiring preserving the right 0 on its underlying set, in which all joins are given by unions.*

4. *Every CCPA complete IL-semiring preserving the right $+$ is isomorphic to a multirelational complete IL-semiring preserving the right $+$ on its underlying set, in which all joins are given by unions.*

By this corollary, it holds that every CCPA quantale is isomorphic to a multirelational quantale on its underlying set, in which all joins are given by unions.

6   Conclusion

We have introduced relational and multirelational complete IL-semirings and shown that

1. every complete IL-semiring is isomorphic to a relational complete IL-semiring, and
2. every complete IL-semiring is isomorphic to a multirelational complete IL-semiring.

1 induces the representation theorem [1, Corollary 3.13] and 2 induces a multirelational representation theorem for quantales. Moreover, we have investigated CCPA complete IL-semirings and shown that

3. a CCPA complete IL-semiring need not be isomorphic to a relational complete IL-semiring in which all joins are given by unions though every quantale is isomorphic to a relational quantale in which all joins are given by unions, and
4. every CCPA complete IL-semiring is isomorphic to a multirelational complete IL-semiring in which all joins are given by unions.

4 induces that

5. every CCPA quantale is isomorphic to a multirelational quantale in which all joins are given by unions.

The following tables summarises these, where CILS stands for complete IL-semiring.

CILS	CCPA CILS	
is	is	isomorphic to a multirelational CILS
is	is	isomorphic to a relational CILS
need not be	is	isomorphic to a multirelational CILS in which all joins are given by unions
need not be	need not be	isomorphic to a relational CILS in which all joins are given by unions

quantale	CCPA quantale	
is	is	isomorphic to a multirelational quantale
is	is	isomorphic to a relational quantale
need not be	is	isomorphic to a multirelational quantale in which all joins are given by unions
need not be	is	isomorphic to a relational quantale in which all joins are given by unions

Acknowledgement

The authors are deeply grateful to Bernhard Möller. He has motivated the authors to address this work by introducing [1] and by posing a question that asks the case of complete IL-semirings instead of quantales. Also, he has given constructive suggestions to the previous version of this paper. We also thank to the anonymous referees for their helpful comments and suggestions. This work was supported in part by Grants-in-Aid for Scientific Research (C) 22500016 from Japan Society for the Promotion of Science (JSPS).

References

1. Brown, C., Gurr, D.: A Representation Theorem for Quantales. Journal of Pure and Applied Algebra 85, 27–42 (1993)
2. Chellas, B.: Modal Logic: An Introduction. Cambridge University Press, Cambridge (1980)
3. Conway, J.: Regular Algebra and Finite Machines. Chapman and Hall, Boca Raton (1971)
4. McCune, W.: Prover9 and Mace4, http://www.cs.unm.edu/~mccune/mace4/
5. Möller, B.: Lazy Kleene Algebra. In: Kozen, D. (ed.) MPC 2004. LNCS, vol. 3125, pp. 252–273. Springer, Heidelberg (2004)

6. Mulvey, C.J.: "&", Supplemento ai Rendiconti del Circolo Matematico di Palermo. Serie II (12), 99–104 (1986)
7. Nishizawa, K., Tsumagari, N., Furusawa, H.: The Cube of Kleene Algebras and the Triangular Prism of Multirelations. In: Berghammer, R., Jaoua, A.M., Möller, B. (eds.) RelMiCS 2009. LNCS, vol. 5827, pp. 276–290. Springer, Heidelberg (2009)
8. Parikh, R.: Propositional logics of programs: new directions. In: Karpinski, M. (ed.) FCT 1983. LNCS, vol. 158, pp. 347–359. Springer, Heidelberg (1983)
9. Rewitzky, I.: Binary multirelations. In: de Swart, H., Orłowska, E., Schmidt, G., Roubens, M. (eds.) Theory and Applications of Relational Structures as Knowledge Instruments. LNCS, vol. 2929, pp. 256–271. Springer, Heidelberg (2003)

Using Bisimulations for Optimality Problems in Model Refinement

Roland Glück

Institut für Informatik, Universität Augsburg,
D-86135 Augsburg, Germany
glueck@informatik.uni-augsburg.de

Abstract. A known generic strategy for handling large transition systems is the combined use of bisimulations and refinement. The idea is to reduce a large system by means of a bisimulation quotient into a smaller one, then to refine the smaller one in such way that it fulfils a desired property, and then to expand this refined system back into a submodel of the original one. This generic algorithm is not guaranteed to work correctly for every desired property; here we show its correctness for a class of optimality problems which can be described in the framework of dioids.

1 Introduction

1.1 General Ideas

In practice one is often confronted with systems containing a large, even infinite number of states and/or transitions, e.g. in control theory, model checking, internet routing and similar cases. If the task is to ensure a certain property (optimality, safety, liveness) by refining, i.e. removing (in practice preventing) transitions, this task can appear to be difficult to solve for the large system. One possible strategy is to reduce the original system into a smaller one using a suitable bisimulation, then to apply a known algorithm to that system, such that a refined subsystem of it fulfils the demanded property, and in a last step to expand that system into a subsystem of the original one. Of course this strategy will not work in all cases. To make sense, the reduction by bisimulation has to decrease the number of states/transitions in a significant way, an algorithm for computing a refined system with the required property has to be known, and the desired property has to be invariant in a certain sense with respect to the chosen bisimulation. As new material in this paper we show the existence of a generic refinement algorithm for a special class of optimisation problems (the second step in the above strategy sketch), and also that this class of problems is treatable by the above bisimulation-based approach (i.e. that the steps one and three are correct in this setting). In contrast to model checking, where bisimulations are commonly used to *check* properties of a system (cf. e.g. [2]) we will use them to *construct* systems with desired properties.

H. de Swart (Ed.): RAMICS 2011, LNCS 6663, pp. 164–179, 2011.
© Springer-Verlag Berlin Heidelberg 2011

1.2 Recent Work

In [14] it was shown how a control policy ensuring a certain optimality property in infinite transition systems can be obtained; that approach worked without the use of bisimulations. However, the iteratively constructed sets (called strata) in that method actually correspond to the equivalence classes of a suitable bisimulation. The successor paper [5] gives an algebraic formulation of bisimulation in general and shows the correctness of the approach for a certain liveness property. The generic algorithm was described in [6].

1.3 Overview

The paper consists of three parts: First, we introduce dioids as a tool for capturing a larger class of model refinement problems. Next we recapitulate the generic method for model refinement via bisimulation quotients. Last we show the correctness of this generic algorithm for the introduced problem class and discuss its efficiency.

2 Dioids and Models

2.1 Dioids and Cumulative Dioids

Our notion of dioid is closely related to the one given in [7]. Like there elements from a dioid will serve as edge labels in graphs to formalise optimality problems. This idea was exploited also for example in [3]. We will use the terms from [7], although there are other namings for the same structures (e.g., a complete dioid corresponds to a quantale). Since we are interested in the construction of refined models (see later) we will deviate from the methods presented in [7].

Definition 2.1. A *complete dioid* is a structure $(D, \Sigma, 0, \cdot, 1)$ such that $(D, \sqsubseteq)$ is a complete lattice with supremum operator Σ and least element 0, where $\sqsubseteq$ is defined by $x \sqsubseteq y \Leftrightarrow \Sigma\{x, y\} = y$, $(D, \cdot, 1)$ is a monoid and $\cdot$ distributes over Σ from both sides. $\sqsubseteq$ is called the *order* of the complete dioid.

In a complete dioid the binary supremum operation is denoted by $+$ and is referred to as *addition*, i.e. $x + y = \Sigma\{x, y\}$. In particular, we have $x + 0 = 0 + x = x$ for all $x \in D$, and $+$ is commutative, associative and idempotent. The operation $\cdot$ is also called *multiplication*. Note that 0 is an annihilator of multiplication (i.e. $0 \cdot x = x \cdot 0 = 0$ for all $x \in D$) due to $\Sigma\emptyset = 0$. Often for readability the $\cdot$ is omitted, so ab stands for $a \cdot b$. As commonly known in this setting, the multiplication is isotone with respect to the order, i.e. $a \sqsubseteq b$ implies $ac \sqsubseteq bc$ as well as $ca \sqsubseteq cb$ for all a, b, c. We use $a \sqsubset b$ as an abbreviation for $a \sqsubseteq b \wedge a \neq b$, and the signs $\sqsupseteq$ and $\sqsupset$ for the converses of the respective relations. In the sequel we will use D to denote the carrier set of a complete dioid when it is clear from the context which complete dioid is considered.

A complete dioid is called *selective* if $a + b \in \{a, b\}$ holds. Obviously this can be extended to the suprema of arbitrary nonempty finite sets. In this case, $\sqsubseteq$ is a linear relation. In the sequel we will consider only selective complete dioids; we call them *s-dioids* for short.

Examples for complete dioids are $(\mathbb{R} \cup \{-\infty, \infty\}, \sup, -\infty, \inf, \infty)$ or $(\mathcal{P}(\mathbb{N}), \cup, \emptyset, \cap, \mathbb{N})$. The order is $\leq$ in the first example and $\subseteq$ in the second one.

A special class of complete dioids are the *cumulative* ones, which are characterised by $a \sqsubseteq 1$ for all a, i.e., 1 is the greatest element with respect to the complete dioid's order. Cumulative dioids are nothing extraordinary, so the well-known sup-inf dioid $(\mathbb{R} \cup \{-\infty, \infty\}, \sup, -\infty, \inf, \infty)$ is cumulative. Under the name *1-bounded* they are also used for language analysis in [4]. The property of being cumulative has equivalent formulations:

Lemma 2.2. *The following statements are equivalent:*

(1) $(D, \Sigma, 0, \cdot, 1)$ *is a cumulative dioid.*
(2) *For all* $a, b, c \in D$ *the implications* $a \sqsubseteq b \Rightarrow ac \sqsubseteq b$ *and* $a \sqsubseteq b \Rightarrow ca \sqsubseteq b$ *hold.*
(3) *For all* $a, b \in D$ *the inequalities* $ab \sqsubseteq a$ *and* $ba \sqsubseteq a$ *hold.*

Proof. (1) $\Rightarrow$ (2): Let $a, b, c \in D$ be arbitrary with $a \sqsubseteq b$. Because of isotony of multiplication wrt. $\sqsubseteq$ and the assumption $c \sqsubseteq 1$ we have $ac \sqsubseteq a \cdot 1 = a$ and hence $ac \sqsubseteq b$. The other implication is shown analogously.
(2) $\Rightarrow$ (3): For arbitrary $a, b \in D$ we have $a \sqsubseteq a$ and due to (2) we have $ab \sqsubseteq a$ (choose $a := a$, $b := a$ and $c := a$). The other inequality follows analogously.
(3) $\Rightarrow$ (1): In (3) we chose an arbitrary b and set $a = 1$. $\qquad\square$

These alternative characterisations will be interpreted and used in the next section.

2.2   Models and Costs

Before defining models we first fix some notation to avoid misunderstandings. A *graph* $G = (V, E)$ is understood as a directed graph, i.e., V is a set of nodes and $E \subseteq V \times V$ is a set of edges. A *walk* in a graph $G = (V, E)$ is a sequence $v_1 v_2 \ldots v_n$ of nodes $v_i \in V$ with $(v_i, v_{i+1}) \in E$. A *path* is a walk with distinct nodes only. A node w_2 is *reachable* from a node w_1 iff there is a walk $v_1 v_2 \ldots v_n$ with $v_1 = w_1$ and $v_n = w_2$ (note the asymmetry of this definition!). The set of all walks beginning at node x and ending at node y is denoted by $W(x, y)$. Analogously, the set of all paths from x to y is denoted by $P(x, y)$. The concatenation $w_1 \circ w_2$ of two walks $w_1 = x_1 x_2 \ldots x_n$ and $w_2 = y_1 y_2 \ldots y_m$ is defined as $x_1 x_2 \ldots x_n y_2 \ldots y_m$ if $x_n = y_1$, and remains undefined else. A walk w' is *subwalk* of a walk w if there are walks w_1 and w_2 such that $w = w_1 \circ w' \circ w_2$. A *cycle* is a walk with identical first and last node. A graph is called *acyclic* if it does not contain cycles. Furthermore, for a function $f : M \to N$ the *image* of f is the set $\{ f(x) \mid x \in M \}$, denoted by $\mathrm{Im}(f)$.

Definition 2.3. A *model* is a pair $M = (G, g)$, where $G = (V, E)$ is a directed graph with node set V and edge set E, and $g : E \to D$ is a mapping from the edge set E into the carrier set D of an s-dioid $(D, \Sigma, 0, \cdot, 1)$. A *target model* is a pair $M_T = (M, T)$, where $M = ((V, E), g)$ is a model, and $T \subseteq V$ is the so called *target set*, where from every $v \in V$ some node $t \in T$ is reachable, and no node $t \in T$ has an outgoing edge. A model is called *finite* iff the associated graph is finite.

Models correspond to edge labelled graphs. The term model was chosen in analogy to model checking. Target models describe the case when one is interested in reaching a certain node set in the underlying graph of a model. Our main interest here lies on target models. The requirement that the target set has to be reachable from every node is motivated by the fact that we will concentrate on walks leading into T.

A (target) model is called *acyclic* if its underlying graph is acyclic. A model $M' = ((V', E'), g')$ is called a *submodel* of a model $M = ((V, E), g)$, written $M' \leq M$, if $V' = V$, $E' \subseteq E$ and $g' = g|_{E'}$ hold ($g|_{E'}$ denotes the restriction of g to the domain E'). A target model $M'_T = (M', T')$ is a *target submodel* of a target model $T_M = (M, T)$ if M' is a submodel of M and $T' = T$. A pair (M, T) where $M = ((V, E), g)$ is a model and $T \subseteq V$ is not reachable from every node in $V \backslash T$ is called a *defect target model*. For a model $M = ((V, E), g)$ and $V' \subseteq V$ the *restriction* of M by V', written $M|_{V'}$, is defined by $M|_{V'} = ((V', E'), g')$ with $V' = V$, $E' = \{(v_1, v_2) \in E | (v_1, v_2) \in V' \times V'\}$ and $g' = g|_{E'}$.

As already mentioned, the edge labels drawn from s-dioids serve to generalise costs of walks in graphs. Concretely this is done in the following manner:

Definition 2.4. Let $M = (G, g)$ with $G = (V, E)$ be a model and $(D, \Sigma, 0, \cdot, 1)$ the associated s-dioid. Then for a walk $w = x_1 x_2 \ldots x_n$ in G the *cost* $c(w)$ of w is defined by $c(w) = \prod_{i=1}^{n-1} g(x_i, x_{i+1})$. For two nodes x and y the *distance* $d(x, y)$ of x and y is defined by $d(x, y) = \sum_{w \in W(x,y)} c(w)$. In a target model $T_M = (M, T)$ the *target distance* $d(x)$ of a node x is defined as $d(x) = \sum_{t \in T} d(x, t)$, where $d(x, t)$ is determined as above in the associated model M. A walk $x_1 x_2 \ldots x_n$ is called *optimal* if $c(x_1 x_2 \ldots x_n) = d(x_1, x_n)$.

Note that the distance is defined for every pair of nodes, even for not reachable nodes: in this case the distance equals the supremum of the empty set, which is 0.

If one chooses the s-dioid $(\mathbb{R}_0^+ \cup \{\infty\}, \inf, \infty, +, 0)$ as codomain of g the cost of a walk corresponds to its length in its classic sense as the sum of the weights of its edges. The distance of two points corresponds to the length of a shortest path connecting these two points, and the target distance $d(x)$ in a target model to the minimal length of a shortest path between x and some node in the target set. Similarly, if the s-dioid $(\mathbb{R} \cup \{-\infty, \infty\}, \sup, -\infty, \inf, \infty)$ is chosen, the cost corresponds to the capacity, and the distance to the maximum capacity of walks.

We represent target models visually as edge labelled graphs, where supremum and multiplication operations are clear from the context or are specified whenever

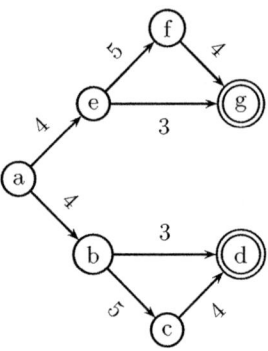

Fig. 1. A Target Model

they are referenced. The nodes of the target set are surrounded by double lines. In the following figure a target model with target set $\{d, g\}$ is shown, but it is not yet clear what the associated s-dioid is.

2.3   Optimality in Models

Note that until now we did not state anything about the existence of optimal walks. E.g., in a graph with edge labels from $\mathbb{R}$ there is no shortest walk between two nodes, if the graph contains a cycle of negative length (note that by definition the distance can be $-\infty$, but there is no walk with this cost). If we use a cumulative s-dioid we are in a much better position: here we can show that in every finite model an optimal walk between two reachable nodes exists, and moreover there always even exists a path with optimal cost between two reachable nodes. This is stated in the next lemma.

Lemma 2.5. *Let* $M = ((V, E), g)$ *be a finite model whose associated s-dioid* $(D, \Sigma, 0, \cdot, 1)$ *is cumulative, and let* x *and* y *be two reachable nodes. Then there is a path* $p \in P(x, y)$ *with* $c(p) = d(x, y)$.

Proof. Let M, x, y be as above and let $w \in W(x, y)$ be an arbitrary walk. Assume that w contains a repeated node, i.e. $w = x_1 x_2 \ldots x_i \ldots x_j \ldots x_n$ with $x_i = x_j$. Consider now the walk $w' = x_1 x_2 \ldots x_{i-1} x_j \ldots x_n$ from x_1 to x_n. Because $(D, \Sigma, 0, \cdot, 1)$ is cumulative we have $c(x_1 x_2 \ldots x_i \ldots x_j) \sqsubseteq c(x_1 x_2 \ldots x_{i-1})$, and together with the isotony of multiplication we obtain $c(w) \sqsubseteq c(w')$. By repeated application of this construction we can obtain from w a path $p \in P(x, y)$ with $c(w) \sqsubseteq c(p)$. So in this case $d(x, y) = \sum_{w \in W(x,y)} c(w) = \sum_{p \in P(x,y)} c(p)$ holds. Since there are only finitely many paths from x to y there is a $p \in P(x, y)$ with $c(p) = d(x, y)$ (this holds, because the order in an s-dioid is linear, so every finite set contains a least element). □

Our goal is to ensure that every walk in a target model leading into the target set is an optimal one. This will happen by constructing a suitable target submodel, which motivates the following definition:

Definition 2.6. For a target model $T_M = (M, T)$ a target submodel $T'_M \leq T_M$ is called an *optimal target submodel*, if for all walks w from x to any node $t \in T$ in T'_M the cost $c(w)$ equals the target distance $d(x)$ in T_M (note that a target submodel is also a target model and therefore in an optimal target submodel the target set has to be reachable from every node outside of it).

If we interpret the labels of Figure 1 in the s-dioid $(\mathbb{R} \cup \{-\infty, \infty\}, \inf, \infty, +, 0)$ then an optimal submodel is given in Figure 2. Assuming the labels to stem from the s-dioid $(\mathbb{R} \cup \{-\infty, \infty\}, \sup, -\infty, \inf, \infty)$ an optimal submodel is depicted in Figure 3.

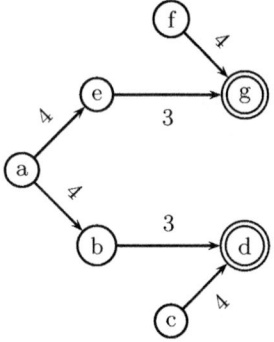

Fig. 2. An Optimal Target Submodel

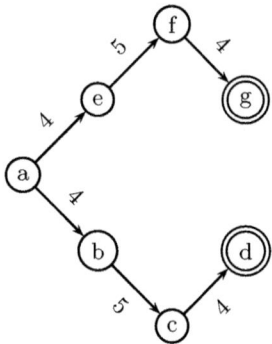

Fig. 3. Another Optimal Target Submodel

2.4   Models with Cumulative S-Dioids

Unfortunately, it turns out that not every target model has an optimal target submodel, as for example shown by the shortest walk problem in the presence of cycles with negative length. So a target model which has an optimal target submodel is called *refinable*. Next we give a necessary and sufficient condition for an s-dioid such that all target models with a distance function based on that s-dioid are refinable.

Theorem 2.7. *Let $\mathcal{D} = (D, \Sigma, 0, \cdot, 1)$ be an s-dioid. Then every finite target model with $\mathcal{D}$ as associated s-dioid is refinable iff $\mathcal{D}$ is cumulative.*

Proof. For $\Rightarrow$ we consider the following algorithm, which is basically a variant of Dijkstras algorithm:

Input:
finite target model $M = (((V, E), g), T)$ with associated cumulative s-dioid
$\mathcal{D} = (D, \Sigma, 0, \cdot, 1)$

initialise dist as an array with indices from V and values from D;
initialise succ as an array with indices from V and values from V;
initialise U as set with elements from V;

```
forall t ∈ T
  dist(t) = 1;
  succ(t) = null;
endfor
forall v ∈ V\T
  dist(v) = 0;
  succ(v) = null;
endfor
U = ∅;
while U ≠ V
  choose v ∉ U with dist(v) = ∑  dist(v');
                              v'∉U
  U = U ∪ {v};
  forall (v', v) ∈ E with v' ∉ U
    if g(v', v) · dist(v) ⊐ dist(v')
      dist(v') = g(v', v)·dist(v);
      succ(v') = v;
    endif
  endfor
endwhile
```

Output:
$\forall v \in V : \mathrm{dist}(v) = d(v)$
succ encodes an optimal submodel

The correctness proof for this algorithm is very similar to the one of the common Dijkstra algorithm. The differences are of course the use of a general cumulative s-dioid instead of $\mathbb{R}_{\geq 0}$, and the fact that we are not interested in optimal paths from a single node to all other nodes but in optimal paths leading into a given set of nodes from every node outside of that set.

The termination of the while-loop is ensured by the termination function $|U|$, since $|U|$ is increased by one after every pass through the while-loop, and it is bounded by $|V|$, since M is assumed to be finite. The for-loops terminate since they run over finite sets only.

To prove the correctness we first introduce the concept of a *maximal path* starting from a node wrt. a given successor array. First, a mapping succ as defined above in the succ-array determines a graph $G_{succ} = (V_{succ}, E_{succ})$ with $V_{succ} = V$ and $(v_1, v_2) \in E_{succ}$ iff $\text{succ}(v_1) = v_2$. In an analogous manner it defines a possibly defect target submodel M_{succ} of M. Then the maximal path $mp(v,\text{succ})$ starting in v wrt. succ is the maximal (i.e. not prolongable) path in M_{succ} starting in v. Intuitively, this describes the maximal path that starts in v and is guided by the instructions encoded in succ.

For the correctness we choose as an invariant for the while-loop the combination of three invariants $I \equiv I_1 \wedge I_2 \wedge I_3$ with

$$I_1 \equiv \forall u \in U : \text{dist}(u) = d(u),$$
$$I_2 \equiv \forall u \in U : c(mp(u, \text{succ})) = \text{dist}(u), \text{ and}$$
$$I_3 \equiv \forall u \notin U : \text{dist}(u) = d(u) \text{ in } M|_{U \cup \{v\}}.$$

Here and in the sequel v denotes the node chosen in the while-loop. Informally this means $\text{dist}(u) = d(u)$ where $d(u)$ is defined on the target model M_{succ}, depending on the progress of the algorithm, whereas $\text{dist}(u) = d(u)$ with a $d(u)$ defined on M holds only for nodes $u \notin U \cup \{v\}$. $\text{succ}(u)$ denotes the successor of u on a walk w with $c(w) = d(u)$ in M_{succ}; it could be null if there is no path in M_{succ} from u leading into some node in the target set.

I_1 and I_2 hold trivially before the first entry into the while-loop because U is empty at this point. I_3 holds before the first entry into the while-loop because at this point M_{succ} has an empty edge set, so $d(u) = 0$ for all $u \notin T$, and $d(u) = 1$ for all $u \in T$ holds in M_{succ}, in consistency with the assignment to dist.

Consider now a pass through the while-loop, and let v be the chosen node (such a node exists, since $\mathcal{D}$ is cumulative, cf. also the proof of lemma 2.5). For nodes $u \in U$ no changes for $\text{dist}(u)$ and $\text{succ}(u)$ are made, so I_1 and I_2 remain valid for such nodes. Next we consider the node v. Assume there is a walk w in M from v into T with $c(w) \sqsupset \text{dist}(v)$. According to I_3 this walk must visit a node u' with $u' \notin U \cup \{v\}$. Let u' be the first node on w with this property. Because of $v, u' \notin U$ and the choice of v we have $\text{dist}(u') \sqsubseteq \text{dist}(v)$. On the other hand, w can be split into two walks as $w = w_1 \circ w_2$ where w_2 is a walk from u' into T, which has, according to I_3, the cost $\text{dist}(u')$. Since the associated dioid is assumed to be cumulativewe have $\text{dist}(u') \sqsupseteq c(w)$, which together with the assumption $c(w) \sqsupset \text{dist}(v)$ contradicts $\text{dist}(v) \sqsupseteq \text{dist}(u')$. So I_1 and I_2 hold after the pass through the while-loop, too.

I_3 remains trivially valid for nodes in $U \cup \{v\}$, so we consider an arbitrary node $v' \notin U \cup \{v\}$. We distinguish two cases. First, assume that every optimal walk in $M|_{U \cup \{v,v'\}}$ starting in v' visits v. Before visiting v such a walk can not visit a node $u' \in U$. To see this let w be the walk from v' to some node $t \in T$, and let w_1 and w_2 be its subpaths from v' to u' and from u' to t, resp. According to I_1 there is a walk w_3 from u' into t with $c(w_3) \sqsupseteq c(w_2)$. But now both $w_1 \circ w_2$ and $w_1 \circ w_3$ are paths from v' into t with $c(w_1 \circ w_3) \sqsupseteq c(w_1 \circ w_2)$, which contradicts the assumption that every optimal walk from v' into T visits v, because $(w_1 \circ w_3)$ does not. So every optimal walk from v' into T consists of the edge (v', v) and an optimal walk from v into T, so it has the cost $g(v', v) \circ \mathrm{dist}(v)$. Hence dist and succ are updated correctly inside the while-loop. The second case is that there is an optimal walk from v' into T in $M|_{U \cup \{v,v'\}}$ which visits nodes in $U \cup \{v'\}$ only. Then dist and succ can remain untouched, so I_3 holds again for all $v' \notin U$.

This showed one direction of Theorem 2.7; for $\Leftarrow$ we consider the following target model with an arbitrary a:

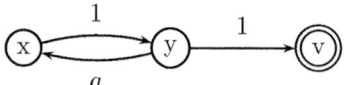

Then the refined graph can either contain the edge (y, x) or not. In the first case we have $1 = c(yv) = c(yxyv) = a$ because every path leading into v has to be optimal, and hence $a \sqsubseteq 1$ for arbitrary a. In the latter case in the original (not refined) target model $d(y) = c(yv)$ holds (because in the refined model only the path yv from y to v exists), and therefore, according to the definition of d, $c(yxyv) \sqsubseteq c(yv)$ holds, which implies $a \sqsubseteq 1$. So $\mathcal{D}$ has to be cumulative according to lemma 2.2. □

2.5   Models with Non-cumulative S-Dioids

Nevertheless there is a possibility to refine models with non-cumulative associated s-dioids if the underlying labelled graph does not contain negative cycles. A *negative cycle* is a cycle $x_1 x_2 \ldots x_n$ with $x_1 = x_n$ and $c(x_1 x_2 \ldots x_n) \sqsupseteq 1$. In the case of absence of negative cycles we can see that between two reachable nodes there is always an optimal walk which also is a path. This can be shown by the same argument as above (removing cycles), since removing a cycle from a walk under this assumption can not decrease its cost (with respect to $\sqsubseteq$). In particular, this means that in a finite model under these circumstances always an optimal walk between two reachable nodes with at most $|V| - 1$ edges exists.

To obtain an optimal submodel in this case, we can apply an algorithm analogous to the Floyd-Warshall algorithm. First we determine by the following algorithm the distances between every pair of nodes and the successor of every node on an optimal path to every other node by the following algorithm:

Input:
finite target model $M = (((V, E), g), T)$ without negative cycles,
nodeset $V = \{1 \ldots n\}$ and associated s-dioid $(D, \Sigma, 0, \cdot, 1)$

initialise dist as an $n \times n$-matrix with entries from D;
initialise succ as an $n \times n$-matrix with entries from V;

```
for (i = 1..n)
  for (j = 1..n)
    if (i, j) ∈ E
      dist(i, j) = g(i, j);
    endif
    else
      dist(i, j) = 0;
    endelse
    succ(i, j) = null;
  endfor
endfor
for (k = 1..n)
  for (i = 1..n)
    for (j = 1..n)
      if dist(i, k) · dist(k, j) ⊐ dist(i, j)
        dist(i, j) = dist(i, k) · dist(k, j);
        succ(i, j) = succ(i, k)
      endif
    endfor
  endfor
endfor
```

Output:
$\forall i, j \in V : \text{dist}(i, j) = d(i, j)$
succ encodes optimal walks

The meaning of $\text{succ}(i, j)$ is analogous to the one in the previous algorithm, with the only difference that not an optimal walk from i into a target set T is encoded but an optimal walk from i to j.

The termination of this algorithm is obvious, and as an invariant for the outermost of the three nested loops we choose the claim that $\text{dist}(i, j)$ equals the cost of an optimal walk from i to j which visits, except for i and j, only nodes v with $v \leq k$, and that $\text{succ}(i, j)$ encodes such a walk. The invariant holds before entering the three nested loops due to the intitialisation of dist and succ and the fact that k is not yet initialised. Consider now a pass through the two inner while-loops. If there is an optimal walk from i to j without visiting the node k than dist and succ remain unchanged and the invariant holds also after the run. Conversely, if there is an optimal walk from some node i to some node j visiting the node k not using nodes k' with $k' > k$ then it has to be composed of an optimal walk from i to k and an optimal walk from k to j without visiting a node k' with $k' > k$. So the update operation preserves the invariant.

Finally, to refine the model to achieve an optimal submodel we have to keep all edges $\{(v, w) \in E \mid \exists t \in T : \text{succ}(v, t) = w \land D(v, t) = \sum_{t' \in T} D(v, t')\}$ and to remove all other edges. Obviously, from every node $v \in V$ a node $t \in T$ is reachable, and every such walk is an optimal one.

3   Bisimulations

The models the previous algorithms are applied to can have large numbers of nodes. One possibility to reduce the problem is the use of bisimulations, which will be introduced in this section. Before getting formal we fix some notation.

For a relation R we denote its converse by R°. For relational composition we use the semicolon $;$. If $E \subseteq X \times X$ is an equivalence relation we write x/E for the equivalence class of an element $x \in X$.

3.1   Basic Definitions

A bisimulation between two relations $R \subseteq X \times X$ and $R' \subseteq X' \times X'$ is a relation $B \subseteq X \times X'$ with the properties $B^\circ ; R \subseteq R' ; B^\circ$ and $B ; R' \subseteq R ; B$. Intuitively this means that if a step from x to y is possible under the relation R then a step from x' to y' is possible under R' where x and x' respectively y and y' are related via B. The analogous property holds for transitions in R' compared to those in R; here the elements are related by B°.

A bisimulation between a relation $R \subseteq X \times X$ and itself is called an *auto-bisimulation*. Since autobisimulations are closed under union, composition and conversion and the identity is an autobisimulation there is a coarsest autobisimulation for a relation $R \subseteq X \times X$, which is an equivalence.

Here we are interested in a special kind of autobisimulations, which also respects the labels of edges of a graph. For a model $M = ((V, E), g)$ we define for every $a \in \mathrm{Im}(g)$ the relation $E_a \doteq \{e \in E \mid g(e) = a\}$. With this convention we define the term bisimulation for models as follows:

Definition 3.1. Let $M = ((V, E), g)$ be a model. A *bisimulation* on M is a relation $B \subseteq V \times V$ such that B is an autobisimulation for E_a for all $a \in \mathrm{Im}(g)$. A *bisimulation* on a target model $M = (((V, E), g), T)$ is a bisimulation B on $((V, E), g)$ such that $B \subseteq T \times T \cup (V - T) \times (V - T)$.

It is easy to see that bisimulations on a model are also closed under union, composition and conversion, and that the identity is a bisimulation on every model, see also [2] for analogous results in an almost identical context and [15] for a relation algebraic treatment. Hence, there is always a coarsest bisimulation on a model, which is an equivalence. For a model M we denote its coarsest bisimulation by $\mathcal{B}_M$. Generally, a bisimulation which is also an equivalence is called a *bisimulation equivalence*.

3.2   Quotient and Expansion

For our purposes bisimulation equivalences can be used to reduce the state numbers of (target) models in a reasonable way. This is done via the *quotient target model*:

Definition 3.2. Let $M = (((V, E), g), T)$ be a target model and $\mathcal{B}$ a bisimulation equivalence on M. The *quotient* $M_{\mathcal{B}}$ of M by $\mathcal{B}$ is the target model $M_{\mathcal{B}} = (((V_{\mathcal{B}}, E_{\mathcal{B}}), g_{\mathcal{B}}), T_{\mathcal{B}})$ with

- $V_{\mathcal{B}} = \{v/\mathcal{B} \mid v \in V\}$,
- $E_{\mathcal{B}} = \{(x/\mathcal{B}, y/\mathcal{B}) \mid (x, y) \in E)\}$,
- $g_{\mathcal{B}}(x/\mathcal{B}, y/\mathcal{B}) = g(x, y)$ and
- $T_{\mathcal{B}} = \{t/\mathcal{B} \mid t \in T\}$.

$g_{\mathcal{B}}$ is well defined due to the requirement that $\mathcal{B}$ is a bisimulation on M.

This construction is analogous to the one of a minimal automaton in automata theory, see for example the classics [11] and [12]. Analogously to there the following lemma about the existence of paths in the quotient holds:

Lemma 3.3. *Let $M = (((V, E), g), T)$ be a target model and $\mathcal{B}$ a bisimulation equivalence on M. If there is a path $x_1 x_2 \cdots x_n$ in M then there is a path $X_1 X_2 \cdots X_n$ in $M/\mathcal{B}$ with $X_i = x/\mathcal{B}$ and $g(x_i, x_{i+1}) = g_{\mathcal{B}}(X_i, X_{i+1})$.*

This shows that in a certain sense dynamics are preserved by quotients.

If we want to reduce the number of states and to preserve the dynamics of a target model at the same time we use the coarsest bisimulation to build the quotient, because it reduces the number of states maximally among all bisimulation equivalences. In this case the resulting quotient is called the *coarsest quotient*.

The coarsest quotient of the target model from Figure 1 is shown in Figure 4. For readability the sets in the nodes' captions are written without parentheses and commas.

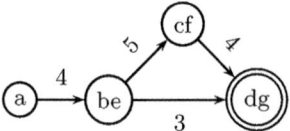

Fig. 4. A Coarsest Quotient

Generally, the coarsest quotient of a target model will have in its associated graph a smaller number of nodes than the original one, especially if the target model is well structured (i.e., it contains identic subgraphs, which cab be merged by the coarsest quotient). If we deal with an infinite target model we should hope that its coarsest quotient is finite.

For the expansion step we need to define an expanding operation, which takes as input a target model and a submodel of one of its quotient models and outputs a target submodel of the original one. The details of the theoretic background can be found in [6]; here we only give the definition.

Definition 3.4. Let $M = (((V, E), g), T)$ be a target model, $\mathcal{B}$ a bisimulation on M and $M'_{\mathcal{B}} = (((V'_{\mathcal{B}}, E'_{\mathcal{B}}), g'_{\mathcal{B}}), T'_{\mathcal{B}})$ a target submodel of the quotient $M_{\mathcal{B}}$. The *expansion* $M'_{\mathcal{B}} \backslash \mathcal{B} = (((V', E'), g'), T')$ of $M'_{\mathcal{B}}$ by $\mathcal{B}$ is given by

- $V' = V$,
- $(x, y) \in E' \Leftrightarrow (x/\mathcal{B}, y/\mathcal{B}) \in E'_\mathcal{B} \wedge (x, y) \in E$,
- $g' = g|_{E'}$ and
- $T' = T$.

Obviously the expansion is a target submodel of the initial target model. Considering a target model M, a bisimulation equivalence $\mathcal{B}$ on M and a target submodel $M'_\mathcal{B}$ of the quotient $M_\mathcal{B}$ the expansion $M'_\mathcal{B} \backslash \mathcal{B}$ is the greatest submodel of M whose quotient by $\mathcal{B}$ equals $M'_\mathcal{B}$. The sign $\backslash$ for the expansion operation indicates that it is a pseudoinverse of the quotient operation based on the notion $/$ for equivalence classes; for details see again [6]. In particular, this means also $M_\mathcal{B} \backslash \mathcal{B} = M$.

Similar to Lemma 3.3 there is also a statement about the existence of paths in the expansion:

Lemma 3.5. *Let $M = (((V, E), g), T)$ be a target model, $\mathcal{B}$ a bisimulation equivalence on M and $M'_\mathcal{B} = (((V'_\mathcal{B}, E'_\mathcal{B}), g'_\mathcal{B}), T'_\mathcal{B})$ a target submodel of the quotient $M_\mathcal{B}$. If there is a path $X_1 X_2 \cdots X_n$ in $M'_\mathcal{B}$ then for every $x_1 \in X_1$ there is a path $x_1 x_2 \cdots x_n$ in $M'_\mathcal{B} \backslash \mathcal{B}$ with $x_i \in X_i$ and $g_\mathcal{B}(X_i, X_{i+1}) = g(x_i, x_{i+1})$.*

Proof. We use induction over n, starting at $n = 2$. Due to $(X_1, X_2) \in E'_\mathcal{B}$ we have also $(X_1, X_2) \in E_\mathcal{B}$. According to the properties of bisimulation equivalences for every $x_1 \in X_1$ there is a $x_2 \in X_2$ with $(x_1, x_2) \in E$ and $g(x_1, x_2) = g_\mathcal{B}(X_1, X_2)$. Together with the definition of the expansion this shows the claim for $n = 2$. Consider now a walk $X_1 X_2 \cdots X_n X_{n+1}$ in $M'_\mathcal{B}$ and an arbitrary $x_1 \in X_1$. Then there is a walk $x_1 x_2 \cdots x_n$ in M with the required properties. An analogous argument as above shows the existence of an $x_{n+1} \in X_{n+1}$ with $g_\mathcal{B}(x_n, x_{n+1}) = g(X_n, X_{n+1})$, which completes the proof. □

There is also a converse version of this lemma:

Lemma 3.6. *Let $M = (((V, E), g), T)$ be a target model, $\mathcal{B}$ a bisimulation equivalence on M and $M'_\mathcal{B} = (((V'_\mathcal{B}, E'_\mathcal{B}), g'_\mathcal{B}), T'_\mathcal{B})$ a target submodel of the quotient $M_\mathcal{B}$. If there is a path $x_1 x_2 \cdots x_n$ in $M'_\mathcal{B} \backslash \mathcal{B}$ then there has to be a path $X_1 X_2 \cdots X_n$ in $M'_\mathcal{B}$ with $x_i \in X_i$ and $g(x_i, x_{i+1}) = g_\mathcal{B}(X_i, X_{i+1})$.*

Proof. It suffices to show that for every edge $(x, y) \in M'_\mathcal{B} \backslash \mathcal{B}$ there is an edge $(X, Y) \in M'_\mathcal{B}$ with $x \in X$, $y \in Y$ and $g_\mathcal{B}(X, Y) = g(x, y)$. But this follows obviously from the fact that $\mathcal{B}$ is a bisimulation equivalence and the definition of the expansion. □

4   Application to Optimality

4.1   Putting the Pieces Together

To make use of the quotient and expansion operations in refinement problems we have to ensure that the desired property is compatible in a certain sense with the chosen bisimulation equivalence. Informally, this means that if the property in the quotient model is ensured by a suitable refinement then the same property does also hold in the expansion of the refined model. Concretely, this is expressed by the next theorem:

Theorem 4.1. *Let $M = (((V,E),g),T)$ be a target model, $\mathcal{B}_M$ its coarsest bisimulation and $M_{\mathcal{B}_M}$ its coarsest quotient. If $M'_{\mathcal{B}_M}$ is an optimal target submodel of $M_{\mathcal{B}_M}$ then $M'_{\mathcal{B}_M} \backslash \mathcal{B}_M$ is an optimal target submodel of M.*

Proof. First we show that for all $x \in V$ the values $d(x)$ in M and $d(x/\mathcal{B}_M)$ in $M_{\mathcal{B}_M}$ coincide. This will be shown by the two inequalities $d(x) \sqsubseteq d(x/\mathcal{B}_M)$ and $d(x/\mathcal{B}_M) \sqsubseteq d(x)$. For the first one we observe according to lemma 3.3 that for every walk $x_1 x_2 \ldots x_n$ in M with $x_1 = x$ and $x_n \in T$ there is a walk $X_1 X_2 \ldots X_n$ in $M_{\mathcal{B}_M}$ with $X_1 = x/\mathcal{B}_M$, $X_n \in T_{\mathcal{B}_M}$ and $g(x_i, x_{i+1}) = g_{\mathcal{B}_M}(X_i, X_{i+1})$. So the set of all costs of paths leading from x to any node $t \in T$ in M is a subset of the costs of all paths leading from $x/\mathcal{B}_M$ to any node $t/\mathcal{B}_M \in T_{\mathcal{B}_M}$, which implies the inequality due to the definition of $d(\cdot)$. Conversely, for every walk $X_1 X_2 \ldots X_n$ in $M_{\mathcal{B}_M}$ with $x \in X_1$ and $X_n \in T_{\mathcal{B}_M}$ according to Lemma 3.5 there exists a walk $x_1 x_2 \cdots x_n$ in M (remember that $M_{\mathcal{B}_M} \backslash \mathcal{B}_M = M$ holds) with $g_{\mathcal{B}_M}(X_i, X_{i+1}) = g(x_i, x_{i+1})$ and, due to the properties of bisimulation, $x_n \in T$. Then an analogous argument shows the claimed inequality.

Let now $w = x_1 x_2 \ldots x_n$ be an arbitrary walk in $M'_{\mathcal{B}_M} \backslash \mathcal{B}_M$. According to Lemma 3.6 there is a walk $W = X_1 X_2 \ldots X_n$ in $M'_{\mathcal{B}_M}$ with $x_i \in X_i$ and $X_n \in T_{\mathcal{B}_M}$, which has the same cost as w. Since $M'_{\mathcal{B}_M}$ is assumed to be an optimal submodel we have $c(W) = d(X_1)$ and $d(X_1) = d(x_1)$ as shown above. Additionally $c(W) = c(w)$ holds, so $M'_{\mathcal{B}_M} \backslash \mathcal{B}_M$ is an optimal submodel of M. $\square$

This result shows the correctness of the following algorithm, if the coarsest quotient of the input target model is finite:

Input:
target model $M = (((V,E),g),T)$ with associated cumulative s-dioid
or without negative cycles

compute the quotient of M;
run a suitable refinement algorithm on the quotient;
expand the refined submodel of the quotient;

Output:
an optimal submodel of M

To demonstrate the algorithm let us consider the target model from Figure 1. Its quotient is shown in Figure 4. If we are interested in shortest paths an optimal submodel of this quotient is the one from Figure 5. Its expansion yields the optimal submodel in Figure 2.

Note also that the proof does not require that the large model is finite; it is only necessary to know that it is refinable at all. So this approach can also be applied to an infinite target model if it is known from other arguments that it is refinable, and its coarsest quotient is finite.

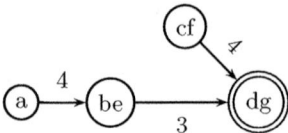

Fig. 5. An Optimal Quotient Target Submodel

4.2 Applicability and Efficiency

After ensuring the correctness of our approach we will take a look at its applicability and efficiency. Let us first assume that the model under consideration is finite. In [13] an algorithm for computing the coarsest bisimulation is presented with a runtime of $\mathcal{O}(|E| \cdot log(|V|))$. The algorithms from subsection 2.3 have a runtime of $\mathcal{O}(|V| \cdot log(|V|) + |E|)$ for the Dijkstra-like one and of $\mathcal{O}(|V|^3)$ for the Floyd-Warshall-like one, and the expansion step can easily be done in $\mathcal{O}(|V| + |E|)$. So for graphs with $|E| \in \mathcal{O}(|V|)$ the immediate application for the Dijkstra-like algorithm and the algorithm using bisimulation have the same asymptotic runtime, whereas for graphs with $|E| \in \Theta(|V|^2)$ the immediate Dijkstra-like algorithm performs better. If one has to solve problems based on non-cumulative dioids there is the possibility that to achieve a better runtime by taking the way over quotient and expansion. The speed-up compared to the immediate application of a refinement algorithm will be the higher the more the state number is reduced by building the quotient. So it appears to be reasonable to apply it especially to well-structured graphs. Examples for this kind of graphs are hierarchical graphs or graphs with a tree-like structure.

Another application field are infinite models. Here the refinement algorithms from 2.3 do not work, but after computing a finite quotient (which is of course not always possible) by for example symbolic algorithms they can be applied and the solution can be expanded. This is what in general happens in [14].

5   Conclusion and Further Work

We have shown that the quotient-refine-expand approach using bisimulations is correct and efficient for a certain class of dioid-based optimality problems. There are two main directions of future work: first, to search and investigate other bisimulation-compatible refinement problems, and second, to place the approach presented here into a more algebraic framework, in a way comparable to [5]. The advantages of the first goal are obvious, as a speed-up for well-structured models can be expected compared to the immediate application of a refinement algorithm. Potential members of this class are flow problems (as for example in [1]), circulation problems (described e.g. in [9] or algebraically in [10]) or problems known from model checking like safety and liveness properties. On the other hand, an algebraic foundation opens the door for deploying automated theorem provers as demonstrated in [8]. A first step into this direction was already done in [5].

Acknowledgments. The ideas in this paper greatly profited from the extensive and pleasant collaboration we had with the late Michel Sintzoff. It is a great pity that he departed so soon and so unepectedly; he will always be remembered in the best way. I am also grateful to Bernhard Möller, Han-Hing Dang, Patrick Roocks, and the anonymous referees for valuable discussions and remarks.

References

1. Ahuja, R.K., Magnati, T.L., Orlin, J.B.: Network Flows. Prentice-Hall, Englewood Cliffs (1993)
2. Baier, C., Katoen, J.-P.: Principles of Model Checking. MIT Press, Cambridge (2008)
3. Billings, J.N., Griffin, T.G.: A model of internet routing using semi-modules. In: Berghammer, R., Jaoua, A.M., Möller, B. (eds.) RelMiCS 2009. LNCS, vol. 5827, pp. 29–43. Springer, Heidelberg (2009)
4. Esparza, J., Kiefer, S., Luttenberger, M.: Derivation tree analysis for accelerated fixed-point computation. In: Ito, M., Toyama, M. (eds.) DLT 2008. LNCS, vol. 5257, pp. 301–313. Springer, Heidelberg (2008)
5. Glück, R., Möller, B., Sintzoff, M.: A semiring approach to equivalences, bisimulations and control. In: Berghammer, R., Jaoua, A.M., Möller, B. (eds.) RelMiCS 2009. LNCS, vol. 5827, pp. 134–149. Springer, Heidelberg (2009)
6. Glück, R., Möller, B., Sintzoff, M.: Model refinement using bisimulation quotients. In: Johnson, M., Pavlovic, D. (eds.) AMAST 2010. LNCS, vol. 6486, pp. 76–91. Springer, Heidelberg (2011)
7. Gondran, M., Minoux, M.: Graphs, Dioids and Semirings. Springer, Heidelberg (2008)
8. Höfner, P., Struth, G.: Automated reasoning in kleene algebra. In: Pfenning, F. (ed.) CADE 2007. LNCS (LNAI), vol. 4603, pp. 279–294. Springer, Heidelberg (2007)
9. Jungnickel, D.: Graphs, Networks and Algorithms, 2nd edn. Springer, Heidelberg (2005)
10. Kawahara, Y.: On the cardinality of relations. In: Schmidt, R.A. (ed.) RelMiCS/AKA 2006. LNCS, vol. 4136, pp. 251–265. Springer, Heidelberg (2006)
11. Myhill, J.: Finite automata and the representation of events. WADD TR-57-624, 112–137 (1957)
12. Nerode, A.: Linear automaton transformations. Proceedings of the American Mathematical Society 9, 541–544 (1958)
13. Paige, R., Tarjan, R.: Three partition refinement algorithms. SIAM Journal for Computing 16(6)
14. Sintzoff, M.: Synthesis of optimal control policies for some infinite-state transition systems. In: Audebaud, P., Paulin-Mohring, C. (eds.) MPC 2008. LNCS, vol. 5133, pp. 336–359. Springer, Heidelberg (2008)
15. Winter, M.: A relation-algebraic theory of bisimulations. Fundam. Inf. 83(4), 429–449 (2008)

Pathfinding through Congruences

Alexander J.T. Gurney and Timothy G. Griffin

Computer Laboratory, University of Cambridge, UK

Abstract. Congruences of path algebras are useful in the definition and analysis of pathfinding problems, since properties of an algebra can be related to properties of its quotient. We show that this relationship can apply even when the algebraic objects involved satisfy weaker forms of the semiring or path algebra axioms. This is useful, since it is just these algebras and their quotients which we need to analyze pathfinding problems characterized by the need to obtain multiple paths even when path preferences are inconsistent, and paths can be filtered out arbitrarily, as in Internet routing.

1 Introduction

For finding optimal paths in graphs and networks, there is a standard theory grounded in linear algebra [2], [3], [4], [11], [15].

But for certain kinds of pathfinding, including some which are important for Internet routing, it seems to be difficult to take advantage of this theory. These situations are problematic because the information sought may not be a single path, because the criteria for path quality may not result in the existence of an optimal solution, or because the routing algorithms are implemented in a distributed and asynchronous fashion. All of these are difficult to incorporate into the theoretical model.

Nonetheless, in recent years there has been an effort to bring the algebraic theory up to speed with the strange and diverse nature of Internet routing. The mathematical language is ultimately not too different, in terms of the signatures of the algebraic objects involved: we will always need some way to compare paths, and some way to compose them out of arcs. The difference comes in the axioms and derived properties that these structures might have. Whereas conditions such as distributivity have historically been assumed for rings and semirings, our new structures may lack distributivity but instead be endowed with other helpful properties [5]. Analogous methods can then be used to treat their structure theory, and in particular the way that important properties are derived compositionally. From a theoretician's perspective, this demonstrates that the unusual features of Internet pathfinding are not so unusual after all, since they are amenable to similar correctness analysis as in the familiar case.

This paper is about the use of congruences as a definitional tool for these new routing algebras. Of course, congruences and quotients are part of the standard abstract algebraic apparatus for familiar structures; the theory of varieties

H. de Swart (Ed.): RAMICS 2011, LNCS 6663, pp. 180–195, 2011.

yields profound insights into how equational properties relate to algebraic constructions. The surprise for our structures is that our most important property is in fact inequational, but relates well with quotient constructions even so.

In particular, we apply congruences to practical problems including the finding of multiple paths, in the presence of filtering, all the while in a world where path preferences do not follow the usual semiring model, but instead satisfy alternative stability criteria.

Much of this material derives from the first author's doctoral thesis [8].

2 Internet Pathfinding in the Abstract

We first summarize the algebraic approach to the analysis of Internet routing, and relate it to the particular features of that problem which differ from more conventional pathfinding.

From a theory perspective, the main issue with interdomain routing is that it is not in fact solving a shortest path problem, and so the usual mathematical apparatus cannot be applied [6]. The use of semirings and related structures for finding best paths in a labelled graph is well understood. The general pattern is that a semiring $(S, \oplus, \otimes)$ can be used to encode path preferences: the elements of S label the arcs of the graph; these are composed with the binary operator $\otimes$ to form path weights, and the best paths emerge when alternatives are summarized with the $\oplus$ operator.

Algorithms based on this pattern solve the best-path problem by computing the closure A^* of the adjacency matrix A of the graph: a matrix whose entries come from S and where matrix addition and multiplication are defined in terms of the operators of S. The facts that the (i, j) entry of

$$A^* = I + A + A^2 + A^3 + \cdots \tag{1}$$

contains the weight of the optimal path from node i to node j, and that this matrix can be computed in finite time, depend on algebraic properties of S. In particular, a distributive law is required for the two operators:

$$\forall a, b, c \in S : c \otimes (a \oplus b) = (c \otimes a) \oplus (c \otimes b). \tag{2}$$

Commonly-considered semirings for this purpose include $(\mathbb{N}, \min, +)$ for computing shortest paths, $(\mathbb{N}, \max, \min)$ for computing widest paths, and $(\mathbb{N}, +, \times)$ for counting paths.

The distributive law encodes the idea that the path choice made by a node (between paths a and b) will be compatible with that made by a neighbour (between $c \otimes a$ and $c \otimes b$). If the first node always behaves as the neighbour would want, then it is algorithmically acceptable for it to make that choice greedily. Therefore, the algorithms of Dijkstra or Bellman-Ford really do compute the best paths, while avoiding the need to enumerate all paths.

Unfortunately, this compatibility of preferences does not hold for Internet pathfinding, where nodes may be controlled by entities in commercial conflict.

In such a situation, it may be that when some node chooses path a over b, its neighbour would have preferred it to take b instead, since $c \otimes b$ could be better than $c \otimes a$. Due to the requirements of hop-by-hop forwarding, a node has no option but to endure its neighbours' choices.

Remarkably, optimal paths can still be computed in this setting, in an efficient manner—as long as we change our definition of optimality. We no longer require a path assignment that is a global optimum, but only a Nash equilibrium, meaning a state from which no node has any incentive to change its current choice of path. A state X in Nash equilibrium can be characterised as a fixed point of

$$X \mapsto AX + I. \tag{3}$$

This operation entails taking each (i, j)-path in X, extending them along the arcs represented in A, and then choosing the best (according to the criteria encoded in the algebra S); the addition of the identity matrix I ensures that the empty path from each node to itself will always be present in the solution. So for a fixed point, we have $X = AX + I$, meaning that when the extension and choice is carried out, each path is the same as it was before. In game-theoretic terms, no (i, j) has any incentive to deviate from its assigned path in X, assuming a game where the only choice is among the paths made available by neighbours. Notably, this equation is the same one which characterises global optimality for shortest paths, if S is the shortest-path semiring. In the wider context, it still represents an optimum: but a local optimum rather than a global one.

To compute such an equilibrium, we simply use the same matrix iteration as in the shortest-path case, with the exception that the underlying algebra is *not* a semiring obeying the distributive law. While this iteration is not guaranteed to terminate in the absence of distributivity, there are other correctness conditions that are sufficient, and are also consistent with the nature of Internet routing. One such is the *strict inflationary* property

$$\forall a, c \in S : a = a \oplus (c \otimes a) \neq c \otimes a, \tag{4}$$

combined with a finite support condition. Even if the distributive law does not hold for a given semiring, this law ensures the existence of a unique fixed point, to which the iterative algorithm converges after finitely many steps, from any starting state [8].

The finite support condition mentioned above is essential for the 'finitely many steps' part of the result. Without this, the possibility remains that the iteration could continue forever, converging towards a state that could never actually be reached. In path computation, it is enough to restrict our set of paths to a some finite subset of the set of all paths in the given graph: so whenever a path arises in the dynamic computation that is outside this set, it should be excluded from consideration. A reasonable choice would be the set of all simple paths in the graph. In terms of weights rather than paths, the condition is that there be only finitely many permitted path weights. Exactly how this kind of condition can be achieved is one of the major topics of this paper, and is explored in Sections 4.1 and 4.2. In brief, the idea is to go from a possibly-infinite semiring-like structure to a finite one, by taking a congruence that identifies all forbidden paths. The convergence theorem can then be stated with the simple precondition that the given algebraic structure be finite.

A further wrinkle is that the semiring multiplication needs to be replaced with function application, if we are to be capable of expressing the diversity of Internet routing configurations. So rather than dealing with semiring-like structures, we are in fact going to use either *order transforms* $(S, \preceq, F)$ or *semigroup transforms* $(S, \oplus, F)$, where S and $\oplus$ are as before; $\preceq$ is a preorder on S; and F is a set of functions from S to S. These respectively generalize ordered semigroups (the semigroup $\otimes$ being replaced by F) and algebras of monoid endomorphisms (except that our functions need not be endomorphisms). In calculations, the functions F are attached to arcs whereas values in S are originated at nodes: path weights are calculated by applying the functions in order to the starting value. The weights can then be compared with $\preceq$ or summarized by $\oplus$, as appropriate. The analogous *strict inflationary* properties here are

$$\forall a \in S, f \in F: \quad a \prec f(a) \tag{5}$$
$$\forall a \in S, f \in F: \quad a = a \oplus f(a) \neq f(a) \tag{6}$$

respectively.

The reason for using these functions is to permit a wide range of possibilities for how path weights can be derived from arc weights. In routing protocols, a multiplicity of attributes are associated with each route: these are calculated in potentially very complex ways, to allow network operators to exercise fine-grained control over the eventual degree of preference each path will receive.

Some options for how functions in F could operate on route data include:

- Adding a numeric arc weight to the path weight.
- Applying 'bottleneck' bandwidth to the bandwidth of a path.
- Adding a node identifier to a list or set.
- Adding a node identifier, *but* also eliminating the path from consideration if that identifier was already present.
- Adding 'community' tags to remotely influence route choice.
- Importing a route from one routing protocol to another, translating route attributes as appropriate.
- Marking routes based on the business relationship between the systems at either end of the arc.

The elements of S may also be *sets* of paths, or other structured collections. In this case, the functions in F apply to the entire set: they can do any of the above operations on a per-path basis, but can also operate on the set as a whole. For example, the set could be reduced to a single best member path, in some way; and that method need not be the same everywhere in the graph.

In structuring the algebraic theory, we have to consider this complexity, and ideally find ways of making it not matter. This involves the development of constructions, that are justified both theoretically and practically, for building algebras from simpler components. With a good choice of constructions, the task of deciding whether a particular algebra has the required correctness conditions should not be too difficult at any stage. This, we believe, is the case for our congruence-based constructions, which are theoretically pleasant, have a good computational interpretation, and are useful for several problems which arise in the modelling of Internet routing.

3   Congruences

The notions of congruence and of quotient are critical to the structure theory of many abstract algebraic objects, including semigroups and semirings. The general picture is that a congruence is an equivalence relation that is compatible with the operations of the object; this makes it possible to lift those operations to deal with equivalence classes rather than elements, thus forming the quotient algebra [7].

Definition 1. *An equivalence relation $\sim$ on semigroup $(S, \oplus)$ is a congruence if*

$$a \sim b \implies (a \oplus c) \sim (b \oplus c) \land (c \oplus a) \sim (c \oplus b)$$

Definition 2. *If $(S, \oplus)$ is a semigroup and $\sim$ is a congruence on S, then the quotient $(S/\sim, \oplus/\sim)$ is also a semigroup. Here, $S/\sim$ is the set of $\sim$-classes. If the class of a is denoted by $[a]$ then the operation of the new semigroup is*

$$[a] \oplus/\sim [b] = [a \oplus b]$$

The fact that $\sim$ is a congruence makes this operation well-defined and associative.

The point of these congruences is that in many cases, properties of $S/\sim$ can be related to properties of S. This is important for understanding Internet routing from an algebraic perspective: it would be convenient if our key correctness properties were preserved under taking congruences, and if congruences turned out to be useful for modelling certain details of Internet pathfinding problems.

Unfortunately, the inequality in our strict inflationary condition means that a quotient algebra is not guaranteed to have that property, even if the starting algebra did. However, there are some important cases where we are able to use congruences to define new algebras with this property being preserved.

We see this most clearly when considering multipath routing; that is, the idea that for each source and destination we want to find as many good paths as possible, as opposed to a single best path. Algebraically, we just need to choose S to contain not path weights, but sets of path weights, and lift the other operations in the obvious way. We prefer to think of this process as a *construction* on S, because that allows us to examine the relationship between the single-path and multipath cases.

There are other ways of dealing with the presence of multiple best paths. One could also use a conventional single-path algorithm, with some rule for discriminating between otherwise equivalent paths. For example, either the oldest or the newest path seen could be selected; though these methods introduce undesirable nondeterminism into the path selection process, making the correctness much less tractable to analyze algebraically. Alternatively, a partial order on paths could be linearized to a total order: this amounts to the introduction of some deterministic tiebreaking method. But this does not suffice even for conventional shortest-path finding, since we can construct order transforms which have one of the required correctness properties (monotonicity), but where no linearization has this property ([8], Theorem 3.1).

In the end, the most serious criticism of any of these ideas is that for some purposes, we want to receive multiple routes. Trying to force the use of a single-path algorithm would be inappropriate: a case of solving the wrong problem. The failure of these strategies should make use of our construction more attractive, provided that it does have the right algebraic properties. So we now need to understand how to define algebras that make use of this idea, and how these behave in terms of the properties we need for correctness.

In the case of an order transform $(S, \preceq, F)$, we want to derive the *algebra of minimal sets* of S, written **minset**(S). The elements will be subsets of S under the condition that everything in a set is either equivalent or incomparable under $\preceq$. We can also define a lifted version of F, and endow this structure with a semilattice join operation (or an equivalent partial order). This amounts to a free distributive lattice construction. In other words, to obtain **minset**(S) we

1. form the power set of S, which is a distributive lattice under inclusion,
2. take a quotient of this lattice by a congruence derived from $\preceq$; this yields the required order or binary operator,
3. and define lifted versions of the functions in F.

Later, we will vary the second step to obtain other useful constructions. These first two steps, taken together, result in the formation of the distributive lattice corresponding to upper sets in S, as in the representation theorem of Birkhoff [1].

Theorem 1 (Birkhoff's theorem). *A finite distributive lattice is isomorphic to the lattice of upper sets of the partial order of its meet-prime elements.*

We quote the theorem in this form (using meet-prime rather than join-prime elements, and upper sets rather than lower sets) because it is the most directly applicable to our purpose, given the conventional interpretation of path preference where $a \prec b$ means that a is *more* preferred than b.

If $(S, \preceq)$ is a partial order, then we can form a corresponding free distributive lattice, whose elements are the upper sets of S, and where the order is the subset order. If the partial order, moreover, has no infinite descending chain, then an equivalent construction takes all sets of the form

$$A = \min(A) \tag{7}$$

where

$$\min(A) = \{x \in A \mid \forall y \in A : \neg(y \prec x)\}. \tag{8}$$

The equivalence comes from the fact that this min operation determines the same congruence as the taking of upper sets [8].

We end up with the same distributive lattice (up to isomorphism). The difference is that using min is a more natural representation of sets for our path problems: $\min(A)$ will (for us) always be a finite set, and its elements have an obvious interpretation as the 'best' things in A. As an alternative, use of a well-quasi-order would guarantee that $\min(A)$ was always a finite set, nonempty unless A were empty [9]. This is because well-quasi-orders, in addition to the lack of infinite descending chains, also lack infinite antichains.

The fact that we have a distributive lattice allows us to deduce immediately that certain computationally useful facts are true of min. In particular, we have

$$\min(A) = \min(\min(A)) \tag{9}$$
$$\min(A \cup B) = \min(\min(A) \cup B) \tag{10}$$

for all A and B. These will influence the implementation of algorithms, by allowing applications of min to be elided in some circumstances.

Essentially the same construction can be carried out if $(S, \preceq)$ is only a preorder. We again obtain 'minimal sets' of elements of S, and a join operator

$$A \oplus B = \min(A \cup B). \tag{11}$$

The functions f in F are lifted to

$$f(A) = \min\{f(a) \mid a \in A\}; \tag{12}$$

note the use of min to put the result set into canonical form. This completes the construction of **minset**(S) for an order transform S: the resulting structure is suitable for use in path algorithms. Sets of paths are combined, via $\oplus$, by finding the best paths out of either set; the f functions now operate on every path in the given set, and only the best paths are allowed to remain.

This idea of *canonicalization* is central to our understanding of congruence-based constructions. Beginning with min, we can derive an equivalence relation

$$A \sim B \iff \min(A) = \min(B) \tag{13}$$

on subsets of S, and so obtain the appropriate distributive lattice by a quotient of the free lattice. The point is that the min operator is a natural one from the perspective of pathfinding algorithms, but it is not the only choice. In general, whenever we have a way of putting elements of S into a canonical form, we would like to be able to derive a congruence so that a version of the above construction can be applied. This is not always possible, but there are sufficient conditions on the canonicalization function which ensure that the derived equivalence relation is a congruence. In fact they are the same as the properties of min from above.

Definition 3. *If $(S, \oplus)$ is a semigroup and r is a function from S to S such that*

1. $r(a) = r(r(a))$
2. $r(a \oplus b) = r(r(a) \oplus b) = r(a \oplus r(b))$

for all a and b in S, then r is a reduction [13],[14].

In the case of a monoid, the first of these axioms is not needed, since we already have $r(a \oplus 1) = r(r(a) \oplus 1)$ from the second axiom, where 1 is the identity for $\oplus$. Similarly, the second axiom can be simplified to a single equality in the case of a commutative semigroup.

A function on a semiring is called a reduction if it is a reduction with respect to both of the semiring operations. Similarly, a reduction on a semigroup transform $(S, \oplus, F)$ is a function r from S to itself, such that r is a reduction on $(S, \oplus)$ and

$$r(f(a)) = r(f(r(a))) \tag{14}$$

for all a in S and f in F. (This replaces the second axiom from Definition 3, for the multiplicative part of the structure.)

A reduction might also be an endomorphism on a semigroup (and similarly, on a semiring), if it additionally satisfies

$$r(a \oplus b) = r(a) \oplus r(b) \tag{15}$$

for all a and b in the carrier set. Furthermore, not every endomorphism of a semigroup will be a reduction, since not all endomorphisms are idempotent.

The min operation with respect to a preorder $(S, \preceq)$ is a reduction on the semigroup $(2^S, \cup)$. Note, however, that it is not a homomorphism. For any function f on S, and any $A \subseteq S$, we also have

$$\min \{f(x) \mid x \in A\} = \min \{f(x) \mid x \in \min(A)\} \tag{16}$$

which demonstrates that min is always a semigroup transform on $(2^S, \cup, F)$, no matter which set of functions F is used.

We now show that a canonicalization or reduction operation defines a congruence, and that conversely every congruence can be used to define a reduction. This also demonstrates that although endomorphisms are not generally reductions, it is always possible to find a reduction that generates the same congruence as a given endomorphism.

Lemma 1. *For any reduction r on $(S, \oplus)$, define a relation $\sim_r$ on S by*

$$a \sim_r b \;\overset{\text{def}}{\Longleftrightarrow}\; r(a) = r(b).$$

This $\sim_r$ is a congruence.

Proof. This is obviously an equivalence relation. To prove that it is a congruence, suppose that $a \sim_r b$, so that $r(a) = r(b)$. Then

$$r(a \oplus c) = r(r(a) \oplus c) = r(r(b) \oplus c) = r(b \oplus c)$$

and likewise for $r(c \oplus a) = r(c \oplus b)$. Hence $\sim_r$ is indeed a congruence. $\qquad \square$

We can also produce a reduction from a congruence. In fact, there will typically be many choices of reduction for a different congruence. Between S and $S/\sim$ there is a homomorphism $\rho^\natural$ called the *natural map*, taking each element of S to its $\sim$-equivalence class. If we choose a function going in the other direction, taking each equivalence class to some representative element within the class, then the composition of these two functions will be a reduction. The choice of representatives means that there may be multiple reduction functions, although they all correspond to the same congruence and define the same equivalence classes.

Lemma 2. *Let $(S, \oplus)$ be a semigroup, $\sim$ a congruence, and $\rho^\natural$ the natural map. If $\theta : S/\sim \longrightarrow S$ is such that $\rho^\natural \circ \theta = id$, then $\theta \circ \rho^\natural$ is a reduction; and $\sim$ is equal to $\sim_{\theta \circ \rho^\natural}$.*

Proof. Note that the condition $\rho^\natural \circ \theta = id$ simply expresses that the representative for a class should be an element of that class. There is always at least one such θ, because there can be no empty classes. This condition also provides that θ must be one-to-one, for if $\theta(P)$ and $\theta(Q)$ are equal, then $(\rho^\natural \circ \theta)(P)$ and $(\rho^\natural \circ \theta)(Q)$ must also be equal; and then $P = Q$.

Now, $\theta \circ \rho^\natural$ satisfies the axioms for a reduction. Firstly, it is idempotent:

$$(\theta \circ \rho^\natural)^2 = \theta \circ (\rho^\natural \circ \theta) \circ \rho^\natural = \theta \circ \rho^\natural.$$

The second reduction axiom is also fulfilled

$$
\begin{aligned}
(\theta \circ \rho^\natural)(a \oplus b) &= \theta(\rho^\natural(a) \oplus \rho^\natural(b)) &&\text{since } \rho^\natural \text{ is a homomorphism} \\
&= \theta(\rho^\natural(\theta(\rho^\natural(a))) \oplus \rho^\natural(b)) &&\text{since } \rho^\natural \circ \theta = id \\
&= (\theta \circ \rho^\natural)((\theta \circ \rho^\natural)(a) \oplus b) &&\text{since } \rho^\natural \text{ is a homomorphism.}
\end{aligned}
$$

and symmetrically for the second equality.

Furthermore, the congruence derived from this reduction is $\sim$ again:

$$
\begin{aligned}
a \sim_{\theta \circ \rho^\natural} b &\iff \theta(\rho^\natural(a)) = \theta(\rho^\natural(b)) \\
&\iff \rho^\natural(a) = \rho^\natural(b) &&\text{since } \theta \text{ is one-to-one} \\
&\iff a \sim b &&\text{by definition of the natural map.}
\end{aligned}
$$

Hence for any congruence there is at least one equivalent reduction. $\qquad\square$

We can therefore choose to represent any reduction r as a pair $(\sim, \theta)$, since this is enough to determine all of the values of the function. The interpretation of reductions in terms of congruences is helpful because it clarifies the true role of a reduction as well as often being more algebraically useful. A reduction is not an arbitrary transformation that fulfils some unusual axioms, but instead arises as the combination of a congruence—to say which distinctions between elements are being ignored—and a choice of representative element from each equivalence class. In some contexts, the reduction function may be the more natural way of thinking about the operations being modelled. This justifies using the reduction idea in the first place, as opposed to making use of congruences throughout. The use of a functional viewpoint rather than a relational one may be more natural from the point of view of implementing a routing protocol, because it provides a direct answer to the question of how to deal with route data. On the other hand, the algebraic theory associated with congruences is much more extensive, which suggests that they should be the preferred representation when trying to prove facts about these algebraic structures.

In terms of algebraic constructions, the picture is that for a given *reduction* on one of our algebraic objects we can define the corresponding *congruence* and therefore the *quotient*.

Specifically, for a given $(S, \oplus, F)$ and reduction $r : S \longrightarrow S$ we can define the quotient S/r as follows.

1. The carrier consists of r-equivalence classes of elements of S; we can choose the canonical representative of each class to be a fixed point of r.
2. The semigroup operation is given by $\rho^\natural(a) \oplus/r \, \rho^\natural(b) = \rho^\natural(a \oplus b)$.
3. The functions in F are lifted: $f(\rho^\natural(a)) = \rho^\natural(f(a))$.

This can be verified to be a semigroup transform. The **minset** construction is clearly a special case, where r is min, S is a set of sets, and $\oplus$ is set union.

4   Applications in Routing

Aside from the use of min-like operations, our main application of congruences is in the handling of pathfinding errors. In practical situations, it is often not enough to have an algorithm simply throw its hands up and declare that no suitable solution exists. Instead, we would like to retain detailed information about what kinds of errors occurred. For example, in interdomain routing there are several reasons why a path might be considered erroneous:

- The same node is visited more than once.
- The path is intended to be filtered out.
- The path violates known economic relationships between networks.
- The path is too long (exceeding a maximum size for routing announcements).
- The origin is unexpected (given neighbours are only anticipated to advertise certain addresses).
- Route data is otherwise malformed.

Any or all of these could be true of a given path.

We believe that from a correctness point of view, it is not enough to sweep all of these under the implementers' rug. Many of the anomalies we observe in Internet routing today can be traced back to the handling of erroneous routes. Error handling is an integral part of the path selection process, and must be dealt with in the algebraic model, just as we deal with ordinary, non-erroneous routes. If not, then the correctness result we obtain is merely 'As long as nothing bad happens, protocol convergence is guaranteed', whereas we would prefer to be in a position to make stronger statements about the resilience of the routing system even in the presence of errors.

A reduction operation is a suitable way to begin encoding error-handling. These functions are all about putting route data into a canonical form: this includes mapping certain routes to error values. In an algebra which includes such values, less preferred than 'ordinary' routes, we obtain the desired behaviour automatically. Erroneous routes are removed from consideration, since they cannot ever be more preferred than a non-erroneous route. Information about the error can still be propagated through the algorithm, enabling diagnosis, but this propagation is suppressed if an alternative route exists. All of this is totally compatible with multipath routing, via **minset** and related constructions.

The safety of these schemes depends on the interaction between

- the nature of path preferences;
- the operations extending paths; and
- the reduction function.

In the remainder of this section, we examine some simple examples of how the language of reductions and congruences can be used to prove required safety properties.

4.1   Forbidden Paths

Presentations of pathfinding algorithms tend to focus on computation of path *weight*, as opposed to returning the identity of each path. In the case of Dijkstra's algorithm, for example, a simple modification allows the recording of path information alongside weight information: this path information is not used while the algorithm is running, but is an additional output. But in our context, the degree of preference associated with a path depends upon the identity of that path—the nodes and arcs that make it up. In particular, we need to exclude, explicitly, paths that are not *simple*, whereas for conventional shortest path problems, this happens automatically. So we will, by default, want to include path information as part of the algebra.

Other paths may also be forbidden, even if they are simple. Network operators are able to make essentially arbitrary decisions about which paths will be unacceptable to them: in protocol implementations, they can be excluded from consideration as soon as they are received. This is equally the case in a multipath context.

Both of these cases can be handled by defining appropriate reductions. The obvious alternative would be to modify each algorithm to have the required behaviour, rather than seeking to encode this within the algebra. The problem with this idea is that it breaks the relationship with the theory of pathfinding based on linear algebra: if this link is not maintained, then we can no longer take advantage of existing theory in understanding the termination or efficiency of algorithms. In terms of convergence proof, our experience has been that it is a great help to make the algorithm as generic as possible, eliminating special cases by putting them into the algebra instead.

The general principle is to define a reduction which will eliminate forbidden paths, by mapping them onto a greatest element. This mirrors the conventional shortest-path model, where non-existent paths are given 'infinite' length. Because any path that is actually present will have finite length, these infinities will only persist in the algorithm if there is no path connecting the nodes in question. Equally, our forbidden paths will be worse than any permitted path, regardless of any of their other merits.

If $(S, \oplus, F)$ is a semigroup transform, with $\oplus$ commutative and having identity 0, and E is a subset of S containing 0, then define a function r_E on S by

$$r_E(x) = \begin{cases} x & x \notin E \\ 0 & x \in E. \end{cases} \tag{17}$$

For this to be a reduction, it is required that E satisfies the property

$$\forall e \in E, x \in S : (x \in E \wedge e \oplus x \in E) \vee (x = e \oplus x). \tag{18}$$

It is then possible to define a new structure based on r_E. This criterion makes operational sense. It states, in effect, that the forbidden paths have to be worse than the non-forbidden paths: if x does not emerge from $e \oplus x$, then all of e, x and $e \oplus x$ are in the error set. So if we forbid a certain path, then we also have to forbid any path for which it is a prefix: once in the error set, we cannot get out.

Definition 4. *Let* $\mathbf{err}(S, E)$ *be the semigroup transform* $(S_E, \oplus_E, F_E)$, *where*

- S_E *consists of those elements of* X *for which* $r_E(x) = x$,
- $x \oplus_E y$ *is* $r_E(x \oplus y)$, *and*
- F_E *consists of functions* f_E *for each* f *in* F, *and* $f_E(x) = r_E(f(x))$.

This $\oplus_E$ can be verified to be associative, since r_E is a reduction. The other properties of $\mathbf{err}(S, E)$ will depend on the choice of S and E.

We have reduced the error set E to a single element in the quotient. Anything in E is mapped to 0, the topmost element of the order; consequently, forbidden paths will be excluded from consideration, in favour of non-forbidden paths of any quality. This mapping is associated with each arc; operationally speaking, this means that on import or export, the forbidden paths are removed from the candidate set.

The congruence associated with such a reduction is related to the notion of a Rees congruence on a semigroup. A subset E of $(S, \oplus)$ is an *ideal* if

$$\forall x \in S, e \in E : (x \oplus e \in E) \wedge (e \oplus x \in E). \tag{19}$$

For a given ideal E, the relation

$$x \sim_E y \xleftrightarrow{\text{def}} x = y \vee (x \in E \wedge y \in E) \tag{20}$$

is a congruence, called the *Rees congruence* with respect to E [7]. In the case of our r_E, the congruence may not be a Rees congruence because E may not satisfy (19). This is in line with our general principle of not enforcing conditions which can be inferred: the definition of **err**(S, E) makes sense even when E is not an ideal, though it may not have desirable properties.

The relationship between reductions and congruences suggests that other representations of **err**(S, E) are possible. Specifically, we could preserve some information about the forbidden path, rather than limiting the available data to merely 'an error occurred'. As long as the correct rules are followed for F and $\oplus$, no difficulty arises. That is, we have to ensure that whatever representation we choose is equivalent under r to the semigroup transform **err**(S, E) above. Instead of mapping everything in E to a single 0, we could have many possibilities, drawn from a subset A of S. This will be acceptable if A is an upper set of S, and if r maps elements of A to elements of A. The correctness argument is the same, but the resulting solution state is perhaps more informative than previously, in the case when the only available path from i to j was forbidden.

4.2   Only Simple Paths

In the multipath setting, a slightly different definition is necessary. We will show an example of how to ensure that only simple paths emerge from the algorithm. The standard algebra of paths is to order them by length: we have a preference relation rather than a semilattice. A variation on the **minset** construction will convert such an algebra into one which can be used in the context of matrix operations.

Let P be the algebra of paths $(N^*, \preceq, C)$, where $p \preceq q$ if and only if $| p | \le | q |$, and C consists of functions c_n for all n in N, which concatenate the node n onto the given path. Let $(S, \le, F)$ be an order transform, which will be responsible for encoding the path weights.

Now, let E be the subset of $S \times N^*$ consisting of those pairs which contain a non-simple path:

$$\{(s, p) \in S \times N^* \mid p \text{ is not simple}\}. \tag{21}$$

The **err** construction cannot be used directly, since E does not satisfy the required property (18). However, there is a reduction which can be used over subsets of $S \times N^*$. Let r be the function

$$r(A) \stackrel{\text{def}}{=} \min(A \setminus E); \tag{22}$$

where min uses the lexicographic order on $S \times N^*$; this satisfies the reduction axioms. It is also operationally consistent with the view of path filtering wherein forbidden paths are removed first, with best-path selection applied to the remainder [12].

Consequently, a semigroup transform can be constructed where

- the elements are those subsets of $S \times N^*$ which are fixed points of r;
- the operation $\oplus$ is given by $A \oplus B \stackrel{\text{def}}{=} r(A \cup B)$; and
- the functions are pairs (f, c_n) for f in F, where

$$(f, c_n)(A) \stackrel{\text{def}}{=} r(\{(f(s), c_n(p)) \mid (s, p) \in A\}).$$

It can be seen that this algebra implements the simple paths criterion in the case of multipath routing: if during the course of computation a non-simple path is computed, it and its associated S-value will be removed from the candidate set.

It is possible to prove that the restriction to simple paths, together with the strict inflationary condition on S, suffice to ensure algorithmic convergence to a unique fixed point [8]. That is, the straightforward algorithm where every node periodically communicates its best paths to its neighbours, and updates its local best path data based on path information received from neighbours, is guaranteed to terminate; moreover, the final state will be a pure Nash equilibrium in the sense of Section 2, and is unique. Indeed, this convergence is guaranteed from *any* starting state, and so the algorithm can be considered to be self-synchronizing to the extent permitted by the nature of the underlying inter-node communication.

5   Algebraic Correctness in Finite Structures

The distinction between the finite and the infinite is of considerable practical importance in network routing. We have already discussed how convergence in a finite number of steps is greatly to be preferred. Another issue in correctness analysis where this distinction arises is in consideration of finite data domains. We almost invariably use the infinite to approximate the finite, working with an idealized, infinite algebraic structure such as $(\mathbb{N}, \min, +)$ for shortest paths, when the actual reality is that routing protocols only allow the expression of a finite number of distinct path lengths. In the case of the Routing Information Protocol (RIP), this finite number is fifteen [10].

The problem for algebraic analysis is that it is much easier to prove results about the infinite structures; indeed, the corresponding 'theorems' for finite structures may even be false. For example, we know that for the lexicographic product **lexprod**(S, T) of two semigroup transforms to be distributive, it suffices for S and T to be individually distributive, and for S to be cancellative, meaning that if $f(a) = f(b)$ then $a = b$, for any f in the function set of S. Addition of integers is a perfectly acceptable cancellative operation. But addition with a finite maximum value is not. On a given graph, our iterative algorithm could fail to reach a global optimum, due to lack of distributivity associated with this upper limit being reached. In particular, the problem would be that some node could be left with the value (∞_S, x) rather than the actual global optimum (∞_S, y), where $y \prec_T x$ according to the order $\preceq_T$ of T, and ∞_S denotes the maximum element of S. This is only a limited form of failure, especially since the termination of the iteration still occurs, but it does seem to undermine the promise of the algebraic method for guaranteeing correctness of pathfinding.

As an aside, the infamous 'counting to infinity' problem of RIP, whereby the protocol could take a long time to adapt to loss of connectivity, is *not* a product of the handling of 'infinity' within RIP. Rather, it derives from the fact that routing information includes the weight of a path but not its identity, and that it is therefore possible for nodes to adopt cyclic paths without realizing. The cycles grow longer and longer, until the limit of sixteen is reached, this 'infinity' denoting the absence of a usable path. If RIP had a more generous notion of infinity, this problem would in fact be even worse, since convergence to the maximum value would take longer.

Returning to proofs of properties, the use of reductions or congruences can ease the difficulty here as well. We can use our **err** operation as part of a larger construction, and trace the correctness properties through. So for an algebra of the form **err**(**lexprod**$(S, T), E)$, we would use our theorems about the lexicographic product to derive properties of **lexprod**(S, T), and then use our theorems about **err** to derive properties of the whole algebra. The existence of these standard constructions allows many cases to be treated uniformly.

In the example above, the real issue is that elements like (∞_S, x) do not in fact denote usable paths: even if the value x is acceptable, the ∞_S is certainly not. Therefore, a way forward is to prohibit such elements from occurring in the computation at all. Take the subset $E = \{\infty_S\} \times T \subseteq S \times T$ and form the algebra **err**(**lexprod**$(S, T), E)$. All of the problematic elements are now identified, meaning that they are no longer barriers to the achieving of a global optimum. We also have a recipe for how to deal with such elements when they crop up in the path computation: map them to a single overall 'infinity' value, effectively by dropping the T component.

It can be shown that an algebra of this form is distributive, if we have a distributivity condition for the appropriate subset of **lexprod**(S, T) (see [8], Theorem 5.9 and Appendix A.5). The condition is that

$$(f, g)(s_1, t_1) \oplus (f, g)(s_2, t_2) = (f, g)\left((s_1, t_1) \oplus (s_2, t_t)\right) \tag{23}$$

for all (f, g) in the function set of **lexprod**(S, T), and all (s_1, t_1) and (s_2, t_2) in the subset $(S \setminus \{\infty_S\}) \times T$ of $S \times T$.

In this way, the required correctness property can be regained, by a modification to the algebra and the use of reduction- or congruence-based theorems.

6   Conclusion

There is an ongoing effort to provide a sound theoretical foundation for Internet routing. While in many cases this task can be tackled on an ad-hoc basis, by writing new definitions and proofs for each proposed routing scheme, a better approach is to provide a general theory which can address several such models. The existing pathfinding theory based on semirings is a sound starting point, but several adaptations need to be made in order to make it applicable to these practical examples.

This paper has demonstrated that several such alterations are more mathematically rich than might be suspected. The apparently awkward 'min' operation has been revealed as having a deep connection with lattice theory and with congruences. Related 'reduction' operations are also susceptible to explanation in terms of congruences. We have also shown that these operations are useful in multipath routing, and for more complex scenarios incorporating route filtering.

The examples in this paper are inspired by interdomain routing. There is considerable scope for applying this theory to the design of future routing systems, so that they can be not only flexible, but also provably correct with reference to an underlying optimization problem.

Acknowledgments

This work was supported by grant EP/F002718/1 from the Engineering and Physical Sciences Research Council (EPSRC). The authors would like to thank Georg Struth and Philip Taylor for their helpful comments.

References

1. Birkhoff, G.: Lattice Theory. American Mathematical Society, Providence (1948)
2. Carré, B.A.: Graphs and networks. Oxford University Press, Oxford (1979)
3. Gondran, M., Minoux, M.: Graphs and algorithms. Wiley, Chichester (1984)
4. Gondran, M., Minoux, M.: Graphes, dioïdes et semi-anneaux: Nouveaux modèles et algorithmes. Tec & Doc, Paris (2001)
5. Griffin, T.G., Gurney, A.J.T.: Increasing bisemigroups and algebraic routing. In: Berghammer, R., Möller, B., Struth, G. (eds.) RelMiCS/AKA 2008. LNCS, vol. 4988, pp. 123–137. Springer, Heidelberg (2008)
6. Griffin, T.G., Shepherd, F.B., Wilfong, G.: The stable paths problem and interdomain routing. IEEE/ACM Trans. Netw. 10(2), 232–243 (2002)
7. Grillet, P.A.: Semigroups: An introduction to the structure theory. Monographs and Textbooks in Pure and Applied Mathematics, vol. 193, Marcel Dekker, New York (1995)
8. Gurney, A.J.T.: Construction and verification of routing algebras. PhD thesis, University of Cambridge (2009)
9. Kruskal, J.B.: The theory of well-quasi-ordering: A frequently discovered concept. J. Combin. Theory Ser. A 13(3), 297–305 (1972)
10. Malkin, G.: RIP version 2. RFC 2453 (1998)
11. Rote, G.: Path problems in graphs. In: Tinhofer, G., Mayr, E.W., Noltemeier, H., Syslo, M. (eds.) Computational Graph Theory. Computing Supplementa, vol. 7, pp. 155–189. Springer, Heidelberg (1990)
12. Wang, Y., Schapira, M., Rexford, J.: Neighbor-specific BGP: More flexible routing policies while improving global stability. In: Douceur, J.R., Greenberg, A.G., Bonald, T., Nieh, J. (eds.) Proceedings of the Eleventh International Joint Conference on Measurement and Modeling of Computer Systems, SIGMETRICS/Performance 2009, pp. 217–228. ACM, New York (2009)
13. Wongseelashote, A.: An algebra for determining all path-values in a network with application to k-shortest-paths problems. Networks 6(4), 307–334 (1976)
14. Wongseelashote, A.: Semirings and path spaces. Discrete Math. 26(1), 55–78 (1979)
15. Zimmermann, U.: Linear and combinatorial optimization in ordered algebraic structures. Annals of Discrete Mathematics, vol. 10. Elsevier North-Holland, Amsterdam (1981)

Towards a Typed Omega Algebra

Walter Guttmann

Department of Computer Science, University of Sheffield, UK
`walter.guttmann@uni-ulm.de`

Abstract. We propose axioms for 1-free omega algebra, typed 1-free omega algebra and typed omega algebra. They are based on Kozen's axioms for 1-free and typed Kleene algebra and Cohen's axioms for the omega operation. In contrast to Kleene algebra, several laws of omega algebra turn into independent axioms in the typed or 1-free variants.

We set up a matrix algebra over typed 1-free omega algebras by lifting the underlying structure. The algebra includes non-square matrices and care has to be taken to preserve type-correctness. The matrices can represent programs in total and general correctness. We apply the typed construction to derive the omega operation on two such representations, for which the untyped construction does not work.

We embed typed 1-free omega algebra into 1-free omega algebra, and this into omega algebra. Some of our embeddings, however, do not preserve the greatest element. We obtain that the validity of a universal formula using only $+$, $\cdot$, $^{+}$, $^{\omega}$ and 0 carries over from omega algebra to its typed variant. This corresponds to Kozen's result for typed Kleene algebra.

1 Introduction

Particular aspects of computations, such as non-termination, are conveniently treated by forming matrices over semirings [14]. A program is represented by a matrix as follows: some of the entries carry information about state transitions and non-terminating executions, while the remaining entries are specific constants. They are chosen and arranged so that matrix multiplication propagates the information as required to model sequential composition.

It is then possible to obtain the Kleene star and omega operations, which underlie the semantics of loops, by standard matrix constructions [3,1,10,13]. Both operations can be derived for the matrices used in total correctness [7], and the Kleene star for the matrices used in general correctness [5]. The approach fails, however, for the omega operation in the latter case: the matrices used in general correctness are not closed under the construction given in [13].

In the present paper we solve this problem by typing the elements of the matrices. As regards the star operation, this means that the underlying structure is a typed Kleene algebra [11]. To deal with the omega operation, we propose a typed omega algebra, based on the untyped axiomatisation of [2].

Section 2 defines the necessary structures. Central to the present paper are (typed) 1-free omega algebras, an extension of the 1-free Kleene algebras of [11],

H. de Swart (Ed.): RAMICS 2011, LNCS 6663, pp. 196–211, 2011.
© Springer-Verlag Berlin Heidelberg 2011

which omit 1 and replace $*$ by $+$. While Kleene algebras are fairly similar to their 1-free variants, we identify several laws of omega algebra as independent axioms of 1-free omega algebra.

In Section 3 we show that finite matrices over typed 1-free omega algebras form (typed) 1-free omega algebras. To this end, we modify the matrix omega operation of [13] to obtain type-correct matrices. Particular subalgebras of the matrix algebra are then used to derive omega for the representation of programs. This works not only for general correctness but also for a recently introduced model that combines it with total correctness [8,6].

In Section 4 we extend results of [11,12], whereby restricted forms of universal statements are valid in the untyped setting if and only if they are valid in the typed setting. In particular, we embed typed 1-free omega algebra into 1-free omega algebra, and the latter into omega algebra. The embeddings require different subsets of axioms, and some do not preserve the greatest element.

Besides the application to program semantics, typed omega algebra can serve the following purposes. The ability to treat non-square matrices is useful for constructions related to automata [10], which indeed motivate typed Kleene algebra, and omega may be used to model infinite executions of the automata. Moreover, typed omega algebra is a subtheory of, and thus may yield insight into, heterogeneous relation algebra [16]; it fits into the hierarchy of [9].

2 Axioms

In this section we give axioms for (typed) (1-free) Kleene and omega algebras. Of these combinations, (typed or 1-free) omega algebras are new.

2.1 Omega Algebra

We recall the axioms of semirings, Kleene algebras and omega algebras. An idempotent semiring is a structure $(S, +, \cdot, 0, 1)$ that satisfies the following axioms:

$$
\begin{array}{lll}
a + (b + c) = (a + b) + c & a(b + c) = ab + ac & a(bc) = (ab)c \\
a + b = b + a & (a + b)c = ac + bc & 1a = a \\
a + a = a & 0a = 0 & a1 = a \\
a + 0 = a & a0 = 0 &
\end{array}
$$

The operation $\cdot$ has higher precedence than $+$ and is frequently omitted by writing ab instead of $a \cdot b$. By $a \leq b \Leftrightarrow a + b = b$ we obtain the partial order $\leq$ on S with join $+$ and least element 0. The operations $+$ and $\cdot$ are $\leq$-isotone.

A Kleene algebra [10] is a structure $(S, +, \cdot, {}^*, 0, 1)$ such that $(S, +, \cdot, 0, 1)$ is an idempotent semiring and the following axioms hold:

$$
\begin{array}{ll}
1 + aa^* = a^* & b + ac \leq c \Rightarrow a^*b \leq c \\
1 + a^*a = a^* & b + ca \leq c \Rightarrow ba^* \leq c
\end{array}
$$

The operation * is $\leq$-isotone and has highest precedence. Every Kleene algebra has the non-empty iteration $a^+ =_{\text{def}} aa^* = a^*a$. It satisfies $a^* = 1 + a^+$ and

$$a + aa^+ = a^+ \qquad\qquad b + ac \leq c \Rightarrow a^+b \leq c$$
$$a + a^+a = a^+ \qquad\qquad b + ca \leq c \Rightarrow ba^+ \leq c$$

The operation $^+$ is $\leq$-isotone and has the same precedence as *.

An omega algebra [2] is a structure $(S, +, \cdot, ^*, ^\omega, 0, 1)$ such that $(S, +, \cdot, ^*, 0, 1)$ is a Kleene algebra and the following axioms hold:

$$aa^\omega = a^\omega \qquad\qquad c \leq ac + b \Rightarrow c \leq a^\omega + a^*b$$

The operation $^\omega$ is $\leq$-isotone and has the same precedence as *. Every omega algebra has a $\leq$-greatest element $\top =_{\text{def}} 1^\omega$. It satisfies

$$a^\omega \top = a^\omega \qquad\qquad a \leq a\top \qquad\qquad \top = \top\top$$
$$a \leq \top \qquad\qquad a \leq \top a$$

We call those axioms of Kleene and omega algebra, which are implications, induction axioms.

2.2   1-Free Omega Algebra

We recall the axioms of 1-free Kleene algebras and introduce 1-free omega algebras. As discussed in Section 5, the restriction to 1-free algebras enables the transfer of universal formulas from the untyped to the typed setting.

A 1-free Kleene algebra [11] is a structure $(S, +, \cdot, ^+, 0)$ that satisfies the idempotent semiring axioms without 1, that is,

$$a + (b + c) = (a + b) + c \qquad\qquad a(b + c) = ab + ac \qquad\qquad a(bc) = (ab)c$$
$$a + b = b + a \qquad\qquad (a + b)c = ac + bc$$
$$a + a = a \qquad\qquad 0a = 0$$
$$a + 0 = a \qquad\qquad a0 = 0$$

and, replacing the *-axioms, the laws about $^+$ mentioned above:

$$a + aa^+ = a^+ \qquad\qquad b + ac \leq c \Rightarrow a^+b \leq c$$
$$a + a^+a = a^+ \qquad\qquad b + ca \leq c \Rightarrow ba^+ \leq c$$

An equivalent structure is obtained by replacing the implications with

$$ac \leq c \Rightarrow a^+c \leq c$$
$$ca \leq c \Rightarrow ca^+ \leq c$$

It follows that the operation $^+$ is $\leq$-isotone.

A 1-free omega algebra is a structure $(S, +, \cdot, ^+, ^\omega, 0, \top)$ such that $(S, +, \cdot, ^+, 0)$ is a 1-free Kleene algebra and the following axioms hold:

$$aa^\omega = a^\omega \qquad\qquad c \leq ac + b \Rightarrow c \leq a^\omega\top + a^+b + b$$

The operation $^{\omega}$ is not $\leq$-isotone in general, but $a \leq b$ implies both $a^{\omega} \leq b^{\omega}\top$ and $a^{\omega}\top \leq b^{\omega}\top$.

Observe the term $a^{\omega}\top$ replacing a^{ω} in the induction axiom to prepare it for the typed setting. We moreover consider the following axioms about $^{\omega}$ and $\top$:

$$
\begin{array}{lll}
(\top 1) \quad a^{\omega}\top = a^{\omega} & (\top 3) \quad a \leq a\top & (\top 5) \quad \top = \top\top \\
(\top 2) \quad\quad a \leq \top & (\top 4) \quad a \leq \top a &
\end{array}
$$

We explicitly state whenever they are used in addition to the axioms of 1-free omega algebra. Except for $(\top 5)$, which follows from $(\top 2)$ and either $(\top 3)$ or $(\top 4)$, these axioms are independent from each other and the axioms of 1-free omega algebra, as counterexamples generated by Mace4 witness.

To improve readability, we use the * notation also in 1-free algebras to abbreviate terms of the form

$$
\begin{array}{ll}
a^*b = a^+b + b & ab^*c = ab^+c + ac \\
ba^* = ba^+ + b & a^*bc^* = a^+bc^+ + a^+b + bc^+ + b
\end{array}
$$

and similar ones, where * occurs in products with at least one 1-free element. For example, the omega induction axiom becomes $c \leq ac + b \Rightarrow c \leq a^{\omega}\top + a^*b$. Due to the semiring axioms, calculations using this notation work as expected. In such contexts * is $\leq$-isotone and the star induction axioms hold.

2.3   Typed 1-Free Omega Algebra

We use the mechanism for typing described in [11]. In particular, we assume a set T of pretypes $s, t, u, v, \ldots$ and obtain the set T^2 of types denoted as $s \to t$. The type of an expression a of omega algebra is denoted by $a : s \to t$ and can be derived using a type calculus with the rules

$$
\frac{a, b : s \to t}{a + b : s \to t} \qquad \frac{a : s \to t \quad b : t \to u}{ab : s \to u} \qquad \frac{a : s \to s}{a^*, a^+, a^{\omega} : s \to s} \qquad \frac{0, \top : s \to t}{1 : s \to s}
$$

The rules for $^{\omega}$ and $\top$ are newly added. Expressions a and b in an equation $a = b$ must have the same type. We also write a_{st} to make clear that a has type $s \to t$.

For example, finite heterogeneous relations are modelled by letting T be the natural numbers. Then $a : s \to t$ denotes that a is a matrix with s rows and t columns. See [11] for further details about the typing mechanism.

A typed Kleene algebra (with pretypes T) is a set S of typed elements $a : s \to t$ $(s, t \in T)$ with polymorphic operators $+$, $\cdot$, *, 0 and 1, typed according to the above inference rules, satisfying all well-typed instances of the Kleene algebra axioms.

Typed 1-free Kleene algebras and typed 1-free omega algebras are defined similarly, using all well-typed instances of the respective axioms in Section 2.2. All well-typed instances of a selection of $(\top 1)$–$(\top 5)$ may be considered besides.

For a typed omega algebra we use all well-typed instances of the omega algebra axioms, except for omega induction, which we replace by the omega induction axiom of 1-free omega algebra $c \leq ac + b \Rightarrow c \leq a^{\omega}\top + a^*b$.

A finitely typed algebra is one with finite T. We denote the set of elements with type $s \to t$ in a typed structure S by S_{st}. An untyped formula is valid in S if all its well-typed instances hold.

Remark. The axiom ($\top 2$) establishes $\top : s \to t$ as the greatest element of type $s \to t$. As in heterogeneous relation algebra, each type has its own greatest element. In the untyped setting, being the greatest element is the main property of $\top$. In the typed setting, emphasis should be on its property to cause a change of types: from $a : s \to t$ we obtain the element $a\top$ of type $s \to u$ by multiplying with $\top : t \to u$. Thus ($\top 5$) decomposes a type cast effected by $\top : s \to u$ into a sequence of two type casts effected by $\top : s \to t$ and $\top : t \to u$.

It is this type changing capacity which is used in the omega induction axiom. This ensures that $a^\omega \top$ is compatible with $a^* b$ also if $b : s \to t$ with $s \neq t$. We have chosen to give a^ω the unique type $s \to s$ for $a : s \to s$, but another approach might give the more general type

$$\frac{a : s \to s}{a^\omega : s \to t}$$

which incorporates the type cast in a (more) polymorphic type of $^\omega$. While this would restore the original form of the omega induction axiom, in the 1-free case additional axioms are required to introduce $\top$. Even if 1 and * are available, it is not possible to derive all well-typed instances of ($\top 5$): Sections 3.2, 3.3 and 5 feature models of typed omega algebra with ($\top 1$)–($\top 4$) but not ($\top 5$).

3   Matrices

In this section we consider finite matrices over typed omega algebras. A matrix algebra is obtained by lifting the underlying structure. This is known for Kleene algebra [3,10], typed Kleene algebra [11] and omega algebra [13]. For typed omega algebra, some modification is required to get the typing right.

The result shows how to obtain the omega operation for typed matrices. We apply it to matrix representations of programs in total and general correctness.

3.1   Matrices over Typed 1-Free Omega Algebra

Fix a typed 1-free omega algebra S with (not necessarily finite) pretypes T. We construct a typed 1-free omega algebra of finite matrices whose entries are elements of S. The pretypes of this matrix algebra are the finite sequences over T. Let $s_1, \ldots, s_m \in T^m$ and $t_1, \ldots, t_n \in T^n$ be pretypes, then a matrix has type $s_1, \ldots, s_m \to t_1, \ldots, t_n$ if and only if its size is $m \times n$ and, for each $1 \leq i \leq m$ and $1 \leq j \leq n$, the entry in row i and column j has type $s_i \to t_j$.

The operations $+$, $\cdot$, 0 and $\top$ are, as usual, the componentwise sum, the matrix product, the 0- and the $\top$-matrix, respectively. The non-empty iteration $^+$ is defined by $(a)^+ = (a^+)$ for 1×1 matrices and, inductively,

$$\begin{pmatrix} a & b \\ c & d \end{pmatrix}^+ = \begin{pmatrix} e^+ & a^* b f^* \\ d^* c e^* & f^+ \end{pmatrix} \quad \text{with} \quad \begin{pmatrix} e \\ f \end{pmatrix} = \begin{pmatrix} a + b d^* c \\ d + c a^* b \end{pmatrix}.$$

This is derived by $A^+ = AA^*$ from the usual matrix * of [3]. It is implicitly used in [11, Lemma 4.1], asserting that the resulting structure of square matrices satisfies the axioms of 1-free Kleene algebra. It is not difficult to check the axioms also for non-square matrices, hence we obtain a typed 1-free Kleene algebra.

The infinite iteration $^\omega$ is given by $(a)^\omega = (a^\omega)$ for size 1×1 and, inductively,

$$\begin{pmatrix} a & b \\ c & d \end{pmatrix}^\omega = \begin{pmatrix} e^\omega & a^*bf^\omega \\ d^*ce^\omega & f^\omega \end{pmatrix} \begin{pmatrix} \top & \top \\ \top & \top \end{pmatrix} \quad \text{with} \quad \begin{pmatrix} e \\ f \end{pmatrix} = \begin{pmatrix} a + bd^*c \\ d + ca^*b \end{pmatrix}.$$

It is instructive to reflect on the type of the involved expressions. By its typing rule, the $^\omega$ operation is applied to a square matrix; its pretype is a finite sequence $t_1, \ldots, t_n, t \in T^{n+1}$. For the inductive step, we take away the last element t of this sequence, denote by $s = t_1, \ldots, t_n$ the remaining ones, and split the matrix accordingly into

$$\begin{pmatrix} a : s \to s & b : s \to t \\ c : t \to s & d : t \to t \end{pmatrix}.$$

Observe that e and f have the same types as a and d, respectively. Hence e^ω, a^*bf^ω, d^*ce^ω and f^ω have the types of a, b, c and d, respectively. The four $\top$ entries of the remaining matrix have these types as well:

$$\begin{pmatrix} a & b \\ c & d \end{pmatrix}^\omega = \begin{pmatrix} e^\omega & a^*bf^\omega \\ d^*ce^\omega & f^\omega \end{pmatrix} \begin{pmatrix} \top_{ss} & \top_{st} \\ \top_{ts} & \top_{tt} \end{pmatrix}$$
$$= \begin{pmatrix} e^\omega \top_{ss} + a^*bf^\omega \top_{ts} & e^\omega \top_{st} + a^*bf^\omega \top_{tt} \\ f^\omega \top_{ts} + d^*ce^\omega \top_{ss} & f^\omega \top_{tt} + d^*ce^\omega \top_{st} \end{pmatrix}.$$

Thus the resulting matrix has the correct type. Note that the columns of the matrix are not identical, as in the untyped case [13], but have their types adjusted. This is not necessary for the * and $^+$ operators.

It remains to show that the $^\omega$ operation thus defined satisfies the axioms. The proof is by induction on the size of the matrix, where the induction step assumes that the omega axioms hold for smaller matrices. The omega unfolding axiom is a consequence of

$$\begin{pmatrix} a & b \\ c & d \end{pmatrix} \begin{pmatrix} e^\omega & a^*bf^\omega \\ d^*ce^\omega & f^\omega \end{pmatrix} = \begin{pmatrix} ae^\omega + bd^*ce^\omega & aa^*bf^\omega + bf^\omega \\ ce^\omega + dd^*ce^\omega & ca^*bf^\omega + df^\omega \end{pmatrix}$$
$$= \begin{pmatrix} (a + bd^*c)e^\omega & (aa^*b + b)f^\omega \\ (c + dd^*c)e^\omega & (ca^*b + d)f^\omega \end{pmatrix} = \begin{pmatrix} ee^\omega & a^*bf^\omega \\ d^*ce^\omega & ff^\omega \end{pmatrix} = \begin{pmatrix} e^\omega & a^*bf^\omega \\ d^*ce^\omega & f^\omega \end{pmatrix},$$

using the induction hypothesis in the last step. For the omega induction axiom, let $u \in T$ and $x, p : s \to u$ and $y, q : t \to u$ such that

$$\begin{pmatrix} x \\ y \end{pmatrix} \le \begin{pmatrix} a & b \\ c & d \end{pmatrix} \begin{pmatrix} x \\ y \end{pmatrix} + \begin{pmatrix} p \\ q \end{pmatrix} = \begin{pmatrix} ax + by + p \\ cx + dy + q \end{pmatrix},$$

hence $y \le d^\omega \top_{tu} + d^*(cx + q)$ by the induction hypothesis. Therefore

$$x \le ax + b(d^\omega \top_{tu} + d^*(cx + q)) + p = (a + bd^*c)x + b(d^\omega \top_{tu} + d^*q) + p$$
$$\le ex + b(f^\omega \top_{tu} + f^*q) + p \,,$$

using $d \leq f$ and $\leq$-isotony. Once more by the induction hypothesis,

$$x \leq e^{\omega} \mathsf{T}_{su} + e^*(bf^{\omega} \mathsf{T}_{tu} + bf^*q + p) .$$

We have to show

$$
\begin{pmatrix} x \\ y \end{pmatrix} \leq \begin{pmatrix} a & b \\ c & d \end{pmatrix}^{\omega} \begin{pmatrix} \mathsf{T}_{su} \\ \mathsf{T}_{tu} \end{pmatrix} + \begin{pmatrix} a & b \\ c & d \end{pmatrix}^* \begin{pmatrix} p \\ q \end{pmatrix}
$$

$$
= \begin{pmatrix} e^{\omega} & a^*bf^{\omega} \\ d^*ce^{\omega} & f^{\omega} \end{pmatrix} \begin{pmatrix} \mathsf{T}_{ss} & \mathsf{T}_{st} \\ \mathsf{T}_{ts} & \mathsf{T}_{tt} \end{pmatrix} \begin{pmatrix} \mathsf{T}_{su} \\ \mathsf{T}_{tu} \end{pmatrix} + \begin{pmatrix} e^+ & a^*bf^* \\ d^*ce^* & f^+ \end{pmatrix} \begin{pmatrix} p \\ q \end{pmatrix} + \begin{pmatrix} p \\ q \end{pmatrix}
$$

$$
= \begin{pmatrix} e^{\omega}(\mathsf{T}_{ss} \mathsf{T}_{su} + \mathsf{T}_{st} \mathsf{T}_{tu}) + a^*bf^{\omega}(\mathsf{T}_{ts} \mathsf{T}_{su} + \mathsf{T}_{tt} \mathsf{T}_{tu}) + e^*p + a^*bf^*q \\ f^{\omega}(\mathsf{T}_{ts} \mathsf{T}_{su} + \mathsf{T}_{tt} \mathsf{T}_{tu}) + d^*ce^{\omega}(\mathsf{T}_{ss} \mathsf{T}_{su} + \mathsf{T}_{st} \mathsf{T}_{tu}) + f^*q + d^*ce^*p \end{pmatrix} .
$$

Consider x: by the above inequality and $e^{\omega} \leq e^{\omega} \mathsf{T}_{ss}$ and $f^{\omega} \leq f^{\omega} \mathsf{T}_{tt}$, it suffices to show $e^*bf^{\omega} \leq a^*bf^{\omega}$ and $e^*bf^*q \leq a^*bf^*q$. By star induction, these reduce to $ea^*bf^{\omega} \leq a^*bf^{\omega}$ and $ea^*bf^*q \leq a^*bf^*q$. Since $e = a + bd^*c$, these follow from $bd^*ca^*bf^{\omega} \leq bf^{\omega}$ and $bd^*ca^*bf^*q \leq bf^*q$. But these are consequences of $d^*ca^*b \leq f^*f = f^+$ since $f^+f^{\omega} \leq f^{\omega}$ and $f^+f^*q \leq f^*q$. The inequality for y follows by swapping $\left(\begin{smallmatrix} x & p & a & b & e & s \\ y & q & d & c & f & t \end{smallmatrix}\right)$.

The same argument applies column-wise to matrices having more than one column (replacing the vectors formed by x, y and p, q, respectively). This shows that the 1-free omega axioms hold for every well-typed instance. Moreover, the matrix algebra satisfies each of (T1)–(T5) if the underlying typed 1-free omega algebra does so. We thus obtain the following result.

Theorem 1. *The finite matrices over a typed 1-free omega algebra form a typed 1-free omega algebra. Each of the axioms (T1)–(T5) is preserved.*

It can be shown that the square matrices of size 2×2 and greater satisfy (T1) vacuously. Only for the 1×1 matrices it is necessary that (T1) holds in the underlying algebra.

For a given dimension n and sequence $(t_i) \in T^n$, the set of $n \times n$ matrices with type $(t_i) \to (t_i)$ is closed under the operations of 1-free omega algebra. We therefore obtain the following consequence of Theorem 1.

Corollary 2. *The $n \times n$ matrices with fixed type over a typed 1-free omega algebra form a 1-free omega algebra. Each of the axioms (T1)–(T5) is preserved.*

By using diagonal matrices with 1-entries and the usual matrix * of [3], the above results also hold for typed omega algebras. Because the omega induction axioms of the typed and the untyped setting differ, the axiom (T1) is needed for the version of Corollary 2 for typed omega algebras.

3.2   Matrices in General Correctness

We now apply the above theory to calculate the omega operation of so-called '(normal) prescriptions', which model programs in general correctness [4]. They are represented by matrices in [14,5].

Let R be an omega algebra. A prescription is a 2×2 matrix

$$\begin{pmatrix} a & b \\ c & d \end{pmatrix} \in R^{2 \times 2}$$

such that $a = \top$ and $b = 0$ and $c = c\top$. Elements of the form $c\top$ model conditions; they are closed under the operations $+$ and $d\cdot$ for any $d \in R$.

The entry d records the terminating executions of a program, and c captures the set of states from which non-terminating executions exist. Fixing $a = \top$ and $b = 0$ ensures that the non-terminating executions of a program x become non-terminating executions of a sequential composition $x \cdot y$, but do not interfere with the terminating executions of $x \cdot y$, as fit for general correctness.

Trying to derive the omega operation on prescriptions using the untyped setting of [13], we obtain $e = \top + 0d^*c = \top$ and $f = d + c\top^*0 = d$, thus

$$\begin{pmatrix} \top & 0 \\ c & d \end{pmatrix}^{\omega} = \begin{pmatrix} e^{\omega} + \top^*0f^{\omega} & e^{\omega} + \top^*0f^{\omega} \\ f^{\omega} + d^*ce^{\omega} & f^{\omega} + d^*ce^{\omega} \end{pmatrix} = \begin{pmatrix} \top^{\omega} & \top^{\omega} \\ d^{\omega} + d^*c\top^{\omega} & d^{\omega} + d^*c\top^{\omega} \end{pmatrix}$$

$$= \begin{pmatrix} \top & \top \\ d^{\omega} + d^*c & d^{\omega} + d^*c \end{pmatrix}.$$

But this is not a prescription due to the entry $\top$ in the first row and the second column.

To solve this problem, let S' be the typed omega algebra with pretypes $T = \{1, 2\}$ such that $S'_{st} = R$ for each $s, t \in T$. The values of well-typed operations are given by calculating in R.

Now consider the substructure S of S' in which $S_{12} = \{0_{12}\}$ and $S_{st} = S'_{st}$ otherwise. Hence we restrict the type $1 \to 2$ to one element, retaining the other types. Then S is closed under the operations of typed omega algebra, except $\top$:

- The sum of two elements has type $1 \to 2$ only if both elements have this type, whence they are both 0_{12} and so is their sum.
- The product of two elements has type $1 \to 2$ only if one of them has this type, whence it is 0_{12} and so is the product.
- The operations $*$, $^{\omega}$, $^+$ and 1 do not apply to the type $1 \to 2$.

The constant 0_{12} is in S_{12} and we take $\top_{12} = 0_{12}$. The instance of the omega induction axiom $c \leq ac + b \Rightarrow c \leq a^{\omega}\top_{12} + a^*b$ holds since c must have the type $1 \to 2$, whence it is 0_{12}. The other axioms of typed omega algebra are satisfied since they hold in S' and S is closed. Therefore S is a typed omega algebra. Moreover, S satisfies (T1)–(T4), but not (T5) since $\top_{12}\top_{21} = 0_{12}\top_{21} = 0_{11} \neq \top_{11}$. Yet we have $\top^{\omega}_{11}\top_{11} = \top_{11}\top_{11} = \top_{11}$.

By Corollary 2 and (T1), the 2×2 matrices over S form an omega algebra. But all prescriptions are elements of this matrix algebra, whence the omega operation is derived as shown in Section 3.1. Again, $e = \top_{11} + 0_{12}d^*c = \top_{11}$ and $f = d + c\top^*_{11}0_{12} = d$, thus

$$\begin{pmatrix} \top_{11} & 0_{12} \\ c & d \end{pmatrix}^{\omega} = \begin{pmatrix} e^{\omega}\top_{11} + \top^*_{11}0_{12}f^{\omega}\top_{21} & e^{\omega}\top_{12} + \top^*_{11}0_{12}f^{\omega}\top_{22} \\ f^{\omega}\top_{21} + d^*ce^{\omega}\top_{11} & f^{\omega}\top_{22} + d^*ce^{\omega}\top_{12} \end{pmatrix}$$

$$= \begin{pmatrix} \top^{\omega}_{11}\top_{11} & \top^{\omega}_{11}0_{12} \\ d^{\omega}\top_{21} + d^*c\top^{\omega}_{11}\top_{11} & d^{\omega}\top_{22} + d^*c\top^{\omega}_{11}0_{12} \end{pmatrix} = \begin{pmatrix} \top_{11} & 0_{12} \\ d^{\omega}\top_{21} + d^*c\top_{11} & d^{\omega} \end{pmatrix}.$$

In $R^{2\times 2}$ this simplifies to

$$\begin{pmatrix} \top & 0 \\ c & d \end{pmatrix}^\omega = \begin{pmatrix} \top & 0 \\ d^\omega \top + d^* c \top & d^\omega \end{pmatrix} = \begin{pmatrix} \top & 0 \\ d^\omega + d^* c & d^\omega \end{pmatrix},$$

since $d^\omega \top = d^\omega$ for $d \in R$ and $c\top = c$. The result is a prescription again.

Corollary 3. *Prescriptions are closed under the following operation $^\omega$, which satisfies the omega axioms:*

$$\begin{pmatrix} \top & 0 \\ c & d \end{pmatrix}^\omega = \begin{pmatrix} \top & 0 \\ d^\omega + d^* c & d^\omega \end{pmatrix}.$$

Note that the set of prescriptions is closed under * but does not form a subalgebra of the matrix algebra because the 0-matrix is not a prescription. Nevertheless, choosing different 0- and 1-elements, they form an omega algebra without right zero (the axiom $a0 = 0$ is omitted) [5].

3.3   Matrices in Total and General Correctness

We can also calculate the omega operation of so-called 'extended designs', which combine certain aspects of total and general correctness [8]. They too can be represented by matrices [6].

Let R be an omega algebra. An extended design is a 3×3 matrix of the form

$$\begin{pmatrix} \top & \top & \top \\ 0 & \top & 0 \\ p & q & r \end{pmatrix} \in R^{3\times 3}$$

such that $p\top = p \le q = q\top$ and $p \le r$.

Here, r records the terminating executions, q captures the non-terminating executions and p the aborting executions of a program. Again, the entries 0 and $\top$ are arranged to propagate this information according to the semantics of extended designs. The constraints $p \le q$ and $p \le r$ are typical for total correctness approaches: in the presence of an aborting execution, any other executions are considered irrelevant and hence absorbed.

Similarly to prescriptions, let S' be the typed omega algebra with pretypes $T = \{1, 2, 3\}$ such that $S'_{st} = R$ for each $s, t \in T$. Consider the substructure S of S' in which $S_{21} = S_{23} = \{0_{12}\}$ and $S_{st} = S'_{st}$ otherwise. Again, S is closed under the operations of typed omega algebra except $\top$. Closure under $\cdot$ is more involved now: for example, it would not suffice to collapse only the type $2 \to 3$, because an element c_{23} of this type may be obtained as the product $c_{23} = a_{21} \cdot b_{13}$. Taking $\top_{21} = 0_{21}$ and $\top_{23} = 0_{23}$, we again establish S as a typed omega algebra with $(\top 1)$–$(\top 4)$.

For the prescription submatrix of an extended design we obtain

$$\begin{pmatrix} \top_{22} & 0_{23} \\ q & r \end{pmatrix}^* = \begin{pmatrix} \top_{22} & 0_{23} \\ r^* q & r^* \end{pmatrix} \quad \text{and} \quad \begin{pmatrix} \top_{22} & 0_{23} \\ q & r \end{pmatrix}^\omega = \begin{pmatrix} \top_{22} & 0_{23} \\ r^\omega \top_{32} + r^* q & r^\omega \end{pmatrix}$$

as for Corollary 3. For the entire matrix we therefore obtain $e = \top_{11}$ and

$$f = \begin{pmatrix} \top_{22} & 0_{23} \\ q & r \end{pmatrix} + \begin{pmatrix} 0_{21} \\ p \end{pmatrix} \top_{11}^{*} \begin{pmatrix} \top_{12} & \top_{13} \end{pmatrix} = \begin{pmatrix} \top_{22} & 0_{23} \\ q & r \end{pmatrix}$$

$$d^{*}ce^{\omega} = \begin{pmatrix} \top_{22} & 0_{23} \\ r^{*}q & r^{*} \end{pmatrix} \begin{pmatrix} 0_{21} \\ p \end{pmatrix} \top_{11}^{\omega} = \begin{pmatrix} \top_{22} & 0_{23} \\ r^{*}q & r^{*} \end{pmatrix} \begin{pmatrix} 0_{21} \\ p\top_{11} \end{pmatrix} = \begin{pmatrix} 0_{21} \\ r^{*}p \end{pmatrix}$$

$$f^{\omega} \begin{pmatrix} \top_{21} \\ \top_{31} \end{pmatrix} = \begin{pmatrix} \top_{22} & 0_{23} \\ r^{\omega}\top_{32} + r^{*}q & r^{\omega} \end{pmatrix} \begin{pmatrix} 0_{21} \\ \top_{31} \end{pmatrix} = \begin{pmatrix} 0_{21} \\ r^{\omega}\top_{31} \end{pmatrix}$$

$$f^{\omega} \begin{pmatrix} \top_{22} & \top_{23} \\ \top_{32} & \top_{33} \end{pmatrix} = \begin{pmatrix} \top_{22} & 0_{23} \\ r^{\omega}\top_{32} + r^{*}q & r^{\omega} \end{pmatrix} \begin{pmatrix} \top_{22} & 0_{23} \\ \top_{32} & \top_{33} \end{pmatrix} = \begin{pmatrix} \top_{22} & 0_{23} \\ r^{\omega}\top_{32} + r^{*}q & r^{\omega} \end{pmatrix}$$

and therefore

$$\begin{pmatrix} \top & \top & \top \\ 0 & \top & 0 \\ p & q & r \end{pmatrix}^{\omega} = \begin{pmatrix} e^{\omega}\top_{11} + a^{*}bf^{\omega}\begin{pmatrix} \top_{21} \\ \top_{31} \end{pmatrix} & e^{\omega}\begin{pmatrix} \top_{12} & \top_{13} \end{pmatrix} + a^{*}bf^{\omega}\begin{pmatrix} \top_{22} & \top_{23} \\ \top_{32} & \top_{33} \end{pmatrix} \\ f^{\omega}\begin{pmatrix} \top_{21} \\ \top_{31} \end{pmatrix} + d^{*}ce^{\omega}\top_{11} & f^{\omega}\begin{pmatrix} \top_{22} & \top_{23} \\ \top_{32} & \top_{33} \end{pmatrix} + d^{*}ce^{\omega}\begin{pmatrix} \top_{12} & \top_{13} \end{pmatrix} \end{pmatrix}$$

$$= \begin{pmatrix} \begin{pmatrix} 0_{21} \\ r^{\omega}\top_{31} \end{pmatrix} + \begin{pmatrix} 0_{21} \\ r^{*}p \end{pmatrix}\top_{11} & \begin{pmatrix} \top_{12} & \top_{13} \end{pmatrix} \\ & \begin{pmatrix} \top_{22} & 0_{23} \\ r^{\omega}\top_{32} + r^{*}q & r^{\omega} \end{pmatrix} + \begin{pmatrix} 0_{21} \\ r^{*}p \end{pmatrix}\begin{pmatrix} \top_{12} & \top_{13} \end{pmatrix} \end{pmatrix}$$

$$= \begin{pmatrix} \top_{11} & \top_{12} & \top_{13} \\ 0_{21} & \top_{22} & 0_{23} \\ r^{\omega}\top_{31} + r^{*}p & r^{\omega}\top_{32} + r^{*}q + r^{*}p\top_{12} & r^{\omega} + r^{*}p\top_{13} \end{pmatrix}.$$

In $R^{3\times3}$ this further simplifies by $r^{\omega}\top = r^{\omega}$ and $r^{*}p\top = r^{*}p \leq r^{*}q$ to the following result. The argument is similar to that for prescriptions, using that extended designs are closed under $*$ [6].

Corollary 4. *Extended designs are closed under the following operation* $^{\omega}$, *that satisfies the omega axioms:*

$$\begin{pmatrix} \top & \top & \top \\ 0 & \top & 0 \\ p & q & r \end{pmatrix}^{\omega} = \begin{pmatrix} \top & \top & \top \\ 0 & \top & 0 \\ r^{\omega} + r^{*}p & r^{\omega} + r^{*}q & r^{\omega} + r^{*}p \end{pmatrix}.$$

4 Typing

In this section we establish a class of theorems that can be transferred from the untyped to the typed setting. The typed setting thus profits from existing theorems, simpler (untyped) proofs of new theorems, and automated theorem provers (such as Prover9) which have no notion of types.

We proceed along the lines of [11] as far as possible. See [15] for a different approach (in Kleene algebra).

4.1   Embedding 1-Free Omega Algebra into Omega Algebra

It is proved in [11, Section 2.2] that every 1-free Kleene algebra can be embedded into a Kleene algebra. We extend that result to omega algebras.

Theorem 5. *Every 1-free omega algebra satisfying* (⊤1) *and* (⊤2) *can be embedded into an omega algebra, except that the embedding need not preserve* ⊤.

Proof. We extend the construction of [11]. Let S be a 1-free omega algebra, and construct the omega algebra $S' =_{\text{def}} \{0, 1\} \times S$ as follows. Intuitively, the element $(0, a)$ represents a, while $(1, a)$ represents $1 + a$. The operations $+$, $\cdot$, *, 0 and 1 on S' are defined as in [11]:

$$(i, a) + (j, b) = (i + j, a + b) \qquad\qquad (i, a)^* = (1, a^+) \qquad\qquad 0 = (0, 0)$$
$$(i, a) \cdot (j, b) = (ij, ab + ib + ja) \qquad\qquad\qquad\qquad\qquad\qquad\quad 1 = (1, 0)$$

This uses $ix = x$ if $i = 1$, and $ix = 0$ if $i = 0$. We add the operation $^\omega$ by

$$(0, a)^\omega =_{\text{def}} (0, a^\omega) \qquad \text{and} \qquad (1, a)^\omega =_{\text{def}} (1, \top) .$$

The omega unfold axiom on S' follows since

$$(0, a) \cdot (0, a)^\omega = (0, a) \cdot (0, a^\omega) = (0, aa^\omega) = (0, a^\omega) = (0, a)^\omega$$
$$(1, a) \cdot (1, a)^\omega = (1, a) \cdot (1, \top) = (1, a\top + \top + a) = (1, \top) = (1, a)^\omega$$

by (⊤2). For the omega induction axiom $c' \le a'c' + b' \Rightarrow c' \le a'^\omega + a'^*b'$ of S' we consider two cases. If $a' = (1, a)$, then $c' \le (1, \top) = a'^\omega$ since $(1, \top)$ is the greatest element of S'. Otherwise, let $a' = (0, a)$ and $b' = (i, b)$ and $c' = (j, c)$. Then

$$(j, c) = c' \le a'c' + b' = (0, a) \cdot (j, c) + (i, b) = (0, ac + ja) + (i, b) = (i, ac + ja + b) .$$

Since the order on S' is componentwise, we have $j \le i$ and $c \le ac + ja + b$. Using the omega induction axiom of S, we obtain

$$c \le a^\omega \top + a^* (ja + b) = a^\omega + ja^+ + a^*b \le a^\omega + ia^+ + a^*b$$

by (⊤1) and since $j \in \{0, 1\}$. Therefore

$$c' = (j, c) \le (i, a^\omega + ia^+ + a^*b) = (0, a^\omega) + (i, a^+b + b + ia^+)$$
$$= (0, a)^\omega + (1, a^+)(i, b) = a'^\omega + (0, a)^* b' = a'^\omega + a'^*b' .$$

The embedding $a \mapsto (0, a)$ is injective and a homomorphism, except that $\top$ is mapped to $(0, \top)$ which is not the greatest element $(1, \top)$ of S'.   □

By using the embedding of Theorem 5 we obtain the following consequence about statements with universally quantified variables.

Corollary 6. *A universal formula using only the operators* $+$, $\cdot$, $^+$, $^\omega$, 0 *is valid in omega algebra if and only if it is valid in 1-free omega algebra with* (⊤1) *and* (⊤2).

If we admit further axioms, we can preserve $\top$ as well. The construction used in the following proof is not required for Kleene algebras.

Theorem 7. *Every 1-free omega algebra satisfying* $(\top 1)$–$(\top 4)$ *can be embedded into an omega algebra.*

Proof. We continue the proof of Theorem 5. Consider the smallest equivalence relation $\cong$ on S' which identifies $(0, \top) \cong (1, \top)$. It is a congruence:

- $(0, \top) + (i, a) = (i, \top + a) = (i, \top) \cong (1, \top) = (1, \top + a) = (1, \top) + (i, a)$ by $(\top 2)$. With commutativity we get congruence with respect to $+$.
- $(0, \top) \cdot (0, a) = (0, \top a) = (0, \top a + a) = (1, \top) \cdot (0, a)$ by $(\top 4)$. Moreover, $(0, \top) \cdot (1, a) = (0, \top a + \top) = (0, \top) \cong (1, \top) = (1, \top a + a + \top) = (1, \top) \cdot (1, a)$ by $(\top 2)$. Congruence in the second argument of $\cdot$ is analogous using $(\top 3)$.
- $(0, \top)^* = (1, \top^+) = (1, \top)^*$.
- $(0, \top)^\omega = (0, \top^\omega) = (0, \top) \cong (1, \top) = (1, \top)^\omega$, since $\top^\omega = \top$ holds in S: by $(\top 3)$ or $(\top 4)$ we have $\top \leq \top\top$, whence $\top \leq \top^\omega \top = \top^\omega \leq \top$ by omega induction, $(\top 1)$ and $(\top 2)$.

We thus obtain the embedding $a \mapsto [(0, a)]_\cong$ by composing the embedding of Theorem 5 with the canonical map h of $\cong$. Observe that h is injective on $S' \setminus \{(1, \top)\}$, thus the new embedding is injective as $(1, \top)$ is not in the image of the previous one. It remains to show that $S'/\cong$ is an omega algebra. The equational axioms follow since $S'/\cong$ is a homomorphic image of the omega algebra S'. We show the conditional equations:

- $ac \leq c \Rightarrow a^*c \leq c$: clear if $c = [(1, \top)]_\cong$ is the greatest element of $S'/\cong$. Otherwise, let $a', c' \in S'$ with $h(a') = a$ and $h(c') = c$ as h is surjective. Since h is a homomorphism, we have

$$ h(a'c' + c') = h(a')h(c') + h(c') = ac + c = c = h(c') . $$

  Because c is not the greatest element, we have $c' \neq (1, \top) \neq a'c' + c'$, hence $a'c' + c' = c'$ since h is injective on $S' \setminus \{(1, \top)\}$. By star induction of S' we obtain $a'^*c' \leq c'$. Hence $a^*c = h(a')^*h(c') = h(a'^*c') \leq h(c') = c$ again since h is a homomorphism.
- $ca \leq c \Rightarrow ca^* \leq c$: symmetrically.
- $c \leq ac + b \Rightarrow c \leq a^\omega + a^*b$: if $ac + b \neq [(1, \top)]_\cong$, apply the previous argument. Otherwise, let $a', b', c' \in S'$ with $h(a') = a$ and $h(b') = b$ and $h(c') = c$. If $c' \leq a'c' + b'$, finish by applying omega induction of S' and the homomorphism h. Otherwise, $a'c' + b' = (0, \top)$ and $c' = (1, c'')$ for some $c'' \in S$. Hence $a' = (0, a'')$ for some $a'' \in S$. By $(\top 3)$,

$$ a'c' \leq (0, a'')(1, \top) = (0, a''\top + a'') = (0, a''\top) = a'(0, \top) . $$

  Therefore $(0, \top) = a'c' + b' \leq a'(0, \top) + b'$, whence $(0, \top) \leq a'^\omega + a'^*b'$ by omega induction of S'. Thus $c \leq [(1, \top)]_\cong = h((0, \top)) \leq h(a'^\omega + a'^*b') = a^\omega + a^*b$ since h is a homomorphism. $\square$

Corollary 8. *A universal formula of 1-free omega algebra is valid in omega algebra if and only if it is valid in 1-free omega algebra with* $(\top 1)$–$(\top 4)$.

Because $(\top 1)$–$(\top 4)$ are independent, these axioms are necessary for Theorem 7 and Corollary 8.

4.2   Embedding Typed 1-Free Omega Algebra into 1-Free Omega Algebra

It is proved in [11, Lemma 4.1] that every typed 1-free Kleene algebra can be embedded into a 1-free Kleene algebra. We extend that result to omega algebras.

As clarified in [12], a typed embedding is required to be injective for each type, but may map elements of distinct types to the same element.

The following result treats the case of finitely typed 1-free omega algebras. It can be generalised to infinitely typed 1-free omega algebras with $(\top 1)$ and $(\top 5)$, though that proof is more involved.

Theorem 9. *Every finitely typed 1-free omega algebra satisfying* $(\top 1)$ *can be embedded into a 1-free omega algebra, except that the embedding need not preserve* $\top$. *Each of the axioms* $(\top 1)$–$(\top 5)$ *is preserved.*

Proof. Let $(S, +, \cdot, {}^{+}, {}^{\omega}, 0, \top)$ be a typed 1-free omega algebra with $(\top 1)$, based on a set of n pretypes T. Arrange the pretypes in a fixed sequence $(t_i) \in T^n$. By Corollary 2, the $n \times n$ matrices with type $(t_i) \to (t_i)$ form a 1-free omega algebra, which satisfies any of $(\top 1)$–$(\top 5)$ if S does so.

We embed S into this matrix algebra by the mapping h defined as follows:

$$h(a_{st})_{uv} =_{\text{def}} \begin{cases} a_{st} & \text{if } u = s \text{ and } v = t \\ a_{st}\top_{tv} & \text{if } u = s \text{ and } v \neq t \\ 0_{uv} & \text{if } u \neq s \end{cases}$$

Thus the element $a : s \to t$ is mapped to a matrix with a in row s and column t, with $a\top$ in any other column of row s, and 0 in any other row. The embedding for 1-free Kleene algebra [11, Lemma 4.1] maps to 0 also in the second case.

Clearly h is injective on each type. We show that h preserves the operations of 1-free omega algebra except $\top$:

- Preservation of $+$ follows since $h(a_{st} + b_{st}) = h(a_{st}) + h(b_{st})$ by

$$h(a_{st} + b_{st})_{uv} = \begin{cases} a_{st} + b_{st} & \text{if } u = s \text{ and } v = t \\ (a_{st} + b_{st})\top_{tv} = a_{st}\top_{tv} + b_{st}\top_{tv} & \text{if } u = s \text{ and } v \neq t \\ 0_{uv} = 0_{uv} + 0_{uv} & \text{if } u \neq s \end{cases}$$
$$= h(a_{st})_{uv} + h(b_{st})_{uv} = (h(a_{st}) + h(b_{st}))_{uv} .$$

- Preservation of $\cdot$ follows if we can show $h(a_{st}b_{tu})_{vw} = (h(a_{st})h(b_{tu}))_{vw}$. If $v \neq s$, then $h(a_{st}b_{tu})_{vw} = 0_{vw}$, but so is

$$(h(a_{st})h(b_{tu}))_{vw} = \sum_{x \in T} h(a_{st})_{vx} h(b_{tu})_{xw} = \sum_{x \in T} 0_{vx} h(b_{tu})_{xw} = 0_{vw} .$$

If $v = s$, then all the summands with $x \neq t$ vanish by $h(a_{st})_{vx}h(b_{tu})_{xw} = h(a_{st})_{vx}0_{xw} = 0_{vw}$, hence

$$(h(a_{st})h(b_{tu}))_{vw} = \sum_{x \in T} h(a_{st})_{sx}h(b_{tu})_{xw} = h(a_{st})_{st}h(b_{tu})_{tw} = a_{st}h(b_{tu})_{tw} \; .$$

If $w = u$, this equals $a_{st}h(b_{tu})_{tu} = a_{st}b_{tu} = h(a_{st}b_{tu})_{vw}$, and if $w \neq u$, it equals $a_{st}b_{tu}\top_{uw} = h(a_{st}b_{tu})_{vw}$ as well.

- For a pretype $s \in T$, let $\top_{s\bar{s}}$ denote the transposed vector of all $\top_{st}$ elements such that $s \neq t \in T$, and similarly for the vector $0_{\bar{s}s}$ and matrix $0_{\bar{s}\bar{s}}$. Preservation of $^+$ follows by

$$h(a_{ss})^+ = \begin{pmatrix} a_{ss} & a_{ss}\top_{s\bar{s}} \\ 0_{\bar{s}s} & 0_{\bar{s}\bar{s}} \end{pmatrix}^+ = \begin{pmatrix} a_{ss}^+ & a_{ss}^*a_{ss}\top_{s\bar{s}}0_{\bar{s}\bar{s}}^* \\ 0_{\bar{s}\bar{s}}^*0_{\bar{s}s}a_{ss}^* & 0_{\bar{s}\bar{s}}^+ \end{pmatrix} = \begin{pmatrix} a_{ss}^+ & a_{ss}^+\top_{s\bar{s}} \\ 0_{\bar{s}s} & 0_{\bar{s}\bar{s}} \end{pmatrix}$$
$$= h(a_{ss}^+) \; .$$

- Preservation of $^\omega$ follows using $(\top 1)$ in

$$h(a_{ss})^\omega = \begin{pmatrix} a_{ss} & a_{ss}\top_{s\bar{s}} \\ 0_{\bar{s}s} & 0_{\bar{s}\bar{s}} \end{pmatrix}^\omega = \begin{pmatrix} a_{ss}^\omega & a_{ss}^*a_{ss}\top_{s\bar{s}}0_{\bar{s}\bar{s}}^\omega \\ 0_{\bar{s}\bar{s}}^*0_{\bar{s}s}a_{ss}^\omega & 0_{\bar{s}\bar{s}}^\omega \end{pmatrix}\begin{pmatrix} \top_{ss} & \top_{s\bar{s}} \\ \top_{\bar{s}s} & \top_{\bar{s}\bar{s}} \end{pmatrix}$$
$$= \begin{pmatrix} a_{ss}^\omega & 0_{s\bar{s}} \\ 0_{\bar{s}s} & 0_{\bar{s}\bar{s}} \end{pmatrix}\begin{pmatrix} \top_{ss} & \top_{s\bar{s}} \\ \top_{\bar{s}s} & \top_{\bar{s}\bar{s}} \end{pmatrix} = \begin{pmatrix} a_{ss}^\omega\top_{ss} & a_{ss}^\omega\top_{s\bar{s}} \\ 0_{\bar{s}s} & 0_{\bar{s}\bar{s}} \end{pmatrix} = \begin{pmatrix} a_{ss}^\omega & a_{ss}^\omega\top_{s\bar{s}} \\ 0_{\bar{s}s} & 0_{\bar{s}\bar{s}} \end{pmatrix}$$
$$= h(a_{ss}^\omega) \; .$$

- Clearly $h(0_{st})$ is the 0-matrix, but $h(\top_{st})$ is not the $\top$-matrix in general. □

Already from this we obtain by modifying the argument of [11, Theorem 1.2] the following consequence. Formulas are implicitly assumed to be finitary.

Corollary 10. *A universal formula using only the operators $+, \cdot, ^+, ^\omega, 0$ is valid in 1-free omega algebra with $(\top 1)$ if and only if it is valid in typed 1-free omega algebra with $(\top 1)$.*

Proof. The backward implication follows since every 1-free omega algebra is a typed 1-free omega algebra (with one type). We prove the forward implication.

The given formula is equivalent to a conjunction of universal implications of the form $\bigwedge_{i \in I} a_i = b_i \Rightarrow \bigvee_{j \in J} c_j = d_j$ with finite index sets I and J and expressions a_i, b_i, c_j, d_j using only the operators $+, \cdot, ^+, ^\omega, 0$. We show the claim for such an implication F.

Assume F holds in 1-free omega algebra with $(\top 1)$. Let S be a typed 1-free omega algebra with $(\top 1)$ and pretypes T. Consider a well-typed instance F' of F. The instance F' only refers to finitely many pretypes $T' \subseteq T$. Let S' be the substructure of S restricted (in types, operations and axioms) to T'. The types which remain in S' keep all of their elements, whence all remaining axioms (equations and implications) still hold. In other words, S' is a finitely typed 1-free omega algebra with $(\top 1)$. Let h be the embedding of S' into a 1-free omega algebra R with $(\top 1)$ according to Theorem 9. In particular, F holds in R.

We show that F' holds in S'. To this end, let v be a valuation of its variables, and assume that the typed instance of each $a_i(v) = b_i(v)$ holds in S'. Then clearly $h(a_i(v)) = h(b_i(v))$ in R. Since h is homomorphic, $a_i(h(v)) = b_i(h(v))$ in R. By F we obtain $c_j(h(v)) = d_j(h(v))$ for some $j \in J$. Since h is homomorphic, $h(c_j(v)) = h(d_j(v))$ in R. Since h is injective on the type of c_j, we obtain $c_j(v) = d_j(v)$ in S'.

Every valuation of the variables of F' in S is a valuation in S', because it must respect the (fixed) types of the variables. Thus F' holds in S, too. □

Because the embedding of Theorem 9 preserves (T2), the same argument works for 1-free omega algebra with (T1) and (T2). We combine this with Corollary 6.

Corollary 11. *A universal formula using only the operators* $+$, $\cdot$, $^+$, $^\omega$, 0 *is valid in omega algebra if and only if it is valid in typed 1-free omega algebra with* (T1) *and* (T2).

Whether the above results can be extended to formulas with $\top$ is open.

5   Conclusion

We conclude with remarks on the condition '1-free'. It is motivated by the counterexample $0 = 1 \Rightarrow a = b$ of [11], which holds in Kleene algebra, but not in typed Kleene algebra under its most general typing $0_{ss} = 1_{ss} \Rightarrow a_{tu} = b_{tu}$. It fails in those and only those typed Kleene algebras, where a type $s \to s$ is collapsed (has only one element) but another type $t \to u$ is not collapsed. An example is given by $T = \{1, 2\}$ and $S_{11} = S_{12} = S_{21} = \{0\}$ and $S_{22} = \{0, 1\}$ with operations defined as usual. In fact, the collapse of a type $s \to s$ triggers the collapse of all types $s \to t$ and $t \to s$: for example, $a_{ts} = a_{ts}1_{ss} = a_{ts}0_{ss} = 0_{ts}$. Further types, such as $t \to t$, are not affected. Sections 3.2 and 3.3 give applications with typed omega algebras where some but not all types are collapsed (though not a square type $s \to s$).

To avoid the above counterexample, further axioms have to be included: we propose (T2) and (T5). Assume a typed Kleene algebra with (T2) and (T5). Then $0_{ss} = 1_{ss}$ implies $\top_{ss} = \top_{ss}1_{ss} = \top_{ss}0_{ss} = 0_{ss}$. Moreover, $\top_{st} = 0_{st}$ for any $s, t \in T$ implies $\top_{uv} = \top_{us}\top_{sv} = \top_{us}\top_{st}\top_{tv} = \top_{us}0_{st}\top_{tv} = 0_{uv}$ by (T5), and hence $a_{uv} = b_{uv}$ for any $a, b \in S_{uv}$ by (T2). Including (T2) and (T5) propagates the collapse of one type to all types. For this reason, it is essential that the typed omega algebras in Sections 3.2 and 3.3 do not satisfy (T5).

There are also models of typed omega algebra with (T1)–(T4) but not (T5), where none of the types is collapsed. An example is given by $T = \{1, 2\}$ where each type contains two Boolean 2×2 matrices under the usual matrix operations: $S_{11} = S_{22} = \{\emptyset, \{(0, 0), (1, 1)\}\}$ and $S_{12} = \{\emptyset, \{(0, 0)\}\}$ and $S_{21} = \{\emptyset, \{(1, 1)\}\}$.

On the other hand, we get another counterexample: $1_{ss} = \top_{ss} \Rightarrow 1_{tt} = \top_{tt}$. While the untyped implication clearly holds, the typed formula is not valid even in heterogeneous relation algebra [16], which is much more restricted than the typed omega algebras discussed in the present paper. Namely, $1 = \top$ holds for relations between one-element sets, but not for relations between larger sets.

Acknowledgement. I thank Georg Struth for helpful discussions and the anonymous referees for valuable comments.

This work was supported by a fellowship within the Postdoc-Programme of the German Academic Exchange Service (DAAD).

References

1. Bloom, S.L., Ésik, Z.: Iteration Theories: The Equational Logic of Iterative Processes. Springer, Heidelberg (1993)
2. Cohen, E.: Separation and reduction. In: Backhouse, R., Oliveira, J.N. (eds.) MPC 2000. LNCS, vol. 1837, pp. 45–59. Springer, Heidelberg (2000)
3. Conway, J.H.: Regular Algebra and Finite Machines. Chapman and Hall, London (1971)
4. Dunne, S.: Recasting Hoare and He's Unifying Theory of Programs in the context of general correctness. In: Butterfield, A., Strong, G., Pahl, C. (eds.) 5th Irish Workshop on Formal Methods. Electronic Workshops in Computing, The British Computer Society (2001)
5. Guttmann, W.: General correctness algebra. In: Berghammer, R., Jaoua, A.M., Möller, B. (eds.) RelMiCS/AKA 2009. LNCS, vol. 5827, pp. 150–165. Springer, Heidelberg (2009)
6. Guttmann, W.: Extended designs algebraically (submitted, 2011)
7. Guttmann, W., Möller, B.: Normal design algebra. Journal of Logic and Algebraic Programming 79(2), 144–173 (2010)
8. Hayes, I.J., Dunne, S.E., Meinicke, L.: Unifying theories of programming that distinguish nontermination and abort. In: Bolduc, C., Desharnais, J., Ktari, B. (eds.) MPC 2010. LNCS, vol. 6120, pp. 178–194. Springer, Heidelberg (2010)
9. Kahl, W.: Refactoring heterogeneous relation algebras around ordered categories and converse. Journal on Relational Methods in Computer Science 1, 277–313 (2004)
10. Kozen, D.: A completeness theorem for Kleene algebras and the algebra of regular events. Information and Computation 110(2), 366–390 (1994)
11. Kozen, D.: Typed Kleene algebra. Tech. Rep. TR98-1669, Cornell University (1998)
12. Kozen, D.: On Hoare logic, Kleene algebra, and types. In: Gärdenfors, P., Woleński, J., Kijania-Placek, K. (eds.) In the Scope of Logic, Methodology, and Philosophy of Science, Synthese Library, vol. 315, pp. 119–133. Kluwer Academic Publishers, Dordrecht (2002)
13. Mathieu, V., Desharnais, J.: Verification of pushdown systems using omega algebra with domain. In: MacCaull, W., Winter, M., Düntsch, I. (eds.) RelMiCS 2005. LNCS, vol. 3929, pp. 188–199. Springer, Heidelberg (2006)
14. Möller, B.: The linear algebra of UTP. In: Uustalu, T. (ed.) MPC 2006. LNCS, vol. 4014, pp. 338–358. Springer, Heidelberg (2006)
15. Pous, D.: Untyping typed algebraic structures and colouring proof nets of cyclic linear logic. In: Dawar, A., Veith, H. (eds.) CSL 2010. LNCS, vol. 6247, pp. 484–498. Springer, Heidelberg (2010)
16. Schmidt, G., Hattensperger, C., Winter, M.: Heterogeneous relation algebra. In: Brink, C., Kahl, W., Schmidt, G. (eds.) Relational Methods in Computer Science, ch. 3, pp. 39–53. Springer, Wien (1997)

Towards an Algebra of Routing Tables

Peter Höfner[1,3] and Annabelle McIver[2,3]

[1] Institut für Informatik, Universität Augsburg, Germany
[2] Department of Computing, Macquarie University, Australia
[3] National ICT Australia Ltd. (NICTA)
peter.hoefner@nicta.com.au, annabelle.mciver@mq.edu.au

Abstract. We use well-known algebraic concepts like semirings and matrices to model and argue about Wireless Mesh Networks. These networks are used in a wide range of application areas, including public safety and transportation. Formal reasoning therefore seems to be necessary to guarantee safety and security. In this paper, we model a simplified algebraic version of the AODV protocol and provide some basic properties. For example we show that each node knows a route to the originator of a message (if there is one).

1 Introduction

Wireless Mesh Networks (WMNs) are currently used in a wide range of application areas, including public safety, transportation, mining, etc. Typically, these networks do not have a central component (router), but each node in the network acts as an independent router, regardless of whether it is connected to another network or not. They allow reconfiguration around broken or blocked paths by hopping from node to node until the destination is reached.

The Ad-hoc On-Demand Distance Vector Routing (AODV) protocol [20] is a routing protocol that finds a route on demand (if needed and if unknown) in WMNs. To guarantee safety and security aspects for WMNs in general and AODV in particular, mathematical analysis should be used. It is our belief that one does not have to develop new theories. We believe that the concepts and tools developed over the last decades are powerful enough to capture new concepts like AODV. For example, Singh et al [21] use process algebra to model a simpler version of AODV. Within this approach processes reflect the behaviour of nodes, i.e., a process describes how nodes react when messages are received. In this paper we will use routing tables as basic elements rather than processes. This allows us to use algebraic concepts like semirings, Kleene algebra and matrices to model some parts of the AODV protocol. The algebraic operations will then be transformers of routing tables. We show how the basic concepts of AODV can be modelled. In particular, we characterise AODV control messages, which are used to distribute knowledge through the network. To achieve this goal, the algebraic operations have to model things like broadcasting (sending to all neighbours) and unicasting (sending to one neighbour) messages.

H. de Swart (Ed.): RAMICS 2011, LNCS 6663, pp. 212–229, 2011.

2   AODV and Its Design Challenges

The AODV protocol [20] supports routing between nodes in a WMN in a demand driven manner. Nodes are able to communicate directly between connected neighbours, but routes between nodes in general comprise a number of "hops". In a dynamic environment, where the network topology can change—typically a consequence of mobile nodes which can cause new connections to be established, or existing ones to disappear—a routing protocol must be able to identify routes which have become obsolete and replace them with valid alternatives.

In order to establish valid routes, information is disseminated throughout the network by the exchange of control messages (requests, replies and error messages) and over time a node essentially "learns" to which distant nodes it is connected, albeit indirectly by multiple hops. One of the challenges presented by a dynamic topology is the management of routes if they become obsolete. Without a careful accounting policy to handle "broken" routes—even if the specific link failure is someway downstream—there is a risk that the routing algorithm establishes so-called "loops": routes which, if followed, will never reach their destination. Reporting broken links to other nodes on a route is one of the functions of AODV and the source of much of its complexity.

In this paper our aim is to describe an algebraic approach to model the behaviour of the network control messages, with a view to modelling some of the principal features of AODV. We begin in the next section by an overview of the underlying network mechanisms and how they fit together.

2.1   Overview of AODV

The basic algorithm underlying AODV disseminates knowledge of local connections or "links" throughout the network; nodes "know" their immediate neighbours (to whom they are connected directly), or "1-hop neighbours" via a network "polling" mechanism known as the *Hello protocol*. Its task is to broadcast a "hello" regularly to any node in its vicinity; any node receiving the broadcast establishes a 1-hop connection to the sender. Any node D which is not directly connected to a source node S requires messages destined for D to be routed via a series of intermediate nodes. We say a *route* from S to D is a sequence of nodes

$$S, N_1, N_2, ..., N_k, D , \tag{1}$$

where each successive node in the sequence is a 1-hop neighbour to its predecessor. The route of Expression (1) has $k+1$ hops. Such a route is said to be *loop-free* when no node occurs more than once.

The objective of AODV is to ensure that, when required, a node is able to identify its next immediate hop on a complete valid route to a given destination D. Nodes maintain a *routing table entry* for each possible destination, which includes the next hop together with the total number of hops required to reach the destination. For example the routing table at node S would have its next hop for destination D registered as N_1 with a hop count $k+1$, whereas each intermediate node N_i $(1 \leq i < k)$ would have its next hop to D registered as N_{i+1} with a hop count $k-i+1$. As we shall see the hop count is necessary for the node to choose between two or more possible valid routes.

Establishing the routing table entries is implemented by flooding information about the 1-hop neighbours throughout the network; it has been shown that the next hop and the total hop count is not sufficient information to avoid the accidental construction of loops. We sketch how that might occur.

Information about the topology, i.e., the 1-hop neighbours, is flooded through-out the network by exchange of control messages, allowing knowledge of routes to be established and next hops to be recorded in the respective routing tables. Consider the scenario in Figure 1. Here there are two valid routes from S to D (via A and B, resp.). Following the specification of the protocol each node is able to store exactly one path to a certain destination. Assume that S has established a path S, A, D. If the link between A and D subsequently breaks, and A initiates a new search for a path to D, its request for a new path would be answered by a response from S that a route can be found to D via A. If fresh requests are not distinguished from previous requests, this would establish a path which is not loop free. To avoid this scenario AODV uses a scheme to distinguish between fresh requests and old routes; in this paper we will concen-trate on setting up the algebra to deal with the simpler case, where freshness is not considered. However, we will discuss extensions which will allow us to deal with the full generality.

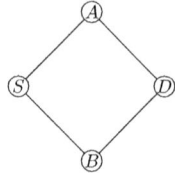

Fig. 1. Illustrating changing topology

2.2   Basic Protocol

As for most routing protocols, the basic idea of AODV is inspired by traditional path search such as Dijkstra's shortest path algorithm [7]—AODV aims to min-imise hop count. Recall that Dijkstra's shortest path algorithm keeps a record of the paths from a source S, extending those paths in each iteration of the algorithm. Loops are avoided by marking nodes that have been visited before. AODV is based on a distributed version of this idea, where the set of visited nodes is extended in several steps by utilising the network control messages. We observe that messages are either *broadcast* or *unicast*, with the difference being that broadcast messages are intended for any node that is connected, whereas unicast messages are addressed to a single specific node. Thus unicast messages are only received by the addressee, and any other node which picks up the message simply ignores it.

The main AODV control messages are as follows:

Route Request (RREQ). When a node requires a route to a (new) destination D, the node broadcasts a RREQ for that destination. A node A that receives the RREQ *either* broadcasts the RREQ again (thereby informing a possibly new set of neighbours) *or* it responds with the "route reply" (see below). The choice to broadcast is made in the case that A itself does not know a route to D; it sends a route reply if it already knows a route to D. Finally if A has already received and broadcasted or responded to a RREQ to D it simply ignores it.[1]

Route Reply (RREP). If R is the destination D or knows a route to D then A unicasts a RREP back to the originator of the corresponding RREQ, with the information about the number of hops to D.

Hello messages are used to establish 1-hop neighbours. A node executing a Hello simply broadcasts a message; any node that picks it up establishes a 1-hop connection.

Route Error (RERR). As mentioned above, WMNs are based on wireless networks with changing topology. Therefore links can be lost. When a link break in an active route is detected, a RERR message is used to notify other nodes that the loss of that link has occurred.[2]

3   Algebraic Approaches for Routing Protocols

We are not the first who want to use algebraic techniques to describe routing procedures. Most of the purely algebraic approaches are done by Sobrinho and Griffin [22,23,10,9]. It seems that Sobrinho was the first who brought algebraic reasoning into the realm of hop-by-hop routing [22]. He uses algebraic properties to argue about the relationship between routing algorithms and Dijkstra's shortest path algorithm. The author states that he follows an algorithmic approach while others use a matrix approach [4]. On the other hand, Griffin uses algebraic structures like semigroups and semirings for the analysis of path vector protocols like the Border Gate Protocol BGP.

In fact, the above mentioned matrix approach is older still and was already used by Backhouse and others [3]. The next evolutionary step of the matrix approach was taken by Kozen [14] who showed that $n \times n$-matrices over Kleene algebras again form a Kleene algebra. He was able to show the correctness of a simple algorithm for reflexive transitive closure purely algebraically, in addition to establishing the relationship between matrix algebra and algorithms for all-pair shortest path algorithms. Later, Möller used test elements—now well-established in semirings and Kleene algebra—to model nodes within the matrix

---

[1] The RREQ handling is more complicated in the case of dealing with broken links.

[2] We include this for completeness.

algebra and to present algebraic versions of the algorithms of Dijkstra and Floyd-Warshall [18]. Another area where matrix algebras can be applied is in static program analysis. This was shown by Fernandes and Desharnais [8].

In this paper, we will extend the matrix approach aiming at a model for the distance vector protocol AODV. In particular, matrices are used as representatives of (local) routing tables, and we formulate AODV control messages as operations on routing tables. To the best of our knowledge, both Sobrinho and Griffin consider static routing protocols, i.e. where routers are required. In this setting link breaks are possible but changes in topology are not, unlike the context for AODV where it is common for links to break and the underlying network topology to change. Moreover, we are interested in modelling concrete actions performed by AODV in an algebraic setting. Again to the best of our knowledge this has not been done before. Interestingly, though path and distance vector protocols are different in practice, the underlying algebraic structures bear strong similarities.

4   Semirings, Kleene Algebras and Extensions

In this section we briefly recapitulate the algebraic structures needed to model AODV control messages.

It is well known that semirings and Kleene algebras model sequential composition, (non-deterministic) choice and finite iteration. They provide the appropriate level of abstraction for modelling actions, programs or state transitions under non-deterministic choice and sequential composition in a first-order equational calculus. A first-order calculus allows the use of first-order automated theorem provers to validate and verify properties. It has been shown that the algebraic structures used are particularly amenable to automated reasoning [13].

An *idempotent semiring (i-semiring)* is a quintuple $(S, +, \cdot, 0, 1)$ such that $(S, +, 0)$ is an idempotent commutative monoid, $(S, \cdot, 1)$ is a monoid, multiplication distributes over addition from the left and right and 0 is a left and right zero of multiplication. The *natural order* $\leq$ on S is given by $a \leq b \Leftrightarrow_{df} a + b = b$. It induces an upper semilattice in which $a + b$ is the supremum of a and b, and 0 is the least element.

A *Kleene algebra* is an idempotent semiring S extended by an operation $^* : S \to S$ for iterating an element an arbitrary but finite number of times. Such an operation has to satisfy the *star unfold* and the *star induction* axioms

$$1 + a \cdot a^* = a^* , \qquad b + a \cdot c \leq c \Rightarrow a^* \cdot b \leq c ,$$
$$1 + a^* \cdot a = a^* , \qquad b + c \cdot a \leq c \Rightarrow b \cdot a^* \leq c .$$

If elements of semirings/Kleene algebras characterise single transitions, whole transition systems are usually encoded in matrices. To calculate with matrices we recapitulate a well known result:

Theorem 4.1. *Standard operations of matrix addition and multiplication turns the family $M(n, K)$ of $n \times n$ matrices over a Kleene algebra K into a Kleene* *algebra again; the zero matrix is neutral w.r.t.* $+$*, and the identity matrix* I *w.r.t. multiplication.*

A proof can e.g. be found in [15], the techniques used are based on preliminary work by Conway [5] and Backhouse [2].

Tests of a program or sets of states of a transition system can also be modelled in this setting. A *test* in an i-semiring S is an element of a Boolean subalgebra test$(S) \subseteq S$ such that test(S) is bounded by 0 and 1 and multiplication $\cdot$ coincides with lattice meet. We will write $a, b, \ldots$ for arbitrary semiring elements and $p, q, \ldots$ for tests. Moreover, $\neg p$ denotes the complement of p in the Boolean subalgebra. An important property (e.g. [6]) of tests is

$$p \cdot a \cdot q \le 0 \Leftrightarrow a \cdot q \le \neg p \cdot a \ . \tag{2}$$

In the matrix model, the maximal test set contains all diagonal matrices with tests of the underlying algebra on its diagonal. Hence Theorem 4.1 also holds for Kleene algebras with tests.

Idempotent semirings admit at least the test algebra $\{0, 1\}$ and can have different test algebras. On test semirings one can define a generalised notion of the weakest liberal precondition wlp (see [19]). This is achieved by defining a box-operator $|\_] : S \mapsto (\text{test}(S) \mapsto \text{test}(S))$ by

$$p \le |a]q \Leftrightarrow p \cdot a \cdot \neg q \le 0 \ , \quad \text{and} \quad |a \cdot b]p = |a](|b]p) \ .$$

The diamond $|\_\rangle$ is the de Morgan dual of this operation, i.e., $|a\rangle p = \neg |a] \neg p$. This operator will be used to determine whether there is a route from p to q in a (checking $p \le |a\rangle q$). An important property of $|a\rangle p$ is that it is the least left preserver of $a \cdot p$. In particular, we have

$$|a\rangle p \cdot a \cdot p = a \cdot p \ . \tag{3}$$

Note that the diamond- and the box-operator bind stronger than addition and multiplication. Test semirings equipped with $|\_]$ and $|\_\rangle$ are called *modal*. Modal semirings can easily be extended to modal Kleene algebras without any further assumptions.

5   Routes and Routing Tables

A routing table is a data structure (often in the form of a table or vector) that lists the routes to network destinations and, most often, also additional information like metrics, sequence numbers or knowledge about the topology.

In classical networks/protocols a routing table is stored in a central component like a router. Reactive network protocols like AODV use a different approach: Each component has its own routing table. This avoids the existence of a central component and is therefore less prone to computer failures like crashes: If a single component crashes the remaining components can still communicate.

In AODV, each component of an arbitrary routing table stores for a known destination, the next node (where to send the packet), the length (hop count) of the route, information about the freshness of a route and some more information. For the moment we only focus on the former two components and omit the remaining (cf. Sections 2 and 7):

- The *next hop* identifies a neighbour where the packet has to be sent to reach a particular destination D.
- The *hop count* specifies the length of the path to D (i.e., number of nodes that have to be visited)

The set of node names is denoted by N; the set H gives a value to compare the quality of two entries (for example the length).

Formally, a *routing table entry* (*entry* for short) is a pair (m, x) over two totally ordered sets $(N, \preceq)$ and $(H, \sqsubseteq)$. On the set $N \times H$ of all entries, we will use the lexicographical ordering (again a total order) to measure the quality of entries.

Next to these "proper" elements, we define two special entries $(\varepsilon, 0)$ and $(\varepsilon, \infty)^3$, both not elements of $N \times H$. The set $\mathbb{M} =_{df} N \times H \cup \{(\varepsilon, \infty), (\varepsilon, 0)\}$ denotes the set of all possible route entries.

In AODV, the names are given as IP addresses; its ordering is straight forward. A proper entry $(m, x) \in N \times H$ states that there is a path (its source and destination will be encoded somewhere else) with next node m and length x; $(\varepsilon, 0)$ states that there is a trivial path of length 0; hence no next hop must be given. In contrast to that, (ε, ∞) denotes an entry that corresponds to a route of infinite length, where also no next hop is given. This construct can be seen as the statement of "no (known) route".

Example 5.1. For the upcoming examples we use $N =_{df} \{A, B, C, \dots\}$ together with the lexicographical order. Moreover, we set $H = \mathbb{N} - \{0\}$. Assume $A \in N$ to be a name, then the pair $(A, 5)$ means that there is a route (its source S and its destination D will be encoded somewhere else) where the next node on the path is A and the length is 5 (cf. Figure 2).   □

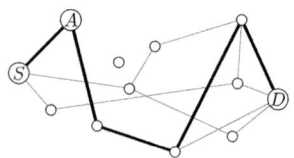

Fig. 2. A route from S to D of length 5

---

^{3} We only use pairs to be consistent with proper elements (m, x). The symbols $(\varepsilon, 0)$ and (ε, ∞) just denote new elements and could be denoted differently.

Next we want to define operations on entries. If there is a choice between two routes going from the same source S to the same destination D, it is obvious that one should prefer the shorter route; if the lengths of two entries are the same, we take the one with the "better name" (using the ordering of names). Again, this is expressed by defining choice (addition) via the lexicographical order[4]:

$$(m, x) + (n, y) =_{df} \begin{cases} (m, x) & \text{if } (x \sqsubset y) \vee (x = y \wedge m \preceq n) \\ (n, y) & \text{otherwise ,} \end{cases}$$

for $(m, x), (n, y) \in N \times H$ being proper entries, where $x \sqsubset y \Leftrightarrow_{df} x \neq y \wedge x \sqsubseteq y$. For the previously defined special elements $(\varepsilon, 0)$ and (ε, ∞) choice is defined by

$$(\varepsilon, \infty) + r =_{df} r + (\varepsilon, \infty) =_{df} r ,$$
$$(\varepsilon, 0) + r =_{df} r + (\varepsilon, 0) =_{df} (\varepsilon, 0) ,$$

for all $r \in \mathbb{M}$. Intuitively, this means that $(\varepsilon, 0)$ is the best entry; an entry of length 0 is favoured over all other entries (and routes). In contrast to that (ε, ∞) is the worst.

Using pairs together with lexicographical ordering in the setting of internet routing is not new; Sobrinho used similar pairs which yield "most-reliable-shortest paths" [22]. However, he does not use the concept of next hops. This concept is mentioned in [10].

Next to choice we also have to combine routes.

Example 5.2. Assume an entry $(A, 5)$ from S to S' and another entry $(B, 7)$ from S' to D, then there is a "route" $(A, 12)$ from S to D with next hop A. A sketch is given in Figure 3.   □

Fig. 3. Route composition

Formally, we assume N, H not only to be ordered sets, but also to offer binary operations $\circ : N \times N \rightarrow N$ and $\bullet : H \times H \rightarrow H$. Entry composition is then defined by pointwise lifting. For $(m, x), (n, y) \in N \times H$ that means

$$(m, x) \cdot (n, y) =_{df} (m \circ n, x \bullet y) .$$

The special elements are handled, for all $r \in \mathbb{M}$, by

$$(\varepsilon, \infty) \cdot r =_{df} r \cdot (\varepsilon, \infty) =_{df} (\varepsilon, \infty) , \qquad (\varepsilon, 0) \cdot r =_{df} r \cdot (\varepsilon, 0) =_{df} r .$$

---

[4] Note that we define the lexicographical order in a right-to-left way. That means that we first compare the second component, whenever the second component is equal, we have a closer look at the first one.

In the examples, the operation $\circ$ coincides with "taking the left element", i.e., $m \circ n = m$ for all $m, n \in N$ and $\bullet$ with ordinary addition on natural numbers.

So far, we have defined a number of useful operations on routing table entries. Under certain circumstances the above operations behave well and form a semiring and a Kleene algebra; in general we get:

Theorem 5.3. *Assume two totally ordered sets $(N, \preceq)$ and $(H, \sqsubseteq)$ with an isotone binary operation $\circ$ on N and a binary operation $\bullet : H \times H \to H$ that is strictly isotone, i.e., $a \sqsubset b \Rightarrow (c \bullet a \sqsubset c \bullet b) \wedge (a \bullet c \sqsubset b \bullet c)$.*

On the set $\mathbb{M} = N \times H \cup \{(\varepsilon, \infty), (\varepsilon, 0)\}$ we define addition and multiplication as above.

(a) *The structure $S =_{df} (\mathbb{M}, +, (\varepsilon, \infty), \cdot, (\varepsilon, 0))$ forms an idempotent semiring, if $x \sqsubset x \bullet y$ for all $x, y \in H$; its natural order coincides on $N \times H$ with the lexicographical order $(n, y) \le (m, x) \Leftrightarrow_{df} (x \sqsubset y) \vee (x = y \wedge m \preceq n)$; (ε, ∞) is the least and $(\varepsilon, 0)$ the greatest element.*

(b) *Under the conditions of Part (a) and setting $r^* =_{df} (\varepsilon, 0)$ for all $r \in \mathbb{M}$ turns S into a Kleene algebra.*

Proof (sketch). The proof is by straightforward calculations. Parts of Part (a) were already stated and proved in [23].

(a) First, the lexicographical order of two total orders also forms a total order. Define (ε, ∞) as least and $(\varepsilon, 0)$ as greatest element, gives yet another total order, where addition calculates the maximum. Hence it is obvious that $(\mathbb{M}, +, (\varepsilon, \infty))$ forms a commutative monoid. Properties of $\cdot$ immediately follow by product construction (e.g. by Birkhoff's HSP theorem of universal algebra). Hence only distributivity has to checked by hand. Here the condition $x \sqsubset x \bullet y$ is needed.

(b) Since in every Kleene algebra $1 \le a^*$, defining Kleene star as $(\varepsilon, 0)$ is necessary. From this, the axioms can easily be checked. $\square$

Note that addition chooses the better route. Therefore the natural order reads as "worse than". In particular, $(n, y) \le (m, x)$ means that the entry (n, y) is worse than (m, x). Most often this implies that (m, x) has a smaller hop count than (n, y).

For the concrete model used in the examples, the condition $x \sqsubset x + y$ obviously holds since $x, y \in \mathbb{N} - \{0\}$. Therefore the structure forms a Kleene algebra and axioms for finite iteration are available. The test algebra is discrete, i.e., it only consists of the two constants. The additional condition is not arbitrary, it even seems to be crucial for routing protocols since in [22] a similar property is used.

So far we only had a look at routing table entries. However, in general we are interested in sets of entries, where each route/entry also has a source and a destination. To model whole sets we use $n \times n$ matrices over entries. An entry at position i, j then corresponds to a route from i to j. To use matrices, we have to assume that the set of names is *finite*. In reality, this is not a restriction at all, since there can only be a finite number of nodes in a given network.[5] Each row of a routing table corresponds to a local routing table of a node. A routing table can therefore be seen as a "snapshot" of the whole network, where all node's routing tables are joined.

Example 5.4. Assume the connectivity graph depicted on the left hand side of Figure 4 and the arbitrary chosen routing table on the right.

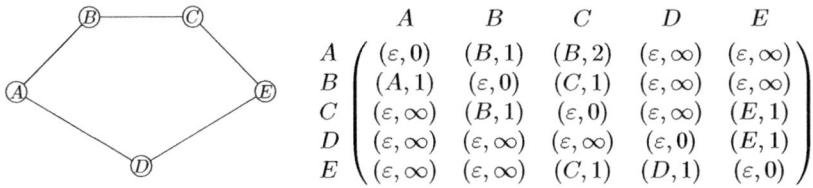

$$
\begin{array}{c}
& A & B & C & D & E \\
A & (\varepsilon,0) & (B,1) & (B,2) & (\varepsilon,\infty) & (\varepsilon,\infty) \\
B & (A,1) & (\varepsilon,0) & (C,1) & (\varepsilon,\infty) & (\varepsilon,\infty) \\
C & (\varepsilon,\infty) & (B,1) & (\varepsilon,0) & (\varepsilon,\infty) & (E,1) \\
D & (\varepsilon,\infty) & (\varepsilon,\infty) & (\varepsilon,\infty) & (\varepsilon,0) & (E,1) \\
E & (\varepsilon,\infty) & (\varepsilon,\infty) & (C,1) & (D,1) & (\varepsilon,0)
\end{array}
$$

Fig. 4. A simple network consisting of 5 nodes

Intuitively, the first entry $(\varepsilon,0)$ states that if A wants to send something to A, there is nothing to do and the known path is "optimal", i.e., it has length 0; the entry $(B,2)$ in the middle of the first row states that A knows a route to C with length 2. The next hop where A has to send the packet to is B. The entry (ε,∞) indicates that no route is known.   □

In general a *routing table* is an element of the set $M(n,K)$ of all $n \times n$ matrices with elements in the Kleene algebra of routing tables presented before. Here, $n = |N|$ is given by the number of possible nodes.

Note that we restrict a routing table in no way. Therefore a node might not even know all its neighbours. Moreover, there is no need of symmetry, i.e., A knows a route to B does not imply necessarily that B knows a route to A. This reflects the reality. However, it seems reasonable to assume that a routing table has $(\varepsilon,0)$-entries on its diagonal. This means that each node at the network knows at least itself. Algebraically this is easily expressed by $1 \le a$.

As stated before tests are subidentities. In our matrix model a test is a matrix that has the entries (ε,∞) or $(\varepsilon,0)$ on its diagonal; all other entries of the matrix are (ε,∞). Typically, a test is used to describe a set of nodes. Premultiplying an arbitrary matrix (routing table) by a test selects certain rows; hence the routing table information about particular nodes is selected. Postmultiplying selects columns, i.e., the snapshopt is "restricted" to information of a particular node or a set of nodes.

Looking at AODV, messages are sent throughout the network. Hence we have to define neighbours in a network, where messages can be sent to. Obviously, this is again done by a routing table.

---

[5] For AODV for example this number is bounded by 2^{32}, the number of possible unique IP addresses (http://en.wikipedia.org/wiki/IP_address).

A *topology* b is a special routing table which contains the identity as submatrix $(1 \leq b)$ *and* also all available links between two nodes.

In our examples a topology contains all available one-hop connections.

Since we have shown that routing tables or, to be more precise sets of routing tables form well-known algebraic structures we are now ready to abstract to algebra and to model parts of AODV.

6   AODV Control Messages Algebraically

AODV is based on control messages sent through the network. As we will see it is less important whether messages are broadcast or unicast. At the algebraic level used, message sending can be expressed by

$$a + p \cdot b \cdot q \cdot (1 + c) , \tag{4}$$

where a, b, c are elements of an i-semiring S and $p, q \in \text{test}(S)$. By distributivity, Expr. (4) consists of three parts: a, $p \cdot b \cdot q$ and $p \cdot b \cdot q \cdot c$. Informally, a describes the system before the message is sent (hence it is not part of the message itself); it can be seen as a snapshot which then, after the message has been delivered, is (if possible) updated by the two other summands. Remember that addition chooses the better route, hence it can be indeed interpreted as update. When a node p receives an AODV control packet from a neighbour q it creates or updates its routing table. If b represents the current topology, the equation $p \cdot b \cdot q$ establishes a 1-hop connection from p to q (if p and q are single nodes), i.e., p knows a path to q. If p, q are sets of nodes, only those paths between p and q are established that really exist in the topology. The third term $(p \cdot b \cdot q \cdot c)$ transmits knowledge (encoded in an element or a routing table c) from q via the topology to p. In the meaning of routing table updates this again reads the other way round. Since p receives a message from q it can update its routing table with information of c.

Before modelling control messages we prove some useful properties about messages in general. For that we define a function

$$\text{msg}(a, b, c) =_{df} a + b \cdot (1 + c)$$

for sending message c via b updating a; here b is either a given topology or it can be restricted by senders and receivers as before. That means that b might have the form $p \cdot b' \cdot q$.

Proposition 6.1

(a) *If the knowledge c and c' is fixed (does not change when sending a message), the order of sending does not matter, i.e.,*

$$\text{msg}(\text{msg}(a, b, c), b', c') = \text{msg}(\text{msg}(a, b', c'), b, c) .$$

(b) *If different messages are sent via a shared topology b, the messages can be sent in parallel, i.e.,*

$$\text{msg}(\text{msg}(a, b, c), b, c') = \text{msg}(a, b, c + c') .$$

(c) *If the same message is sent via different connections, connections can be joined, i.e.,*

$$\mathtt{msg}(\mathtt{msg}(a,b,c),b',c) = \mathtt{msg}(a,b+b',c) \ .$$

(d) *Sending c via b using different sets of senders p and p' to receivers q and q' is not interchangeable. We only have*

$$\mathtt{msg}(\mathtt{msg}(a,p \cdot b \cdot q,c),p' \cdot b \cdot q',c) \le \mathtt{msg}(a,(p+p') \cdot b \cdot (q+q'),c) \ .$$

(e) *It is interchangeable if p is not connected to q' via b ($q' \le |b|\neg p$) and p' not to q, i.e.,*

$$q' \le |b|\neg p \ \wedge \ q \le |b|\neg p'$$
$$\Rightarrow \mathtt{msg}(\mathtt{msg}(a,p \cdot b \cdot q,c),p' \cdot b \cdot q',c) = \mathtt{msg}(\mathtt{msg}(a,p' \cdot b \cdot q',c),p \cdot b \cdot q,c) \ .$$

Morover, if $q' \le |b|\neg p$ and $q \le |b|\neg p'$ then

$$\mathtt{msg}(\mathtt{msg}(a,p \cdot b \cdot q,c),p' \cdot b \cdot q',c) = \mathtt{msg}(\mathtt{msg}(a,p' \cdot b \cdot q',c),p \cdot b \cdot q,c) \ ,$$

meaning that sending can be done in parallel.

In general, Part (d) cannot be strengthened to an equation since p might be able to to send information to q' and p' to q. This behaviour is excluded when p is not connected to q' and p' not to q (cf. Part (e)).

The proofs are straightforward algebraic calculations. Moreover, they can be automated using first-order theorem provers—another advantage of an algebraic approach. We used Prover9 [16] to verify these properties. The input files can be found at [12].

A node following the rules of the AODV protocol often forwards messages. When a content c is forwarded, it is also changed. For example, if p receives knowledge c via b, it will not forward c but $b \cdot c$. Sending a message c via the topology b and forwarding it via another topology b' can be encoded by $\mathtt{msg}(\mathtt{msg}(a,b,c),b',b \cdot c)$. Assuming $1 \le b'$, we get

$$
\begin{aligned}
& \mathtt{msg}(\mathtt{msg}(a,b,c),b',b \cdot c) \\
= \ & a+b+b \cdot c+b'+b' \cdot b \cdot c \\
\le \ & a+b'+b' \cdot b+b' \cdot b \cdot c \\
= \ & a+b'(1+b+b \cdot c) \\
= \ & \mathtt{msg}(a,b',b+b \cdot c) \ .
\end{aligned}
$$

Intuitively, this means that the knowledge after forwarding a message once can be approximated by sending a single message via b' with knowledge of the first topology b and the learnt component $b \cdot c$. Forwarding and broadcasting a message through an entire network, is now forwarding the message again and again. In algebra this yields long, but simple expressions (cf. the calculation above). In the situation where the network topology does not change things become much easier. Forwarding once yields now the equation

$$\mathtt{msg}(\mathtt{msg}(a,b,c),b,b \cdot c) = \mathtt{msg}(a,b,b \cdot c) \ .$$

Hence, by simple fixpoints arguments, broadcasting a message can be modelled by (a single message)

$$\mathtt{msg}(a, b, b^* \cdot c) = a + b \cdot (1 + b^*c) = a + b + b \cdot c + b \cdot b \cdot c + b \cdot b \cdot b \cdot c + \ldots .$$

The snapshot a is first updated with information from the topology b, then by information c sent via the topology by a 1-hop connection, then by the information c sent via a 2-hop connection and so on.

Before turning to the core messages of AODV, we look at a special case of forwarding messages. We assume that there is a sender p which broadcasts the empty message ($c = 1$) and that only those nodes that have actually received information forward the messages . The first message is given by $\mathtt{msg}(a, b \cdot p, 1) = a + b \cdot p$. The receivers of this message are characterised by $|b\rangle p$. After forwarding the received message once, the system looks like

$$
\begin{aligned}
&\mathtt{msg}(\mathtt{msg}(a, b \cdot p, 1), b \cdot |b\rangle p, b \cdot p) \\
={}& a + b \cdot p + b \cdot p + b \cdot |b\rangle p + b \cdot |b\rangle p \cdot b \cdot p \\
={}& a + b \cdot (|b^0\rangle p + |b^1\rangle p) + b \cdot p + b \cdot b \cdot p \\
={}& a + b \cdot (|b^0 + b^1\rangle p) + b \cdot (p + b \cdot p) .
\end{aligned}
$$

The first step is by definition and distributivity. The second step holds since $p = |1\rangle p = |b^0\rangle p$, by commutativity, distributivity, and Equation (3). The third is again by distributivity and additivity of $|_-\rangle$. Propagating the message through the whole network can therefore again be expressed by using Kleene star.

$$a + b \cdot |b^*\rangle p + b^* \cdot p . \tag{5}$$

Using the assumption $1 \le a$, this is equivalent to $\mathtt{msg}(a, b \cdot |b^*\rangle p, b^* \cdot p)$.

Let us now formalise the core messages of AODV.

Hello Messages do not send any information ($c = 1$) except the one of the connection. Moreover, if p sends such a message, every neighbour will receive it. Hence it has the form $a + b \cdot p$. In the matrix model terms like $b \cdot p$ restrict matrices column-wise. In particular this means that only information about routes towards p can be learned.

Example 6.1. Assume that the topology T is given by the graph of Figure 5; we further assume that D is the test element representing the single node D and that no node has any information about routes (except the trivial one to themselves), i.e., we start with the identity matrix as snapshot. Then $I + T \cdot D$ models that D sends a hello message. The result is given on the right hand side of Figure 5.

The resulting matrix shows that now the nodes B, C and E have established routes to D.   □

Route Requests are a bit more complicated. On the one hand a node that receives a message establishes a 1-hop connection as in the case of saying "Hello". This is the first iteration of the broadcast. However, if the neighbours do not have information about the destination, they forward the message by broadcasting it to their neighbours.

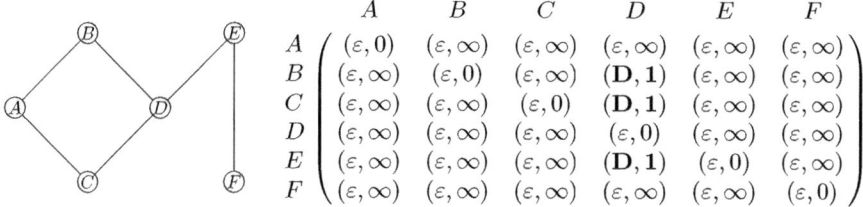

$$
\begin{array}{c@{\quad}c@{\quad}c@{\quad}c@{\quad}c@{\quad}c@{\quad}c}
 & A & B & C & D & E & F \\
\end{array}
$$

$$
\begin{array}{c}
A \\ B \\ C \\ D \\ E \\ F
\end{array}
\left(
\begin{array}{cccccc}
(\varepsilon,0) & (\varepsilon,\infty) & (\varepsilon,\infty) & (\varepsilon,\infty) & (\varepsilon,\infty) & (\varepsilon,\infty) \\
(\varepsilon,\infty) & (\varepsilon,0) & (\varepsilon,\infty) & (\mathbf{D},\mathbf{1}) & (\varepsilon,\infty) & (\varepsilon,\infty) \\
(\varepsilon,\infty) & (\varepsilon,\infty) & (\varepsilon,0) & (\mathbf{D},\mathbf{1}) & (\varepsilon,\infty) & (\varepsilon,\infty) \\
(\varepsilon,\infty) & (\varepsilon,\infty) & (\varepsilon,\infty) & (\varepsilon,0) & (\varepsilon,\infty) & (\varepsilon,\infty) \\
(\varepsilon,\infty) & (\varepsilon,\infty) & (\varepsilon,\infty) & (\mathbf{D},\mathbf{1}) & (\varepsilon,0) & (\varepsilon,\infty) \\
(\varepsilon,\infty) & (\varepsilon,\infty) & (\varepsilon,\infty) & (\varepsilon,\infty) & (\varepsilon,\infty) & (\varepsilon,0)
\end{array}
\right)
$$

Fig. 5. Sending Hello Messages

Example 6.2. Assume the topology T and the snapshot of Figure 5—denoted by X. Now we want to assume that A sends a RREQ looking for a path to D. As first step the snapshot is updated by $X + T \cdot A$, where A is the test representing the single node A; the result is

$$
\begin{array}{c@{\quad}c@{\quad}c@{\quad}c@{\quad}c@{\quad}c@{\quad}c}
 & A & B & C & D & E & F \\
\end{array}
$$

$$
\begin{array}{c}
A \\ B \\ C \\ D \\ E \\ F
\end{array}
\left(
\begin{array}{cccccc}
(\varepsilon,0) & (\varepsilon,\infty) & (\varepsilon,\infty) & (\varepsilon,\infty) & (\varepsilon,\infty) & (\varepsilon,\infty) \\
(\mathbf{A},\mathbf{1}) & (\varepsilon,0) & (\varepsilon,\infty) & (D,1) & (\varepsilon,\infty) & (\varepsilon,\infty) \\
(\mathbf{A},\mathbf{1}) & (\varepsilon,\infty) & (\varepsilon,0) & (D,1) & (\varepsilon,\infty) & (\varepsilon,\infty) \\
(\varepsilon,\infty) & (\varepsilon,\infty) & (\varepsilon,\infty) & (\varepsilon,0) & (\varepsilon,\infty) & (\varepsilon,\infty) \\
(\varepsilon,\infty) & (\varepsilon,\infty) & (\varepsilon,\infty) & (D,1) & (\varepsilon,0) & (\varepsilon,\infty) \\
(\varepsilon,\infty) & (\varepsilon,\infty) & (\varepsilon,\infty) & (\varepsilon,\infty) & (\varepsilon,\infty) & (\varepsilon,0)
\end{array}
\right)
$$

We see that the nodes B and C have learned about the connection to A. Now these nodes should forward the request if they do not know D. However, both already know a connection to D, hence nothing happens.

Using I as snapshot instead of the one of Fiure. 5 yields another outcome.

$$
\begin{array}{c@{\quad}c@{\quad}c@{\quad}c@{\quad}c@{\quad}c@{\quad}c}
 & A & B & C & D & E & F \\
\end{array}
$$

$$
\begin{array}{c}
A \\ B \\ C \\ D \\ E \\ F
\end{array}
\left(
\begin{array}{cccccc}
(\varepsilon,0) & (B,1) & (C,1) & (\varepsilon,\infty) & (\varepsilon,\infty) & (\varepsilon,\infty) \\
(A,1) & (\varepsilon,0) & (\varepsilon,\infty) & (\varepsilon,\infty) & (\varepsilon,\infty) & (\varepsilon,\infty) \\
(A,1) & (\varepsilon,\infty) & (\varepsilon,0) & (\varepsilon,\infty) & (\varepsilon,\infty) & (\varepsilon,\infty) \\
(\mathbf{B},\mathbf{2}) & (B,1) & (C,1) & (\varepsilon,0) & (\varepsilon,\infty) & (\varepsilon,\infty) \\
(\varepsilon,\infty) & (\varepsilon,\infty) & (\varepsilon,\infty) & (\varepsilon,\infty) & (\varepsilon,0) & (\varepsilon,\infty) \\
(\varepsilon,\infty) & (\varepsilon,\infty) & (\varepsilon,\infty) & (\varepsilon,\infty) & (\varepsilon,\infty) & (\varepsilon,0)
\end{array}
\right)
$$

The original message is forwarded and broadcasted until it reaches D.   □

Following the specification of AODV, the request is only forwarded if no route to the destination is known. In other words only those nodes have to forward the request who (at the beginning) do not have information about the destination q. To achieve this, we modify the topology b and use $b \cdot \lfloor a \rfloor \neg q$ instead.

Due to this, forwarding and broadcasting a whole request can be modelled by

$$
a + b' \cdot |b'^{*}\rangle p + b'^{*} \cdot p \; ,
$$

where $b' = b \cdot \lfloor a \rfloor \neg q$.

Of course this compact algebraic expression is again only possible if the topology does not change. However, from a practical perspective it is reasonable to assume that the topology ceases to change after some time.

Route Reply. For unicasting a route reply back, we restrict the topology or more precisely the snapshot in such a way that it contains only the single path back to the originator. At the moment we think that we need domain specific knowledge for this task.

Example 6.3. For the previously presented matrix, the path from D back to A can easily be constructed from the matrix: the information of D (the 4-th row) shows that the next hop towards A is B. A similar argument shows that the next hop from B towards A is A. Hence we would use a new topology that only contains these single 1-hop connections:

$$
\begin{array}{c}
 \\ A \\ B \\ C \\ D \\ E \\ F
\end{array}
\begin{array}{cccccc}
A & B & C & D & E & F \\
\left(\varepsilon,0\right) & (\varepsilon,\infty) & (\varepsilon,\infty) & (\varepsilon,\infty) & (\varepsilon,\infty) & (\varepsilon,\infty) \\
(\mathbf{A},1) & (\varepsilon,0) & (\varepsilon,\infty) & (\varepsilon,\infty) & (\varepsilon,\infty) & (\varepsilon,\infty) \\
(\varepsilon,\infty) & (\varepsilon,\infty) & (\varepsilon,0) & (\varepsilon,\infty) & (\varepsilon,\infty) & (\varepsilon,\infty) \\
(\varepsilon,\infty) & (\mathbf{B},1) & (\varepsilon,\infty) & (\varepsilon,0) & (\varepsilon,\infty) & (\varepsilon,\infty) \\
(\varepsilon,\infty) & (\varepsilon,\infty) & (\varepsilon,\infty) & (\varepsilon,\infty) & (\varepsilon,0) & (\varepsilon,\infty) \\
(\varepsilon,\infty) & (\varepsilon,\infty) & (\varepsilon,\infty) & (\varepsilon,\infty) & (\varepsilon,\infty) & (\varepsilon,0)
\end{array}
$$

□

From an algebraic perspective, a route reply then becomes the same as a route request; just using the modified topology. In other words we restrict the topology to links that actually forward the request. In AODV each node that receives a route request from p and has information about a route to the destination q sends a reply. At least this set can be characterised purely algebraically by $|a\rangle q; |(b \cdot |a]\neg q)^*\rangle p$. The first part selects all nodes that have information about the destination q; the second one selects all nodes receiving information from p. Nodes only receive a route request if there is a path from p where no intermediate node has information about the destination.

Route Error messages are also spread throughout the network. Hence we can use the same mechanism of message sending discussed before. However, as a reaction of an incoming error message the routing table of a node has to be modified. More precisely, an entry has to be invalidated or removed. On the level of entries this can easily achieved by annihilation $((m, x) \cdot (\varepsilon, \infty) = (\varepsilon, \infty))$. However, matrices select the best known route and any existing route would be preferred over (ε, ∞). There are two possible solutions, which we can only sketch (due to lack of space). One is to define a meet operation $\sqcap$ which selects the worst route. This would require a Boolean algebra as underlying structure. Another solution is to use some indication of freshness. In AODV this is achieved by sequence numbers (see Section 2). As soon as sequence numbers are embedded, invalidating route entries will come for free without changing the underlying algebraic structure.

To conclude this section, we show the feasibility of the algebraic approach and show an important property of the AODV protocol.

Theorem 6.4. *If a message (e.g. a route request) is broadcasted via a topology b with $1 \leq b$ and is not stopped (it is forwarded by all intermediate nodes), every node (connected to the originator) knows all its neighbours and each node knows a route to the originator of the message (if there is one). Mathematically this means for a topology b,*

$$q \leq |b^*\rangle p \;\Rightarrow\; b \cdot q + q \leq a + b \cdot |b^*\rangle p \cdot (1 + b^*p) \ .$$

The proof can again be automated; it takes less than a second.

In particular this means that the backward routes are loop free. The same argument then holds for the route request; hence whenever messages are forwarded through a topology (which might be changed for sending RREPs) the outcome is loop free.

However, as discussed in Section 2, loop freeness in general does not hold in our simplified setting. It holds in Theorem 6.4 since we assume that the message is broadcasted and forwarded throughout the entire network. To overcome this deficit, the protocol AODV provides sequence numbers to indicate the freshness of a route and information about validity of routes. Therefore a next step towards a real characterisation of AODV is to integrate sequence numbers in our algebra; however one crucial point is how to compose different routes (cf. Figure 3) with different sequence numbers.

7   Conclusion and Outlook

The aim of the present paper is to present first steps towards an algebraic characterisation for AODV. At the moment we are able to model the core parts of AODV like sending route requests. However the model excludes substantial details like sequence numbers.

The presented approach has several advantages: First, in its purely algebraic form simple algebraic calculations and reasoning with off-the-shelf theorem provers is feasible; secondly, on the model (matrix) level, standard and well established algorithms (like algorithms for matrix multiplication) can be used for model checking or for determining case studies[6]; thirdly, since in the model elements are routing tables, the connection between abstraction, "reality" and implementation (e.g., [1]) can be seen quite easily.

The ultimate aim for future work is of course to completely specify AODV, to verify properties like loop freeness and maybe even improve the protocol. To achieve this goal, one has to add sequence numbers and route validity. However, when extending the underlying algebra the axioms of Kleene algebra should still be satisfied. In particular, the condition of Theorem 5.3 should hold. For

---

[6] In fact, we used a simple Haskell implementation to produce the example of the present paper.

reasoning about concurrency in AODV (e.g., when sending a message) one might extend the given algebra to a concurrent Kleene algebra [11].

So far we have modelled changing topology by separate matrices. Another approach would equip connections by probability. This would not only allow modelling changing topology, but also message losses during transmission. A first thought is to use probabilistic Kleene algebra [17,24] instead of standard Kleene algebra. However as pointed out by Takai and Furusawa in the erratum to [24] probabilistic Kleene algebra is not closed under forming matrices.

A different approach to AODV might even be to look at the protocol from another point of view. In the present paper we used a classical approach; however if one interprets nodes as individuals or agents, their routing tables as the local knowledge, reactive protocols become multiagent systems where knowledge about routes is distributed. Hence one should be able to adapt yet another well-known theory to protocols.

Acknowledgement. We thank Peter Jipsen, Bernhard Möller and the anonymous referees for useful comments and suggesting improvements to the presentation.

References

1. AODV-UU: An implementation of the AODV routing protocol (IETF RFC 3561), http://sourceforge.net/projects/aodvuu/ (accessed February 26, 2011)
2. Backhouse, R.: Closure Algorithms and the Star-Height Problem of Regular Languages. Ph.D. thesis, Imperial College, London (1975)
3. Backhouse, R., Carré, B.A.: Regular algebra applied to path-finding problems. Journal of the Institute of Mathematics and Applications (1975)
4. Carré, B.A.: Graphs and Networks. Oxford Applied Mathematics & Computing Science Series. Oxford University Press, Oxford (1980)
5. Conway, J.H.: Regular Algebra and Finite Machines. Chapman & Hall, Boca Raton (1971)
6. Desharnais, J., Möller, B., Struth, G.: Modal Kleene algebra and applications — A survey. Journal of Relational Methods in Computer Science 1, 93–131 (2004)
7. Dijkstra, E.W.: A note on two problems in connexion with graphs. Numerische Mathematik 1, 269–271 (1959)
8. Fernandes, T., Desharnais, J.: Describing data flow analysis techniques with Kleene algebra. SCP 65, 173–194 (2007)
9. Griffin, T.G., Gurney, A.J.T.: Increasing bisemigroups and algebraic routing. In: Berghammer, R., Möller, B., Struth, G. (eds.) RelMiCS/AKA 2008. LNCS, vol. 4988, pp. 123–137. Springer, Heidelberg (2008)
10. Griffin, T.G., Sobrinho, J.: Metarouting. SIGCOMM Comp. Com. Rev. 35, 1–12 (2005)
11. Hoare, C.A.R., Möller, B., Struth, G., Wehrman, I.: Concurrent Kleene algebra. In: Bravetti, M., Zavattaro, G. (eds.) CONCUR 2009. LNCS, vol. 5710, pp. 399–414. Springer, Heidelberg (2009)
12. Höfner, P.: Database for automated proofs of Kleene algebra, http://www.kleenealgebra.de (accessed February 26, 2011)

13. Höfner, P., Struth, G.: Automated reasoning in kleene algebra. In: Pfenning, F. (ed.) CADE 2007. LNCS (LNAI), vol. 4603, pp. 279–294. Springer, Heidelberg (2007)

14. Kozen, D.: The Design and Analysis of Algorithms. Springer, Heidelberg (1991)

15. Kozen, D.: A completeness theorem for Kleene algebras and the algebra of regular events. Information and Computation 110(2), 366–390 (1994)

16. McCune, W.W.: Prover9 and Mace4, http://www.cs.unm.edu/~mccune/prover9 (accessed February 26, 2011)

17. McIver, A.K., Gonzalia, C., Cohen, E., Morgan, C.C.: Using probabilistic Kleene algebra pKA for protocol verification. J. Logic and Algebraic Programming 76(1), 90–111 (2008)

18. Möller, B.: Dijkstra, Kleene, Knuth. Talk at WG2.1 Meeting, slides available online at, http://web.comlab.ox.ac.uk/jeremy.gibbons/wg21/meeting61/MoellerDijkstra.pdf (accessed February 26, 2006)

19. Möller, B., Struth, G.: WP is WLP. In: MacCaull, W., Winter, M., Düntsch, I. (eds.) RelMiCS 2005. LNCS, vol. 3929, pp. 200–211. Springer, Heidelberg (2006)

20. Perkins, C., Belding-Royer, E., Das, S.: Ad hoc on-demand distance vector (AODV) routing. RFC 3561 (Experimental) (July 2003), http://www.ietf.org/rfc/rfc3561.txt

21. Singh, A., Ramakrishnan, C.R., Smolka, S.A.: A process calculus for mobile ad hoc networks. SCP 75, 440–469 (2010)

22. Sobrinho, J.: Algebra and algorithms for QoS path computation and hop-by-hop routing in the internet. IEEE/ACM Trans. Networking 10(4), 541–550 (2002)

23. Sobrinho, J.: Network routing with path vector protocols: Theory and applications. In: Applications, Technologies, Architectures, and Protocols for Computer Communications. SIGCOMM 2003, pp. 49–60. ACM Press, New York (2003)

24. Takai, T., Furusawa, H.: Monodic tree kleene algebra. In: Schmidt, R.A. (ed.) RelMiCS/AKA 2006. LNCS, vol. 4136, pp. 402–416. Springer, Heidelberg (2006) (accessed February 26, 2011) Errata available at, http://www.sci.kagoshima-u.ac.jp/ furusawa/person/Papers/correct_ monodic_kleene_algebra.pdf

Dependently-Typed Formalisation of Relation-Algebraic Abstractions

Wolfram Kahl

McMaster University, Hamilton, Ontario, Canada
kahl@cas.mcmaster.ca

Abstract. We present a formalisation in the dependently-typed programming language Agda2 of basic category and allegory theory, and of generalised algebras where function symbols are interpreted in a parameter category. We use this nestable algebra construction as the basis for nestable category and allegory constructions, ultimately aiming at a formalised foundation of the algebraic approach to graph transformation, which uses constructions in categories of graph structures considered as unary algebras.

The features of Agda permit strongly-typed programming with these nested algebras and with relational homomorphisms between them in a natural mathematical style and with remarkable ease, far beyond what can be achieved even in Haskell.

Keywords: Dependently typed programming, algebras as data, allegories of relational algebra morphisms, nested algebras.

1 Introduction

In the context of computation, algebras are frequently seen as models of data *types*, with computations implementing their primitive and derived operations. However, algebras also have uses as data *values*, with computations producing new algebras from old. Examples for this are not only the Abstract State Machines (originally "Evolving Algebras") of Gurevich [13,14], but also any graph data structures, which can be considered as (typically unary) algebras. The "algebraic approach" to graph transformation [6] in particular takes that point of view, and applies abstractions from category theory to define and reason about graph transformation systems.

In this paper, we explore a flexible formalisation of aspects of relational categories and universal algebra in the dependently-typed programming language (and proof checker) Agda2 [26], leading up to allegories of "relational homomorphisms" between algebras, technically also known as bisimulations.

We start with an introduction to essential features of Agda2 (in the following just referred to as Agda) and its current standard library, and then (Sect. 3) summarise our formalisation of relation algebraic operations for the standard concept of relations in Agda. In Sect. 4 we turn to fine-grained, universe-polymorphic formalisations of categories and allegories, and elaborate more on the topics of domain (Sect. 5) and restricted residuals (Sect. 6). Our generalised formalisation of algebras is summarised in Sect. 7. We discuss some related work in Sect. 8.

H. de Swart (Ed.): RAMICS 2011, LNCS 6663, pp. 230–247, 2011.

The Agda theories discussed in this paper are available on-line at the URL
http://RelMiCS.McMaster.ca/~kahl/RATH/Agda/.

2 Introduction to Agda: Types, Sets, Equality

The Agda home page[1] states:

Agda is a dependently typed functional programming language.
It has inductive families, i.e., data types which depend on values, such as the
type of vectors of a given length. It also has parametrised modules, mixfix
operators, Unicode characters, and an interactive Emacs interface which can
assist the programmer in writing the program.

Agda is a proof assistant. It is an interactive system for writing and
checking proofs. Agda is based on intuitionistic type theory, a foundational
system for constructive mathematics developed by the Swedish logician Per
Martin-Löf. It has many similarities with other proof assistants based on
dependent types, such as Coq, Epigram, Matita and NuPRL.

Syntactically and "culturally", Agda is quite close to Haskell. However, since Agda
is strongly normalising and has no $\perp$ values, the underlying semantics is quite different. Also, since Agda is dependently typed, it does not have Haskell's distinction between terms, types, and kinds (the "types of the types"). The Agda constant
Set corresponds to the Haskell kind *; it is the type of all "normal" datatypes. For
example, the Agda standard library defines the type Bool as follows:

data Bool : Set **where** true : Bool
                        false : Bool

Since Set needs again a type, there is Set_1, with Set : Set_1, etc., resulting in a
hierarchy of "universes". Since Version 2.2.8, Agda supports *universe polymorphism*, with universes Set i where i is an element of the following special-purpose
variant of the natural numbers:

data Level : Set **where** zero : Level
                        suc : (i : Level) → Level

With this, the conventional usage turns into syntactic sugar, so that Set is now
Set zero, and Set_1 = Set (suc zero). The standard library uses "⊔" for maximum
on Level; in our development, we systematically rename this to "ʊ", so that we
use "⊔" as join in the inclusion order of morphisms, as customary in abstract
relation algebra [28,27].

With universe polymorphism enabled, we may quantify over Level-typed variables that occur as Level arguments of Set. Universe polymorphism is essential
for being able to talk about both "small" and "large" categories or relation algebras, or, for another example, also for being able to treat diagrams of graphs and
graph homomorphisms as graphs again. We therefore use universe polymorphism
throughout this paper.

---

[1] http://wiki.portal.chalmers.se/agda/

For example, the standard library includes the following definition for the universe-polymorphic parameterised Maybe type:

```
data Maybe {a : Level} (A : Set a) : Set a where just : (x : A) → Maybe A
                                                nothing : Maybe A
```

Maybe has two parameters, a and A, where dependent typing is used since the type of the second parameter depends on the first parameter. The use of {...} flags a as an *implicit parameter* that can be elided where its type is implied by the call site of Maybe. This happens in the occurrences of Maybe A in the types of the data constructors just and nothing: In Maybe A, the value of the first, implicit parameter of Maybe can only be a, the level of the set A.

The same applies to implicit function arguments, and in most cases, implicit arguments or parameters are determined by later arguments respectively parameters. Frequently, implicit arguments correspond quite precisely to the implicit context of mathematical statements, so that the reader may be advised to skip implicit arguments at first reading of a type, and return to them for clarification where necessary for understanding the types of the explicit parameters.

While the Hindley-Milner typing of Haskell and ML allows function definitions without declaration of the function type, and type signatures without declaration of the universally quantified type variables, in Agda, all types and variables need to be declared, but implicit parameters and the type checking machinery used to resolve them alleviate that burden significantly. For example, the original definition writes only Maybe {a} (A : Set a) : Set a, since the type of a will be inferred from a's use as argument to Set. In this paper, we will rarely use this possibility to elide types of named arguments, since we estimate that the clarity of explicit typing is worth the additional "optical noise" especially for readers who are less familiar with Agda or dependently-typed theories.

The "programming types" like Maybe can be freely mixed with "formula types", inspired by the Curry-Howard-correspondence of "formulae as types, proofs as programs". The formula types of true formulae contain their proofs, while the formula types of false formulae are empty.

The standard library type of propositional equality has (besides two implicit parameters) one explicit parameter and one explicit argument; the definition therefore gives rise to types like the type "$2 \equiv 1 + 1$", which can be shown to be inhabited using the definition of natural numbers and natural number addition $+$, and the type "$2 \equiv 3$", which is an empty type, since it has no proof[2].

```
data _≡_ {a : Level} {A : Set a} (x : A) : A → Set a where refl : x ≡ x
```

The definition introduces types $x \equiv y$ for any x and y of type A, but only the types $x \equiv x$ are inhabited, and they contain the single element refl {a} {A} {x}.

In Agda, as in other type theories without quotient types, sets with equality are typically modelled as *setoids*, that is, carrier types equipped with an

---

[2] In Agda, almost all lexemes are separated by spaces, since almost all symbol combinations form legal names. Underscores as part of names indicate positions of explicit arguments for mixfix operators.

equivalence. This closely corresponds to the non-primitive nature of the "equality" test $(\equiv)$: Eq a $\Rightarrow$ a $\rightarrow$ a $\rightarrow$ Bool in Haskell.

The standard library defines the following type of homogeneous relations:

Rel : {a : Level} $\rightarrow$ Set a $\rightarrow$ (l : Level) $\rightarrow$ Set (a ⊔ suc l)
Rel A l = A $\rightarrow$ A $\rightarrow$ Set l

A proof that $\_\approx\_$ is an equivalence relation is a record containing the proofs of reflexivity, symmetry, and transitivity:

record IsEquivalence {a l : Level} {A : Set a} ($\_\approx\_$: Rel A l) : Set (a ⊔ l) **where**
  field refl : {x : A} $\rightarrow$ x $\approx$ x
       sym : {x y : A} $\rightarrow$ x $\approx$ y $\rightarrow$ y $\approx$ x
       trans : {x y z : A} $\rightarrow$ x $\approx$ y $\rightarrow$ y $\approx$ z $\rightarrow$ x $\approx$ z

A setoid is a dependent record consisting of a Carrier set, a relation $\_\approx\_$ on that carrier, and a proof that that relation is an equivalence relation:

record Setoid c l : Set (suc (c ⊔ l)) **where**
  field Carrier : Set c
     $\_\approx\_$: Rel Carrier l
     isEquivalence : IsEquivalence $\_\approx\_$
  open IsEquivalence isEquivalence **public**

An Agda record is also a module that may contain other material besides its **field**s; the "**open**" clause makes the fields of the equivalence proof available as if they were fields of Setoid. This language feature enables incremental extension of smaller theories to larger theories at very low notational cost.

The Preorder type of the Agda standard library adds a second preorder relation to a Setoid, with reflexivity with respect to the setoid equality; for a Poset, that preorder also needs to be antisymmetric with respect to the setoid equality.

3   Generalised Heterogeneous Concrete Relations

Concrete relations from a set A to another set B are normally defined to be the subsets of the Cartesian product A × B. Equivalently, they can be seen as characteristic functions of type (A × B) $\rightarrow$ Bool, or, in the curried variant, of type A $\rightarrow$ B $\rightarrow$ Bool (function type construction associates to the right). In Agda, it is more natural to replace Bool with a Set universe, with the understanding that R a b is the type of proofs that the pair (a, b) is in R, that is, R a b is empty if (a, b) is not in R, and inhabited if (a, b) is in R.

Therefore, we will use R : A $\rightarrow$ B $\rightarrow$ Set for "small concrete" relations. This relation type, A $\rightarrow$ B $\rightarrow$ Set, can also serve to represent other structures, for example, G : N $\rightarrow$ N $\rightarrow$ Set could represent a graph G with node type N, and G n_1 n_2 would be the set of all edges from n_1 to n_2. Utilities defined for relations with types like A $\rightarrow$ B $\rightarrow$ Set therefore can be applied in many different contexts.

Since categories can be seen as graphs with additional structure, we obviously need full universe polymorphism, and define:

$Rel : \{i\,j : Level\} \to (k : Level) \to Set\,i \to Set\,j \to Set\,(i \cup j \cup suc\,k)$
$Rel\,k\,A\,B = A \to B \to Set\,k$

The order of parameters is a matter of taste; the standard library, which until recently defined only types of homogeneous relations, now defines the same types, but with a different argument order in the parameters of their constant REL. However, we have, by the definitions, type equality $Rel\,k\,A\,B = REL\,A\,B\,k$, so that our Rel-based library is fully interoperable with the standard library, which does not provide typical relation-algebraic operations and laws. (The AoPA library of Mu *et al.* [25] does provide relation-algebraic operations and laws, but supports heterogeneous binary relations only at the levels 0 and, to a lesser degree, 1).

From this definition of relation types, we completely follow the standard procedure to formalisation of concrete relations in dependent type theory; inclusion, and equality of relations as derived from inclusion are defined as follows:

$\_\subseteq\_ \; : \{i\,j\,k_1\,k_2 : Level\}\,\{A : Set\,i\}\,\{B : Set\,j\}$
$\qquad \to Rel\,k_1\,A\,B \to Rel\,k_2\,A\,B \to Set\,(i \cup j \cup k_1 \cup k_2)$
$P \subseteq Q = \forall\,\{x\,y\} \to P\,x\,y \to Q\,x\,y$

$\_\doteq\_ \; : \{i\,j\,k_1\,k_2 : Level\} \to \{A : Set\,i\} \to \{B : Set\,j\}$
$\qquad \to Rel\,k_1\,A\,B \to Rel\,k_2\,A\,B \to Set\,(i \cup j \cup k_1 \cup k_2)$
$R \doteq S = (R \subseteq S) \times (S \subseteq R)$ -- $\times$ encodes logical conjunction $\wedge$

Due to universe polymorphism, we could also have declared (equivalently):

$\_\subseteq\_ \; : \{i\,j\,k_1\,k_2 : Level\}\,\{A : Set\,i\}\,\{B : Set\,j\}$
$\qquad \to Rel\,(i \cup j \cup k_1 \cup k_2)\,(Rel\,k_1\,A\,B)\,(Rel\,k_2\,A\,B)$

These relations on Rel-relations are then used to define Setoids and Posets of Rel-relations. However, since Rel deals directly with Sets, not with setoids, the identity relation has to be based on propositional equality, but that provided by the standard library, $\_\equiv\_$ presented in Sect. 2, forces the Level of its arguments onto its result, so we provide our own fully universe-polymorphic variant:

data $\_\equiv\equiv\_$ $\{k\,a\}\,\{A : Set\,a\}\,(x : A) : A \to Set\,k$ **where** $\equiv\equiv$-refl $: x \equiv\equiv x$

With this, we can define universe-polymorphic identity relations:

$idR : \{k\,i : Level\}\,\{A : Set\,i\} \to Rel\,k\,A\,A$
$idR = \_\equiv\equiv\_$

In the module hierarchy Relation.Binary.Heterogeneous, we define standard relation-algebraic operations and properties, and prove the relevant laws in particular to be able to implement the abstract theories presented in Sect. 4.

4   Semigroupoids, Categories, Allegories, Collagories

We present a relatively fine-grained modularisation of sub-theories of distributive allegories, following our work on using semigroupoids to provide the theory of finite relations between infinite types, as they frequently occur as data structures in programming [19], and on collagories as foundation for relation-algebraic graph transformation [20].

Semigroupoids are to categories as semigroups are to monoids — no identities are assumed. The following definition is taken from [19], and will probably appear to be quite conventional to readers familiar with the basics of category theory (except perhaps for the argument order of composition):

Definition 4.1 A **semigroupoid** $(\mathsf{Obj}, \mathsf{Mor}, \mathsf{src}, \mathsf{trg}, \,\text{;}\,)$ is a graph with a set Obj of *objects* as vertices, a set Mor of *morphisms* as edges, with $\mathsf{src}, \mathsf{trg} : \mathsf{Mor} \to \mathsf{Obj}$ assigning source and target object to each morphism (we write "$f : \mathcal{A} \to \mathcal{B}$" instead of "$f \in \mathsf{Mor}$ and $\mathsf{src}\ f = \mathcal{A}$ and $\mathsf{trg}\ f = \mathcal{B}$"), and an additional partial operation "$\_ \,\text{;}\, \_$" of composition such that the following hold:

- For $f : \mathcal{A} \to \mathcal{B}$ and $g : \mathcal{B}' \to \mathcal{C}$, the composition $f \,\text{;}\, g$ is defined iff $\mathcal{B} = \mathcal{B}'$, and if it is defined, then $(f \,\text{;}\, g) : \mathcal{A} \to \mathcal{C}$.
- Composition is associative, i.e., if one of $(f \,\text{;}\, g) \,\text{;}\, h$ and $f \,\text{;}\, (g \,\text{;}\, h)$ is defined, then so is the other and they are equal.

For two objects $\mathcal{A}$ and $\mathcal{B}$, the collections of morphisms $f : \mathcal{A} \to \mathcal{B}$ is also called the *homset* from $\mathcal{A}$ to $\mathcal{B}$, and written $\mathsf{Hom}(\mathcal{A}, \mathcal{B})$.

A morphism is called an *endomorphism* iff its source and target objects coincide; an endomorphism R is called *idempotent* if $R \,\text{;}\, R = R$.   □

From semigroupoids, we obtain more specialised theories by adding in particular identities (to obtain categories), converse, domain, and local ordering of homsets, and their combinations. This continues with semi-lattice and lattice properties for the homsets, and various coherence properties; the following is the inclusion graph of most of our current theories, the most important of which will be discussed in more detail in the remainder of this and the next two sections.

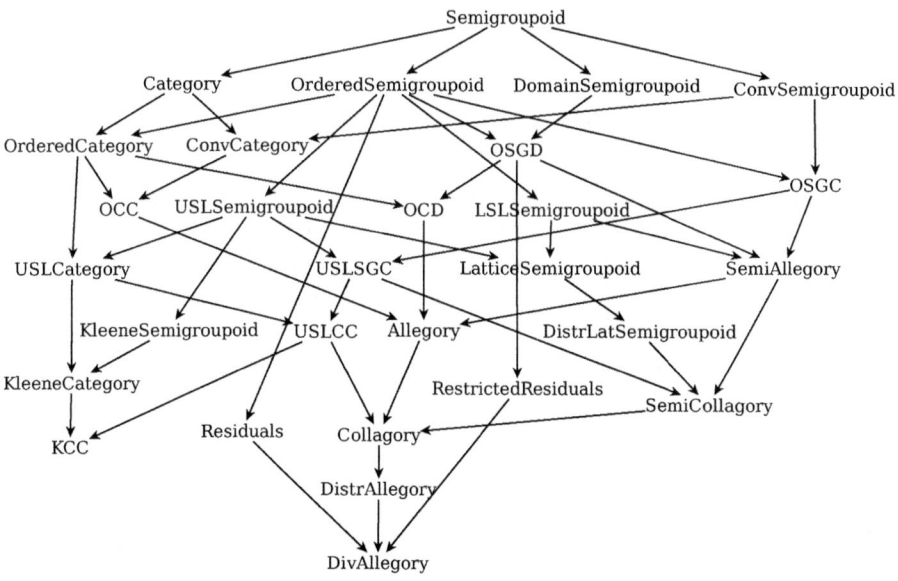

Such a fine-grained modularisation automatically comes with some overhead in comparison with larger, monolithic theories as used for example by Gonzalía [12]. It is interesting to see how the namespace management by Agda's module system, which includes nested modules and records considered as modules, minimises that cost, and enables different approaches to achieving essentially the effect of subtyping with extensible records, which are not available.

Using universe-polymorphism throughout our development gives us the flexibility to use both "small" explicitly constructed examples, like the two-object two-morphism category $\bullet \rightrightarrows \bullet$, standard next-level "large" categories like *Set* and *Rel*, and even larger categories. We use variable names i, j, k, k_1, etc. for universe levels.

As our way to deal with homsets (i.e., sets of morphisms from one object to another) we choose to use the standard-library Setoid theory for individual homsets, starting from a function Hom : Obj → Obj → Setoid j k, and derive from this the underlying function Mor : Obj → Obj → Set j exposing the types of the morphisms, and lifting the equality (i.e., the Setoid equivalence) into the global view:

module HomSetoid {i j k : Level} {Obj : Set i} (Hom : Obj → Obj → Setoid j k) **where**
  Mor : Rel Obj j
  Mor = Setoid.Carrier $\circ_2$ Hom   -- using: f $\circ_2$ g = λ x y → f (g x y)

  infix 4 $\_ \approx \_$
  $\_ \approx \_$: {A B : Obj} → Rel (Mor A B) k
  $\_ \approx \_$ {A} {B} = Setoid.$\_ \approx \_$ (Hom A B)

This HomSetoid module also includes the lifted properties of the equality, since for using the underlying Setoid properties, one would always need to identify the relevant hom-setoid:

  $\approx$-refl : {A B : Obj} → {R : Mor A B} → R $\approx$ R
  $\approx$-refl   {A} {B} = Setoid.refl (Hom A B)
  $\approx$-sym : {A B : Obj} → {R S : Mor A B} → R $\approx$ S → S $\approx$ R
  $\approx$-sym   {A} {B} = Setoid.sym (Hom A B)
  $\approx$-trans : {A B : Obj} → {Q R S : Mor A B} → Q $\approx$ R → R $\approx$ S → Q $\approx$ S
  $\approx$-trans {A} {B} = Setoid.trans (Hom A B)

Equivalently, one could have derived the individual setoids from the global morphism equality, which is essentially the path chosen by Gonzalía [12], and closer in spirit to Def. 4.1. However, we believe the approach chosen here, with Hom primitive instead of src and trg, is more natural in the dependently-typed context, where src and trg are hardly ever needed since their results are typically already available once their argument is, since they are part of the argument's type:

  src : {A B : Obj} → Mor A B → Obj
  src {A} {_} _ = A
  trg : {A B : Obj} → Mor A B → Obj
  trg {_} {B} _ = B

Morphism composition is defined in the context of such a HomSetoid. Thanks to universe-polymorphism together with the formulae-as-types view of relations, the standard-library type Transitive also provides the typing for composition: the composition operator is a proof for higher-level transitivity of Mor. This typing also turns composition into a total function, so that the definedness discussion in Def. 4.1 does not need to be reflected here.

Due to the setoid setup, we need, besides associativity, also demand a congruence property (defined using the standard-library type _ Preserves$_2$ _ → _ → _).

We also show here one of the definitions of derived concepts that we include in this module (many others are suppressed in this short summary): A morphism I is a *left identity* if for all compatible morphisms R we have I ⨾ R ≈ R:

record CompOp {i j k : Level} {Obj : Set i} (Hom : Obj → Obj → Setoid j k)
    : Set (i ⊔ j ⊔ k) **where**

 open HomSetoid Hom

 infixr 9 _⨾_
 field _⨾_ : Transitive Mor
   ⨾-cong : {A B C : Obj}
      → ((_⨾_ {A} {B} {C}) Preserves$_2$ _≈_ → _≈_ → _≈_)
   ⨾-assoc : {A B C D} {f : Mor A B} {g : Mor B C} {h : Mor C D}
      → ((f ⨾ g) ⨾ h) ≈ (f ⨾ (g ⨾ h))
 isLeftIdentity : {A : Obj} → Mor A A → Set (i ⊔ j ⊔ k)
 isLeftIdentity {A} I = {B : Obj} {R : Mor A B} → I ⨾ R ≈ R

A *semigroupoid* is fully defined by the parameters and fields of a CompOp, and we represent it as a dependent (record) product — the type of the second field depends on the first field Hom.

record Semigroupoid {i : Level} (j k : Level) (Obj : Set i) : Set (i ⊔ suc (j ⊔ k)) **where**
 field Hom : Obj → Obj → Setoid j k
   compOp : CompOp Hom
 open HomSetoid i j k Hom **public**
 open CompOp compOp **public**

The last two **open** ... **public** lines serve to re-export everything defined in the (record) modules HomSetoid and CompOp. Since these re-exports do not contribute to the essence of the formalisation, we suppress their rendering from now on. However, these re-exports are essential in that they effectively hide the fact that we modularised the definition of Semigroupoid — a future "**open** Semigroupoid" brings also all the items defined in HomSetoid and CompOp into scope.

Following relation-algebraic terminology, we call the involution of self-dual semigroupoids "*converse*", and define it again in an independent building block ConvOp, including the converse operator and its axioms as **fields**. Combining a Semigroupoid with a ConvOp then produces a ConvSemigroupoid (not shown).

For illustrating the flavour of these developments, we also show here one immediate consequence of the axioms (un˘-cong), the derived concept of symmetry, and a proof that left identities are symmetric. Both proofs are presented in the calculational style, using the mixfix operators ≈-begin_, _≈⟨_⟩_ and _□ which are variants of the calculational reasoning operators provided by the standard library, equipped with two additional implicit object parameters (similar to ≈-sym etc.) to enable calculational reasoning in homsets without having to specify the homset explicitly.

```
record ConvOp {i j k : Level} {Obj : Set i}
                  (SG : Semigroupoid {i} {j} {k} Obj) : Set (i ⊔ j ⊔ k) where
  open Semigroupoid SG

  field  ˘          : {A B : Obj} → Mor A B → Mor B A
         ˘-cong     : {A B : Obj}  {R S : Mor A B} → R ≈ S → R ˘ ≈ S ˘
         ˘˘         : {A B : Obj}  {R : Mor A B}   → (R ˘) ˘ ≈ R
         ˘-involution : {A B C : Obj} {R : Mor A B} {S : Mor B C}
                      → (R ⨾ S) ˘ ≈ (S ˘ ⨾ R ˘)

  un˘-cong : {A B : Obj} {R S : Mor A B} → R ˘ ≈ S ˘ → R ≈ S
  un˘-cong {A} {B} {R} {S} R˘≈S˘ = ≈-begin
      R    ≈⟨ ≈-sym ˘˘ ⟩          (R ˘) ˘
           ≈⟨ ˘-cong R˘≈S˘ ⟩      (S ˘) ˘
           ≈⟨ ˘˘ ⟩                S                                □

  isSymmetric : {A : Obj} → Mor A A → Set k
  isSymmetric R = R ˘ ≈ R

  isLeftIdentity-isSymmetric : {A : Obj} {R : Mor A A}
                            → isLeftIdentity R → isSymmetric R
  isLeftIdentity-isSymmetric {A} {R} left = ≈-begin
     R ˘ ≈⟨ ≈-sym left ⟩              R ⨾ R ˘
         ≈⟨ ⨾-cong₁ (≈-sym ˘˘) ⟩     (R ˘) ˘ ⨾ R ˘
         ≈⟨ ≈-sym ˘-involution ⟩     (R ⨾ R ˘) ˘
         ≈⟨ ˘-cong left ⟩            (R ˘) ˘
         ≈⟨ ˘˘ ⟩                     R                              □
```

For locally ordered semigroupoids, we replace the Setoid in the result type of Hom with the stronger Poset, and therefore need to adapt this in the instantiations of HomSetoid and CompOp. The local poset ordering relations are again collected into a global parameterised relation, ⊑. The following is a flattened presentation of the OrderedSemigroupoid definition — the original is composed from several modules and records:

```
record OrderedSemigroupoid {i : Level} (j k₁ k₂ : Level) (Obj : Set i)
                        : Set (i ⊔ suc (j ⊔ k₁ ⊔ k₂)) where
  field Hom : Obj → Obj → Poset j k₁ k₂
        compOp : CompOp i j k₁ (posetSetoid ∘₂ Hom)

  semigroupoid : Semigroupoid Obj
  semigroupoid = record {Hom = posetSetoid ∘₂ Hom; compOp = compOp}
  open Semigroupoid semigroupoid hiding (Hom; compOp)

  infix 4 _⊑_
  _⊑_ : {A B : Obj} → Rel (Mor A B) k₂   -- The morphism ordering.
  _⊑_ {A} {B} = Poset._≤_ (Hom A B)
  field ⨾-monotone   : {A B C : Obj} {f f' : Mor A B} {g g' : Mor B C}
                     → f ⊑ f' → g ⊑ g' → (f ⨾ g) ⊑ (f' ⨾ g')
```

The definition and **open**ing of semigroupoid here produces the illusion that an OrderedSemigroupoid is a "Semigroupoid with a local order on each homset", even though the definition has been structured in a different way. From now on we will suppress the definitions of such subtheories, like semigroupoid here, if they are not used inside the shown part of the enclosing definition.

Also included in OrderedSemigroupoid (but not shown) are proofs of the ordering properties expressed in terms of $\sqsubseteq$, one-sided monotonicity properties of composition, and numerous derived concepts and laws, including properties and types of idempotent subidentities.

An *ordered semigroupoid with converse (OSGC)* adds to its constituents the monotonicity law for converse (and contains a large number of derived concepts and lemmas, not shown here):

record OSGC $\{i\ :\ \mathsf{Level}\}\ (j\ k_1\ k_2\ :\ \mathsf{Level})\ (\mathsf{Obj}\ :\ \mathsf{Set}\ i)\ :\ \mathsf{Set}\ (i \sqcup \mathsf{suc}\ (j \sqcup k_1 \sqcup k_2))$ **where**
    field orderedSemigroupoid : OrderedSemigroupoid $j\ k_1\ k_2$ Obj
    open OrderedSemigroupoid orderedSemigroupoid

    field convOp : ConvOp semigroupoid
    open ConvOp convOp

    field ˘-monotone : $\{A\ B\ :\ \mathsf{Obj}\}\ \{R\ S\ :\ \mathsf{Mor}\ A\ B\} \rightarrow R \sqsubseteq S \rightarrow (R\ ˘) \sqsubseteq (S\ ˘)$

When defining *upper* and *lower semilattice semigroupoids*, we do not add the join respectively meet operators directly to the range of Hom, since the standard library currently only provides lattices, but not semilattices. Otherwise, these definitions are straightforward and not shown.

Semi-allegories are, by analogy with semigroupoids, "allegories without identity morphisms", i.e., lower semilattice semigroupoids with converse and domain (see Sect. 5) satisfying the Dedekind rule:

record SemiAllegory $\{i\ :\ \mathsf{Level}\}\ (j\ k_1\ k_2\ :\ \mathsf{Level})\ (\mathsf{Obj}\ :\ \mathsf{Set}\ i)$
                                      $:\ \mathsf{Set}\ (i \sqcup \mathsf{suc}\ (j \sqcup k_1 \sqcup k_2))$ **where**
    field osgc   : OSGC $j\ k_1\ k_2$ Obj
    open OSGC   osgc
    field meetOp   : MeetOp   orderedSemigroupoid
          domainOp   : OSGDomainOp orderedSemigroupoid
    open MeetOp   meetOp
    open OSGDomainOp domainOp

    field Dedekind : $\{A\ B\ C\ :\ \mathsf{Obj}\}\ \{Q\ :\ \mathsf{Mor}\ A\ B\}\ \{R\ :\ \mathsf{Mor}\ B\ C\}\ \{S\ :\ \mathsf{Mor}\ A\ C\}$
                       $\rightarrow (Q \,\mathbin{\fatsemi}\, R \sqcap S) \sqsubseteq (Q \sqcap S \,\mathbin{\fatsemi}\, R\ ˘) \,\mathbin{\fatsemi}\, (R \sqcap Q\ ˘ \,\mathbin{\fatsemi}\, S)$

We now turn to categories, where the new ingredient is the operation assigning an identity morphism to each object:

record IdOp $(i\ j\ k\ :\ \mathsf{Level})\ \{\mathsf{Obj}\ :\ \mathsf{Set}\ i\}\ (\mathsf{Hom}\ :\ \mathsf{Obj} \rightarrow \mathsf{Obj} \rightarrow \mathsf{Setoid}\ j\ k)$
            $(\_\,\mathbin{\fatsemi}\,\_\ :\ \mathsf{Transitive}\ (\mathsf{Setoid.Carrier} \circ_2 \mathsf{Hom}))\ :\ \mathsf{Set}\ (i \sqcup j \sqcup k)$ **where**
    open HomSetoid $i\ j\ k$ Hom

    field Id   : $(A\ :\ \mathsf{Obj}) \rightarrow \mathsf{Mor}\ A\ A$
          leftId   : $\{A\ B\ :\ \mathsf{Obj}\} \rightarrow \{f\ :\ \mathsf{Mor}\ A\ B\} \rightarrow (\mathsf{Id}\ A \,\mathbin{\fatsemi}\, f) \approx f$
          rightId : $\{A\ B\ :\ \mathsf{Obj}\} \rightarrow \{f\ :\ \mathsf{Mor}\ A\ B\} \rightarrow (f \,\mathbin{\fatsemi}\, \mathsf{Id}\ B) \approx f$

This identity module can now be added easily to (ordered) semigroupoids to produce the corresponding categories:

- A Category consists of a Semigroupoid and an IdOp
- An OrderedCategory consists of an OrderedSemigroupoid and an IdOp.
- A ConvCategory consists of a ConvSemigroupoid and an IdOp; preservation of identities by converse follows from the lemma shown in ConvSemigroupoid:

$Id^\smile$: {A : Obj} → (Id {A}) $^\smile$ ≈ Id {A}
$Id^\smile$ {A} = isLeftIdentity-isSymmetric leftId

- An OCC (*ordered category with converse*) consists of an OSGC and an IdOp.
- An Allegory is an OCC with a MeetOp satisfying Dedekind, and can be shown to contain a SemiAllegory by defining a DomainOp based on:

dom : {A B : Obj} → Mor A B → Mor A A
dom R = Id ⊓ R ⨾ R $^\smile$

Since the OCC theory exports so much material, it appears more natural to use OCCs as starting point for defining allegories analogously to semiallegories — the alternative would have been to start from semiallegories and work analogously to the OCC definition, which in this case would have required careful re-export of the OCC material since Agda currently does not permit re-export of the same item via more than one interface.

- Similarly, a (Semi-)Collagory is a (Semi-)Allegory with a JoinOp and lattice distributivity.
- A DistrAllegory is a Collagory with least morphisms and zero laws.

In OCCs, we also have the standard relation-algebraic way of defining properties like univalence (($R^\smile$ ⨾ R) ⊑ Id B), totality (Id A ⊑ (R ⨾ $R^\smile$)), injectivity, etc., and deriving laws for them. However, since most reasoning with these properties immediately uses the identity laws of composition, the corresponding definitions in OSGC typically make for shorter proofs, and we use those to define a proof-carrying type of Mappings (all in OSGC):

isUnivalent : {A B : Obj} → Mor A B → Set ($i ⊔ j ⊔ k_2$)
isUnivalent R = isSubidentity ($R^\smile$ ⨾ R)

isTotal : {A B : Obj} → Mor A B → Set ($i ⊔ j ⊔ k_2$)
isTotal R = isSuperidentity (R ⨾ $R^\smile$)

isMapping : {A B : Obj} → Mor A B → Set ($i ⊔ j ⊔ k_2$)
isMapping R = isUnivalent R × isTotal R

record Mapping (A B : Obj) : Set ($i ⊔ j ⊔ k_2$) **where field** mor : Mor A B
                                                          prf : isMapping mor

The semigroupoid of mappings in an OSGC and the category of mappings in an OCC are easily constructed (in module Categoric.MapSG):

MapSG : {i j k_1 k_2 : Level} {Obj : Set i}
        → OSGC j k_1 k_2 Obj → Semigroupoid ($i ⊔ j ⊔ k_2$) k_1 Obj
MapCat : {i j k_1 k_2 : Level} {Obj : Set i}
        → OCC j k_1 k_2 Obj → Category ($i ⊔ j ⊔ k_2$) k_1 Obj

5   Domain

Domain can be defined in allegories as shown above; for weaker theories, Desharnais *et al.* [9] axiomatised domain operators essentially in an ordering context (in semirings and Kleene algebras), where the domain operator produces subidentities. A more recent alternative is the purely equational approach of Desharnais *et al.* [7], which starts just from semigroups. We have formalised both approaches, concentrating on the aspects that do not require complements.

Most of [7, Section 3] has been generalised to *left closure semigroupoids*:

field
    dom : $\{A\ B : Obj\} \to Mor\ A\ B \to Mor\ A\ A$
    dom-cong : $\{A\ B : Obj\}\ \{R\ S : Mor\ A\ B\} \to R \approx S \to dom\ R \approx dom\ S$
    D1 : $\{A\ B : Obj\}\ \{R : Mor\ A\ B\} \to (dom\ R)\ \sqamp\ R \approx R$
    L2 : $\{A\ B : Obj\}\ \{R : Mor\ A\ B\} \to dom\ (dom\ R) \approx dom\ R$
    L3 : $\{A\ B\ C\}\ \{R : Mor\ A\ B\}\ \{S : Mor\ B\ C\} \to (dom\ R)\ \sqamp\ dom\ (R\ \sqamp\ S) \approx dom\ (R\ \sqamp\ S)$
    D4 : $\{A\ B\ C\}\ \{R : Mor\ A\ B\}\ \{S : Mor\ A\ C\}$
        $\to (dom\ R)\ \sqamp\ (dom\ S) \approx (dom\ S)\ \sqamp\ (dom\ R)$

    $\_\preccurlyeq\_$: $\{A\ B : Obj\}\ (R\ S : Mor\ A\ B) \to Set\ k$ -- The *"fundamental order"*
$R \preccurlyeq S = R \approx (dom\ R)\ \sqamp\ S$

For the fundamental order $\preccurlyeq$, we have been able to show some additional properties, namely that it is preserved by multiplication with domain elements from the left, and that the domain semigroup axiom D3 implies monotonicity of dom with respect to $\preccurlyeq$ (proofs not shown):

dom$_\sqamp$-$\preccurlyeq$monotone   : $\{A\ B\ C : Obj\}\ \{Q : Mor\ A\ B\}\ \{R\ S : Mor\ A\ C\}$
        $\to R \preccurlyeq S \to ((dom\ Q)\ \sqamp\ R) \preccurlyeq ((dom\ Q)\ \sqamp\ S)$
dom-D3-$\preccurlyeq$monotone : $\{A\ B : Obj\}\ \{R\ S : Mor\ A\ B\}$
        $\to (dom\ ((dom\ R)\ \sqamp\ S) \approx (dom\ R)\ \sqamp\ (dom\ S)) \to R \preccurlyeq S \to dom\ R \preccurlyeq dom\ S$

However, $\preccurlyeq$-monotonicity of dom does not imply D3; the model searcher Mace4 [24] finds a four-element counter-example.

For the subidentity-based approach, we adapt the definitions of [9] to the ordered semigroupoid setting:

record OSGDomainOp $\{i\ j\ k_1\ k_2 : Level\}\ \{Obj : Set\ i\}$
    (base : OrderedSemigroupoid $j\ k_1\ k_2\ Obj$) : $Set\ (i \sqcup j \sqcup k_1 \sqcup k_2)$ **where**
  open OrderedSemigroupoid base
  field dom : $\{A\ B : Obj\} \to Mor\ A\ B \to Mor\ A\ A$
      domSubIdentity   : $\{A\ B : Obj\}\ \{R : Mor\ A\ B\} \to isSubidentity\ (dom\ R)$
      dom-$\sqamp$-idempotent : $\{A\ B : Obj\}\ \{R : Mor\ A\ B\} \to (dom\ R)\ \sqamp\ (dom\ R) \approx dom\ R$
      domPreserves$\sqsubseteq$: $\{A\ B : Obj\}\ \{Q\ R : Mor\ A\ B\} \to Q \sqsubseteq R \to Q \sqsubseteq (dom\ R)\ \sqamp\ Q$
      domLeastPreserver : $\{A\ B : Obj\}\ \{R : Mor\ A\ B\}\ \{d : Mor\ A\ A\}$
               $\to isSubidentity\ d \to (d\ \sqamp\ d \approx d) \to (R \sqsubseteq d\ \sqamp\ R) \to dom\ R \sqsubseteq d$
      domLocality   : $\{A\ B\ C : Obj\}\ \{R : Mor\ A\ B\}\ \{S : Mor\ B\ C\}$
               $\to dom\ (R\ \sqamp\ dom\ S) \sqsubseteq dom\ (R\ \sqamp\ S)$

Here, we show that this satisfies all the conditions of a domain semigroupoid, and define *domain minimality*, which has been proposed by Desharnais and Möller [8] for characterising determinacy in Kleene algebras.

6 Restricted Residuals

Motivated by application to relations between infinite sets, where residuals (with respect to composition) of finite relations typically have an infinite "uninteresting part", [19] introduced *restricted residuals* that characterise the finite "interesting part"; they have also found applications to substitutions [22]. For ordered semi-groupoids with domain dom and range ran, restricted residuals are distinguished from standard residuals by the additional restr axiom:

field $\_\,/\,\_$ $\qquad$: $\{A\ B\ C\} \to$ Mor $A\ C \to$ Mor $B\ C \to$ Mor $A\ B$
$\quad$ /-cancel-outer : $\{A\ B\ C\}\ \{S : $ Mor $A\ C\}\ \{R : $ Mor $B\ C\} \to (S \,/\, R) \,\mathbin{\mathrm{\circ}}\, R \sqsubseteq S$
$\quad$ /-restr $\qquad$: $\{A\ B\ C\}\ \{S : $ Mor $A\ C\}\ \{R : $ Mor $B\ C\} \to$ ran $(S \,/\, R) \sqsubseteq$ dom R
$\quad$ /-universal $\quad$: $\{A\ B\ C\}\ \{S : $ Mor $A\ C\}\ \{R : $ Mor $B\ C\}\ \{Q : $ Mor $A\ B\}$
$\qquad\qquad\qquad\qquad \to Q \mathbin{\mathrm{\circ}} R \sqsubseteq S \to$ ran $Q \sqsubseteq$ dom $R \to Q \sqsubseteq S \,/\, R$

From the many properties of standard residuals (see for example [11]), a remarkable number carries over to restricted residuals; we list these derived properties without their proofs and without their implicit arguments:

/-cancel-inner	: ran $T \sqsubseteq$ dom S	$\to T \sqsubseteq (T \mathbin{\mathrm{\circ}} S) \,/\, S$
/-monotone	: $S_1 \sqsubseteq S_2$	$\to S_1 \,/\, R \sqsubseteq S_2 \,/\, R$
/-antitone	: $R_2 \sqsubseteq R_1 \to$ dom $R_1 \sqsubseteq$ dom R_2	$\to S \,/\, R_1 \sqsubseteq S \,/\, R_2$
/-cancel-middle	:	$(S \,/\, R) \mathbin{\mathrm{\circ}} (R \,/\, T) \sqsubseteq S \,/\, T$
/-cancel-$\mathbin{\mathrm{\circ}}$	: ran $(S \,/\, R) \sqsubseteq$ dom $(R \mathbin{\mathrm{\circ}} T)$	$\to S \,/\, R \sqsubseteq (S \mathbin{\mathrm{\circ}} T) \,/\, (R \mathbin{\mathrm{\circ}} T)$
/-outer-$\mathbin{\mathrm{\circ}}$	:	$F \mathbin{\mathrm{\circ}} (S \,/\, R)\ \sqsubseteq (F \mathbin{\mathrm{\circ}} S) \,/\, R$
dom-/	:	dom $(S \,/\, R) \sqsubseteq$ dom S
domS/S$\approx$domS	:	dom $(S \,/\, S) \approx$ dom S
ranS/S$\approx$domS	:	ran $(S \,/\, S)\ \approx$ dom S
S/S-$\mathbin{\mathrm{\circ}}$-S	:	$(S \,/\, S) \mathbin{\mathrm{\circ}} S\ \approx S$
S/S-isTransitive	:	isTransitive $(S \,/\, S)$

(The property /-cancel-middle has first been shown in [15].) Restricted right residuals are defined dually, and the following laws hold for combining the two:

/-twist	: dom $(S \,/\, R) \sqsubseteq$ ran $(T \,/\, S) \to S \,/\, R \sqsubseteq (T \,/\, S) \mathbin{\backslash} (T \,/\, R)$
/-twist-down	: dom $(S \,/\, R) \sqsubseteq$ ran $(R \,/\, S) \to S \,/\, R \sqsubseteq (R \,/\, S) \mathbin{\backslash} (R \,/\, R)$
/-twist-up	: $\qquad\qquad\qquad\qquad\qquad\qquad S \,/\, R \sqsubseteq (S \,/\, S) \mathbin{\backslash} (S \,/\, R)$

7 Generalised Algebras

In the context of many-sorted algebras, a signature consists of a set of sorts and a set of function symbols, each equipped with information about its argument and result sorts. An algebra consists of interpretations of the syntactic elements of its signature. Typically, sorts are interpreted as sets, and function symbols as functions from the Cartesian products of the argument sort interpretations to the target sort interpretation.

For a signature Σ, the type of Σ-algebras is then not a "small" Set, but a "large" Set_1. i.e., a member of the next universe encompassing Set. essentially since there are at least as many Σ-algebras as there are "small" Sets. This implies that we cannot, at the same universe level, directly define an algebra which has a carrier that is some set of algebras of some possibly different signature. A practical example where this is desirable is algebras of graphs with graph operations, where the graphs themselves are considered as unary algebras. Therefore, we need a universe-polymorphic concept of Σ-algebras.

In addition to universe polymorphism, we also require *shape polymorphism*, where a shape specifies the allowed arities of function symbols. The resulting allegories of algebras satisfy different properties depending on these shapes (as indicated):

```
data SHAPE : Set where LIST    : SHAPE   -- arbitrary arities: Allegory
                       MAYBE : SHAPE   -- only 0-ary and unary: Collagory
                       NELIST : SHAPE   -- not 0-ary: Allegory with zero laws
                       ONE    : SHAPE   -- only unary: Distributive allegory
```

We introduce auxiliary definitions ShapedList, ShapeProduct and ShapeFunctor etc. to deal with shape polymorphism, and define FunSig as type for the signatures of single function symbols, where src is the argument sort list and has its possible lengths determined by Shape:

```
record FunSig (Shape : SHAPE) (Sort : Set) : Set where
   field src : ShapedList Shape Sort
         trg : Sort
```

A signature of a given Shape and with Sort as set of sorts and FSymb as set of function symbols provides for each function symbol an individual signature (FunSig):

```
Sig : (Shape : SHAPE) → (Sort : Set) → (FSymb : Set) → Set
Sig Shape Sort FSymb = FSymb → FunSig Shape Sort
```

Furthermore, we also need *category polymorphism*, since we want to be able to interpret our signatures over categories different from *Set* — in the graph operation example mentioned above, we might want to move to some category of graph homomorphisms.

For defining just an algebra, we need neither semigroupoid composition nor ShapeProduct parallel composition of morphisms, and therefore declare only the necessary parameters, which include the mapping objProd that takes a Shaped list of semigroupoid objects and maps them to the corresponding product object:

```
record Algebra {Shape : SHAPE} {Sort FSymb} (sig : Sig Shape Sort FSymb)
   {i j : Level} (Obj : Set i) (Mor : Rel Obj j) (objProd : ShapedList Shape Obj → Obj)
   : Set (i ⊔ j) where
   field carrier : Sort → Obj   -- interpretation of sorts
         op : (f : FSymb)       -- interpretation of function symbols
            → Mor (objProd (mapShapedList Shape carrier (FunSig.src (sig f))))
            (                                carrier (FunSig.trg (sig f)))
```

Declaring **open** Algebra makes the field selectors carrier and op available unqualified. This allows a quite concise definition of bisimulations, or relational algebra homomorphisms, where the morphisms used as function symbol interpretations are restricted to be mappings in an OSGC. Such a bisimulation is a sort-indexed family of morphisms between the respective interpretations of each sort, together with a proof of the bisimulation property:

```
record AlgBiSim
  {Shape : SHAPE} {Sort : Set} {FSymb : Set}
  (sig : Sig Shape Sort FSymb)
  {i j k₁ k₂ : Level} {Obj : Set i}
  (base : OSGC j k₁ k₂ Obj)
  (P : ShapeProductSGFunctor Shape (OSGC.semigroupoid base))
  (A : Algebra sig Obj (OSGC.Mapping base) (ShapeProductSGFunctor.objProd P))
  (B : Algebra sig Obj (OSGC.Mapping base) (ShapeProductSGFunctor.objProd P))
  : Set (i ⊔ j ⊔ k₁ ⊔ k₂)
  where
    open OSGC base
    open ShapeProductSGFunctor P
    field
      hom        : (s : Sort) → Mor (carrier A s) (carrier B s)
      commutes : (f : FSymb)
               → morProd hom (FunSig.src (sig f)) ⨾ Mapping.mor (op B f)
               ⊑ Mapping.mor (op A f) ⨾ hom (FunSig.trg (sig f))
```

As a conventional mathematical definition, this might be expressed as follows:

Definition 7.1 Let a signature $\Sigma = (\mathcal{S}, \mathcal{F}, \mathrm{src}, \mathrm{trg})$, an OSGC **C** with sufficient direct products, and two abstract Σ-algebras $\mathcal{A}$ and $\mathcal{B}$ over **C** be given.

A Σ-*bisimulation* Φ *from* $\mathcal{A}$ *to* $\mathcal{B}$ is an $\mathcal{S}$-indexed family of **C**-morphisms $\Phi_s : s^{\mathcal{A}} \to s^{\mathcal{B}}$ such that for every function symbol $f \in \mathcal{F}$ with $f : s_1 \times \cdots \times s_n \to t$ the following inclusion holds:

$$(\Phi_{s_1} \times \cdots \times \Phi_{s_n}) \, ; f^{\mathcal{B}} \sqsubseteq f^{\mathcal{A}} \, ; \Phi_t \, . \qquad \square$$

As usual in such conventional mathematics, many parameters are left implicit, and even where they are technically turned into implicit parameters in Agda, they need to be explicitly listed in the definition. For example, in Def. 7.1 the Shape of the signature is not mentioned explicitly at all, but, together with the ShapeProductSGFunctor, subsumed in the phrase "with sufficient direct products".

Composition of bisimulations is component-wise composition of the morphism families; the necessary correctness proof is, due to explicit associativity steps etc., a bit longer than in usual mathematical presentations, but quite readable — prodComp is distributivity of the morphism part morProd of the Shaped list product functor P over composition:

```
let homComp = λ s → hom R s ⨾ hom S s in record
  {hom = homComp
  ; commutes = λ f → let open FunSig (sig f) using (src; trg) in ⊑-begin
      morProd homComp src ⨾ Mapping.mor (op C f)
```

$\approx\langle$ ⨾-cong$_1$ prodComp $\rangle$
  (morProd (hom R) src ⨾ morProd (hom S) src) ⨾ Mapping.mor (op C f)
$\approx\langle$ ⨾-assoc $\rangle$
  morProd (hom R) src ⨾ morProd (hom S) src ⨾ Mapping.mor (op C f)
$\sqsubseteq\langle$ ⨾-monotone$_2$ (commutes S f) $\rangle$
  morProd (hom R) src ⨾ Mapping.mor (op B f) ⨾ hom S trg
$\approx\langle$ ⨾-assocL $\rangle$
  (morProd (hom R) src ⨾ Mapping.mor (op B f)) ⨾ hom S trg
$\sqsubseteq\langle$ ⨾-monotone$_1$ (commutes R f) $\rangle$
  (Mapping.mor (op A f) ⨾ hom R trg) ⨾ hom S trg
$\approx\langle$ ⨾-assoc $\rangle$
  Mapping.mor (op A f) ⨾ hom R trg ⨾ hom S trg
$\approx\langle$ $\approx$-refl $\rangle$
  Mapping.mor (op A f) ⨾ homComp trg   □}

From there, it is relatively straightforward to define the instances of the semigroupoid and category types introduced in Sect. 4.

8 Related Work

Our approach to categories with setoids of morphisms, but not of objects, derives essentially from Kanda's "effective categories" [23]; it is also used by Huet and Saïbi [16] for their formalisation of category theory in Coq, and by Gonzalía [12], who produced formalisations of concrete heterogeneous binary relations and of Freyd and Scedrov's allegory hierarchy [10] in Alf, a predecessor of Agda.

Mu et al. [25] have contributed Agda2 theories inspired by Bird and de Moor's *Algebra of Programming* [4]; they note the advantages that Agda2 brought to their formalisations of concrete relations over the Alf formalisations of Gonzalía.

Jackson [17] formalised abstract algebra as used in computer algebra systems in Nuprl, which uses a variant of type theory that provides sets, and therefore does not need setoids. This work also does not include a general approach using signatures.

Capretta [5] formalised universal algebra in Coq, with fixed encoding of the sets of sorts and function symbols as finite natural number sets. Barthe et al. [2] provide an in-depth discussion of different treatments of setoids.

9 Conclusion

Our extension of a treatment of binary heterogeneous relations similar to that of Mu et al. [25] to full universe polymorphism as mentioned in Sect. 3 is a minor, technical contribution which nonetheless constitutes a significant generalisation.

The semigroupoids and categories of Sect. 4 not only bring formalisations similar to Gonzalía's [12] into a current system, and into fully universe-polymorphic shape; they also reflect recent developments towards finer granularity of these theories. They are also a significant advance over the Isabelle theories of [18] both in scope and in style of exposition: Besides of the more natural formalisation of categories in a dependently-typed system, Agda also enables more flexibility with structuring a theory hierarchy through arbitrary (sub-)module opening, whereas the records underlying locales in Isabelle allow only extension at predefined extension points. As a result, the Agda formalisation appears to be much more maintainable.

Finally, our way of constructing allegories (etc.) of algebras from underlying allegories seems to not have been formalised in a mechanised theorem prover before. Doing this in Agda has proven quite satisfactory, since the language combines natural mathematical expressiveness with a programming attitude. Future work will continue to formalise material required for powerful relation-algebraic graph transformation concepts [21], and will explore to use Agda's foreign-function interface to Haskell to combine verified graph transformation algorithms in Agda with graphical user interfaces [29], or to use it in code generation back-ends [1].

References

1. Anand, C.K., Kahl, W.: An optimized Cell BE special function library generated by Coconut. IEEE Transactions on Computers 58(8), 1126–1138 (2009)
2. Barthe, G., Capretta, V., Pons, O.: Setoids in type theory. J. Funct. Program. 13(2), 261–293 (2003)
3. Berghammer, R., Jaoua, A.M., Möller, B. (eds.): RelMiCS 2009. LNCS, vol. 5827. Springer, Heidelberg (2009)
4. Bird, R.S., de Moor, O.: Algebra of Programming. International Series in Computer Science, vol. 100. Prentice-Hall, Englewood Cliffs (1997)
5. Capretta, V.: Universal algebra in type theory. In: Bertot, Y., Dowek, G., Hirschowitz, A., Paulin, C., Théry, L. (eds.) TPHOLs 1999. LNCS, vol. 1690, pp. 131–148. Springer, Heidelberg (1999)
6. Corradini, A., Montanari, U., Rossi, F., Ehrig, H., Heckel, R., Löwe, M.: Algebraic approaches to graph transformation, part I: Basic concepts and double pushout approach. In: Rozenberg, G. (ed.) Handbook of Graph Grammars and Computing by Graph Transformation, Foundations, vol. 1, ch. 3, pp. 163–245. World Scientific, Singapore (1997)
7. Desharnais, J., Jipsen, P., Struth, G.: Domain and antidomain semigroups. In: Berghammer et al. [3], pp. 73–87
8. Desharnais, J., Möller, B.: Characterizing determinacy in Kleene algebras. Information Sciences 139, 253–273 (2001)
9. Desharnais, J., Möller, B., Struth, G.: Kleene algebra with domain. ACM Transactions on Computational Logic 7(4), 798–833 (2006)
10. Freyd, P.J., Scedrov, A.: Categories, Allegories. North-Holland Mathematical Library, vol. 39. North-Holland, Amsterdam (1990)
11. Furusawa, H., Kahl, W.: A study on symmetric quotients. Tech. Rep. 1998-06, Fakultät für Informatik, Universität der Bundeswehr München (December 1998)
12. Gonzalía, C.: Relations in Dependent Type Theory. Ph.D. thesis, also as Technical Report No. 14D, Department of Computer Science and Engineering, Chalmers University of Technology, Göteborg University (2006)
13. Gurevich, Y.: Evolving Algebras: An attempt to discover semantics. In: Rozenberg, G., Salomaa, A. (eds.) Current Trends in Theoretical Computer Science, pp. 266–292. World Scientific, Singapore (1993)

14. Gurevich, Y.: Sequential abstract state machines capture sequential algorithms. ACM Transactions on Computational Logic 1(1), 77–111 (2000)
15. Han, J.: Proofs of Relational Semigroupoids in Isabelle/Isar. M.Sc. thesis, McMaster University, Department of Computing and Software (2008)
16. Huet, G., Saïbi, A.: Constructive category theory. In: Plotkin, G.D., Stirling, C., Tofte, M. (eds.) Proof, language, and interaction: Essays in honour of Robin Milner. Foundations of Computing Series, pp. 239–275. MIT Press, Cambridge (2000)
17. Jackson, P.B.: Enhancing the Nuprl Proof Development System and Applying it to Computational Abstract Algebra. Ph.D. thesis, Cornell University (1995)
18. Kahl, W.: Calculational relation-algebraic proofs in Isabelle/Isar. In: Berghammer, R., Möller, B., Struth, G. (eds.) RelMiCS 2003. LNCS, vol. 3051, pp. 178–190. Springer, Heidelberg (2004)
19. Kahl, W.: Relational semigroupoids: Abstract relation-algebraic interfaces for finite relations between infinite types. J. Logic and Algebraic Programming 76(1), 60–89 (2008)
20. Kahl, W.: Collagories for relational adhesive rewriting. In: Berghammer et al [3], pp. 211–226
21. Kahl, W.: Amalgamating pushout and pullback graph transformation in collagories. In: Ehrig, H., Rensink, A., Rozenberg, G., Schürr, A. (eds.) ICGT 2010. LNCS, vol. 6372, pp. 362–378. Springer, Heidelberg (2010)
22. Kahl, W.: Determinisation of relational substitutions in ordered categories with domain. J. Logic and Algebraic Programming 79, 812–829 (2010)
23. Kanda, A.: Constructive category theory (no. 1). In: Gruska, J., Chytil, M.P. (eds.) MFCS 1981. LNCS, vol. 118, pp. 563–577. Springer, Heidelberg (1981)
24. McCune, W.: Prover9 and Mace4, version LADR-2009-11A (2009), http://www.prover9.org/
25. Mu, S.C., Ko, H.S., Jansson, P.: Algebra of programming using dependent types. In: Audebaud, P., Paulin-Mohring, C. (eds.) MPC 2008. LNCS, vol. 5133, pp. 268–283. Springer, Heidelberg (2008)
26. Norell, U.: Towards a Practical Programming Language Based on Dependent Type Theory. Ph.D. thesis, Department of Computer Science and Engineering, Chalmers University of Technology (September 2007)
27. Schmidt, G., Hattensperger, C., Winter, M.: Heterogeneous relation algebra. In: Brink, C., Kahl, W., Schmidt, G. (eds.) Relational Methods in Computer Science. Advances in Computing Science, ch. 3, pp. 39–53. Springer, Wien (1997)
28. Schmidt, G., Ströhlein, T.: Relations and Graphs, Discrete Mathematics for Computer Scientists. EATCS-Monographs on Theoret. Comput. Sci. Springer, Heidelberg (1993)
29. West, S., Kahl, W.: A generic graph transformation, visualisation, and editing framework in Haskell. In: Boronat, A., Heckel, R. (eds.) Proceedings of the Eighth International Workshop on Graph Transformation and Visual Modeling Techniques (GT-VMT 2009). Electronic Communications of the EASST, vol. 18, pp. 12.1–12.18 (September 2009)

Omega Algebras and Regular Equations

Michael R. Laurence and Georg Struth

Department of Computer Science
The University of Sheffield, UK
{m.laurence,g.struth}@dcs.shef.ac.uk

Abstract. We study a weak variant of omega algebra, where one of the usual star induction axioms is absent, in the context of recursive regular equations. We present abstract conditions for explicitly defining the omega operation and use them for proving an algebraic variant of Arden's rule for solving such equations. We instantiate these results in concrete models—languages, traces and relations—showing, for instance, that the omega captures precisely the empty word property in regular languages. Finally, we derive Salomaa's axioms for the algebra of regular events. This yields a sound and complete axiomatisation in which the "regular" axioms are weaker than Kleene algebra.

1 Introduction

More than thirty years ago, Arden, Brzozowski, Salomaa and others developed a beautiful and intriguingly simple algebraic approach to automata, regular languages and regular expressions. Arden associated automata with systems of recursive language equations and provided a rule—*Arden's rule*—for solving these equations [1]. Brzozowski introduced residuals and language derivatives as a simple algebraic tool for obtaining (minimal deterministic) automata from regular expressions [3]. Salomaa combined these techniques in an algebra of regular events in which all valid identities of regular expressions can be derived. In the derivation process, characteristic recursive equations for regular expressions are constructed by equivalence preserving transformations within the algebra. Arden's rule is then used for solving these equations [14]. This approach had considerable influence, for instance, on Milner's development of CCS and the π-calculus (cf. [12]). Today, however, it seems to be unknown except to a small circle of specialists.

Continuing our work on termination and nontermination analysis in variants of Kleene algebras [5,9], we revisit Arden and Salomaa's approach to solving regular equations from the point of view of omega algebras. These were introduced as algebras of omega-regular events [4] to capture the equational theories of omega-regular expressions and languages. They expand the regular operations of union, concatenation and finite iteration, as axiomatised by Kleene algebras [11], by an operation for infinite iteration. More specifically, we investigate a weak variant of omega algebras in the context of *regular* languages; of linear recursive equations over regular or Kleene algebra terms. We call these *left omega algebras* since the usual right star induction axiom of Kleene algebra is absent.

H. de Swart (Ed.): RAMICS 2011, LNCS 6663, pp. 248–263, 2011.

It is well known that Kleene algebra is sound and complete for the equational theory of regular expressions [11], but the greater expressivity of omega algebra yields new insights. It allows us to round up previous research on (non)termination and put it into a broader perspective. It motivates some interesting research questions, as outlined at the end of this paper. Our main results are as follows:

- We provide sufficient conditions for explicitly defining the omega operation in left Kleene algebras and building left omega algebras as conservative extensions (more precisely, extensions by definition) of left Kleene algebras.
- We show that Arden's rule from formal language theory, which cannot be expressed algebraically in Kleene algebra, is derivable both as a quasi-identity and as an inference rule in left omega algebra. This abstract version of Arden's rule states that certain recursive regular equations, that is, linear recursive equations in the language of Kleene algebras, have unique solutions. We also show that the applicability condition for Arden's rule (that a certain omega term vanishes) has meaningful interpretations in various models.
- We specialise our abstract results to regular language models. Here, the omega operation is boolean-valued, hence a predicate. It characterises precisely the empty word property, which holds if a language contains the empty word. Arden's rule for regular languages can be obtained from its abstract relative via a simple length-increase argument. Analogous results are obtained for trace and path algebras.
- We show that Salomaa's axioms are derivable in left omega algebras. Consequently, left omega algebras are sound and complete for the algebra of regular events. The omega-free fragment of left omega algebras thus coincides with the equational fragment of regular languages, which is decidable.
- Finally, we show that in relation semirings, the omega of an element vanishes if and only if that element is wellfounded. Arden's rule now specialises to a unique extension property (cf. [7]), which we can derive, for the first time, in a first-order setting.

Our results show that (left) omega algebras play an interesting role in formal language theory and computational modelling, whenever a system is specified in terms of recursive equations. They allow us to obtain explicit definitions of the omega operator and to derive generic conditions for unique solutions of regular equations by equational first-order reasoning. As so often with Kleene algebras and related approaches, results that were previously fragmented across concrete models can be obtained in an abstract, uniform and very simple way. The proof of Arden's rule, in particular, is almost trivial in this setting.

All calculational results in this paper have been formally verified by automated reasoning within the Isabelle/HOL theorem prover [13]. In this paper, therefore, we only show proofs that we find interesting. The remaining ones can be found online in the omega algebra file of our Isabelle repository for algebraic methods [16].

2 Left Omega Algebras

The algebras studied in this paper are based on idempotent semirings or dioids. Dioids expanded by an operation of finite iteration are known as Kleene algebras. Omega algebras are obtained by further expanding these by an operation of infinite iteration.

Formally, a *semiring* is a structure $(S, +, \cdot, 0, 1)$ over a set S such that $(S, +, 0)$ is a commutative monoid, $(S, \cdot, 0)$ is a monoid, multiplication distributes over addition from the left and right,

$$x \cdot (y + z) = x \cdot y + x \cdot z, \qquad (x + y) \cdot z = x \cdot z + y \cdot z,$$

and zero is a left and right annihilator with respect to multiplication,

$$0 \cdot x = 0, \qquad x \cdot 0 = x.$$

An *idempotent semiring* or *dioid* is a semiring in which the following additive idempotency law holds:

$$x + x = x.$$

Every dioid is ordered by the usual order $\leq$ on the semilattice reduct $(S, +)$. The operations of addition and multiplication are isotone with respect to that order and 0 is the least element.

Semirings and dioids satisfy a duality principle. The *opposite* of a semiring or dioid can be formed by swapping the order of multiplication. Since all axioms of semirings and dioids are transformed into axioms under opposition, the opposite of a semiring or dioid is again a semiring or dioid and theorems are preserved under opposition as well.

A *left Kleene algebra* is a dioid K expanded by a star operation $^* : K \to K$ which satisfies the unfold axiom and induction axiom

$$1 + xx^* = x^* \qquad \text{and} \qquad z + xy \leq y \Rightarrow x^*z \leq y.$$

A *Kleene algebra* is a left Kleene algebra where the dual axioms $1 + x^*x = x^*$ and $z + yx \leq y \Rightarrow zx^* \leq y$ hold, too. While the opposite of a Kleene algebra is also a Kleene algebra, left Kleene algebras need not be closed under opposition [10].

A *left omega algebra* is a left Kleene algebra K expanded by an omega operation $^\omega : K \to K$ which satisfies the unfold axiom and the coinduction axiom

$$x^\omega = xx^\omega \qquad \text{and} \qquad y \leq xy + z \Rightarrow y \leq x^\omega + x^*z.$$

An *omega algebra* is a left omega algebra that is also a Kleene algebra.

The operations of star and omega are intended to model finite and strictly infinite iteration. Formally, the star is defined as a least fixpoint and the star induction law provides the corresponding induction principle. Similarly, the omega is defined as a greatest fixpoint with a corresponding law for coinduction. Every omega algebra has a maximal element with respect to the natural order, namely 1^ω, whereas Kleene algebras need not possess maximal elements. We write $\top = 1^\omega$. Omega algebras need not be closed under opposition. An important property is that all operations of (left) omega algebras are isotone with respect to the natural order.

A hierarchical axiomatisation of all these structures and a large number of facts, including a formalisation of the most important models, can be found in our Isabelle repository [16]. More information about this repository and automated theorem proving in algebras with Isabelle can be found in a tutorial paper [8].

3   Languages, Relations and Traces

In this paper, we mainly focus on the regular language model, the regular trace model and the regular relational model of left Kleene algebras and left omega algebras, which of course are also models of Kleene algebras and omega algebras. By "regular" we mean that only the regular operations, which are given by the signature of Kleene algebras, are used and only finite words and traces are considered. These models have been studied extensively in the literature. We follow the definitions in [9] and refer to that publication for additional information. An important distinction, however, is that the language, relation and trace models considered in this text are not necessarily complete, that is, their boolean reducts need not be closed under arbitrary infima and suprema.

To obtain strong statements, while proving theorems for left Kleene algebras and left omega algebras, we construct counterexamples usually for Kleene algebras and omega algebras.

Example 1. Regular languages form Kleene algebras. Let Σ be a finite set or alphabet and let Σ^* be the free monoid (set of all finite words) over Σ. A *language* is a subset X of Σ^*. The structure $(2^{\Sigma^*}, \cup, \circ, {}^*, \emptyset, \{\epsilon\})$ forms a Kleene algebra under the standard regular operations of formal language theory:

- $X \cup Y$ is the set-theoretic union of the languages X and Y,
- $XY = \{xy \in \Sigma^* : x \in X \text{ and } y \in Y\}$,
- $X^* = \bigcup_{i \geq 0} X^i$, where the powers of X are inductively defined.

This Kleene algebra is called the *full language Kleene algebra* over Σ. Each subalgebra—not necessarily complete—of a full language Kleene algebra is a Kleene algebra. We call these subalgebras *language Kleene algebras* over Σ. A language is *regular* if it can be inductively generated from the empty language, the empty string language $\{\epsilon\}$ and the singleton languages $\{a\}$ for each $a \in \Sigma$ by applying the regular operations.

It is well known that regular languages can be represented by regular expressions, and we usually do not distinguish between the two. Regular expressions and Kleene algebras have the same signature. A classical result in Kleene algebra shows that the above axioms for Kleene algebras are not only sound, but also complete for the equational theory of regular languages or regular expressions, which is also called the *algebra of regular events* [11]. Hence identities between Kleene algebra terms are equivalent to equivalences between regular expressions, and they can be decided by finite automata. The proof of this theorem is based on the fact that Kleene algebras are closed under matrix formation and these matrices can represent finite automata. The standard constructions of automata theory, automata from regular expressions, ϵ-elimination, determinisation and minimisation, can then be carried out within Kleene algebra and Kleene's theorem becomes a theorem of Kleene algebra. These constructions, however, require the presence of both the left and the right star induction axioms. Whether left Kleene algebras are complete for the algebra of regular events is, as far as we know, open.

Example 2. Binary relations form Kleene algebras. Let A be a set. The structure $(2^{A \times A}, \cup, \circ, {}^*, \emptyset, 1_A)$ forms a Kleene algebra, where $2^{A \times A}$ denotes the set of all binary relations on A, $\cup$ is set union, $\circ$ is relative product, * is the reflexive-transitive closure operation, $\emptyset$ is the empty relation and 1_A the identity relation on A. This Kleene algebra is called the *full relation Kleene algebra* over A. Each subalgebra of a full relation Kleene algebra is a Kleene algebra; a *relation Kleene algebra* over A.

Example 3. Sets of traces form Kleene algebras. A *trace* over a (finite) set P and a (finite) set A is a finite sequence over $(P \cup A)^*$, in which the first and last letter are in P and in which letters from P and A alternate. $(P, A)^*$ denotes the set of all traces over P and A. The product of traces is a partial operation:

$$p_0 a_0 \ldots a_{m-1} p_m \cdot q_0 b_0 \ldots b_{n-1} q_n = p_0 a_0 \ldots a_{m-1} p_n b_0 \ldots b_{n-1} q_n$$

if $p_m = q_0$, and it is undefined otherwise. The product $T_1 \cdot T_2$ on sets of traces, which is total, and the remaining regular operations can be defined as in the language case (c.f. [9]). This turns $2^{(P,A)^*}$ into a Kleene algebra, the *full trace Kleene algebra* over P and A. In particular, P is the multiplicative unit of this algebra. Again, every subalgebra of a full trace Kleene algebra is a Kleene algebra; a *trace Kleene algebra*.

Path algebras can be obtained from trace algebras by "forgetting" the elements of A. They are very similar to trace algebras [9] and we therefore do not discuss them any further in this paper.

Complete language, relation, trace and path Kleene algebras all have complete boolean algebras as reducts, and since the regular operations are isotone, the map $\lambda y.xy$ has a greatest fixed point by the Knaster-Tarski theorem.

Theorem 4. *Every complete language, relation, trace or path Kleene algebra can be expanded uniquely to an omega algebra.*

4   Conditions for Defining Omega

This section provides conditions for explicitly defining the omega operation in left omega algebras. These conditions extend, simplify and generalise previous work on trace semirings [9].

We call an element x of a dioid *dense* if $x \le xx$ holds. In particular, every multiplicatively idempotent element is dense, and every element above 1 is dense, since $1 \le x$ if and only if $x = x + 1$. Therefore $xx = (x+1)x = xx + x \ge x$. Subidentities in dioids, however, need not be dense—Nitpick, a counterexample generator which is part of the Isabelle system, found a counterexample with three elements—but they are dense in many models of interest.

We call an element x of a (left) omega algebra ω-*trivial* if $x^\omega = 0$. The following facts have been verified with Isabelle.

Lemma 5. *Let x be a dense element of a left omega algebra. Then*

(a) $x^\omega = x\top$,
(b) $(x+y)^\omega = y^\omega + y^*x\top$.

The following facts are immediate from Lemma 5(a) and because $x\top = \top$ for all $x \ge 1$.

Lemma 6. *In every left omega algebra,*

(a) $0^\omega = 0$,
(b) $1 \le x \Rightarrow x^\omega = \top$.

Lemma 5(a) and Lemma 6 show that the omega of many elements of left omega algebras can be explicitly defined, in particular, that of all dense elements and of all elements greater than 1. Lemma 5(b) can be turned into an abstract sufficient condition for explicitly defining omega.

Proposition 7. *Let x be an element of a left omega algebra that can be split as $x = x_0 + x_1$, where x_0 is dense and x_1 ω-trivial. Then*

$$x^\omega = x_1^* x_0 \top.$$

Hence in every left omega algebra in which every element can be split into a dense and an ω-trivial part, omega can be explicitly defined.

Of course, 0 is both dense and ω-trivial, hence in the above definition both x_0 and x_1 can be zero.

The following sections show that this splitting works in many important models, in particular language, trace and path models. However, it fails in the relational model. In the cases we consider, splitting is associated with a length-increase argument. A more general investigation of this association and other mechanisms for splitting elements is left for future research.

The property of ω-triviality can be related with an equivalent condition that holds in arbitrary dioids. We call an element x of a dioid *deflationary* if

$$\forall y.y \le xy \Rightarrow y = 0.$$

This is motivated by the fact that, in Bourbaki-Witt fixpoint theory, functions on posets that satisfy $y \le f(y)$ for all y are often called *inflationary*[1].

---

[1] Backhouse and Carré [2] write *definite* instead of *deflationary*.

Lemma 8. *An element of a left omega algebra is deflationary if and only if it is ω-trivial.*

This fact is already known for omega algebras [9]. The following fact rules out that elements greater than one are deflationary.

Lemma 9. *For every element $x \geq 1$ in a dioid, the map $\lambda y.xy$ is inflationary.*

It is well known that extensions of theories by definition are conservative extensions, that is, every statement of the restricted theory holds in its extension, and every statement in the language of the restricted theory which holds in the extended theory holds already in the restriction. In other word, extensions by definition do not add expressive power.

Section 6 and Section 8 show that language and trace omega algebras are conservative extensions of the respective Kleene algebras.

5 Arden's Rule Abstractly

Arden's rule is a fundamental tool of formal language theory [1]. To determine, for instance, the language accepted by the automaton

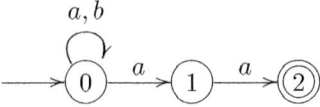

it can be translated into a system of recursive language equations

$$x_0 = (a+b)x_0 + ax_1$$
$$x_1 = ax_2$$
$$x_2 = 1$$

Arden's rule yields a way of solving this system. The solution to the first equation—which is recursive—is $x_0 = (a+b)^*ax_1$. Solutions for x_1 and then x_2—which are not recursive—are obtained by substitution. In particular, we obtain the solution $x_0 = (a+b)^*aa1 = (a+b)^*aa$, which is, of course, the regular expression corresponding to the automaton.

More generally, Arden's rule states that, whenever a language denoted by a regular expression x does not contain the empty word, if $y = xy + z$ is valid, then $y = x^*z$ is valid. In other words, if a language x does not have the empty word property, then the recursive equation $y = xy + z$ has the unique solution $y = x^*z$.

It should be evident that Arden's rule is of general interest for modelling and reasoning about computing systems in terms of systems of recursive equations. The work of Salomaa [14] shows that Arden's rule yields a simple algebraic proof of one direction of Kleene's theorem (the other one can be obtained nicely and algebraically by using Brzozowski's language derivatives [3]). Apart from

Milner's work mentioned in the introduction, also Backhouse and Carré's study of matrix algebras over regular algebras [2] rely heavily on Arden's rule.

To provide a more general context for Arden's rule we prove it abstractly in left omega algebras and discuss some of its consequences and variants at the algebraic level.

Theorem 10 (Arden's rule). *Let x be an ω-trivial element of a left omega algebra. Then*

$$y = xy + z \Rightarrow y = x^* z.$$

Proof. $xy + z \leq y \Rightarrow x^* z \leq y$ is the induction axiom of left Kleene algebra. So it remains to show that $y \leq xy + z \Rightarrow y \leq x^* z$. By omega coinduction, $y \leq xy + z$ implies $y \leq x^\omega + x^* z$, so $y \leq x^* z$ holds whenever $x^\omega \leq x^* z$, hence in particular when $x^\omega = 0$. $\qquad\square$

We display this proof only as an example for the simplicity and genericity of algebraic calculations. A formal proof can be found in our repository. We are indebted to one referee for pointing out that instead of x being ω-trivial we could have used the weaker condition $x^\omega \leq x^* z$ in the statement of Arden's rule. We have adapted our proof accordingly. While this weakening might be interesting in its own right, ω-triviality corresponds more closely to the statement of Arden's rule in language theory.

Arden's rule is often used as an inference rule which is also derivable in left omega algebras due to standard properties of universal quantification.

Corollary 11. *In left omega algebras, the following inference rule is derivable.*

$$\frac{x^\omega = 0 \qquad y = xy + z}{y = x^* z}$$

The next proposition shows that left omega algebras are very useful for deriving consequences and variations of Arden's lemma.

Proposition 12

(a) *In left Kleene algebra, $y = x^* z \Rightarrow y = xy + z$.*
(b) *In left omega algebra, $y = xy + z \Leftrightarrow y = x^* z$ if x is deflationary (ω-trivial).*
(c) *In left Kleene algebra (left omega algebra), $\forall y, z.(y \leq xy + z \Rightarrow y \leq x^* z)$ implies that x is deflationary (ω-trivial).*
(d) *In left omega algebra, $\forall y, z.(y = xy + z \Leftrightarrow y = x^* z)$ if and only if x is deflationary (ω-trivial).*
(e) *In left omega algebra, the equation $y = xy + z$ has the unique solution $y = x^* z$ if and only if x is deflationary (ω-trivial).*

Again, all proofs can be found in our repository. They are fully automated, which further underpins the simplicity of the algebraic approach.

It is interesting to contrast Arden's rule with the star induction rule which, in left Kleene algebras, is equivalent to

$$xy + z = y \Rightarrow x^* z \leq y.$$

Accordingly, x^*z is the *least* solution of the equation $y = xy + z$ in every left Kleene algebra. But, as the next lemma shows, it need not be the only solution.

Lemma 13

(a) *Every left Kleene algebra contains elements x, y and z such that $y = xy + z$, but $y \neq x^*z$.*

(b) *In every left Kleene algebra, if $1 \leq x$, then $y = xy + z$ has solutions $y = x^*w$ for all elements $w \geq z$.*

(c) *In some left Kleene algebras, for some elements x and z, the equation $y = xy + z$ has more than one solution.*

Proof.

(a) Let $x = y = 1$ and $z = 0$. Then $1 = 1 \cdot 1 + 0$, but $1 \neq 0 = 1^* \cdot 0$.

(b) If $1 \leq x$, then $\lambda y.xy$ is inflationary, hence $z \leq xz \leq x^*w$ and $x^* = xx^*$. Therefore $xx^*w + z = x^*w + z = x^*w$.

(c) Every bounded distributive lattice is a Kleene algebra bounded by 0 and 1. Join is addition, meet is multiplication and $x^* = 1$ for each element of x. In this model, for every element w, $xw + z$ solves the equation $y = xy + z$ since $y = xy + z = x(xw + z) + z = xxw + xz + z = xw + z$. It remains to find w_1 and w_2 for which $xw_1 + z \neq xw_2 + z$. These constraints are met, for instance, by any chain where $w_1 > x > z > w_2$.   □

These last statements consider arbitrary fixpoints between the least and the greatest one.

An interesting open question is whether deflationarity, instead of ω-triviality, implies Arden's rule already in (left) *Kleene* algebras, that is, for a signature that contains only those operations that occur in the rule itself.

6   Omega and Regular Languages

The results of the previous sections immediately specialise to regular language omega algebras over an alphabet Σ.

In this setting, the elements above 1 are those languages that contain the empty word, that is, which have the empty word property.

Lemma 14. *If a language L has the empty word property, then*

(a) *L is dense,*

(b) *$L^\omega = \Sigma^*$.*

This is an immediate instance of Lemma 5 and the fact that all $x \geq 1$ in a dioid are dense.

Lemma 15. *Languages that do not satisfy the empty word property are deflationary.*

Proof. Assume that L does not have empty word property and let $L' \neq \emptyset$. Then any word of minimal length in LL' must be strictly greater than any word of minimal length in L and therefore $L' \not\leq LL'$, that is, L is deflationary.   □

This algebraic characterisation of the empty word property has already been used by Backhouse and Carré [2]. Their Theorem 5.2 proves the following corollary (with respect to deflationary elements) for the special case of matrices over regular languages, from which the statement for regular languages follows as a subcase.

Corollary 16. *Languages do not have the empty word property if and only if they are deflationary (ω-trivial).*

Proof. The only-if-direction is immediate from Lemma 15 and Lemma 8. The if-direction is immediate from Lemma 9.   □

Now, obviously, every language L does or does not have the empty word property. Hence the splitting condition of Proposition 7 is trivially applicable. The following fact is then obvious from our abstract results.

Proposition 17. *Every language (left) Kleene algebra can be uniquely expanded into a language (left) omega algebra. For every language L,*

$$
L^\omega = \begin{cases} \Sigma^* & \text{if } L \text{ has the empty word property,} \\ \emptyset & \text{otherwise.} \end{cases}
$$

Hence language omega algebras are extensions by definition of language Kleene algebras.

It is important that "extension" means that the omega operation is defined on *regular languages*, not that regular languages are expanded to omega-regular languages. Unfortunately, the uses of "extension" in model theory and universal algebra, where it means adding elements to models, and proof theory, where conservative extensions expand signatures, may be confusing. As a consequence of Proposition 17, the absence of the empty word property can be *defined* as an identity in left omega algebra, whereas this is not possible in Kleene algebra, where deflationarity, which is a quasi-identity, hence a universally quantified equational Horn formula, could be used. A second consequence is important for the following section.

Corollary 18. *Language (left) omega algebras are conservative extensions of language (left) Kleene algebras.*

Therefore, if a theorem in the language of Kleene algebras holds in all language omega algebras, it already holds in all language Kleene algebras.

Finally, Arden's rule of formal language theory arises as an immediate consequence of Corollary 16 and Theorem 10.

Lemma 19. *If the language L does not have the empty word property, then the equation $Y = LY + L'$ has the unique solution $Y = L^*L'$.*

Formally, of course, it can be written as a quasi-identity, which is not possible in Kleene algebra. All the other abstract results from Section 4 and Section 5 hold in the language model, too.

7 Soundness and Completeness for Regular Languages

An abstract variant of Arden's rule plays a prominent role as an axiom in Salomaa's sound and complete axiomatisation for the algebra of regular events [14]. In this setting, soundness means that regular languages are models of Salomaa's axioms, hence all equations derived from these axioms hold when interpreted in regular languages. Completeness means that all identities between regular expressions—that is, identities denoting the same regular language—can be derived from the axioms.

Salomaa essentially expands dioids by a star operation that satisfies the axioms

$$1 + x^*x = x^*, \qquad (1+x)^* = x^*$$

and by the following inference rule for solving equations:

$$\frac{y = yx + z}{y = zx^*}.$$

It can be applied whenever x does not have the empty word property. This last rule is the opposite of Arden's rule.

Proposition 20. *Salomaa's (dual) axioms are theorems of left omega algebra.*

But because theorems of Kleene algebras and valid regular identities are preserved by opposition, the duals of Salomaa's axioms yield a complete axiomatisation for the algebras of regular events, too. Technically this means that Salomaa's representation of regular expressions (or automata) in terms of recursive equation must be dualised in order to use the axioms of weak omega algebras. Algebras satisfying the duals of Salomaa's axioms have been called *regular algebras* by Backhouse and Carré [2].

Saloma's axioms were criticised by Kozen for not being algebraic: Due to the meta-level side condition "x does not have the empty word property", Salomaa's variant of Arden's rule is not preserved under substitution [11]: "Another way to say this is that [Salomaa's variant of Arden's rule] must not be interpreted as a universal Horn formula". While this is certainly true in the context of Kleene algebras, left omega algebras yield an algebraic axiomatisation.

Theorem 21. *Left omega algebra is complete for the algebra of regular events: Let s and t be two regular expressions over an alphabet Σ that denote the same regular language. Then $s = t$ is a theorem of left omega algebra.*

Proof. Salomaa's axioms are complete for the algebra of regular events [14], and these axioms (more precisely their duals) are derivable in left omega algebra. Hence left omega algebra is also complete.   □

Theorem 22. *Left omega algebra is sound for the algebra of regular events.*

Proof. This holds because language omega algebras are conservative extensions of language Kleene algebras.   □

Corollary 23. *The equational theory of regular (omega-free) terms in left omega algebra is decidable.*

By soundness and completeness, the free algebras of omega-free terms in the class of omega algebras are isomorphic to regular languages. Equality between these can be decided, for instance, by using automata, or by Salomaa's procedure for solving equations.

The completeness result in this section might seem bizarre, because it holds for terms in a restricted signature and an algebra considered over a rather unnatural model. But it is situated between Salomaa's result (since it is algebraic) and Kozen's result (since it uses weaker Kleene algebra axioms); and it raises the question whether a completeness result for the algebra of regular events and *left Kleene algebras* is possible.

8   Omega and Traces

Trace omega algebras have, to some extent, been studied in [9]. The arguments are similar to, but slightly different from language omega algebras.

In trace models, the elements between 0 and 1 are the subsets of P, that is, sets of traces of length one. They form a boolean subalgebra and are therefore multiplicatively idempotent, hence dense. A set of traces has been called *test-free* if it does not contain a subset of P [9]. Each set of traces can be split in a subset of P and a test-free subset (both possibly empty).

Lemma 24 ([9]). *If a set of traces is test-free, then it is deflationary (ω-trivial).*

A length-increase argument is again the key to proving this fact. It follows that the omega operation can once more be explicitly defined.

Proposition 25. *Every trace (left) Kleene algebra can be uniquely expanded to a trace (left) omega algebra. For every set T of traces,*

$$T^\omega = \begin{cases} T \circ \top & \text{if } T \subseteq P, \\ \emptyset & \text{otherwise,} \end{cases}$$

where $\top$ denotes the set of all traces.

So trace omega algebras are extensions by definition of trace Kleene algebras and therefore conservative extensions.

It is also obvious that a variant of Arden's rule can be obtained for the trace model which can be used for solving recursive trace equations, for instance in the context of reactive system verification, where trace models are important.

As already mentioned, a special case of trace omega algebras are path omega algebras (cf. [2,9]). In path algebras, the elements between 0 and 1 are the sets of paths of length one. Sets of paths can again be split into subsets of P and test-free paths. The test-free paths are deflationary, and the omega of a set of paths is that set composed with the set of all paths, if the set is a subset of P and it is empty otherwise. All further results that hold of trace left omega algebras also hold of path left omega algebras. Additional results about paths dioids and their relationship to trace dioids can be found in [9].

9   Omega and Relations

Relation omega algebras differ from trace and language omega algebras in that a length-increase argument for showing that an element is deflationary does not work anymore.

In relational models, all elements above 1 are reflexive relations. Their omega is, of course, ⊤ (the full cartesian product). Also, as in the case of trace models, all elements below 1 are multiplicatively idempotent, hence dense.

Lemma 26. *Each subidentity R of a relation left omega algebra satisfies*

$$R^{\omega} = R \circ \top.$$

Hence deflationary or ω-trivial elements must be irreflexive. It is also clear that each relation can again be split into a subidentity and an irreflexive part. But will R^{ω} vanish for all irreflexive relations?

Lemma 27. *There exists a relation dioid in which some irreflexive relation is not deflationary.*

Proof. Consider the full relation dioid over the booleans $\mathbb{B} = \{0, 1\}$. The relation R is depicted in the left-hand diagram below whereas the right-hand diagram shows $\top = \mathbb{B}^2$. It is easy to see that $R \circ \top = \top$ and, obviously, $\top \neq \emptyset$.

□

Hence the situation is more complex than in trace semirings.

However it turns out that in relation dioids, being deflationary means being wellfounded, and we can prove this abstractly. The concept of wellfoundedness has already been investigated in the context of domain semirings and Kleene algebras with domain [6,5]. Formally, a *domain semiring* [6] is a semiring S expanded by a domain operation $d : S \to S$ that satisfies

$$x \leq d(x)x, \qquad d(xy) = d(xd(y)), \qquad d(x) \leq 1,$$
$$d(0) = 0, \qquad d(x + y) = d(x) + d(y).$$

For relation semirings over a set a A, the domain operation models

$$d(R) = \{(p, p) \in A \times A : (p, q) \in R \text{ for some } q \in A\},$$

which corresponds to the set of all states p at which the relation R is enabled. It can be shown that the set $d(S)$ of all domain elements forms a bounded distributive lattice with minimal element 0 and maximal element 1. In the relation semiring, these elements can be identified with sets of states (formally, they are subidentities). State spaces that form Boolean algebras can be obtained from an alternative axiomatisation which entails the present one [6].

A Kleene star and omega operator can be added to the signature without any need of modifying the domain axioms.

An element x of a domain semiring S is *wellfounded* if

$$p \leq d(xp) \Rightarrow p = 0$$

holds for all $p \in d(S)$. The expression $d(xp)$ models the preimage of the set p under the (abstract) action x, that is, the set of all elements in S which are related by x with some element in p. If $p \leq d(xp)$, then the set p is closed under x-actions, hence no element in p can have x-maximal elements. By the above formula, therefore, only the empty set can (vacuously) have x-maximal elements. But this means that, in the case of relations, x is wellfounded in the set-theoretic sense (cf. [5] for further discussion and [16] for a formalisation).

The formula that expresses wellfoundedness is very similar to that expressing deflation, and we will now show that under some abstract conditions, which can easily be verified for relation semirings (and falsified for trace and language semirings) [9], the two conditions are equivalent.

We assume a domain semiring S with a maximal element $\top$, that is, $x \leq \top$ holds for all $x \in S$. In fact, such elements can be adjoined to any dioid. In relation semirings over a set A, $\top = A \times A$. We further assume that the domain semiring satisfies

$$d(x)\top = x\top,$$

which in relation dioids immediately follows from the definition of the relative product. This condition has been called the *taming condition* in [9].

Lemma 28. *In every domain semiring with $\top$,*

$$x\top = 0 \Leftrightarrow x = 0.$$

Proof. $x\top = 0 \Leftrightarrow xd(\top) = 0 \Leftrightarrow x1 = 0 \Leftrightarrow x = 0$. □

We can now show the main statement of this section.

Proposition 29. *Let $d(x)\top = x\top$ hold in a domain semiring with $\top$. An element is wellfounded if and only if it is deflationary.*

Proof. Assume that x is deflationary, that is, $y \leq xy \Rightarrow y = 0$ holds for all y and suppose that $p \leq d(xp)$. Then $p\top \leq d(xp)\top = xp\top$, hence $p\top = 0$ follows from the assumption. But, by Lemma 28, this is the case if and only if $p = 0$.

Conversely, assume that x is wellfounded, that is, $p \leq d(xp) \Rightarrow p = 0$ holds for all $p \in d(S)$, and suppose that $y \leq xy$. Then $d(y) \leq d(xy) = d(xd(y))$ holds because domain is isotone [6], and therefore $d(y) = 0$ by the assumption of wellfoundedness. This is the case if and only if $y = 0$ [6]. □

Corollary 30. *An element of a relation left omega algebra is wellfounded if and only if it is ω-trivial.*

Formally verified proofs of all the results in this sections can again be found in our repository. More generally, it can be shown [5,9] that, in domain semirings,

$$R^\omega = \nabla(R)\top,$$

where $\nabla(R)$ is an element of $d(A \times A)$ that characterises all those elements of A from which infinite R-chains emanate. Hence R^ω can again be defined explicitly in this setting.

Arden's lemma, of course, holds in the relational setting.

Corollary 31. *In relation left omega algebras, if R is wellfounded, then the equation $Y = R \circ Y + S$ has the unique solution $Y = R^* \circ S$.*

This fact has already been proved in [7], but in a higher-order setting and using fixed point fusion. To our knowledge, ours is the first proof that is entirely within first-order logic and which could be obtained by automated reasoning.

10   Conclusion and Future Work

We have studied left omega algebras both abstractly and on *regular* models given by languages, traces, paths and relations. We derived abstract criteria generalising Arden's rule for solving systems of recursive regular equations, and for defining the omega operation explicitly on interesting classes of models. We linked left omega algebras with Salomaa's axioms for the algebra of regular events and showed that left omega algebras are sound and complete for this class. As so often with Kleene algebra, a main achievement is certainly generality and simplicity.

An important model that could not be discussed in this paper is formed by the matrices over omega algebras, which themselves form omega algebras. Our abstract results are, of course, valid in this setting, but particular criteria for unique solvability certainly deserve further investigation.

In addition, the results obtained motivate further interesting research questions: Can an abstract version of Arden's lemma be proved already in (left) Kleene algebra, for instance by using deflationarity? Can the length-increase argument for language, trace and path arguments be generalised? Can more general conditions for splitting elements into dense and deflationary parts be obtained? Is left Kleene algebra complete for the algebra of regular events? Is omega algebra complete for the algebra of omega-regular events?

At least the last question has a positive answer: In a forthcoming paper we show that Wagner's complete axiomatisation of omega-regular languages [17] is derivable in variants of left omega algebras [15]—see our repository for a derivation of Wagner's axioms—and that these algebras are sound for omega-regular languages as well. The results in this paper are instrumental for dealing with the regular parts of languages in the absence of the right induction axiom, as in Wagner's approach.

References

1. Arden, D.N.: Delayed-logic and finite-state machines. In: Annual IEEE Symposium on Foundations of Computer Science, pp. 133–151 (1961)
2. Backhouse, R.C., Carré, B.A.: Regular algebras applied to path-finding problems. IMA J. Appl. Math. 15(2), 161–186 (1975)
3. Brzozowski, J.A.: Derivatives of regular expressions. J. ACM 11(4), 481–494 (1964)
4. Cohen, E.: Separation and reduction. In: Backhouse, R., Oliveira, J.N. (eds.) MPC 2000. LNCS, vol. 1837, pp. 45–59. Springer, Heidelberg (2000)
5. Desharnais, J., Möller, B., Struth, G.: Algebraic notions of termination. Logical Methods in Computer Science 7(1:1), 1–29 (2011)
6. Desharnais, J., Struth, G.: Internal axioms for domain semirings. Science of Computer Programming 76(3), 181–203 (2011)
7. Doornbos, H., Backhouse, R.C., van der Woude, J.: A calculational approach to mathematical induction. Theor. Comput. Sci. 179(1-2), 103–135 (1997)
8. Foster, S., Struth, G., Weber, T.: Automated engineering of relational and algebraic methods in isabelle/hol. In: de Swart, H. (ed.) RAMiCS 2011, vol. 6663, pp. 52–67. Springer, Heidelberg (2011)
9. Höfner, P., Struth, G.: Algebraic notions of nontermination: Omega and divergence in idempotent semirings. J. Logic and Algebraic Programming 79(8), 794–811 (2010)
10. Kozen, D.: On Kleene algebras and closed semirings. In: Rovan, B. (ed.) MFCS 1990. LNCS, vol. 452, pp. 26–47. Springer, Heidelberg (1990)
11. Kozen, D.: A completeness theorem for Kleene algebras and the algebra of regular events. Information and Computation 110(2), 366–390 (1994)
12. Milner, R.: A complete inference system for a class of regular behaviours. J. Computer and System Sciences 28, 439–466 (1982)
13. Paulson, L., Nipkow, T., Wenzel, M.: Isabelle, http://www.cl.cam.ac.uk/research/hvg/Isabelle/index.html
14. Salomaa, A.: Two complete axiom systems for the algebra of regular events. J. ACM 13(1), 158–169 (1966)
15. Struth, G.: Modal tools for separation and refinement. Electronic Notes in Theoretical Computer Science 214, 81–101 (2008)
16. Struth, G., et al.: Isabelle algebraic methods repository (2011), http://www.dcs.shef.ac.uk/~georg/isa
17. Wagner, K.W.: Eine Axiomatisierung der Theorie der regulären Folgenmengen. Elektronische Informationsverarbeitung und Kybernetik 12(7), 337–354 (1976)

On Probabilistic Kleene Algebras, Automata and Simulations

Annabelle McIver[1], Tahiry M. Rabehaja[2], and Georg Struth[2]

[1] Department of Computing, Macquarie University, Australia
annabelle.mciver@mq.edu.au
[2] Department of Computer Science, The University of Sheffield, UK
{t.rabehaja,g.struth}@dcs.shef.ac.uk

Abstract. We show that a class of automata modulo simulation equivalence forms a model of probabilistic Kleene algebra. We prove completeness of this model with respect to continuous probabilistic Kleene algebras. Hence an identity is valid in continuous probabilistic Kleene algebras if and only if the associated automata are simulation equivalent.

1 Introduction

Kleene algebras are a family of mathematical structures that are fundamental to many computing applications. Variants for specific models and tasks including processes (cf. [1]), probabilistic protocol analysis [16], program refinement [21] or grainless concurrency [10] have been developed. The best understood variant, which initiated this line of research, has been introduced by Kozen [12]. A classical result relates Kozen's Kleene algebras to regular languages and the regular expressions that represent them. Kozen has shown that the free algebras in this class are isomorphic to regular languages [11]. In other words, regular languages are models of this algebra—a soundness result—and every valid identity between regular expressions can be derived from its axioms—a completeness result. Consequently, the equational theory of Kozen's Kleene algebras is decidable using the standard procedures for regular expressions, for instance, finite automata. Since this variant is also sound and complete with respect to the equational theory of binary relations under the regular operations provided by Kleene algebra, the decision procedure is also relevant for relation-based program analysis. It has already been implemented in theorem provers such as Coq [2] and Isabelle [13].

Much less is known, however, about other variants of Kleene algebras, where completeness results and decision procedures would be of comparable interest. At least for variants without the star, which correspond to variants of near-semirings, the situation is quite clear (cf.[9]). The free dioids, which are reducts of Kozen's Kleene algebras, are isomorphic to sets of strings. From the point of view of process algebra, equality in this variant corresponds to trace equivalence. It is easy to obtain the elements of the free algebras syntactically by normal form computation, using the left distributivity axiom $x(y + z) = xy + xz$ to rewrite arbitrary term-trees into "polynomials", that is, sums of products. It is also easy

H. de Swart (Ed.): RAMICS 2011, LNCS 6663, pp. 264–279, 2011.
© Springer-Verlag Berlin Heidelberg 2011

to see that terms can be interpreted as automata, and that determinising and minimising them preserves equality with respect to the axioms. For reducts of other variants, where the left distributivity law is absent, term normal forms correspond to proper trees and equivalence is induced by simulation or bisimulation. This yields a more fine-grained resolution of the nondeterministic choices in computations.

In the presence of the Kleene star, the situation changes drastically and simple term normal forms no longer characterise the free algebras. In the case of Kozen's variant, minimal deterministic automata are used as "normal forms" for Kleene algebra terms. Kozen's completeness proof uses the fact that Kleene algebras are closed under matrix formation and that finite automata can be encoded as matrices. The operations of epsilon-elimination, determisation and minimisation of automata are shown to preserve equivalence within Kleene algebra. For variants without the left distributivity law, Fokkink and Zantema prove a completeness results for a variant of near-semirings with iteration with respect to bisimulation equivalence [6]. Furusawa and Takai adapt the matrix construction to prove completeness of probabilistic Kleene algebras (without one of the zero law) with respect to a class of regular tree languages [20]. Their result, however, contains a gap[1]. Cohen uses a modification of Brzozowski's technique of language derivatives to construct a term model of probabilistic Kleene algebra [5]. Coalgebraically, these terms represent automata and the associated notion of equivalence is induced by simulation. He also presents a tentative completeness proof depending on Furusawa and Takai's result. Finally, in these proceedings, Furusawa and Nishizawa present a soundness and completeness result for the complete semiring reduct of probabilistic Kleene algebras with respect to a certain class of multirelations [7]. It is well known that the Kleene star can be defined explicitly in this setting, but the relationship with the star axioms of probabilistic Kleene algebras is not further explored.

This paper adds another piece to the puzzle. In contrast to Cohen, we use an explicit construction à la Kleene's theorem to obtain an automata-theoretic model of probabilistic Kleene algebra. We then use this model to prove completeness with respect to continuous or complete probabilistic Kleene algebras and simulation equivalence. In contrast to Furusawa and Nishizawa's approach, particular emphasis is on representing the star. Also, automata yield a somewhat more fine-grained model than multirelations; the correspondence between the two remains an interesting open question. Finally, through the automata-correspondence, decision procedures for process algebras based on partition refinement [4] become available for continuous probabilistic Kleene algebras.

We believe that the techniques developed in this paper may be useful for proving completeness in the non-continuous case. However, continuity is not an unnatural restriction, since two of the most important models of probabilistic Kleene algebras, namely expectation transformers [15] and up-closed multireations [8] have this property.

---

[1] http://www.sci.kagoshima-u.ac.jp/~furusawa/person/Papers/
correct_monodic_kleene_algebra.pdf

2   Probabilistic Kleene Algebras

Probabilistic Kleene algebras [16,14] have been introduced for resolving nondeterministic choices as they occur, for instance, in probabilistic protocols that
involve adversarial scheduling. They are very similar to process algebras like
CCS or ACP, but do not consider parallelism and communication. Simulation
equivalence instead of bisimilarity is the underlying notion of equivalence. In
addition, a variant of the Kleene star axioms is used for modelling iteration.

Formally, a *probabilistic Kleene algebra* is a structure $(K, +, \cdot, *, 0, 1)$, where

- $(K, +, 0)$ is a commutative idempotent monoid,
- $(K, \cdot, 1)$ is a monoid, where $x \cdot y$ will be simply denoted xy,
- 0 is a left and right annihilator $(0x = 0 = x0)$,
- multiplication is right distributive and left subdistributive,

$$(x + y)z = xz + yz, \qquad xy + xz \le x(y + z),$$

- the star satisfies the left unfold and left induction axiom

$$1 + xx^* \le x^*, \qquad xy \le y \Rightarrow x^*y \le y,$$

where the order is defined as usual in Kleene algebras, $x \le y$ iff $x + y = y$.
The right induction axiom $y(x + 1) \le y \Rightarrow yx^* \le y$ is usually added, but our
completeness result does not depend on it. The main purpose of the right induction axiom is to express x^* as the supremum of the sequence $((x + 1)^\alpha)_\alpha$,
where α ranges over ordinals. Left subdistributivity is equivalent to left isotonicity $x \le y \Rightarrow zx \le zy$ (right isotonicity follows from right distributivity). Basic
process algebras are obtained from probabilistic Kleene algebras essentially by
dropping left subdistributivity, whereas (left) Kleene algebras are obtained by
replacing it by a left distributivity axiom. As already explained these variations
account for the difference between bisimilarity, simulation equivalence and trace
equivalence and for tree-based versus trace-based models.

A probabilistic Kleene algebra K is *continuous* if multiplication is continuous
from the left and the right, that is, if it distributes over left and right directed
joins:

$$x(\sup D) = \sup\{xy \mid y \in D\} \quad \text{and} \quad (\sup D)x = \sup\{yx \mid y \in D\}$$

hold for all $x \in K$ and directed sets $D \subseteq K$. In fact, we only need *conditional
continuity*, that is, only existing suprema of directed sets need to be preserved.

The star unfold and left induction law ensure that the least fixpoint of $f(x) =
1 + ux$ exists. By continuity, it can be reached by iteration at the first infi-
nite ordinal, $u^* = \sup_{n \in \mathbb{N}} f^n(0)$. Moreover, $vu^*v' = \sup_n vf^n(0)v'$ because the
sequence $f^n(0)$ is directed.

Lemma 1. *In every continuous probabilistic Kleene algebra*

$$x^* = \sup_{n \in \mathbb{N}} (1 + x)^n.$$

Proof. First, $x^* = \sup_n f^n(0)$. Second, the sequence $(f^n(0)|n)$ is increasing since f is isotone. Hence $f^n(0) = (1+x)^{n-1}$ holds by induction.   □

Lemma 2. *The operation $+$ is completely additive.*

Proof. We prove a slightly more general fact for existing suprema. Consider a family y_i such that $\sup_i y_i = y$. Then $\sup_i(x + y_i) \leq x + y$ by isotonicity. Conversely, if $z \geq x + y_i$ for all i, then $z + x + y \geq x + y_i + x + y = x + y$, hence $\sup_i(x + y_i) \geq x + y$.   □

3 An Automata-Based Model

This section presents a Kleene-style construction of an automata-theoretic model for probabilistic Kleene algebras, hence a soundness result. It makes Cohen's coalgebraic construction, which is perhaps more elegant, more explicit. The link between the two approaches is that the terms constructed in Cohen's approach can directly be interpreted as automata. Brzozowski's original paper [3] and Rutten's review from a coalgebraic point of view [19] provide excellent introductions for the case of bisimilarity. A minor difference to Cohen's approach is that we also need to check continuity.

As usual, a *nondeterministic finite automaton* (NFA) is a tuple $(G, \Sigma, \delta, i, F)$ where G is a (finite) set of states, Σ a finite alphabet and $\delta \subseteq G \times (\Sigma \cup \{\varepsilon\}) \times G$ a transition relation. ε denotes the empty word, $i \in G$ the initial state and $F \subseteq G$ the set of final states. An automaton is *accessible* if every state is reachable from the initial state and every state except the initial one reaches some final state. We will only consider accessible automata.

We follow the standard construction in Kleene's theorem and inductively interpret terms of probabilistic Kleene algebras by NFAs. We do not explicitly define terms, but we assume that the set of constants from which they are constructed yields the alphabet Σ of the associated automaton. Hence we use Σ also to denote the set of all constants from which terms are built. We denote the initial state by • and final states by ∘. We write ⊙ for a state which is both initial and final.

Formally, we write $G(s)$ for the automaton associated with term s. We inductively define

- $G(0) = \bullet$,
- $G(1) = \odot$,
- for $a \in \Sigma$, $G(a) = \bullet \xrightarrow{\ a\ } \circ$,
- Given $G(s)$ and $G(t)$, $G(s+t)$ is

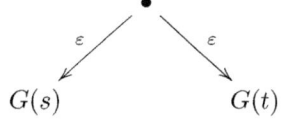

Here, the only initial state is • and the targets of the ε-transitions are the former initial states of $G(s)$ and $G(t)$.

- Given $G(s)$ and $G(t)$, $G(st) = G(s) \xrightarrow{\varepsilon} G(t)$, where each final state of $G(s)$ is linked to the initial state of $G(t)$ by an epsilon transition.
- Given $G(s)$, $G(s^*)$ is

$$\odot \overset{\varepsilon}{\underset{\varepsilon}{\rightleftarrows}} G(s)$$

that is, each final state of $G(s)$ is linked to $\odot$ via some ε-transition. $\odot$ is, in turn, linked by an ε-transition to the initial state of $G(s)$.

G is a homomorphism from probabilistic Kleene algebra terms to automata under the operations corresponding to the constructions given. For instance, $G(s + t)$ induces the operation $G(s) + G(t)$ on the automata $G(s)$ and $G(t)$. But the operations constructed do not depend on the fact that the operation was inductive. They are well-defined for arbitrary (accessible) automata. We henceforth write $G + H$, GH and G^* for arbitrary (accessible) automata G, H.

The automata corresponding to non-zero terms $s + t$, st and s^* are accessible by construction, as the diagrams show. If one term is 0, then a node, from which no final state is reachable, could be added. We always remove this node and the edge leading to it. This preserves accessibility and we will show that it does not influence soundness and completeness. Again, this applies to arbitrary accessible automata as well; hence we may assume without loss of generality that the operations $G + H$, GH and G^* preserve accessibility.

In the construction, ε-transitions have been introduced. It is well known from automata theory that such transitions can be removed without affecting acceptance. Here, the situation is different since the absence of left distributivity induces a different notion of equivalence.

So we first define notions of simulation and simulation equivalence on the accessible automata with ε constructed. We then verify that these automata under the regular operations satisfy the axioms of continuous probabilistic Kleene algebra with respect to simulation equivalence. Finally, we demonstrate that standard ε-elimination from automata theory preserves simulation equivalence.

First we define $\varepsilon(x)$, the ϵ-closure of state x, as the set of states which are reachable by ϵ-transitions only from x. In particular, $x \in \varepsilon(x)$. We also define

$$\overline{\delta_a}x = \{x' \mid \exists y \cdot y \in \varepsilon(x) \wedge x' \in \delta_a y\} \text{ and } \overline{F} = \{x \mid \varepsilon(x) \cap F \neq \emptyset\}$$

where $\delta_a x = \{x' \mid (x, a, x') \in \delta\}$. This extended transition $\overline{\delta}$ is the ε-closure of δ and $\overline{F}$. It contains those states leading to some final states by ε-transitions only.

Let G and H be automata. A relation $R \subseteq G \times H$ is a *simulation* if

- $(i_G, i_H) \in R$,
- for all $a \in \Sigma$, if $(x, y) \in R$ and $x' \in \overline{\delta_a}x$ then $(x', y') \in R$ for some $y' \in \overline{\delta_a}y$,
- if $(x, y) \in R$ and $x \in \overline{F_G}$, then $y \in \overline{F_H}$.

We write $G \preceq H$ and say that H *simulates* G whenever there is a simulation $R \subseteq G \times H$. Simulations need not be total. There can be $x \in G$ on which a simulation is undefined. The reason is that in the second condition above, ε transitions have not been considered. But every simulation R can be totalised by setting $R' = R \cup \{(x,y) \mid R.x = \emptyset \wedge \exists x' \cdot (x \in \varepsilon(x') \wedge (x',y) \in R)\}$.

It is well known that simulations on $G \times H$ are closed under union and composition. It follows that all simulations can be extended to maximal ones. It is also well known that simulations induce preorders and equivalences. Two automata G and H are *simulation equivalent*, $G \cong H$, if $G \preceq H$ and $H \preceq G$.

Theorem 1 (Soundness). *The accessible automata under the regular operations form a probabilistic Kleene algebra (without the right induction law) with respect to simulation equivalence.*

Proof. Let G, H, K be accessible automata.

- $(G+H)+K \cong G+(H+K)$. The relation $R \subseteq ((G+H)+K) \times (G+(H+K))$ defined by the following diagram is a simulation.

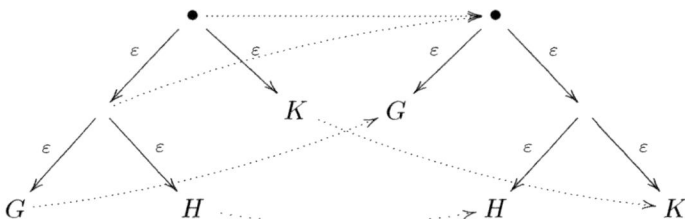

  We can use a similar construction for the converse direction.
- The simulations for $G + H \cong H + G$ and $G + G \cong G$ are trivial.
- $0 + G \cong G$. We can use the identity simulation because $0 + G$ is transformed into G by making the automaton accessible.
- $(GH)K \cong G(HK)$. The automata corresponding to the left-hand and the right-hand side are identical by construction.
- $1G \cong G$. We use the simulation

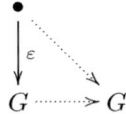

  and its converse. The simulation for $G1 \cong G$ is similar.

- $G0 \cong 0$. Both automata have no final state. By making the left-hand side accessible, it becomes identical to the right-hand side.

- $0G \cong 0$. In the left-hand side, G is not reachable and the automaton becomes 0 by making it accessible.

- $GH + GK \preceq G(H + K)$. The simulation is essentially shown in the following diagram.

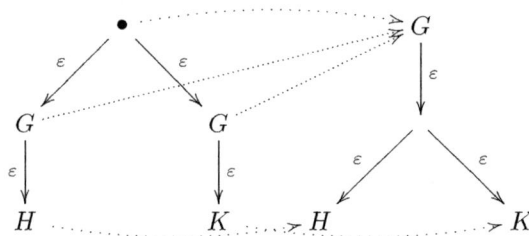

The initial state • is mapped to the initial state of G and the dotted arrows from G to G (resp. H to H and K to K) are essentially the identity relation.

- $GK + HK \cong (G + H)K$. The following figure gives a simulation that works both ways.

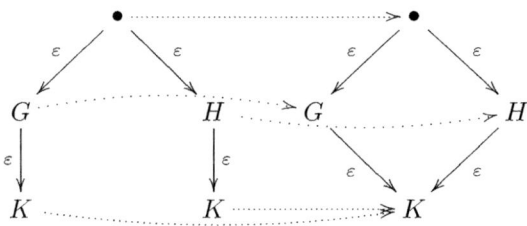

- $1 + GG^* \preceq G^*$. The following construction yields a simulation for one unfold of G^*.

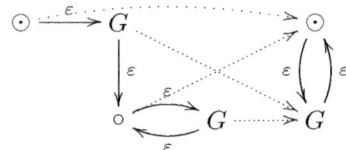

- $GH \preceq H \Rightarrow G^*H \preceq H$. Let $R \subseteq GH \times H$ be the maximal simulation. We write $G^n = G_n \cdots G_1$, where G_n is the n-th copy of the automaton G. We write x_n for the copy of state x in G_n. The notation in the following diagram has been changed to show the construction more clearly.

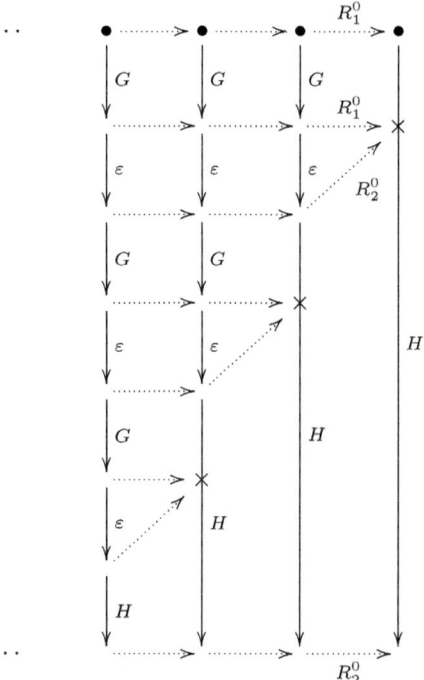

We define $R = R^0 = R_1^0 \cup R_2^0$ where $R_1^0 \subseteq G \cdot \varepsilon \times H$ and $R_2^0 \subseteq H \times H$ as in the diagram, so that R^0 is a simulation. We inductively define $R^n = R_1^n \cup R_2^n$ where $R_1^n = R_1^0$ and $R_2^n = R^{n-1} \circ \cdots \circ R^0 \circ R_2^0$. It follows by induction that each R^n is a simulation for $G^n H \preceq H$.

We now define $R^* \subseteq G^* H \times H$ and show that it is a simulation for $G^* H \preceq H$. For $x, y \in G \cup H$ we define

$$(x,y) \in R^* \text{ iff } \begin{cases} (x_i, y) \in R^n \text{ for some } i, n \text{ with } i \leq n, & \text{if } x \in G, \\ (x,y) \in R, & \text{if } x \in H. \end{cases}$$

The initial state ι of G^* is mapped to every state of H in the image of a copy of i_G under R^n. The final states of $G^* H$ are related to those of H. We now prove that R^* is a simulation by inspecting transitions in the automata.

First, $(\iota, i_H) \in R^*$ since $(i_G, i_H) \in R^0$.

Next, suppose $x \in \overline{F_{G^* H}}$ and $(x, y) \in R^*$. There are two cases: (i) If $x \in H$, then (x, y) is already in R. (ii) If $x \in G \cup \{\iota\}$ then $1 \preceq H$, so $x \in \overline{F_G} \cup \{\iota\}$. By definition $(x_i, y) \in R^n$ for some i and n. Therefore $x_i \in \overline{F_{G^n H}}$ and consequently $y \in \overline{F_H}$ because R^n is a simulation (consider the diagram).

Next, suppose $(x, y) \in R^*$ and $x' \in \overline{\delta}_a x$ is the result of a transition in the automaton $G^* H$, that is, it is either a transition in H or a transition in G^* passing through i_{G^*} or a transition from some final state of G^* to some state in H. We distinguish three cases. (i) If $x \in H$ then, then we are done since the simulation used for the step is R by definition. Otherwise, there exists i, n such that $(x_i, y) \in R^n$. (ii) $x' \in \overline{\delta}_a x$ is obtained by a transition in G.

Then R^n is a simulation and there exists $y' \in \bar{\delta}_a y$ such that $(x'_i, y') \in R^n$. Hence $(x', y') \in R^*$. (iii) $x' \in \bar{\delta}_a x$ is obtained by a transition of the form $x \xrightarrow{\varepsilon} \iota \xrightarrow{\varepsilon a} x'$. In the diagram, by ε-closure, there will therefore be additional edges that can either loop back into G or lead into H. That is, $x'_{i-1} \in \bar{\delta}_a x_i$ or $x' \in H$ and $i = 1$. In the first case, when we loop back into G, there exists a state y' such that $(x'_{i-1}, y') \in R^n$. Therefore, by definition, $(x', y') \in R^*$. In the second case, when the transition leads into H, there exists a state y' such that $(x', y') \in R$, by definition. Again, $(x', y') \in R^*$.   □

Proposition 1. *The accessible automata under the regular operations form a continuous probabilistic Kleene algebra with respect to simulation equivalence.*

Proof. It remains to show that multiplication of automata from the left and right is continuous. We first define a notion of residuation on automata. We then establish a Galois connection between residuation and multiplication, from which continuity follows.

For automata G and $H \neq 0$ we define the automaton G/H with initial state $i_{G/H} = i_G$, final states $F_{G/H} = \{x \in G \mid H \preceq G_x\}$, where G_x is constructed from G by making its initial state into x. We make the resulting automaton accessible by discarding all states and edges that do not lead to a final state.

We now show that $KH \preceq G$ iff $K \preceq G/H$. Assume R is a simulation from KH to G. That means R is in particular a simulation from H to G_x for some x. By definition of G/H, therefore, R is a simulation from K to G/H, since the state x becomes final in G/H and an image of a final state of K under R.

For the converse direction, suppose that R is a simulation from K to G/H. By Theorem 1, multiplication is isotone, hence $KH \preceq (G/H)H$, and it remains to show that $(G/H)H \preceq G$.

First, if $F_{G/H}$ is empty then $G/H = 0$ and the result follows.

Otherwise, assume that R' is the simulation from KH to $(G/H)H$. By construction of G/H, we also know that there exists a simulation S_x from H to G_x for all final states x of G/H and that there is a simulation (except for the final state property) between G/H and G, namely the identity relation id. Hence $S' = (\cup_x S_x) \cup id$ is indeed a simulation from $(G/H)H$ to G and $R' \circ S'$ is a simulation from KH to G.

It then follows from general properties of Galois connections that $(\cdot H)$ is (conditionally) completely additive, hence right continuous.

It remains to show left continuity. Let $(G_i)_i$ be a directed set of automata such that $\sup_i G_i = G$ and let H be any automaton. Then $\sup_i (HG_i) \preceq HG$ because multiplication is isotone and it remains to show $HG \preceq \sup_i (HG_i)$. Let us assume that $\sup_i (HG_i) \preceq K$. We will show that $HG \preceq K$.

By definition of supremum, $HG_i \preceq K$ for all i, hence there is a set of states $X_i = \{x \in K \mid G_i \preceq K_x\}$, that is, the set of all those states in K from which G_i is simulated. Obviously, $X_i \subseteq X_j$ if $G_j \preceq G_i$ in the directed set. But since K has only finitely many states, there must be a minimal set X in that directed set such that all G_i are simulated by K_x for $x \in X$. By definition, therefore $G = \sup_i G_i \preceq K_x$ for all $x \in X$. There exists a simulation $S_X \subseteq HG_i \times K$ for some i such that the residual automaton K/G_i has precisely X as its set of final states. We take the union of S_X restricted to H with all simulations yielding $G \preceq K_x$ for all $x \in X$ and verify that this is a simulation from HG to K.   □

The following examples show that the axioms for automata under simulation equivalence can neither be weakened nor strengthened.

Example 1

(a) It is clear by considering the diagram for subdistributivity in the proof of Theorem 1 that a simulation from the right-hand automaton to the left-hand automaton is impossible. This refutes left distributivity for our model.
(b) The left star unfold axiom can be strengthened to $1 + xx^* = x^*$, but the inequality $x^* \leq 1 + x^*x$ is not valid.

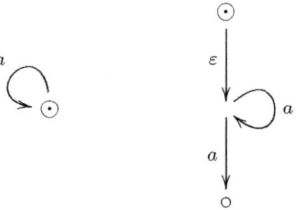

It is easy to see that there cannot be a simulation between these automata. But the following diagram shows that the inequality $x^* \leq 1 + x^*(x+1)$ holds.

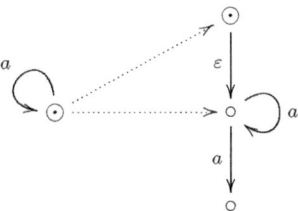

(c) A right star induction law $yx \leq y \Rightarrow yx^* \leq y$ does not hold.

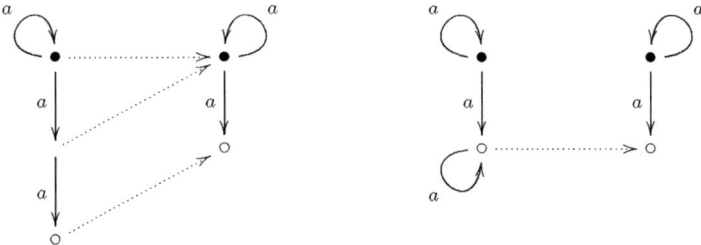

The left diagram shows a simulation from a^*aa to a^*a though there is no simulation in the right figure since the displayed arrow could not be satisfied.

(d) Kozen's counterexample [12] on Kleene algebras possessing a least fixpoint for $1 + ax$ but not for $1 + xa$ still holds in our setting (i.e. for $1 + x(a + 1)$). Therefore the right induction axiom of probabilistic Kleene algebras is independent.

We close this section by showing that ε-elimination is possible in our automata model, using the standard technique. The general idea is shown in the following diagram.

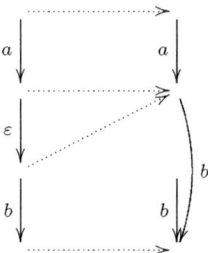

Proposition 2. *Each accessible automaton is simulation equivalent to an accessible ϵ-free one.*

Proof. Let G be an automaton and let G_ε be an automaton with transition relation $\delta = (\delta_G \cup \bigcup_{a \in \Sigma}\{(x, a, z) |\ \exists y \in \varepsilon(x) \cdot z \in \delta_a y\}) \setminus \varepsilon$, with initial state i and set of final states $F = \{f \in \overline{F_G} \mid f$ accessible in $G_\varepsilon\}$. We also assume that there are no non-accessible states in G_ε.

First, we show that $G \preceq G_\varepsilon$. Consider the relation $R \subseteq G \times G_\varepsilon$ such that $(x, y) \in R$ iff $x \in \varepsilon(y)$ in G. We must show that R is a simulation. We have $(i_G, i_{G_\varepsilon}) \in R$ because $i_G \in \varepsilon(i_G)$. Assume $(x, y) \in R$ and $x \in \overline{F_G}$, by definition $\varepsilon(x) \cap F_G \neq \emptyset$ but $\varepsilon(x) \subseteq \varepsilon(y)$ so $y \in F_{G_\varepsilon}$.

Let $a \in \Sigma$. Since $x \in \varepsilon(y)$ we know that $\overline{\delta}_a x \subseteq \overline{\delta}_a y$ (transition in G). Let $x' \in \overline{\delta}_a x$. Then $y \xrightarrow{a} z \in \delta$ and then $z \in G_\varepsilon$ since G_ε is accessible.

We now show that $G_\varepsilon \preceq G$. The simulation relation in G is defined with respect to the ε-closure of δ_G. Hence G has at least as many states and non-ε-transitions as G_ε. Hence it can simulate G_ε.   □

4   Automata Labelled by Terms

We now extend the construction of automata from terms in Section 2 to include labels for states. This is needed in the next section. It is again by induction.

  - Base cases: $G(a) = a \xrightarrow{a} 1$, $G(0) = 0$ and $G(1) = 1$.
  - Addition: $G(s + t)$ is

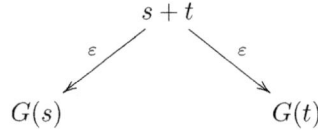

- Multiplication: $G(st)$ is $G(s).t \xrightarrow{\varepsilon} G(t)$
- Kleene star: $G(s^*)$ is

$$s^* \underset{\varepsilon}{\overset{\varepsilon}{\rightleftarrows}} G(s).s^*$$

In these diagrams final states labelled with a term greater than or equal to 1. $G(s).t$ is the automaton where the label of each state is multiplied by t.

Final states of $G(t)$ can be defined by the following endofunction [3].

- $o(1) = 1$, $o(0) = 0$, and $o(a) = 0$ for each $a \in \Sigma$.
- $o(s + t) = o(s) + o(t)$, $o(st) = o(s)o(t)$ and $o(s^*) = 1$.

The following proposition adapts a similar statement and proof by Milner [18]. A similar technique has been used by Cohen [5].

Proposition 3. *Every term t of probabilistic Kleene algebra is equivalent to a sum $\sum_{a \in \Sigma} \sum_{s \in \bar{\delta}_a t} as + o(t)$, where δ is the transition relation of $G(t)$.*

Proof. By structural induction. We only consider the induction steps.

- If the term is of the form $s + t$, then

$$s = \sum_{a \in \Sigma} \sum_{s' \in \bar{\delta}_a s} as' + o(s) \qquad \text{and} \qquad t = \sum_{a \in \Sigma} \sum_{t' \in \bar{\delta}_a t} at' + o(t).$$

Since $\bar{\delta}_a(s + t) = \bar{\delta}_a(s) \cup \bar{\delta}_a(t)$, it follows that

$$\sum_a \sum_{u \in \bar{\delta}_a s \cup \bar{\delta}_a t} au + o(s + t) = \sum_a \sum_{s' \in \bar{\delta}_a s} as' + o(s) + \sum_a \sum_{t' \in \bar{\delta}_a t} at' + o(t).$$

- In the product case st, we assume the same sums for s and t as before. We have $\bar{\delta}_a(st) = \bar{\delta}'_a(s) \cup o(s)\bar{\delta}_a(t)$ where $\bar{\delta}'_a(s) = \bar{\delta}_a(s).t$.

$$st = \left(\sum_a \sum_{s' \in \bar{\delta}_a s} as' + o(x) \right) t = \sum_a \sum_{s' \in \bar{\delta}'_a s} as' + o(s) \left(\sum_a \sum_{t' \in \bar{\delta}_a t} at' + o(t) \right)$$

$$= \sum_a \sum_{u \in \bar{\delta}'_a s \cup o(s)\bar{\delta}_a t} au + o(st)$$

The first step uses the induction hypothesis. The second step uses the induction hypothesis again, right distributivity and the identity $\bar{\delta}'_a(x) = \bar{\delta}_a(x).y$. The third step uses $o(st) = o(s)o(t)$ and the fact that $o(s) \in \{0, 1\}$.

- Finally, for the case of $*$, we can assume without loss of generality that $o(s) = 0$ because $(x + 1)^* = x^*$ in probabilistic Kleene algebras. Therefore

$$s^* = ss^* + 1 = \left(\sum_a \sum_{s' \in \bar{\delta}_a s} as' \right) s^* + 1 = \sum_a \sum_{s' \in \bar{\delta}_a s^*} as' + o(s^*).$$

The second step uses the induction hypothesis. The third step uses the identities $\bar{\delta}_a(s^*) = \bar{\delta}_a(s).s^*$ and $o(s^*) = 1$. □

5   A Completeness Result

In this section we simply call *tree* an automaton whose graph is a tree. In the case of accessible automata, all leaves are final states, but there may also be some internal final states. In contrast to Furusawa and Takai's approach, we are dealing with standard automata rather than tree-automata. Our approach is therefore more similar to Cohen's.

It is obvious from our construction of automata that each tree is the interpretation of some $*$-free term in probabilistic Kleene algebra. If T is a tree then t_T denotes a $*$-free term such that $T \cong G(t_T)$.

Proposition 4. *Let T and T' be trees and s a term.*

(a) If $T \preceq G(s)$, then $t_T \leq s$ is valid in probabilistic Kleene algebras.
(b) If $T \preceq T'$, then $t_T \leq t_{T'}$ is valid in probabilistic Kleene algebras.

Proof. Let $T \preceq G(s)$, without loss of generality we assume the automata are ε-free. Assume that $T \cong G(t)$ for some $*$-free term t and consider a leaf in $G(t)$. It is labelled by the term 1 and related by the simulation to $G(s)$ to final states of $G(s)$. Hence the label of these must be greater than or equal to 1. By induction, we assume that $x \leq y$ for all subterm x of t of size n and y such that $G(t)_x \preceq G(s)_y$. We must show this property for all subterms of size $n+1$. By Proposition 3, $x = \sum_a \sum_{x' \in \delta_x} ax' + o(x)$. That is, x is a sum of monomials ax' where all x' come from the level below x in the tree. Similarly, every $y \in G(s)$ can be written that way. It follows that $x \leq y$ because multiplication and addition are isotone. The case where both automata are trees is an instance.   □

As a consequence of this proposition, we will denote a $*$-free term and a tree by the same notation, usually t. For each automaton G, consider the set of trees

$$\tau(G) = \{t \mid t \preceq G \wedge t \text{ is a tree}\}.$$

This set is stable under addition and down-closed. We define the operations

$$\tau + \tau' = \{t + t' \mid t \in \tau \wedge t' \in \tau'\},$$
$$\tau\tau' = \downarrow \{tt' \mid t \in \tau \wedge t' \in \tau'\},$$
$$\tau^* = \downarrow \{(t+1)^n \mid t \in \tau\},$$

where $\downarrow \tau$ denotes the down-closure of τ. All these sets are again stable under addition and down-closed.

The previous proposition implies that $\tau(G(u)) = \{G(t) \mid t \leq u \wedge t \text{ is } *\text{-free}\}$. We denote $\tau(u) = \{t \mid t \leq u \wedge t \text{ is } *\text{-free}\}$.

Proposition 5. *τ is a homomorphism to sets of $*$-free terms.*

Proof. Obviously, $\tau(u) + \tau(v) \subseteq \tau(u + v)$. For the converse inclusion suppose $t \in \tau(u + v)$, that is, $t \leq u + v$ so $G(t) \preceq G(u + v) \cong G(u) + G(v)$. Since the automata are disjoint, we can decompose t into $t_u + t_v$ such that $t_u \leq u$ and $t_v \leq v$ (this is possible because t is a tree). Therefore $t \in \tau(u) + \tau(v)$.

We have $\tau(u)\tau(v) \subseteq \tau(uv)$ by isotonicity. If $t \in \tau(uv)$ then $G(t) \preceq G(uv) \cong G(u)G(v)$. Then $t = t_u(t_{v1}, \ldots, t_{vn})$ for some $t_u \in \tau(u)$ and $t_{vi} \in \tau(v)$. So $t \leq t_u(\sum_i t_{vi}) \in \tau(u)\tau(v)$.

By isotonicity of $*$, $\tau(u)^* \subseteq \tau(u^*)$. Let $t \in L(u^*)$, then $G(t) \preceq G(u)^*$ so $t \leq (t'+1)^n$ for some $t' \in \tau(u)$ and $n \in \mathbb{N}$. In fact, since t has finite depth, we may unfold u^* finitely many times and reason as in the case of sequential composition to construct t'. Ones need to be added because each intermediate state after each iteration of $G(u)$ is a final state. □

Next we show that $\preceq$ corresponds to tree language inclusion.

Theorem 2. $G \preceq H$ iff $\tau(G) \subseteq \tau(H)$.

Proof. The forward implication is obvious by transitivity of $\preceq$. For the converse implication, suppose $\tau(G) \subseteq \tau(H)$ and consider the relation $R \subseteq G \times H$ such that $(x,y) \in R$ iff $L(G_x) \subseteq L(H_y)$. We show that R is a simulation (the maximal simulation, in fact). We must check the three defining conditions. (i) $(i_G, i_H) \in R$ is immediate. (ii) If $(x,y) \in R$ and $x \in F_G$, then $1 \in \tau(G_x) \subseteq \tau(H_y)$, so $y \in F_H$. (iii) Assume, by contradiction, that $(x,y) \in R$, $x \xrightarrow{a} x'$ and for every $y_i' \in \delta_a y$ there exists $t_i \in \tau(x')$ such that $t_i \notin \tau(y_i')$. Since $\delta_a y$ is a finite set, we define $t = \sum_i t_i \in \tau(x')$ and therefore, $at \in \tau(x) \subseteq \tau(y)$. Therefore, there exists $y_i' \in \delta_a y$ such that $t \preceq G_{y_i'}$ which is impossible since $\tau(G_{y_i'})$ is down-closed. □

We now characterise terms in continuous probabilistic Kleene algebras by the set of ($*$-free) terms or trees that approximate them from below.

Proposition 6. *Every term u of a continuous probabilistic Kleene algebra satisfies $u = \sup \tau(u)$.*

Proof. By structural induction. We already know that $\sup \tau(u) \leq u$ and that $\tau(u)$ is directed.

- For the base case, if u is a tree then $u \in \tau(u)$ so we are done.
- If $u = u_1 + u_2$ then by Proposition 5, $\tau(u) = \tau(u_1) + \tau(u_2)$ so $\sup \tau(u) = \sup\{t_1+t_2 \mid t_1 \in \tau(u_1) \wedge t_2 \in \tau(u_2)\}$. Let $t_1 \leq u_1$, by continuity and induction hypothesis $t_1+u_2 = \sup\{t_1+t \mid t \leq u_2\} \leq \sup \tau(u)$. Therefore, by continuity again $u \leq \sup\{t_1 + u_2 \mid t_1 \leq u_1\} \leq \sup \tau(u)$. Hence $u = \sup \tau(u)$.
- Let $u = u_1 u_2$, we have $\tau(u) = \tau(u_1)L(u_2)$ and we use the same reasoning as before. Let $t_1 \in \tau(u_1)$, then by continuity and induction hypothesis, $t_1 u_2 = \sup\{t_1 t \mid t \leq u_2\} \leq \sup \tau(u)$. By continuity again, $u_1 u_2 = \sup\{t_1 u_2 \mid t_1 \leq u_1\} \leq \sup \tau(u)$. we conclude $u \leq \sup \tau(u)$. Hence $u = \sup \tau(u)$.
- Let $u = v^*$. Then by Proposition 5, $\tau(u) = \tau(v)^*$ and we have to show $u \leq \sup\{(t+1)^n \mid t \leq v \wedge n \in \mathbb{N}\}$ by definition of τ^*. But $\sup\{(t+1)^n \mid t \leq v\} = (v+1)^n \leq \sup \tau(u)$ (induction on n and using the case of multiplication). So, $\sup\{(v+1)^n \mid n \in \mathbb{N}\} \leq \tau(u)$ and therefore, by continuity (existence of $\sup_n (v+1)^n$), $v^* \leq \tau(u)$. □

Finally, we can prove our completeness result; the main result of this paper.

278 A. McIver, T.M. Rabehaja, and G. Struth

Theorem 3 (Completeness). *If $G(u) \preceq G(v)$, then $u \leq v$ is derivable in continuous probabilistic Kleene algebra.*

Proof. If $G(u) \preceq G(v)$, then $\tau(u) \subseteq \tau(v)$ by Theorem 2. It follows from Proposition 6 that $u = \sup \tau(u) \leq \sup \tau(v) = v$. □

Note that G is a continuous mapping due to Proposition 6 and Theorem 2.

6   Conclusion

The main contribution of this paper is a completeness result for continuous probabilistic Kleene algebras with respect to a class of automata. We also provided a variant of Cohen's soundness result for these classes where the automata constructions have been made explicit. The techniques developed in this paper may be usefull for proving a completeness result in the non-continuous case.

The results in this paper motivate some further investigations. First, to associate the automata obtained with canonical ones, for instance, by minimisation. It seems that the automata obtained by Cohen's adaptation of Brzozowski's translation procedure are minimal. Determinisation of automata is not possible in this context since it does not preserve simulation equivalence.

Second, the completeness result should be turned into an effective decision procedure based on checking simulations between automata [17]. Such a procedure would be very useful when applying probabilistic Kleene algebras in the context of automated protocol verfication.

Third, our completeness result should be related to the representability result of Furusawa and Nishizawa. In the context of Kozen's Kleene algebras, the equational theories of Kleene algebras, continuous Kleene algebras and relation Kleene algebras coincide. It would be interesting to investigate whether a similar result holds for probabilistic Kleene algebras, continous probabilistic Kleene algebras and (weak) Kleene algebras of multirelations.

Acknowledgement. We would like to thank Ernie Cohen and Hitoshi Furusawa for helpful discussions, and to Jeff Sanders for valuable comments. We are especially grateful to the reviewers for helping us to improve the presentation of this paper.

References

1. Bergstra, J.A., Ponse, A., Smolka, S.A. (eds.): Handbook of Process Algebra. Elsevier, Amsterdam (2001)
2. Braibant, T., Pous, D.: An efficient coq tactic for deciding kleene algebras. In: Kaufmann, M., Paulson, L.C. (eds.) ITP 2010. LNCS, vol. 6172, pp. 163–178. Springer, Heidelberg (2010)
3. Brzozowski, J.A.: Derivatives of regular expressions. J. ACM 11(4), 481–494 (1964)
4. Bustan, D., Grumberg, O.: Simulation-based minimization. ACM Trans. Comput. Logic 4(2), 181–206 (2003)

5. Cohen, E.: Weak Kleene algebra is sound and (possibly) complete for simulation. CoRR abs/0910.1028 (2009)
6. Fokkink, W., Zantema, H.: Basic process algebra with iteration: Completeness of its equational axioms. Comput. J. 37(4), 259–268 (1994)
7. Furusawa, H., Nishizawa, K.: Relational and multirelational representation theorems for complete idempotent left semirings. In: de Swart, H. (ed.) RAMICS 2011. LNCS, vol. 6663, Springer, Heidelberg (2011)
8. Furusawa, H., Tsumagari, N., Nishizawa, K.: A non-probabilistic relational model of probabilistic kleene algebras. In: Berghammer, R., Möller, B., Struth, G. (eds.) RelMiCS/AKA 2008. LNCS, vol. 4988, pp. 110–122. Springer, Heidelberg (2008)
9. van Glabbeek, R.G.: The linear time-branching time spectrum (extended abstract). In: Baeten, J.C.M., Klop, J.W. (eds.) CONCUR 1990. LNCS, vol. 458, pp. 278–297. Springer, Heidelberg (1990)
10. Hoare, C.A.R.T., Möller, B., Struth, G., Wehrman, I.: Concurrent kleene algebra. In: Bravetti, M., Zavattaro, G. (eds.) CONCUR 2009. LNCS, vol. 5710, pp. 399–414. Springer, Heidelberg (2009)
11. Kozen, D.: A completeness theorem for Kleene algebras and the algebra of regular events. Information and Computation 110(2), 366–390 (1994)
12. Kozen, D.: On Kleene algebras and closed semirings. In: Rovan, B. (ed.) MFCS 1990. LNCS, vol. 452, pp. 26–47. Springer, Heidelberg (1990)
13. Krauss, A., Nipkow, T.: Proof pearl: Regular expression equivalence and relation algebra. Journal of Automated Reasoning (to appear, 2011)
14. McIver, A.K., Cohen, E., Morgan, C.C.: Using probabilistic kleene algebra for protocol verification. In: Schmidt, R.A. (ed.) RelMiCS/AKA 2006. LNCS, vol. 4136, pp. 296–310. Springer, Heidelberg (2006)
15. McIver, A.K., Morgan, C.C.: Abstraction, Refinement and Proof for Probabilistic Systems. Springer, Heidelberg (2005)
16. McIver, A.K., Weber, T.: Towards automated proof support for probabilistic distributed systems. In: Sutcliffe, G., Voronkov, A. (eds.) LPAR 2005. LNCS (LNAI), vol. 3835, pp. 534–548. Springer, Heidelberg (2005)
17. Milner, R.: An algebraic definition of simulation between programs. Tech. rep., Stanford University, Stanford, CA, USA (1971)
18. Milner, R.: A complete inference system for a class of regular behaviours. J. Comput. Syst. Sci. 28(3), 439–466 (1984)
19. Rutten, J.J.M.M.: Automata and coinduction (an exercise in coalgebra). In: Sangiorgi, D., de Simone, R. (eds.) CONCUR 1998. LNCS, vol. 1466, pp. 194–218. Springer, Heidelberg (1998)
20. Takai, T., Furusawa, H.: Monodic tree kleene algebra. In: Schmidt, R.A. (ed.) RelMiCS/AKA 2006. LNCS, vol. 4136, pp. 402–416. Springer, Heidelberg (2006)
21. von Wright, J.: Towards a refinement algebra. Sci. Comput. Program. 51(1-2), 23–45 (2004)

Ampersand
Applying Relation Algebra in Practice

Gerard Michels[1], Sebastiaan Joosten[1], Jaap van der Woude[1],
and Stef Joosten[1,2]

[1] Open Universiteit Nederland,
Postbus 2960,
6401 DL HEERLEN
[2] Ordina NV
Nieuwegein
{gerard.michels,jaap.vanderwoude,stef.joosten}@ou.nl

Abstract. Relation algebra can be used to specify information systems
and business processes. It was used in practice in two large IT projects
in the Dutch government. But which are the features that make relation algebra practical? This paper discusses these features and motivates
them from an information system designer's point of view. The resulting language, Ampersand [1], is a syntactically sugared version of relation
algebra. It is a typed language, which is supported by a compiler. The
design approach, also called Ampersand, uses software tools that compile
Ampersand scripts into functional specifications. This makes Ampersand
interesting as an application of relation algebra in the industrial practice. The purpose of this paper is to define Ampersand and motivate its
features from a practical perspective.

This work is part of the research programme of the Information Systems & Business Processes (IS&BP) department of the Open University.

Keywords: heterogeneous relation algebra – domain-specific feedback
– specification language – business rules – requirements engineering –
type system – rule based design.

1 Introduction

Ampersand is a simple requirements specification language with relational semantics. It is a syntactically sugared version of relation algebra. It has been
designed for students and practitioners with minimal mathematical background,
who use it for designing business processes. In the sequel, we shall call these
users 'requirements engineers', because Ampersand enables them to design by
eliciting requirements. The purpose of this paper is to describe the features of

---

[1] It is named after the ampersand symbol (&), which means "and". The name hints
at the desire to have it all: getting the best from both business and IT, achieving
results from theory and practice alike, and realising the desired results effectively
and more efficiently than ever before.

H. de Swart (Ed.): RAMICS 2011, LNCS 6663, pp. 280–293, 2011.
© Springer-Verlag Berlin Heidelberg 2011

Ampersand that make it practical. Ampersand is currently used in teaching at the Open University of the Netherlands (OUNL) and it has already been used to design several large scale IT-systems in the Dutch government[2].

A challenge in designing Ampersand was to obtain a useful requirements specification language that is faithful to Tarski's axioms. This sets Ampersand apart from Codd's relational model [3], which has been criticized for being unfaithful to its mathematical origins [4]. In this paper Ampersand is defined with semantics in a homogeneous relation algebra which is faithful to Tarski's axioms. But only relations with a type are considered useful. Ampersand with semantics in a heterogeneous relation algebra is discussed in [15]. Generalization (subtyping) is exploited in that paper to embed the relations appropriately such that the composition and union operations are total and the Tarski axioms are valid to a large extent.

Ampersand features rules, relations and concepts. It has a type system that blocks compilation as long as a script contains type incorrect expressions. In essence, a specification in Ampersand is a set of rules and a set of relation symbol declarations with a concept-based type. Each rule is an expression (a relation term) in a relation algebra that must be kept true throughout time. Thus, each relation that is a rule represents an invariant requirement of the business. A requirements engineer takes responsibility for a correspondence between requirements (in natural language) and rules (i.e., constraints in relation algebra).

Ampersand features a compiler that produces requirements specifications in natural language in the form of a PDF-document. A type system ensures the absence of relation terms that are predictably nonsensical to the end user. Relation terms that are type incorrect (i.e., make no sense) are reported back to the requirements engineer as a type error. This feedback explains why an erroneous relation term makes no sense within the chosen context. The feedback generator is implemented as an attribute grammar of untyped relation terms with a type function as one of its attributes. This is described in section 4. This implementation allows for future extensions with new attributes to analyze and report in even more detail.

The relevance of relation algebra in practice has surprised the authors. Formalizing requirements results in compact scripts that allow substantial pieces of the design process to be automated. Students at the OUNL now use an online tool in which they can study the impact of their Ampersand scripts, analyse them and generate requirements specifications from them. For the future we plan to implement editing of rule models within the tool. This will give students a working prototype of their design, which is expected to improve their learning curve.

The work presented here advances the state-of-the-art of designing business processes and the IT-systems that support them. The innovation Ampersand

---

[2] In 2007 a new information system to support the Dutch Immigration Service in managing all immigration requests was designed. In 2010 a platform independent design for all courts of administrative law in the Netherlands was generated.

brings is to generate design artifacts, such as class diagrams, directly from requirements. The idea to use algebra for knowledge representation [1] lies at the heart of our research. Ideas from heterogeneous relation algebra [13,7] were borrowed to construct a type system. Furthermore, the existence of a well studied body of knowledge called relation algebra [10] eliminated the need to invent a new (possibly poorer) language. Ampersand has the distinct advantage of tapping into an existing, well studied body of knowledge. Besides, relation algebras are being used (e.g. [2]) and taught (e.g. [5]).

This research is relevant for requirements engineers and students alike, who specify business processes by means of constraints. They can rely on the feedback from their design tool. This audience consists mostly of students and practitioners in business information systems. The attention given to feedback is justified by the difficulty of learning how to formalize business rules in Ampersand.

The paper starts in section 2 with an informal introduction on requirements and the way an information system satisfies them. Then the syntax and semantics of Ampersand are defined. Next the type function is introduced. The Ampersand compiler does type checking and type deduction simultaneously. It has been implemented as an Haskell attribute grammar. The feedback system implementation is described and demonstrated.

2   Language for Requirements

Requirements engineers need a language that is shared by all stakeholders. For example, let us assume we are working for an auction house in Dendermonde. The sentence "Peter presides over auction room 3" makes sense in this context, if there is a person called Peter, there is an auction room 3, and the auction house is familiar with the idea that a person presides over an auction room. In this context the sentence "Vehicle 06-GNL-3 presides over auction room 3" does not make sense, because vehicles never preside over auction rooms. That is non-sense. The type system is an instrument by which the requirements engineer can exclude a significant class of nonsense. The declaration *presides* : Person $\sim$ AuctionRoom introduces a (heterogeneous) relation with the name *presides*, in which only pairs of persons and auction rooms reside. By specifying relations, a requirements engineer introduces the basic sentences of a language. More complex sentences are made by means of the operators of relation algebra. If stakeholders agree to use that language to express requirements, we call this language *shared*.

Ampersand sees an information system as a collection of data that represents facts in a given context. If, for example, Peter presides over auction room 3, one expects a tuple $\langle$ 'Peter', 'auction room 3'$\rangle$ in a relation *presides* to represent this fact. Some time later that relation might contain the tuple $\langle$ 'Sue-Ellen', 'auction room 3'$\rangle$. In an information system, data content changes as facts in the given context change. Rules constrain that data content. For example, the auction house might require that a bid on a lot at an auction may be placed only by persons who are registered as a bidder for that particular auction. Using relations to represent parts of that sentence, a requirements engineer assembles relation terms and declares them as logical expressions with the keyword RULE.

$$bid : \mathtt{Person} \sim \mathtt{Lot} \tag{1}$$

$$at : \mathtt{Lot} \sim \mathtt{Auction} \tag{2}$$

$$registeredfor : \mathtt{Person} \sim \mathtt{Auction} \tag{3}$$

$$\mathtt{RULE}\ bid \ \fatsemi\ at \Rightarrow registeredfor \tag{4}$$

This fragment assembles relations into a rule by means of operators from relation algebra, in this example ⨟ (compose), and ⇒ (implication). The full set of operators is introduced in definition 3. In this manner, a requirements engineer formalizes all requirements that need to be maintained by an information system. He can do this one rule at a time, adding requirements and nuance to the specification incrementally. In the fragment above the requirements engineer may add nuance by adding a rule covering that *at* is at least a partial function, which is likely to be assumed by the reader of the original requirement.

An information system contains all relations and their data. The task of the information system is to keep all rules satisfied. If a bid is placed by an unregistered bidder, the computer might simply block the transaction, explaining that you must be registered in order to place a bid. However, it might also signal this event to a registrar, to register the new bidder. Or, it might automatically look up this person, conclude that he is a subscribed member and register him automatically. For the purpose of this paper, we restrict the large number of options to two: One option is to block with an error message while the data remain in the old state (rollback). The other option is to proceed with one or more violations. In that case signals are raised, which persist as long as the violation persists. This mechanism allows an information system to control business processes, i.e., collaborations between people and computers. Blocking rules enforce consistent data throughout the business process. Signal rules trigger people to take some action in order to restore the situation that all rules are satisfied. Such rules may be satisfied with some delay, because people need some time to act.

Summarizing, Ampersand is a language built on binary relations, providing a natural model for data and business rules. This differs from the idea to use relations as a model for nondeterministic programs, which became common in the eighties [12]. There are other relational languages for constraint-based modelling, like Alloy [9] or Z [14]. Semantically, Ampersand, Z, and Alloy bear great resemblance [8]. All three are declarative specification languages with a type system. Underlying Alloy, one will find relations and atoms in a similar way Ampersand has. Both Alloy and Ampersand do automatic analysis, albeit in different ways. Ampersand uses notations of relation algebra, where Alloy and Z are more into set-notations. More significant differences are found in what Ampersand does with a specification. Unlike Alloy and Z, Ampersand generates data structures and code for a database that can hold every model of an Ampersand script. Alloy is a model checker, whereas Ampersand derives an implementation. Alloy tries to find a model, whereas Ampersand verifies any particular model at runtime and reports violations. Alloy supports consistency checks at design time, as the user can look for counterexamples on specific assertions. Ampersand checks for consistency at runtime, and does so automatically.

3   Ampersand

This section defines the core of Ampersand. Its abstract syntax is defined in section 3.1. The semantics of relation terms are given in section 3.2. Ampersand chooses an interpretation that allows the representation of requirements and their satisfaction by means of data. A type function is introduced which is partially defined. All expressions with an undefined type are deemed incorrect and therefore rejected by the type checker. The type system is discussed in section 3.3.

3.1   Syntax

The Ampersand syntax consists of constant symbols for (business) concepts, (business) elements, relations and relation operators. Relation terms can be constructed with relations and relation operators.

Let $\mathbb{C}$ be a set of concept symbols. A concept is represented syntactically by an alphanumeric character string starting with an upper case character. In this paper, we use A, B, and C as concept variables.

Let $\mathbb{U}$ be a set of atom symbols. An atom is represented syntactically by an ASCII string within single quotes. All atoms are an element of a concept, e.g. 'Peter' is an element of Person. We use a, b, and c as atom variables.

Let $\mathbb{D}$ be a set of relation symbols. A relation symbol is represented syntactically by an alphanumeric character string starting with a lower case character. For every $A, B \in \mathbb{C}$, there are special relation symbols, I_A and $V_{A \times B}$. We use r, s, and t as relation variables.

Let $\neg$, $^{\sqcup}$, $\sqcup$, and $\mathbin{\text{\scriptsize ⨾}}$ be relation operators of arity 1, 1, 2, and 2 respectively. The binary relation operators $\sqcap, \Rightarrow$ and $\equiv$ are cosmetic and only defined on the interpretation function (see definition 3).

Let $\mathbb{R}$ be the set of relation terms. We use R, S, and T as variables that denote relation terms.

Definition 1 (relation terms)
$\mathbb{R}$ *is recursively defined by*

$$I_A, V_{A \times B}, r, \neg R, R^{\sqcup}, R \sqcup S, R \mathbin{\text{\scriptsize ⨾}} S \in \mathbb{R}$$

$$\textit{provided that } R, S \in \mathbb{R}, r \in \mathbb{D}, \textit{ and } A, B \in \mathbb{C}$$

Definition 2 (statements)
An Ampersand design of context $\mathfrak{C}$ is a user-defined collection of statements where

- RUL $\subseteq \mathbb{R}$ *is a collection of relation terms called rule statements.*
- REL *is a collection of* $r : A \sim B$ *for all* $r \in \mathbb{D}$ *such that* $A, B \in \mathbb{C}$. *The instances of* REL *are called relation declarations.*
- POP *is a collection of* $a \ r \ b$ *such that* $a \in A, b \in B$ *and* $(r : A \sim B) \in$ REL. *The instances of* POP *are called relation elements.*

The relation declarations define the conceptual structure and scope of $\mathfrak{C}$. Relation elements define facts in $\mathfrak{C}$. *POP* is called the *population* of $\mathfrak{C}$. Rules are constraints on that population.

An Ampersand script is a user-defined collection of relation declarations, relation elements and rule statements. It describes a (business) context $\mathfrak{C}$, in compliance with the OMG standard *Semantics of Business Vocabulary and Business Rules (SBVR)*. The Ampersand compiler contains a parser, which extracts *RUL*, *REL*, and *POP* from an Ampersand script that describes $\mathfrak{C}$.

3.2   Semantics

The previous section defines the syntactic structure of Ampersand. This section introduces an interpretation $\mathfrak{I}(R)$ that defines the semantics of a relation term R. This function interprets relation terms based on *POP* and a relation algebra [10] $\langle \mathbb{R}, \cup, \overline{}, ; , \check{}, \mathbb{I} \rangle$ where $\mathbb{R} \subseteq \mathcal{P}(\mathbb{U})$. All relation symbols used in a relation term are either declared by the user in *REL* or I_A or $V_{A \times B}$. So, the relation algebra on $\mathbb{R}$ is configured by the user through *REL*. The interpretation of all relation symbols in $\mathbb{D}$ is completely user-defined through *POP*. Thus, given some *REL* and some *POP*, $\mathfrak{I}(R)$ determines whether some relation holds between two elements.

Definition 3 (interpretation function)
Given some $\mathfrak{C}$, *the interpretation function of relation terms is defined by*

$$\text{relation} \qquad \mathfrak{I}(r) = \{\langle a, b \rangle \mid a \ r \ b \in \text{POP}\} \tag{5}$$

$$\text{identity} \qquad \mathfrak{I}(I_A) = \{\langle a, a \rangle \mid a \in A\} \tag{6}$$

$$\text{universal} \qquad \mathfrak{I}(V_{A \times B}) = \{\langle a, b \rangle \mid a \in A, \ b \in B\} \tag{7}$$

$$\text{complement} \qquad \mathfrak{I}(\neg R) = \overline{\mathfrak{I}(R)} \tag{8}$$

$$\text{converse} \qquad \mathfrak{I}(R^{\sqcup}) = \mathfrak{I}(R)^{\check{}} \tag{9}$$

$$\text{union} \qquad \mathfrak{I}(R \sqcup S) = \mathfrak{I}(R) \cup \mathfrak{I}(S) \tag{10}$$

$$\text{composition} \qquad \mathfrak{I}(R \,\fatsemi\, S) = \mathfrak{I}(R); \mathfrak{I}(S) \tag{11}$$

(the interpretation of the mentioned cosmetic relation operators)

$$\text{intersection} \qquad \mathfrak{I}(R \sqcap S) = \mathfrak{I}(\neg(\neg R \sqcup \neg S)) \tag{12}$$

$$\text{implication} \qquad \mathfrak{I}(R \Rightarrow S) = \mathfrak{I}(\neg R \sqcup S) \tag{13}$$

$$\text{equivalence} \qquad \mathfrak{I}(R \equiv S) = \mathfrak{I}((R \Rightarrow S) \sqcap (S \Rightarrow R)) \tag{14}$$

Section 2 informally described that relations need to have a type. A relation term $R \in \mathbb{R}$ with a type $\mathfrak{T}(R)$ has an interpretation. If it has no type, it is said to have a *type error* or to be *(semantically) incorrect*. An end user might call a relation term without a type *nonsense*. Relation terms with a type error are rejected with a proper feedback message.

By a relation declaration $r : A \sim B$ the user declares the existence of a relation symbol r with a type denoted $\mathfrak{T}(r) = A \sim B$. We use X, Y as type variables. By Definition 2, given some $r : A \sim B$, the user can only define a relation element $a\ r\ b \in POP$ if $a \in A$ and $b \in B$. $\mathfrak{T}(R)$ is inspired on Hattensperger's typing function for heterogeneous relation algebra [7].

Definition 4 (typing function)
Given some $\mathfrak{C}$, the partial typing function of relation terms is defined by

$$\mathfrak{T}(r) = A \sim B \qquad , \text{if } r : A \sim B \in \text{REL} \tag{15}$$

$$\mathfrak{T}(I_A) = A \sim A \qquad , \text{if } A \in \mathbb{C} \tag{16}$$

$$\mathfrak{T}(V_{A \times B}) = A \sim B \qquad , \text{if } A, B \in \mathbb{C} \tag{17}$$

$$\mathfrak{T}(\neg R) = \mathfrak{T}(R) \qquad , \text{if } \mathfrak{T}(R) \text{ is defined} \tag{18}$$

$$\mathfrak{T}(R^{\sqcup}) = B \sim A \qquad , \text{if } \mathfrak{T}(R) = A \sim B \tag{19}$$

$$\mathfrak{T}(R \sqcup S) = \mathfrak{T}(R) \qquad , \text{if } \mathfrak{T}(R) = \mathfrak{T}(S) \tag{20}$$

$$\mathfrak{T}(R \mathbin{\raise0.3ex\hbox{\tiny$\circ$}\kern-0.1em\lower0.3ex\hbox{\tiny$\circ$}} S) = A \sim C \qquad , \text{if } \mathfrak{T}(R) = A \sim B, \mathfrak{T}(S) = B \sim C \tag{21}$$

3.3 Type System

Only type correct expressions are allowed in Ampersand scripts. Ampersand allows overloading of relation symbols. This means that relation term names need not be unique. Overloading is necessary for practical reasons only, in order to give users more freedom to choose names. In particular, requirements engineers may choose short uniform names, such as *aggr*, *has*, *in*, to represent different relations for different types. The type system deduces all possible types for any given relation term name. If there are multiple possibilities, a type error is given. The script writer can disambiguate any relation term name by adding type information explicitly.

Ampersand also allows that two different concepts overload an atom symbol, i.e., an $a \in A$ has an interpretation different from $a \in B$. Atoms are only interpreted in the context of typed relation terms, preventing an ambiguous identity of atoms. In practice, business concepts may be overlapping, e.g., 'Peter' the Person has the same practical identity as 'Peter' the Auctioneer. In the companion paper, Van der Woude and Joosten [15] embed generalization of concepts in Ampersand. They introduce supertype relations (ϵ), called embeddings, to define total extensions of the partial (heterogeneous) relational operators except for the complement. These extensions give requirements engineers a type-controlled freedom to express themselves in relations on sub- or superconcepts. The use of negation in business rules with generalization requires further investigation.

The requirements engineer declares the supertype relation between subconcept A and superconcept B as $\epsilon : A \sim B \in REL$.

Altogether, the problem that the type system must solve is twofold. The type system must deduce one relation term from a relation term name, and check the type of relation terms simultaneously. We use R', S', T' as variables to denote relation term names.

The type function $\mathfrak{T}'(R')$ is based on the partial typing function $\mathfrak{T}(R)$. The type system examines $\mathfrak{T}'(R')$ to mark the name R' as bound to a relation, ambiguous, undeclared or undefined.

- If some r is not declared with a type in REL, then r is said to be *undeclared*.
- If some R' can only refer to one type correct relation term R, then R' is said to *bind to R*.
- If some R' does not refer to any type correct relation term, then R' is said to be *undefined*.
- If some R' can refer to more than one type correct relation term, i.e., an *alternative*, then R' is said to be *ambiguous*.

Definition 5 (type function)
Given some $\mathfrak{C}$, the type function of relation term names is defined by

$$\mathfrak{T}'(r) = \{A \sim B \mid (r : A \sim B) \in \text{REL } \} \tag{22}$$
$$\mathfrak{T}'(I_A) = \{A \sim A\} \tag{23}$$
$$\mathfrak{T}'(V_{A \times B}) = \{A \sim B\} \tag{24}$$
$$\mathfrak{T}'(R'_X) = \mathfrak{T}'(R') \cap \{X\} \tag{25}$$
$$\mathfrak{T}'(\neg R') = \mathfrak{T}'(R') \tag{26}$$
$$\mathfrak{T}'(R'^{\sqcup}) = \{B \sim A | A \sim B \in \mathfrak{T}'(R')\} \tag{27}$$
$$\mathfrak{T}'(R' \sqcup S') = \mathfrak{T}'(R') \cap \mathfrak{T}'(S') \tag{28}$$
$$\mathfrak{T}'(R' \, \S \, S') = \{A \sim C \mid \sharp \{B \mid A \sim B \in \mathfrak{T}'(R'), B \sim C \in \mathfrak{T}'(S')\} = 1\} \tag{29}$$

Rule 25 enables the requirements engineer to disambiguate an ambiguous relation term name with type information. Rule 29 ensures that if $R' \, \S \, S'$ yields more than one alternative with type $A \sim C$, then $A \sim C \notin \mathfrak{T}'(R' \, \S \, S')$.

This type system is not monotone, i.e., a well typed program can become ill-typed by adding rules for different, independent relations. When this happens to a requirements engineer, he will notice that the introduction of a new relation forces him to add more type information to other parts in his script.

4   Feedback System

The type system is embedded in the feedback system of the Ampersand compiler. The feedback system must give an error message if and only if a relation term name cannot be bound to a relation term. The error messages must be concise, i.e., correct and kept short, precise and relevant. The quality of the feedback deserves attention, because it lets students focus on learning rule-based design.

The feedback system is implemented in Haskell [11] based on an attribute grammar [6] Haskell library developed at Utrecht University. The feedback system checks all relation term names R' used in any statement in RUL or POP, given the conceptual structure of $\mathfrak{C}$ represented by REL. REL is an inherited attribute called *context structure*. Within the scope of a statement, given the context structure, R' is bound, ambiguous, undeclared or undefined.

A bound R' implies the existence of one suitable alternative for any subname S' of R'. The reader may verify this surjective function from subname S' to bound R' in rule 22-29. However, $\mathfrak{T}'(S')$ may yield more than one alternative. The feedback system uses two attributes to determine the type of any S' within the scope of R'. $\mathfrak{T}'(S')$ is a synthesized attribute called *pre-type*. Some $X \in \mathfrak{T}'(S')$ is an inherited attribute called *automatic type directive*. The automatic type directive is based on Definition 5.

For example, consider two statements a rule and a relation element, both yielding relation *rel1*:

$$rel1 \sqcup rel2 \in RUL$$

$$\text{'atom1'} \ rel1 \ \text{'atom2'} \in POP$$

where

$$REL = \{rel1 : \mathsf{Cpt1} \sim \mathsf{Cpt2}, rel1 : \mathsf{Cpt1} \sim \mathsf{Cpt3}, rel2 : \mathsf{Cpt1} \sim \mathsf{Cpt2}\}$$

The rule binds to a typed relation term, although $\mathfrak{T}'(rel1)$ yields more than one alternative:

$$
\begin{aligned}
\mathfrak{T}'(rel1) &= \{\mathsf{Cpt1} \sim \mathsf{Cpt2}, \ \mathsf{Cpt1} \sim \mathsf{Cpt3}\} && \text{(rule 22, ambiguous)} \\
\mathfrak{T}'(rel2) &= \{\mathsf{Cpt1} \sim \mathsf{Cpt2}\} && \text{(rule 22, bound)} \\
\mathfrak{T}'(rel1 \sqcup rel2) &= \mathfrak{T}'(rel1) \cap \mathfrak{T}'(rel2) && \text{(rule 28)} \\
&= \{\mathsf{Cpt1} \sim \mathsf{Cpt2}, \ \mathsf{Cpt1} \sim \mathsf{Cpt3}\} \cap \{\mathsf{Cpt1} \sim \mathsf{Cpt2}\} \\
&= \{\mathsf{Cpt1} \sim \mathsf{Cpt2}\} && \text{(bound)} \\
&= \mathfrak{T}'((rel1 \sqcup rel2)_{\mathsf{Cpt1} \sim \mathsf{Cpt2}}) && \text{(auto)} \\
&= \mathfrak{T}'((rel1_{\mathsf{Cpt1} \sim \mathsf{Cpt2}} \sqcup rel2_{\mathsf{Cpt1} \sim \mathsf{Cpt2}})_{\mathsf{Cpt1} \sim \mathsf{Cpt2}}) && \text{(auto)} \\
\mathfrak{T}'(rel1_{\mathsf{Cpt1} \sim \mathsf{Cpt2}}) &= \{\mathsf{Cpt1} \sim \mathsf{Cpt2}\} && \text{(rule 22, bound)} \\
\mathfrak{T}'(rel2_{\mathsf{Cpt1} \sim \mathsf{Cpt2}}) &= \{\mathsf{Cpt1} \sim \mathsf{Cpt2}\} && \text{(rule 22, bound)} \\
&&& \text{(done)}
\end{aligned}
$$

The relation element is ambiguous, because $\mathfrak{T}'(rel1)$ yields more than one alternative:

$$
\begin{aligned}
\mathfrak{T}'(rel1) &= \{\mathsf{Cpt1} \sim \mathsf{Cpt2}, \ \mathsf{Cpt1} \sim \mathsf{Cpt3}\} && \text{(rule 22, ambiguous)} \\
&&& \text{(done)}
\end{aligned}
$$

The requirements engineer should have specified a type directive

$$\text{'atom1'} \ rel1_{\mathsf{Cpt1} \sim \mathsf{Cpt2}} \ \text{'atom2'} \in POP$$

or

$$\texttt{'atom1'}\ rel1\,_{\texttt{Cpt1}\sim\texttt{Cpt3}}\ \texttt{'atom2'} \in POP$$

Detailed information can be obtained through synthesized attributes of the attribute grammar. Three synthesized attributes are defined in our implementation. One attribute holds either $\mathfrak{T}(R)$ if R' is bound to R and $\mathfrak{T}(R)$ is defined, or an error message as described in Section 4.1 otherwise. Another attribute holds a fully typed relation term name $R_{\mathfrak{T}(R)}$ if the previous attribute holds $\mathfrak{T}(R)$, and is undefined otherwise. The third attribute holds an extensive LaTeX report for complete detail on type errors or binding relation term names. Such a report contains equational traces like presented for the examples in this section. This report is generated by the Ampersand compiler.

4.1 Error Messages

If a relation term has an error, then an error message must be composed. We have designed templates for error messages which relate to the type system rule at which an error first occurs. The error messages are defined short but complete and specific. The templates are presented in the same order of precedence as the operators they relate to. If subterms of a relation term have error messages, then the error message of the relation term is the union of all the error messages of subterms.

relation undeclared If $\mathfrak{T}'(r)$ yields no types, then there is no relation declaration for r.

relation undeclared: r

relation undefined Requirements engineers may use the unique name of some R e.g. R'_X. This may cause different kinds of errors which are checked in the same chronological order as described here.

If $X = A \sim B$ and A or B does not occur in the type signature of any relation declaration, then there is probably a typo in the concept name.

unknown concept: A, or

unknown concept: B, or

unknown concepts: A and B

If $X \notin \mathfrak{T}'(R')$, then there is no R such that $\mathfrak{T}(R) = X$. If $R' = r$ then r_X is undeclared.

relation undeclared: r_X

In all other cases:

relation undefined: r_X

possible types are: $\mathfrak{T}'(R')$

incompatible/ambiguous composition Let $\mathfrak{T}'(R')$ and $\mathfrak{T}'(S')$ yield alternatives. If $\mathfrak{T}'(R' \, \mathbf{;} \, S')$ yields no types, then there is no alternative for $R' \, \mathbf{;} \, S'$. In case $\sharp\{B \mid A \sim B \in \mathfrak{T}'(R'), B \sim C \in \mathfrak{T}'(S')\} = 0$, then R' and S' are incompatible for composition.

$$\textit{incompatible composition: } R' \, \mathbf{;} \, S'$$

$$\textit{possible types of } R' \textbf{:} \; \mathfrak{T}'(R')$$

$$\textit{possible types of } S' \textbf{:} \; \mathfrak{T}'(S')$$

In case $\sharp\{B \mid A \sim B \in \mathfrak{T}'(R'), B \sim C \in \mathfrak{T}'(S')\} > 1$, then the composition of R' and S' is ambiguous.

$$\textit{ambiguous composition: } R' \, \mathbf{;} \, S'$$

$$\textit{possible types of } R' \textbf{:} \; \{A \sim B \mid A \sim B \in \mathfrak{T}'(R'), B \sim C \in \mathfrak{T}'(S')\}$$

$$\textit{possible types of } S' \textbf{:} \; \{B \sim C \mid A \sim B \in \mathfrak{T}'(R'), B \sim C \in \mathfrak{T}'(S')\}$$

incompatible comparison Let $\mathfrak{T}'(R')$ and $\mathfrak{T}'(S')$ yield alternatives. If $\mathfrak{T}'(R' \sqcup S')$ yields no types, then R' and S' are incompatible for comparison.

$$\textit{incompatible comparison: } R' \sqcup S'$$

$$\textit{possible types of } R' \textbf{:} \; \mathfrak{T}'(R')$$

$$\textit{possible types of } S' \textbf{:} \; \mathfrak{T}'(S')$$

ambiguous type If R' is ambiguous within the scope of a statement, then the ambiguity is reported as an error.

$$\textit{ambiguous relation: } R'$$

$$\textit{possible types: } \mathfrak{T}'(R')$$

4.2   Demonstration

Let us demonstrate a typical constraint that some relation is contained within another relation.

$$(rel1 \sqcap rel2 \Rightarrow rel0) \in RUL$$

where

$$REL = \{rel1 : \mathsf{Cpt}1 \sim \mathsf{Cpt}2, rel2 : \mathsf{Cpt}3 \sim \mathsf{Cpt}4\}$$
$$R \sqcap S = \neg(\neg R \sqcup \neg S)$$
$$R \Rightarrow S = \neg R \sqcup S$$

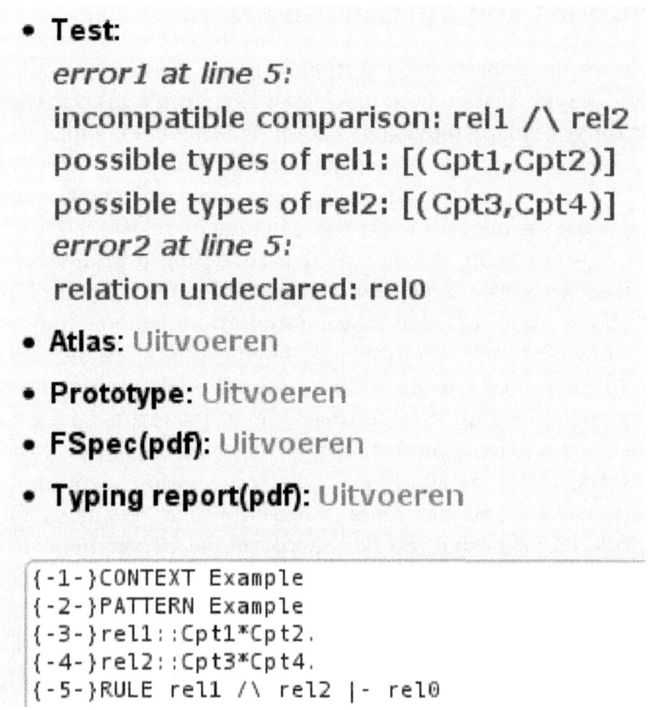

- **Test:**

  error1 at line 5:

  incompatible comparison: rel1 /\ rel2
  possible types of rel1: [(Cpt1,Cpt2)]
  possible types of rel2: [(Cpt3,Cpt4)]
  error2 at line 5:

  relation undeclared: rel0

- **Atlas:** Uitvoeren

- **Prototype:** Uitvoeren

- **FSpec(pdf):** Uitvoeren

- **Typing report(pdf):** Uitvoeren

```
{-1-}CONTEXT Example
{-2-}PATTERN Example
{-3-}rel1::Cpt1*Cpt2.
{-4-}rel2::Cpt3*Cpt4.
{-5-}RULE rel1 /\ rel2 |- rel0
```

Fig. 1. compiler screen snippet with type error (Dutch)

The rule contains two errors. Both will be mentioned in the error message as depicted in Figure 1.

These messages provide the requirements engineer with relevant and sufficient information which they understand. For complete detail, the requirements engineer requests a LaTeXreport containing a trace like:

$$\mathfrak{T}'(rel1) = \{\text{Cpt1} \sim \text{Cpt2}\} \qquad \text{(rule 22, bound)}$$
$$\mathfrak{T}'(\neg rel1) = \mathfrak{T}'(rel1) \qquad \text{(rule 26)}$$
$$= \{\text{Cpt1} \sim \text{Cpt2}\} \qquad \text{(bound)}$$
$$\mathfrak{T}'(rel2) = \{\text{Cpt3} \sim \text{Cpt4}\} \qquad \text{(rule 22, bound)}$$
$$\mathfrak{T}'(\neg rel2) = \mathfrak{T}'(rel2) \qquad \text{(rule 26)}$$
$$= \{\text{Cpt3} \sim \text{Cpt4}\} \qquad \text{(bound)}$$
$$\mathfrak{T}'(\neg rel1 \sqcup \neg rel2) = \mathfrak{T}'(\neg rel1) \cap \mathfrak{T}'(\neg rel2) \qquad \text{(rule 28)}$$
$$= \{\} \qquad \text{(undefined)}$$
$$\mathfrak{T}'(rel0) = \{\} \qquad \text{(rule 22, undeclared)}$$
$$\text{(done)}$$

5   Conclusions and Further Research

In this paper we have introduced Ampersand by means of an abstract syntax, semantics, and a type system, and motivated this from a practical use of specifying information systems. Ampersand allows requirements engineers to represent a situation specific language by relation declarations and express truths in it. True facts are represented by relation elements and requirements are represented by rules. Rules are terms in a syntactic calculus of relations based on Tarski's axioms. The relation terms are enriched with typing arguments to enable the filtering of rules without a type, and relation elements of undeclared relations. Ampersand allows overloading of relation and atom symbols and generalization of concepts, in order to give requirements engineers a more natural syntax. The type system filters relation terms with an undefined or ambiguous type.

Ampersand generates precise feedback on type errors. This feature keeps requirements engineers from implementing rules which will never hold. The feedback is generated by an attribute grammar on relation term names. The concreteness and relevance of the generated feedback is meant to focus students on learning rule-based design. So far, the practical experience with students is encouraging. Systematic experimentation and evaluation of the learning process is scheduled in the nearby future.

For the purpose of systematic research and learning, Ampersand development is moving from an ASCII text editor towards an Integrated Development Environment (IDE) for education. We consider this IDE as an information system for the business process of system development with Ampersand. As such it is described within Ampersand, and a prototype web-based information system is generated with the compiler.

The current information system for system development with Ampersand, called *Atlas*, is read-only. The Atlas is the prototype generated from an Ampersand script describing the Ampersand language. The compiler loads type correct Ampersand scripts, and derived data like rule violations and pictures into the database of this system. Through the generated web-interface students can explore the design of their context $\mathfrak{C}$ by clicking on related elements. Student behaviour is stored in the database and may serve as input to prove intended didactical improvements on the system.

The Atlas will become editable in controlled phases. First students will be able to edit the *POP* of their context. Certain rules, e.g., syntax rules, may be violated by changes in *POP*. These rules need to be implemented as business rules such that the generated IDE gives feedback on violations of these rules. After *POP* has successfully become editable, *REL* followed by *RUL* will become editable resulting in a rule-based educational environment to develop with Ampersand.

The use of Ampersand in practice is also encouraging. Large information system projects in the Dutch government have already been designed in Ampersand, and one has been realized.

Further research in the Ampersand project focuses on:

- Publishing the software and disclosing further results in the open source domain (ampersand.sourceforge.net).

- Refining the didactics of teaching Ampersand as a rule based design method.
- Design for large IT projects in industry.
- Generate software prototypes from Ampersand scripts.
- Automate the design of web services.

References

1. Brink, C., Schmidt, R.A.: Subsumption computed algebraically. Computers and Mathematics with Applications 23(2-5), 329–342 (1992)
2. Brink, C., Kahl, W., Schmidt, G. (eds.): Relational methods in computer science. Advances in Computing. Springer, New York (1997)
3. Codd, E.F.: A relational model of data for large shared data banks. Communications of the ACM 13(6), 377–387 (1970)
4. Date, C.J.: What not how: the business rules approach to application development. Addison-Wesley Longman Publishing Co., Inc., Boston (2000)
5. Desharnais, J.: Basics of relation algebra, http://www2.ift.ulaval.ca/~Desharnais/Recherche/Tutoriels/TutorielRelMiCS10.pdf
6. Dijkstra, A., Swierstra, S.D.: Typing haskell with an attribute grammar. In: Vene, V., Uustalu, T. (eds.) AFP 2004. LNCS, vol. 3622, pp. 1–72. Springer, Heidelberg (2005)
7. Hattensperger, C., Kempf, P.: Towards a formal framework for heterogeneous relation algebra. Inf. Sci. 119(3-4), 193–203 (1999)
8. Jackson, D.: A comparison of object modelling notations: Alloy, UML and Z. Tech. rep. (1999), http://sdg.lcs.mit.edu/publications.html
9. Jackson, D.: Software Abstractions: Logic, Language, and Analysis. The MIT Press, Cambridge (2006)
10. Maddux, R.D.: Relation Algebras. Studies in logic, vol. 150. Elsevier, Iowa (2006)
11. Peyton Jones, S. (ed.): Haskell 98 Language and Libraries – The Revised Report. Cambridge University Press, Cambridge (2003)
12. Sanderson, J.G.: A Relational Theory of Computing. LNCS, vol. 82. Springer, New York (1980)
13. Schmidt, G., Hattensperger, C., Winter, M.: Heterogeneous Relation Algebra. In: Relational Methods in Computer Science. Advances in Computing, ch. 3, pp. 39–53. Springer, New York (1997)
14. Spivey, J.M.: The Z Notation: A reference manual, 2nd edn. International Series in Computer Science. Prentice Hall, New York (1992)
15. van der Woude, J., Joosten, S.: Relational heterogeneity relaxed by subtyping, (submitted 2011)

Programming from Galois Connections

Shin-Cheng Mu[1] and José Nuno Oliveira[2]

[1] IIS, Academia Sinica, Taiwan, Taiwan
[2] High Assurance Software Lab, University of Minho, Portugal

Abstract. Problem statements often resort to superlatives such as in eg. "...the smallest such number", "...the best approximation", "...the longest such list" which lead to specifications made of two parts: one defining a broad class of solutions (the *easy* part) and the other requesting the optimal such solution (the *hard* part).

This paper introduces a binary relational combinator which mirrors this linguistic structure and exploits its potential for calculating programs by optimization. This applies in particular to specifications written in the form of Galois connections, in which one of the adjoints delivers the optimal solution being sought.

The framework encompasses re-factoring of results previously developed by Bird and de Moor for greedy and dynamic programming, in a way which makes them less technically involved and therefore easier to understand and play with.

1 Introduction

Computer programming is admittedly a challenging intellectual activity, calling for experience and training under a *read-understand-repeat* learning cycle. By acquiring good practices, relying on experienced teachers, the learning curve eventually bends, but reliability cannot be fully ensured. If one asks a student in programming about *why* she/he programs in that way (whatever this is) the answer is likely to be: *I don't know — my teachers used to do it this way.*

Why is this so? Isn't programming a *scientific discipline*? Surely it is, as several landmark textbooks show[1]. But, perhaps the question

Why *and* in what measure *is programming difficult?*

is yet to be given a satisfactory answer. By satisfactory we mean one which should unravel the ingredients of problem solving in a structured way, thus identifying which skills one should acquire to become a good programmer.

Abstraction is one such skill [12]. Abstracting from the programming language and underlying technology is generally accepted as mandatory in the early stages of thinking about a software problem. This has lead to *abstract modeling*, which has become a discipline in itself [10,9]. However, handling abstractions is not

---

[1] See eg. the following (by no means exhaustive) list of widely acclaimed references: [11,5,21,20,4,3].

H. de Swart (Ed.): RAMICS 2011, LNCS 6663, pp. 294–313, 2011.

easy either (many will say it is harder) and the question persists: *why* and *in what measure* is abstract modeling difficult?

Induction is another such skill, to which programmers unconsciously appeal whenever solving a complex problem by (temporarily) imagining some (smaller) parts of it already solved (the *divide-and-conquer* strategy). However, where and how does *induction* crop up in the design of a program? For instance, where exactly in the design of *quicksort* from its specification: *yield an ordered permutation of the input sequence*, does the *doubly recursive* strategy of the algorithm show up? The starting specification does not look inductive at all.

This paper tries to answer the questions above by splitting algorithmic specifications generically in two parts, to be addressed in different stages. Let us see where these come from.

In program construction one often encounters specifications asking for the "best" solution among a collection of solution candidates. Such specifications may have the form "the smallest such number . . .", "the best approximation such that . . .", "the longest prefix of a list satisfying . . .", etc. A typical example is the definition of whole number division $x \div y$, for natural numbers x and (positive) y. A specification in words would say that $x \div y$ is the *largest* natural number that, when multiplied by y, is at most x. The standard function *takeWhile p*, as another example, returns the *longest* prefix of the input list such that all elements satisfy predicate p.

Many other, less classroom-like problem statements share the same linguistic pattern in their use of superlatives. For instance, the computation of the "best" schedule for a collection of tasks, given their time spans and an acyclic *Gantt* graph (describing which tasks depend upon completion of which other tasks) is another problem of the same kind. Such a schedule is "best" (among other schedules paying respect to the given graph of dependencies) in the sense that its tasks start as early as possible.

It is often relatively easy to construct a program that meets half of such specifications: returning or enumerating the feasible solution candidates, such as a natural number, or prefixes of the input list. This is the easy part. The hard part of the specification, however, demands that we return a candidate that is "best" in some sense (eg. some ordering): the largest integer, or the longest prefix, that satisfies the first, easy part of the specification.

In this paper we propose a new relational operator mirroring this "easy/hard" dichotomy of problem statements into mathematics. The operator is of the form

$$E \upharpoonright H,$$

where E specifies the easy part — the collection of solution candidates — while H specifies the hard part — criteria under which a best solution is chosen.

One might wonder how to come up with the easy/hard split in the first place. In this paper we aim at characterizing problem specifications in terms of *Galois connections* [6], in which one of the adjoints specifies the easy part (usually a known function) and the other specifies the one at target (the hard one). For instance, the (easy) adjoint of whole division is multiplication. This setting, which suggests that *"mathematics comes in easy/hard pairs"*, provides a natural way to split a problem in its components, as seen below.

Paper structure. In Section 2 we argue why Galois connections are suitable as calculational specifications, before motivating and introducing the ($\restriction$) operator in Section 3. If some components in the Galois connection are inductively defined, as reviewed in Section 4, the two theorems presented in Section 5, demonstrated by two examples, allow us to calculate the wanted adjoint. A larger example, scheduling a collection of tasks given a Gantt graph, is presented in Section 6, before we conclude in Section 7. A minimal review of relational program calculation is given in Appendix A and B.

2   Galois Connections as Program Specifications

Let us take the problem of writing the algorithm of *whole division* as starting example[2]. Its specification has already been stated above, informally:

> $x \div y$ *is the largest natural number that, if multiplied by* y, *is at most* x.

Which mathematics should we write to capture the text above? One possibility is to write a "literal" one,

$$x \div y = \langle \bigvee z \; :: \; z \times y \leq x \rangle, \tag{1}$$

encoding superlative *largest* explicitly as a supremum. Handling suprema, however, is not easy in general. A second version will circumvent this difficulty,

$$z = x \div y \; \equiv \; \langle \exists r \; : \; 0 \leq r < y : \; x = z \times y + r \rangle, \qquad \begin{array}{c|c} x & y \\ \hline r & z \end{array} \tag{2}$$

at the cost of existentially quantifying over remainders.

A third alternative is surprisingly simpler [18]: an equivalence

$$z \times y \leq x \; \equiv \; z \leq x \div y \qquad (y > 0) \tag{3}$$

universally quantified in all its variables. Pragmatically, it expresses a "shunting" rule which enables one to exchange between a whole division in the upper side of a ($\leq$) inequality and a multiplication in the lower side, very much like in handling equations in school algebra.

Equivalences such as (3) are known as Galois connections [1,6,18]. In general, a Galois connection (GC) is a pair of functions f and g satisfying the equivalence $f \, z \leq x \; \equiv \; z \sqsubseteq g \, x$, for all z and x, given preorders ($\leq$) and ($\sqsubseteq$) (which can be the same). Functions f and g are said to be *adjoints* of each other — f is the *lower* adjoint and g the *upper* adjoint. In the case of (3) the adjoints are

$$z \underbrace{(\times y)}_{f} \leq x \; \equiv \; z \leq x \underbrace{(\div y)}_{g}.$$

---

[2] This example is taken from [18].

Why can one be so confident of the adequacy of (3) in the face of the given requirements? Do substitution $z := x \div y$ in (3) and obtain $(x \div y) \times y \leq x$: this tells that $x \div y$ is a candidate solution. Now read (3) from left to right, that is, focus on the implication $z \times y \leq x \;\Rightarrow\; z \leq x \div y$: conclude that $x \div y$ is largest among all other candidate solutions z.

So (3) means the same as (1). What are the advantages of the former over the latter? It turns up that (3) is far more generous with respect to inference of properties of $x \div y$, some of which are mere instantiations:

$$0 \leq x \div y, \qquad\qquad (z := 0)$$
$$y \leq x \;\equiv\; 1 \leq x \div y. \qquad\qquad (z := 1)$$

Other facts, for instance $x \div 1 = x$, call for properties of the lower adjoint:

$$z \leq x \div 1$$

$\equiv \qquad$ { Galois connection (3), for $y := 1$ }

$$z \times 1 \leq x$$

$\equiv \qquad$ { 1 is the unit of $\times$ }

$$z \leq x.$$

That is, every natural number z which is at most $x \div 1$ is also at most x. We conclude that $x \div 1$ and x are the same. The rationale behind this style of reasoning is known as the principle of *indirect equality*[3]:

$$a = b \;\equiv\; \langle \forall x \;::\; x \leq a \equiv x \leq b \rangle. \qquad\qquad (4)$$

More elaborate properties can be inferred from (3) using indirect equality and basic properties of the "easy" adjoint, for instance $(n \div m) \div d \;=\; n \div (d \times m)$, for $m, d > 0$. Again GC (3) blends well with indirect equality in an easy proof:

$$z \leq (n \div m) \div d$$

$\equiv \qquad$ { Galois connection (3), twice }

$$(z \times d) \times m \leq n$$

$\equiv \qquad$ { $\times$ is associative }

$$z \times (d \times m) \leq n$$

$\equiv \qquad$ { Galois connection (3) again, in the opposite direction }

$$z \leq n \div (d \times m)$$

$:: \qquad$ { indirect equality (4) }

$$(n \div m) \div d = n \div (d \times m).$$

---

[3] See [1]. Readers unaware of this way of indirectly establishing algebraic equalities will recognize that the same pattern of indirection is used when establishing set equality via the membership relation, cf. $A = B \;\equiv\; \langle \forall x \;::\; x \in A \equiv x \in B \rangle$ as opposed to, e.g. circular inclusion: $A = B \;\equiv\; A \subseteq B \wedge B \subseteq A$.

Readers are challenged to compare this with alternative proofs of the same result using (1) or (2) instead of (3), not to mention the inductive proof required if relying on the obvious recursive implementation of $x \div y$ [18].

This strategy is applicable to arbitrarily complex problem domains, provided candidate solutions are ranked by a partial order such as $\leq$ above. This is shown in our next example, in which the underlying partial order is the *prefix* relation $\sqsubseteq$ on finite sequences and what is being specified is *take*, the function which yields the longest prefix of its input sequence up to some given length n[4]:

$$length\ z \leq n\ \wedge\ z \sqsubseteq x\ \equiv\ z \sqsubseteq take(n,x). \tag{5}$$

The property being sought,

$$take(n, take(m,x))\ =\ take(min(n,m),x), \tag{6}$$

will rely on another GC — that of defining the minimum of two numbers,

$$x \leq n\ \wedge\ x \leq m\ \equiv\ x \leq min(n,m), \tag{7}$$

in a way which shows how effectively GCs compose with each other[5]:

$$z \sqsubseteq take(n, take(m,x))$$

$\equiv$ { Galois connection (5), twice }

$$length\ z \leq n \wedge length\ z \leq m \wedge z \sqsubseteq x$$

$\equiv$ { Galois connection of *min* of two numbers (7) }

$$length\ z \leq min(n,m) \wedge z \sqsubseteq x$$

$\equiv$ { (5) again, now folding }

$$z \sqsubseteq take(min(n,m),x)$$

$::$ { indirect equality over prefix partial ordering $\sqsubseteq$ }

$$take(n, take(m,x)) = take(min(n,m),x).$$

Once again, the inductive proof of the same property performed over the recursive definition of *take* can but be regarded as an over-kill in face of such a simple calculation relying on the Galois connection concept.

One may wonder about the extent to which such a calculational style carries over to supporting the actual *synthesis* of the implementation of *take* given its specification (5) in the form of a Galois connection. This brings us to the core subject of the current paper: *how calculational is programming from Galois*

---

[4] See [16]. The authors would like to thank Roland Backhouse for spotting this Galois connection, whose upper adjoint $g = take$ is *specified* in terms of a lower adjoint involving *id* and *length*: $f\ z = (length\ z, z)$. Thus the lower ordering is the product partial order $(\leq) \times (\sqsubseteq)$, defined pointwise in the obvious way.

[5] For a detailed account of the algebra of Galois connections see eg. [1,18,16].

connections? Reference [18] shows how the defining Galois connection of ($\div$) provides most of what is required for calculating its implementation. Reference [16] does the same for *take*, but Galois connection (5) is productive only after an inductive definition of prefix ($\sqsubseteq$) is given explicitly, at point level. This somehow suggests that similar, but more economic and generic reasoning could be performed at the pointfree level of the algebra of programming [4], capitalizing on the pointfree definition of partial orderings such as prefix as relational folds.

Presenting such a generic, pointfree style of *programming from Galois connections* is the main aim of the current paper and leads us into the core of the research being reported.

3   Calculating Galois Adjoints

Recall the definition of a GC: given two preorders ($\leq$) on A and ($\sqsubseteq$) on B, we say that two functions $f : A \leftarrow B$ and $g : B \leftarrow A$ form a GC if they satisfy the following equivalence:

$$f \, x \leq y \;\equiv\; x \sqsubseteq g \, y \qquad \text{cf. diagram:} \qquad (8)$$

It is quite common in GCs to have adjoints of disparate complexity. In GC (3) relating multiplication ($\times y$) and whole division ($\div y$), for example, the former is easier to define than the latter. A common scenario is that of one being given the two preorders and an easy adjoint, thereupon targeting at calculating the other adjoint.

Recall the easy/hard split discussed in Section 1. We will propose in this section a relational operator that manifests the split: by $E \restriction H$ we denote a problem specification where the easy part E is "shrunk" by the requirements of the hard part H. It will then be shown that given ($\leq$), ($\sqsubseteq$), and lower adjoint f in a Galois connection, the upper adjoint can be expressed by:

$$g = (f^{\circ} \cdot (\leq)) \restriction (\sqsupseteq). \qquad (9)$$

We will then discuss, in this section and the next, some properties of ($\restriction$) that help us to calculate g. The operator ($\restriction$) is similar to, and shares many properties of, the *min* operator of Bird and de Moor [4], with the significant advantage of not requiring a power allegory.

3.1   The "Shrink" Operator

From now on we will be using a number of definitions and rules of the pointfree calculus of relations. For the reader's convenience, minimal review is given in Appendix A. For a thorough introduction, the reader is referred to Aarts et al. [1], and to Bird and de Moor [4] for a categorical perspective.

The first step toward manifesting the easy/hard split is to rewrite (8) to pointfree style by turning both sides into relations between x and y. Since partial orders such as $(\leq)$ and $(\sqsubseteq)$ are relations that map "larger" elements to "smaller" ones, the right hand side trivially translates to $(\sqsubseteq) \cdot g$. The left hand side, noting that $(x, f\,x) \in f^\circ$ and that $f\,x \leq y$ is another way of writing $(f\,x, y) \in (\leq)$, translates to $f^\circ \cdot (\leq)$. The equivalence means that the two relations are equal:

$$f^\circ \cdot (\leq) \;=\; (\sqsubseteq) \cdot g. \tag{10}$$

Such equality splits into two inclusions to be dealt with separately:

$$(\sqsubseteq) \cdot g \;\subseteq\; f^\circ \cdot (\leq) \;\wedge\; f^\circ \cdot (\leq) \;\subseteq\; (\sqsubseteq) \cdot g. \tag{11}$$

We show that the first inclusion in (11) is equivalent to $g \subseteq f^\circ \cdot (\leq)$ provided that f is monotonic, that is, $x \sqsubseteq y \Rightarrow f\,x \leq f\,y$, which can be written pointfree as $(\sqsubseteq) \cdot f^\circ \subseteq f^\circ \cdot (\leq)$. That it implies $g \subseteq f^\circ \cdot (\leq)$ is easy to see — since $(\sqsubseteq)$ is a preorder, $g \subseteq id \cdot g \subseteq (\sqsubseteq) \cdot g$. For the other direction, we reason:

$$g \subseteq f^\circ \cdot (\leq)$$
$$\Rightarrow \quad \{ \text{ monotonicity of } (\cdot) \}$$
$$(\sqsubseteq) \cdot g \;\subseteq\; (\sqsubseteq) \cdot f^\circ \cdot (\leq)$$
$$\Rightarrow \quad \{ \text{ assumption: } f \text{ monotonic } \}$$
$$(\sqsubseteq) \cdot g \;\subseteq\; f^\circ \cdot (\leq) \cdot (\leq)$$
$$\Rightarrow \quad \{ \leq \text{ transitive: } (\leq) \cdot (\leq) \subseteq (\leq) \}$$
$$(\sqsubseteq) \cdot g \;\subseteq\; f^\circ \cdot (\leq).$$

Concerning the second condition in (11), by taking converses of both sides and using the function shunting (23) rule, we transforming it to $g \cdot (f^\circ \cdot (\leq))^\circ \subseteq (\sqsupseteq)$. All in all, we have just factored Galois connection (11) into two parts,

$$f^\circ \cdot (\leq) \;=\; (\sqsubseteq) \cdot g \quad \equiv \quad \underbrace{g \subseteq f^\circ \cdot (\leq)}_{\text{"easy"}} \;\wedge\; \underbrace{g \cdot (f^\circ \cdot (\leq))^\circ \subseteq (\sqsupseteq)}_{\text{"hard"}}. \tag{12}$$

uncovering the easy/hard blend which is implicit in the original formulation. To see this, let us first abbreviate $f^\circ \cdot (\leq)$ to R. The left hand operand of the conjunction, $g \subseteq R$, states that g must return a result permitted by R — the "easy" part. The right hand operand $g \cdot R^\circ \subseteq (\sqsupseteq)$, on the other hand, states that if R maps x to y (therefore $(x, y) \in R^\circ$), it must be the case that $g\,x \sqsupseteq y$. That is, g returns a maximum result, under $(\sqsupseteq)$, among those results allowed by R. This is the "hard" part of the connection.

This is in fact nothing surprising: we have merely reconstructed an equivalent definition of a Galois connection [1, Theorem 5.29, page 66]: (1) f is monotonic, (2) $(f \cdot g)\,x \leq x$, (3) $(f\,x) \leq y \Rightarrow x \sqsubseteq (g\,y)$. The calculation above, however, inspires us to capture this pattern by a new relational operator. Given $R :: A \leftarrow B$ and $S :: A \leftarrow A$, define $R \upharpoonright S :: A \leftarrow B$, pronounced "$R$ shrunk by S", by

$$X \subseteq R \upharpoonright S \;\equiv\; X \subseteq R \wedge X \cdot R^\circ \subseteq S, \tag{13}$$

The definition states that X must be at most R, and that if X yields an output for an input x, it must be a maximum, with respect to S, among all possible outputs of x. In terms of the easy/hard split, R is the easy part and S defines the (optimization) criterion to be taken into account in the hard part. Using the properties of relational intersection and division, one may come up with a closed form for $R \upharpoonright S$:

$$R \upharpoonright S \;=\; R \cap S/R^{\circ}. \tag{14}$$

With the new notation we can go back to (12) and rephrase the right hand side of the equivalence in terms of $(\upharpoonright)$:

$$g \subseteq (f^{\circ} \cdot (\leq)) \upharpoonright (\sqsupseteq). \tag{15}$$

3.2 Properties of Shrinking

From the definition (13), it is clear that $R \upharpoonright S \subseteq R$. It is easy to find out under what condition the other direction of inclusion holds: $R \subseteq R \upharpoonright S$ iff $R \cdot R^{\circ} \subseteq S$, and so

$$R = R \upharpoonright S \;\equiv\; img\ R \subseteq S. \tag{16}$$

since $img\ R = R \cdot R^{\circ}$. Since $\top$ is above anything, we have $R \upharpoonright \top = R$, that is, R stays the same if we put no constraints in the "hard" part. When $S = \bot$, no maximum exists, and thus $R \upharpoonright \bot$ yields nothing for any input: $R \upharpoonright \bot = \bot$.

The following rule shows how $(\upharpoonright Q)$ distributes into relational union:

$$(R \cup S) \upharpoonright Q \;=\; ((R \upharpoonright Q) \cap Q/S^{\circ}) \cup ((S \upharpoonright Q) \cap Q/R^{\circ}). \tag{17}$$

This arises from (14) and distribution of intersection over union. A most important consequence of (17) is that $(\upharpoonright Q)$ distributes into joins,

$$[R, T] \upharpoonright S \;=\; [R \upharpoonright S, T \upharpoonright S], \tag{18}$$

— recalling that $[R, S] = (R \cdot inl^{\circ}) \cup (S \cdot inr^{\circ})$ — and therefore conditionals,

$$(p \rightarrow R, T) \upharpoonright S \;=\; (p \rightarrow (R \upharpoonright S), (T \upharpoonright S)). \tag{19}$$

The following two rules allow us to distribute a function in and out of $(\upharpoonright)$:

$$(R \cdot f) \upharpoonright S \;=\; (R \upharpoonright S) \cdot f, \qquad (f \cdot R) \upharpoonright S \;=\; f \cdot (R \upharpoonright (f^{\circ} \cdot S \cdot f)).$$

The first equality can be proved using shunting and indirect equality, while the second generalizes a similar result in [4].

A number of results of the $(\upharpoonright)$ combinator relate to simplicity. Recall that the image of a simple relation R is coreflexive, that is, $img\ R \subseteq id$. Then, from (16) we draw $R = R \upharpoonright S$ if R is simple and S is reflexive, since $img\ R \subseteq id$ and $id \subseteq S$ entail $img\ R \subseteq S$.

Very often, S in (13) is anti-symmetric: $S \cap S^{\circ} \subseteq id$. In this case it can be shown that $R \upharpoonright S$ is always simple [7]. An application of this result concerns (15), ensuring $(f^{\circ} \cdot (\leq)) \upharpoonright (\sqsupseteq)$ simple for $(\sqsupseteq)$ a partial order. Thus equality (9) holds in such a situation.

The special case $S = id$ in (13) deserves some attention. In this situation, each output in the shrunk relation can relate only to itself. Thus $(y, x) \in R \upharpoonright id$ only when y is the sole value that x is mapped to by R. When more than one such y exists, x cannot be in the domain of $R \upharpoonright id$. Therefore, $R \upharpoonright id$ is the largest deterministic fragment of R. Formally,

$$X \subseteq R \upharpoonright id \quad \equiv \quad X \sqsubseteq R \land X \cdot X^{\circ} \subseteq id. \tag{20}$$

where $X \sqsubseteq R$ means $R \cdot dom\ X = X$, that is, X is less defined than R but as non-deterministic as R where defined[6].

4   Inductive Relations

A question was raised in Section 1: where and how does induction crop up in the design of a program? An answer is provided in the remainder of this paper, in two steps. First, we recall that the "natural" way of ordering inductively defined data (such as eg. lists and trees) is through inductive relations defined using well-known combinators of the algebra of programming known as *folds* and *unfolds* [4]. Second, we show how specifications written as GCs on such inductive orderings "naturally" lead to inductive implementations, by calculation.

Functional programmers are familiar with inductive definitions of datatypes such as the natural number $\mathbb{N}$ and finite lists $List\ A$, and the fold function defined on them. The notion can be generalised to relations. For a review, the reader is referred to Appendix B. While functional folds are often used to define operations on inductively defined datatypes, it is often overlooked that many relations between inductively defined data can also be inductively defined as relational folds. The $(\geq)$ ordering on $\mathbb{N}$, for example, is nothing but the *least* relation satisfying

$$x \geq 0 \qquad \land \qquad x \geq y \Rightarrow (x+1) \geq (y+1).$$

The two conditions respectively translate to $\top \cdot zero^{\circ} \subseteq (\geq)$ and $(\geq) \subseteq suc^{\circ} \cdot (\geq) \cdot suc$ in pointfree style. It turns out that $(\geq)$ is a fold:

$$\top \cdot zero^{\circ} \subseteq (\geq) \land (\geq) \subseteq suc^{\circ} \cdot (\geq) \cdot suc$$

$\equiv$ { shunting, since $R \subseteq T \land S \subseteq T \equiv R \cup S \subseteq T$ }

$$(\top \cdot zero^{\circ}) \cup (suc \cdot (\geq) \cdot suc^{\circ}) \subseteq (\geq)$$

$\equiv$ { by (25): $[R, S] \cdot [T, U]^{\circ} = (R \cdot T^{\circ}) \cup (S \cdot U^{\circ})$ }

---

[6] This is the $\vdash_{pre}$ ordering of [17], where it is shown to be a factor of the standard refinement ordering. The proof of (20), omitted for space economy, essentially shows that the right hand sides of (13) and (20) coincide, for $S = id$.

$$[\top, suc \cdot (\geq)] \cdot [zero, suc]^\circ \subseteq (\geq)$$
$$\equiv \quad \{ \text{ absorption (24) } \}$$
$$[\top, suc] \cdot (id + (\geq)) \cdot [zero, suc]^\circ \subseteq (\geq)$$
$$\equiv \quad \{ \text{ (26) } \}$$
$$(\geq) = (\![\top, suc]\!).$$

Note that $(+)$ in the penultimate line denotes the sum functor (see Appendix A) rather than numerical sum.

This not the only way the ordering on natural numbers can be defined, however. If we instead perform case analysis on the lesser side of the ordering, we come up with:

$$0 \leq y \quad \wedge \quad x \leq y \Rightarrow (x+1) \leq (y+1).$$

The first line translates to $zero \cdot \top \subseteq (\leq)$, where $\top$, having type $A \leftarrow \mathbb{N}$, is equivalent to $[zero, suc]^\circ$. By a similar calculation, we come up with a definition of $(\leq)$ as a fold (see Appendix A for the definition of $(\![\_, \_]\!)$):

$$(\leq) = (\![zero, zero \cup suc]\!).$$

Given two finite lists xs and ys, let $xs \sqsubseteq ys$ mean that xs is a prefix of ys. Natural numbers and finite lists are similar in structure and, through a similar calculation, one comes up with the following definition of $(\sqsubseteq)$ as a fold:

$$(\sqsubseteq) = (\![nil, nil \cup cons]\!). \tag{21}$$

Since lists are special cases of binary trees, one can generalise $(\leq)$ and $(\geq)$ to regular datatypes such that trees "grow larger" by substituting of empty nodes to other (sub)trees, and prove generically that $(\leq)^\circ = (\geq)$ (see [14]). The two orderings above are enough for our purposes of showing their role in calculating implementations of adjoints of Galois connections, as is shown in the sequel.

5 Program Calculation by Optimization — "Shrinking Specs into Programs"

Given a Galois connection $f\ x \leq y \equiv x \sqsubseteq g\ y$, recall the conclusion of Section 3.1 that g can be expressed as $g = (f^\circ \cdot (\leq)) \upharpoonright (\sqsupseteq)$. The next step is triggered by a question: what can we do wherever $(\leq)$ and/or $(\sqsubseteq)$ are inductive relations?

In this section we will see two examples that follow a standard scheme we propose: (1) fusion, in the easy part, of the inner ordering $(\leq)$ with f°, to form either a fold or a restricted form of a hylomorphism (a fold followed by the converse of a fold); (2) shrinking the easy part using the hard part ($\upharpoonright(\sqsupseteq)$), hence the *motto*: "shrinking specs into programs".

We present two theorems to perform the shrinking: the *Greedy Theorem*, which applies when the easy part is a fold, and the *Dynamic Programming (DP)*

Theorem, when it is a hylomorphism where the folding phase is a function. The Greedy Theorem is a simplification of that of Bird and de Moor [4]: it does not need a power allegory, and thus is applicable in more categories and, we believe, easier to comprehend. The DP-Theorem is similar to that of Bird and de Moor, with a different precondition, arising from its more general setting.

Both theorems are datatype-generic, and applicable not only for Galois connections, but also for optimisation problems in general. Due to space constraints we are unable to cover this aspect and will defer the discussion to a later work.

5.1   Example of Greedy Programming

Given a predicate p, *takeWhile* p xs yields the longest prefix of xs whose elements all satisfy p:

$$all\ p\ xs\ \wedge\ xs \sqsubseteq ys\quad \equiv\quad xs \sqsubseteq takeWhile\ p\ ys. \tag{22}$$

This expresses a Galois connection between the set of all finite sequences ys and that of the ones (xs) whose elements all satisfy p. The upper adjoint is *takeWhile* p and the lower adjoint is the embedding of all such sequences into the larger set. To see this we rewrite (22) into the pointfree equality

$$map\ p?\cdot(\sqsubseteq)\quad =\quad (\sqsubseteq)\cdot takeWhile\ p$$

by expressing *all p* by coreflexive relation *map p?*. Recall that $(a,a) \in p? \equiv p\ a$. Therefore, $(xs, xs) \in map\ p? \equiv all\ p\ xs$.

Note how *map p?* captures the lower-adjoint of the connection, as it is simple and entire over the set of all sequences satisfying p. Since $(map\ p?)^\circ$ is the same as *map p?* (coreflexives are symmetric) we have that *takeWhile* p can be defined in terms of $(\upharpoonright)$:

$$takeWhile\ p\ =\ (map\ p?\cdot(\sqsubseteq))\upharpoonright(\sqsupseteq).$$

What to do now? If we manage to transform the easy part $map\ p?\cdot(\sqsubseteq)$ into a fold, the following *Greedy Theorem* gives us conditions under which we may promote $(\upharpoonright(\sqsupseteq))$ into a fold:

Theorem 1. $(\!|\,R\upharpoonright S\,|\!) \subseteq (\!|\,R\,|\!)\upharpoonright S$ *if* S *is transitive and* R *is monotonic with respect to* S°, *that is,* $R\cdot \mathsf{F}S^\circ \subseteq S^\circ\cdot R$.

Proof: *see appendix C.*   □

The "monotonic condition" $R\cdot \mathsf{F}S^\circ \subseteq S^\circ\cdot R$ states that if x_1 is no worse than x_2 under S, at least one output of R on x_1 is no worse than any output on x_2. Thus we lose nothing if we compute only the locally optimal answers, that is, doing $(\upharpoonright S)$ in the fold.

Transforming $map\ p?\cdot(\sqsubseteq)$ into a fold turns out to be easy because, as shown in (21), $(\sqsubseteq)$ is already a fold. By a standard fold-fusion we get:

$$map\ p?\cdot(\sqsubseteq) = (\!|\,nil, nil \cup (cons\cdot(p?\times id))\,|\!),$$

that is, in every step we may choose between taking an empty prefix (nil) and, if the current element satisfies p, attach it to the previously computed prefix

$(cons \cdot (p? \times id))$. The monotonicity condition basically says that a longer prefix remains longer after such an operation. Its formal proof makes use of (21), the fact that $(\sqsubseteq)$ is a fold.

By Theorem 1 we may choose $(\![\, [nil, nil \cup (cons \cdot (p? \times id))] \upharpoonright (\sqsupseteq) \,]\!)$ as a candidate for $takeWhile\ p$. By (18), we may distribute $(\upharpoonright (\sqsupseteq))$ into the join. The relation $(nil \cup (cons \cdot (p? \times id))) \upharpoonright (\sqsupseteq)$ returns a longer list whenever possible, that is, whenever the current element satisfies p. Thus the fold refines to $(\![\, nil, ((p \cdot fst) \to nil, cons) \,]\!)$, which translates to the usual definition of $takeWhile$:

$$
\begin{aligned}
takeWhile\ p\ [] \quad &=\ [] \\
takeWhile\ p\ (x : xs)\ \mid\ p\ x\ &=\ x : takeWhile\ p\ xs \\
\mid\ otherwise\ &=\ [].
\end{aligned}
$$

5.2   Example of DP-Programming

Given GC (3) between $(\times y)$ and $(\div y)$, this can be expressed in terms of $(\upharpoonright)$:

$$(\div y)\ =\ ((\times y)^{\circ} \cdot (\leq)) \upharpoonright (\geq).$$

To calculate $(\div y)$, one may proceed the same way as in the previous section and fuse $(\times y)^{\circ}$ into $(\leq)$ to form a fold, and attempt to apply Theorem 1. This time, however, we can not prove the monotonicity condition. Fortunately, for this and many other examples, the following Dynamic Programming Theorem applies:

Theorem 2. $\mu(\lambda X \to (in \cdot \mathsf{F}X \cdot T^{\circ}) \upharpoonright S)\ \subseteq\ (\![\, T \,]\!)^{\circ} \upharpoonright S$ *if in is monotonic with respect to* S, *that is,* $in \cdot \mathsf{F}S \subseteq S \cdot in$, *and* $dom\ T \subseteq dom\ \mathsf{F}((\![\, T \,]\!)^{\circ} \upharpoonright S)$.

Proof: *see [14].*   □

The notation μf denotes the least fixed point of f. The constructor in can in fact be an arbitrary function, a generalisation we do not need here.

To apply Theorem 2, we aim at turning $(\times y)^{\circ} \cdot (\leq)$ to converse of a fold or, equivalently, turning $(\geq) \cdot (\times y)$ into a fold. It is known that $(\times y) = (\![\, zero, (+y) \,]\!)$: starting with 0, and add y in each step. By fold fusion, we get $(\geq) \cdot (\times y) = (\![\, \top, (+y) \,]\!)$: the base case can be any number.

The monotonicity condition in Theorem 2 instantiates to: $[zero, suc] \cdot (id + (\geq)) \subseteq (\geq) \cdot [zero, suc]$, which can be proved formally using the definition of $(\geq)$ as a fold. Theorem 2 is thus applicable and we get:

$$\mu(\lambda X \to ([zero, suc] \cdot (id + X) \cdot [\top, (+y)]^{\circ}) \upharpoonright (\geq))\ \subseteq\ (\![\, zero, (+y) \,]\!)^{\circ} \upharpoonright (\geq).$$

Denote $(+y)^{\circ}$, a partial function that applies only to input no less than y, by $(-y)$, and note that $zero \cdot \top = zero$. By (25), the left hand side simplifies to $\mu(\lambda X \to (zero \cup (suc \cdot X \cdot (-y))) \upharpoonright (\geq))$. It is a recursive definition where, in every step, we may choose to simply return 0 or, if possible, subtract y from the input and add 1 to the recursively computed result.

We have yet to simplify $(zero \cup (suc \cdot X \cdot (-y))) \upharpoonright (\geq)$. Not having space for the formal detail, we simply note here that since the result of $suc \cdot R$ is strictly larger than 0, to maximise the output, we shall just choose the right branch whenever possible, that is, when the input is no less than y. This results in the usual program for division:

$$\begin{aligned} x \div y \mid x \geq y &= 1 + ((x - y) \div y) \\ \mid \mathbf{otherwise} &= 0. \end{aligned}$$

6   Case Study: Scheduling as a Galois Connection

As our closing case study, we will be looking at a more complex problem related to task scheduling. The full detail cannot be covered in this paper, and we will be proceeding in a less formal manner, sketching only an outline of the development.

Let A be a set of tasks, and let $g :: Gantt = \mathsf{P}A \leftarrow A$ such that for each $x \in A$, $g\ x$ is the set of tasks that have to wait for x to complete before commencing, while the spans, time need by each task, is given by a function $Spans = \mathbb{N} \leftarrow A$ where $\mathbb{N}$ models discrete time intervals (eg. days, months). $Gantt$ and $Spans$ form an acyclic graph, known as a $Gantt\ graph$, coined after Henry Gantt (1861-1919) who introduced them. A time schedule associating starting times to tasks (optimal or not), is also modelled by a function of type $Schedule = \mathbb{N} \leftarrow A$. The types $Spans$ and $Schedule$ will be refined later. We use variables sp for $Spans$, sh for $Schedule$, x, y, etc. for tasks, and s, t for time.

Given $g :: Gantt$, the goal is to calculate a function $bsch_g :: Schedule \leftarrow Spans$ that computes the "best" schedule for the tasks — "best" in the sense that tasks start as early as possible. Take, for instance, $A = \{a, b, c, d\}$, for task spans $sp = \{(1, a), (5, b), (10, c), (20, d)\}$ and graph $g = \{(\{b\}, a), (\{c\}, b), (\{\}, c), (\{c\}, d)\}$, the best schedule will be $bsch_g\ sp = \{(0, a), (1, b), (20, c), (0, d)\}$.

How do we specify $bsch_g$? Note that "best" means smallest and that $bsch_g\ sp$ should be monotonic in both arguments: more dependencies in g and/or longer tasks in sp can only defer tasks start-up times into the future. This suggests specifying $bsch_g$ as adjoint of a Galois connection between schedules and spans. Let $lazy_g :: Spans \leftarrow Schedule$ be a function that, given a schedule, computes for each task the maximum time it is allowed to take (hence the name), we have

$$lazy_g\ sh \overset{.}{\geq} sp \;\equiv\; sh \overset{.}{\geq} bsch_g\ sp,$$

where $(\overset{.}{\geq})$ denotes $(\geq)$ lifted to functions: $f \overset{.}{\geq} h \equiv \langle \forall x : x \in A : f\ x \geq h\ x \rangle$.

The function $lazy_g$ appears to be easier to define than $bsch_g$. In the definition below, $(t \hookleftarrow x) \uplus sh$ denotes a function sh, whose domain does not include x, extended with a mapping from x to t.

$$\begin{aligned} lazy_g\ \{\} &= \{\} \\ lazy_g\ ((t \hookleftarrow x) \uplus sh) \mid g\ x \subseteq dom\ sh &= (s \hookleftarrow x) \uplus sp \\ \mathbf{where}\ sp &= lazy_g\ sh \\ s &= \sqcap\{sh\ y \mid y \in g\ x\} - t. \end{aligned}$$

The $\sqcap$ operator in the non-empty case takes the minimum of a set, thus the span allowed for each task x is the difference between the earliest scheduled time among tasks that follow x and t, the time scheduled for x. The non-deterministic pattern $(t \leftarrow x) \uplus sh$ does not explicitly specify an order in which tasks are picked. However, the guard $g\ x \subseteq dom\ sh$, needed because we want to look up all the y's in sh, implicitly enforces the topological order — x is processed before all tasks that depend on it. Equivalently, we could have treated the schedule as a list of pairs sorted in topological order: $Schedule = Spans = [(\mathbb{N}, A)]$. One may thus drop the domain check and come up with the following definition for $lazy_g$:

$$
\begin{aligned}
lazy_g\ [] \quad &= []\\
lazy_g\ ((t,x) : sh) &= (s, x) : lazy_g\ sh\\
\textbf{where}\ s &= \sqcap\{sh\ y \mid y \in g\ x\} - t.
\end{aligned}
$$

For brevity we still use the syntax $sh\ y$ for looking up.

To calculate $bsch_g = ((lazy\ g)^\circ \cdot (\dot{\geq})) \restriction (\dot{\leq})$, we have to construct the converse of $lazy_g$. Consider, in $s = \sqcap\{sh\ y \mid y \in g\ x\} - t$, what t could be given s and x. If $g\ x$ is empty, $s = \infty$, and t could be any finite value. With $g\ x$ non-empty, we have $t = \sqcap\{sh\ y \mid y \in g\ x\} - s$. However, $t :: \mathbb{N}$ must be non-negative. So we are putting an constraint on sh: $\sqcap\{sh\ y \mid y \in G\ x\}$ must be no smaller than s. That gives us a very non-deterministic program for $(lazy_g)^\circ$: we go through the graph in topological order until we reach a task say y, for which $g\ y$ is empty, guess a possible time to schedule it, and go back to some task x that must be done before y. If y is scheduled late enough that x can finish, that's fine. Otherwise this trial fails and we backtrack.

We can refine $(lazy_g)^\circ$ to a more deterministic program that explicitly pass the constraint $\sqcap\{sh\ y \mid y \in g\ x\} \geq s$ down through the recursive calls, so that the choice of t for when $g\ x = \{\}$ is guaranteed to be late enough. We use an extra argument, a mapping from tasks to time, that records the earliest time each task must be scheduled. Initially it is all zero, meaning that there is no constraint yet: $(lazy_g)^\circ\ sp = sche_g\ (sp, \{(z,0) \mid z \in dom\ sp\})$. In pointfree style, let $init\ sp = (sh, \{(z,0) \mid z \in dom\ sp\})$, we have $(lazy\ g)^\circ = sche_g \cdot init$.

The main computation happens in $sche$, the name suggesting that it returns a scheduling, but not always the best one. It can be defined as:

$$
\begin{aligned}
sche_g\ ([], \_) \quad &= []\\
sche_g\ ((s,x) : sp, c) &= (t, x) : sche_g\ (sp, c')\\
\textbf{where}\ t &= \textbf{if}\ null\ (g\ x)\ \textbf{then}\ (\text{something no less than } c\ x)\ \textbf{else}\ c\ x\\
c'\ y &= \textbf{if}\ y \notin g\ x\ \textbf{then}\ c\ y\ \textbf{else}\ (t+s) \sqcup (c\ y).
\end{aligned}
$$

This is an unfold, that is, converse of a fold, on lists. In each step, the next task in topological order is scheduled, and the constraint set c is updated to c' to schedule the rest of the tasks.

Now that we have $bsch_g = (sche_g \cdot init \cdot (\dot{\geq})) \restriction (\dot{\leq})$, the next steps are to fuse $(\dot{\geq})$ into $sche_g \cdot init$ to form an unfold, and to promote $(\restriction(\dot{\leq}))$ into the unfold. Fusing $(\dot{\geq})$ with $sche_g$ merely makes the value of t more non-deterministic: we are left with only $t \geq c\ x$. To promote $(\restriction(\dot{\leq}))$ we need a theorem related to

Theorem 2 that needs a stronger antecedent. It confines the value of t to the smallest possible: $c\ x$. The development concludes with the following program:

$$
\begin{aligned}
bsch_g\ ([\,],\_) \quad &= [\,] \\
bsch_g\ ((s,x):sp,c) &= (t,x):bsch_g\ (sp,c')
\end{aligned}
$$
$$
\textbf{where } c'\ y = \textbf{if } y \notin g\ x \textbf{ then } c\ y \textbf{ else } (c\ x + s) \sqcup (c\ y).
$$

7   Conclusions and Future Work

Poor scalability is often pointed out as a problem of the mathematics of program construction. By contrast, Galois connections are a well-known example of mathematical device which scales up from trivial to complex problem domains. The research programme which embodies this paper starts from the conjecture that the latter could help the former to scale up.

In this context, "programming from Galois connections" is proposed as a way of calculating programs from specifications which take the form of Galois connections. This (emerging) discipline is beneficial in several respects. In particular, the specification of a "hard" operation as adjoint of a GC provides early insight on its properties, well before the actual implementation is derived. This is granted by the rich algebra of GCs, which compose which each other in several ways (thus growing larger and larger) and offer a powerful framework for reasoning about suprema without making these explicit in the calculations.

It should be noted that Galois connections are ubiquitous in mathematics and computer science [13]. In the latter case, they have been shown to offer a powerful way to structure the allegory calculus of Freyd and Ščedrov [8,4], of which Tarski's relation algebra may in retrospect be seen as an instance [19]. Several examples of such GCs are given in the current paper (see eg. [1,6,15] for a detailed account). At the other side of the spectrum, GCs have even been proposed (together with the principle of indirect equality) as the building block of a new brand of theorem provers [18].

In this context, the main contribution of the current paper is to be found in the proposed process of deriving, using the algebra of programming [4], the algorithmic implementation of Galois adjoints, expressed in closed formulæ which record what is "easy" and "hard" to implement. However, instead of resorting to explicit, point-level suprema, as is usual in textbooks, a new pointfree relational combinator (named *shrinking*) is proposed. Thanks to the rich algebra of this combinator, already sketched in [7], one is able to express and generalize previous results on dynamic and greedy programming [4], in a way which dispenses with the heavy artillery of power-allegories [8]. Such results thus become accessible to a wider audience and easier to apply.

The *whole division* example provides a measure of progress: the *verification* of a given algorithm against the given GC (3), carried out in [18], gives place in the current paper to its *construction* from the connection itself.

So much for *pros*. Future work is concerned with a number of *cons*, namely the fact that not every problem casts into a GC. The typical counter-example arises from the (false) lower adjoint being an embedding (or even the identity) and lacking monotonicity.

Still on the negative side, we feel that the conceptual economy of the overall approach is still unmatched by the effort needed to carry out particular examples. A body of knowledge around these results needs to be developed, structured in corollaries, special cases, etc. The general result concerning checking monotonicity in the side conditions of Theorems 1 and 2 given in [14] is an example of what is required.

Last but not least, we find that the *shrinking* combinator has a lot more to offer to algorithmic refinement, in particular with respect to its two-dimensional factorization: either increasing definition or reducing non-determinism [17]. As discussed in Section 3.1, $R \upharpoonright id$ is the largest deterministic fragment of a specification R, that is, that part of R which cannot be further refined. So, in a sense, all effort should go into refining the complement of $R \upharpoonright id$ with respect to R. Embodying this intuition in the greedy and dynamic programming theorems is clearly a subject for future research.

Acknowledgements. Special thanks go to Roland Backhouse for spotting the Galois connection of *take*, which triggered talk [16] and interesting discussions at IFIP WG2.1 thereupon. Thanks are also due to Jeremy Gibbons for his comments on an earlier draft of this paper.

This research was partly supported by the MONDRIAN Project funded by the Portuguese NSF under contract PTDC/EIA-CCO/108302/2008.

References

1. Aarts, C., Backhouse, R., Hoogendijk, P., Voermans, E., van der Woude, J.: A relational theory of datatypes (December 1992), http://www.cs.nott.ac.uk/~rcb
2. Backhouse, R.: Chapter 4 Galois connections and fixed point calculus. In: Blackhouse, R., Crole, R.L., Gibbons, J. (eds.) Algebraic and Coalgebraic Methods in the Mathematics of Program Construction. LNCS, vol. 2297, pp. 89–148. Springer, Heidelberg (2002)
3. Backhouse, R.: Program Construction: Calculating Implementations from Specifications. John Wiley & Sons, Inc., New York (2003)
4. Bird, R., de Moor, O.: Algebra of Programming. Series in Computer Science. Prentice-Hall International, Englewood Cliffs (1997) C.A.R. Hoare, series editor
5. Dijkstra, E.: A Discipline of Programming. Prentice-Hall, Englewood Cliffs (1976)
6. Doornbos, H., Backhouse, R., van der Woude, J.: A calculational approach to mathematical induction. Theor. Comp. Science 179(1-2), 103–135 (1997)
7. Ferreira, M., Oliveira, J.: Variations on an Alloy-centric tool-chain in verifying a journaled file system model. Technical Report DI-CCTC-10-07, Univ. of Minho (January 2010)
8. Freyd, P., Scedrov, A.: Categories, Allegories, Mathematical Library, vol. 39. North-Holland, Amsterdam (1990)
9. Jackson, D.: Software Abstractions: Logic, Language, and Analysis, 9th edn. The MIT Press, Cambridge (2006) ISBN 0-262-10114-9
10. Jones, C.: Software Development — A Rigorous Approach. Prentice-Hall International, Englewood Cliffs (1980)

11. Knuth, D.: The Art of Computer Programming, 2nd edn. Addison/Wesley, Amsterdam (1997)
12. Kramer, J.: Is abstraction the key to computing? Commun. ACM 50(4), 37–42 (2007)
13. Melton, A., Schmidt, D.A., Strecker, G.E.: Galois connections and computer science applications. In: Poigné, A., Pitt, D.H., Rydeheard, D.E., Abramsky, S. (eds.) Category Theory and Computer Programming. LNCS, vol. 240, pp. 299–312. Springer, Heidelberg (1986)
14. Mu, S.C., Oliveira, J.: Programming from Galois Connections — Principles and Applications. Tech. Report TR-IIS-10-009, Academia Sinica (December 2010)
15. Oliveira, J.: Extended static checking by calculation using the pointfree transform. In: Bove, A., Barbosa, L.S., Pardo, A., Pinto, J.S. (eds.) Language Engineering and Rigorous Software Development. LNCS, vol. 5520, pp. 195–251. Springer, Heidelberg (2009)
16. Oliveira, J.: A Look at Program "Galculation". Presentation at the IFIP WG 2.1 #65 Meeting (January 2010)
17. Oliveira, J.N., Rodrigues, C.J.: Pointfree factorization of operation refinement. In: Misra, J., Nipkow, T., Karakostas, G. (eds.) FM 2006. LNCS, vol. 4085, pp. 236–251. Springer, Heidelberg (2006)
18. Silva, P., Oliveira, J.: 'Galculator': functional prototype of a Galois-connection based proof assistant. In: PPDP 2008, pp. 44–55. ACM, New York (2008)
19. Tarski, A., Givant, S.: A Formalization of Set Theory without Variables, vol. 41. A. M. Society, AMS Colloquium Publications (1987)
20. Ullman, J.: Principles of Database Systems. Computer Science Press, Rockville (1981)
21. Wirth, N.: Algorithms + Data Structures = Programs. Prentice-Hall, Englewood Cliffs (1976)

A   Relational Calculus

Relations. A relation R from set B to set A, written $R :: A \leftarrow B$, is a subset of the set $\top = \{(a,b) \mid a \in A \wedge b \in B\}$. When $(a,b) \in R$, we say R maps b to a. Set operations such as union, intersection, etc., apply to relations as well. The largest relation (with respect to set inclusion ($\subseteq$)) of its type is $\top$, while the empty relation is denoted by $\bot$. Given $R :: A \leftarrow B$ and $S :: B \leftarrow C$, their composition $R \cdot S :: A \leftarrow C$ is defined by:

$$(a,c) \in (R \cdot S) \;\equiv\; \langle \exists b \;::\; (a,b) \in R \;\wedge\; (b,c) \in S \rangle.$$

Composition is monotonic with respect to ($\subseteq$). The identity relation $id_A :: A \leftarrow A$ defined by $\langle \forall a \; : \; a \in A : \; (a,a) \in id_A \rangle$ is the unit of composition. We often omit the subscript when it is clear from the context. Given a relation $R :: A \leftarrow B$, its *converse* $R^\circ :: B \leftarrow A$ is defined by $(b,a) \in R^\circ \equiv (a,b) \in R$.

A relation that is a subset of id is said to be *coreflexive*, often used to filter results satisfying certain conditions. Given a predicate p, the coreflexive relation $p?$ is defined by: $(a,a) \in p? \;\equiv\; p\,a$. The domain and range of a relation R are given respectively by $dom\ R = id \cap (R^\circ \cdot R)$ and $ran\ R = id \cap (R \cdot R^\circ)$. A relation R is said to be (1) *simple*, if $(a,b) \in R$ and $(a',b) \in R$ implies $a = a'$, or $R \cdot R^\circ \subseteq id$; (2) *entire*, if every $b \in B$ is mapped to some a, or $id \subseteq R^\circ \cdot R$. A (total) function is a relation that is both simple and entire. As a convention, single small-case letters refer to functions. One nice property of functions is that inclusion equivalues equality: $f \subseteq g \equiv f = g$. The following *shunting* rules allows us to move functions to the other side of inclusion:

$$f \cdot R \subseteq S \equiv R \subseteq f^\circ \cdot S, \qquad R \cdot f^\circ \subseteq S \equiv R \subseteq S \cdot f. \qquad (23)$$

The relation $R \cdot R^\circ$ is called the *image* of R, denoted by $img\ R$.

Given $R :: A \leftarrow B$, $S :: B \leftarrow C$, and $T :: A \leftarrow C$, the relation $T/S :: A \leftarrow B$ is defined by the Galois connection:

$$R \cdot S \subseteq T \equiv R \subseteq T/S.$$

If $(\cdot S)$ is like multiplication, $(/S)$ is like division: T/S is the largest relation such that $T/S \cdot S \subseteq T$.

Relators, Sum, and Product. A *relator* is an extension of a *functor* in category theory. For the purpose of this paper it suffices to know that a relator F consists of an operation on types that takes a type A to another type FA, and an operation on relations, denoted by the same symbol F, that takes $R :: A \leftarrow B$ to $FR :: FA \leftarrow FB$. A relator is supposed to preserve identity ($F id_A = id_{FA}$) and composition ($FR \cdot FS = F(R \cdot S)$), and is monotonic with respect to ($\subseteq$) ($R \subseteq S \Rightarrow FR \subseteq FS$). The unit relator **1** takes any type to the unit type (with one element denoted by ()), and any relation to id.

A bi-relator is a relator generalised to having two arguments. We will need two bi-relators: sum ($+$) and product ($\times$). For ($\times$), the operation on types is the Cartesian product $A \times B$, defined by $\{(a, b) \mid a \in A \wedge b \in B\}$. The projections are $fst\ (a, b) = a$ and $snd\ (a, b) = b$. Given $R :: A \leftarrow C$ and $S :: B \leftarrow C$, the "split" $\langle R, S \rangle :: (A \times B) \leftarrow C$ is defined by:

$$((a, b), c) \in \langle R, S \rangle \equiv (a, c) \in R \wedge (b, c) \in S.$$

Equivalently, $\langle R, S \rangle = (fst^\circ \cdot R) \cap (snd^\circ \cdot S)$. The operation on relations is defined using split:

$$(R \times S) = \langle R \cdot fst, S \cdot snd \rangle.$$

Functional programmers may be more familiar with the special case for functions: $\langle f, g \rangle\ a = (f\ a, g\ a)$, and $(f \times g)\ (a, b) = (f\ a, g\ b)$.

The disjoint sum of two sets A and B is defined by $A + B = \{inl\ a \mid a \in A\} \cup \{inr\ b \mid b \in B\}$, with inl and inr being two injections. Given two relations $R :: A \leftarrow B$ and $S :: A \leftarrow C$, their "join" $[R, S] :: A \leftarrow (B + C)$ is defined by:

$$(a, inl\ b) \in [R, S] \equiv (a, b) \in R \qquad (a, inr\ c) \in [R, S] \equiv (a, c) \in S.$$

Equivalently, $[R, S] = (R \cdot inl^\circ) \cup (S \cdot inr^\circ)$. This gives rise to the relator operation on relations:

$$R + S = [inl \cdot R, inr \cdot S].$$

Note the symmetry between the definitions for sum and product. We will often need this absorption law:

$$[R, S] \cdot (T + U) = [R \cdot T, S \cdot U]. \tag{24}$$

One of the applications of the join is to define the branching operator $(P \rightarrow R, S)$, corresponding to the **if** P **then** R **else** S construct in many programming languages:

$$
\begin{aligned}
(p \rightarrow R, S) &= [R, S] \cdot ((inl \cdot p?) \cup (inr \cdot (\neg p)?)) \\
&= (R \cdot p?) \cup (S \cdot (\neg p)?).
\end{aligned}
$$

More generally, a common programming pattern is to use the converse of a join $[T, U]^\circ = (inl \cdot T^\circ) \cup (inr \cdot U^\circ)$ to simulate possibly non-deterministic case analysis, and process the two cases by another join. In such situations the following rule comes in handy:

$$[R, S] \cdot [T, U]^\circ = (R \cdot T^\circ) \cup (S \cdot U^\circ). \tag{25}$$

B   Inductively Defined Datatypes and Catamorphisms

Inductively defined datatypes. Natural numbers are often inductively defined to be the smallest set $\mathbb{N}$ such that (a) $0 \in \mathbb{N}$; (b) if $n \in \mathbb{N}$, so does $1 + n$. Let $\mathsf{F}_\mathbb{N}$ be a function from sets to sets defined by $\mathsf{F}_\mathbb{N} X = \{0\} \cup \{1 + n \mid n \in X\}$. The two conditions together are equivalent to saying that $\mathsf{F}_\mathbb{N} \mathbb{N} \subseteq \mathbb{N}$, and the requirement that $\mathbb{N}$ being the smallest means that $\mathbb{N}$ is the *least prefix-point*, and also the *least fixed-point* of $\mathsf{F}_\mathbb{N}$[7].

If we abstract over 0 and $(1+)$, representing them respectively by inl () and inr, F can be expressed as the type operation of relator $\mathsf{F}_\mathbb{N} X = \mathbf{1} + X$, where $\mathbf{1}$ is the unit type. Letting $in_\mathbb{N} :: \mathbb{N} \leftarrow \mathsf{F}_\mathbb{N}\mathbb{N}$ be the isomorphism between $\mathsf{F}_\mathbb{N}\mathbb{N}$ and $\mathbb{N}$, the successor function $(1+)$ can be encoded by $suc = in_\mathbb{N} \cdot inr$. The number 0 is encoded by $in_\mathbb{N}$ (inl ()). In calculations, however, we often find the constant function $zero = in_\mathbb{N} \cdot inl \cdot \top$ (that always yields 0 for any input) more useful.

Many inductively defined datatypes can be encoded this way. A finite list of elements of type A, for example, can be defined as the least fixed-point of $\mathsf{F}_{List} X = \mathbf{1} + A \times X$, with constructors $nil :: List\ A \leftarrow B$ defined by $in_{List} \cdot inl \cdot \top$ and $cons :: List\ A \leftarrow (A \times List\ A)$ by $in_{List} \cdot inr$. The type of leaf-valued binary trees, as defined in Haskell notation by **data** $Tree\ A = Tip\ A \mid Bin\ (Tree\ A)\ (Tree\ A)$, is the least fixed-point of $\mathsf{F}_{Tree} X = A + X \times X$.

Catamorphisms. To design programs on these inductively defined datatypes, one is often encouraged to define them over the input inductive structure. The so-called *catamorphism*, also known as *fold*, is one such useful pattern of induction.

Folds exist for all datatypes defined as least fixed-points of so-called *regular* relators: those defined in terms of $\mathbf{1}$, $(+)$, $(\times)$, constants, and type relators. Let

---

[7] For f monotonic on $(\leq)$, x is a prefix-point of f if $f\ x \leq x$, and a fixed-point if $f\ x = x$. The least prefix-point is also the least fixed-point [2].

T denote the least fixed-point of the type operation of relator F. Given a relation $R :: B \leftarrow \mathsf{F}B$, the catamorphism $(\!(R)\!)_\mathsf{F} :: B \leftarrow \mathsf{T}$ is the least prefix point, and also the least fixed-point, of $\lambda X \to R \cdot \mathsf{F}X \cdot in_\mathsf{T}^\circ$. Thus it is the least relation satisfying:

$$(\!(R)\!)_\mathsf{F} \supseteq R \cdot \mathsf{F}(\!(R)\!)_\mathsf{F} \cdot in_\mathsf{T}^\circ, \tag{26}$$
$$(\!(R)\!)_\mathsf{F} = R \cdot \mathsf{F}(\!(R)\!)_\mathsf{F} \cdot in_\mathsf{T}^\circ.$$

Take $\mathsf{F}X = 1 + A \times X$ as an example, and note that every relation $R :: B \leftarrow (1 + A \times B)$ can be factored to $[R_1, R_2]$ with $R_1 :: B \leftarrow 1$ and $R_2 :: B \leftarrow (A \times B)$. By taking $in_\mathsf{T} = [nil, cons]$ and instantiating R_1 and R_2 respectively to a constant and a function we recover *foldr* above.

The fold fusion rule is one of the most important properties of folds:

$$(\!(T)\!) \subseteq S \cdot (\!(R)\!)_\mathsf{F} \quad \Leftarrow \quad T \cdot \mathsf{F}S \subseteq S \cdot R.$$

It states conditions under which we may promote relations into the body of the fold.

C   Proof of Theorem 1

Proof:

$$(\!(S \upharpoonright R)\!) \subseteq (\!(S)\!) \upharpoonright R$$
$\equiv$ { universal property of ($\upharpoonright$) }
$$(\!(S \upharpoonright R)\!) \subseteq (\!(S)\!) \ \wedge \ (\!(S \upharpoonright R)\!) \cdot (\!(S)\!)^\circ \subseteq R$$
$\equiv$ { monotonicity of $(\!(\_)\!)$ and $X \upharpoonright R \subseteq R$ }
$$(\!(S \upharpoonright R)\!) \cdot (\!(S)\!)^\circ \subseteq R$$
$\equiv$ { hylomorphism: $(\!(R)\!) \cdot (\!(S)\!)^\circ = \langle \mu X :: R \cdot \mathsf{F}X \cdot S^\circ \rangle$ }
$$\langle \mu X :: (S \upharpoonright R) \cdot \mathsf{F}X \cdot S^\circ \rangle \subseteq R$$
$\Leftarrow$ { least prefix point }
$$(S \upharpoonright R) \cdot \mathsf{F}R \cdot S^\circ \subseteq R$$
$\Leftarrow$ { monotonic condition: $S \cdot \mathsf{F}R^\circ \subseteq R^\circ \cdot S$ }
$$(S \upharpoonright R) \cdot S^\circ \cdot R \subseteq R$$
$\Leftarrow$ { since $S \upharpoonright R \subseteq R/S^\circ$ }
$$(R/S^\circ) \cdot S^\circ \cdot R \subseteq R$$
$\Leftarrow$ { division: $R/S \cdot S \subseteq R$ }
$$R \cdot R \subseteq R$$
$\equiv$ { R transitive }
$$true. \qquad \square$$

Constructions around Partialities

Gunther Schmidt

Fakultät für Informatik, Universität der Bundeswehr München
85577 Neubiberg, Germany
gunther.schmidt@unibw.de

Abstract. That matrices of relations also obey the rules of relation algebra is well known. When a suitable ordering relation is given, partialities may be conceived as their lattice-continuous mappings — corresponding to existential images which are often studied independently. Matrices of partialities would considerably improve the possibility to study non-strictness, streams, partial evaluation, and net properties in a compact relation-algebraic form. They seem, however, to lead inevitably to some borderline cases as the Boolean lattice $\mathbb{B}^0$ and row-less matrices, that shall here be dealt with.

Keywords: relation, existential image, partiality.

1 Introduction

It was a much-remembered achievement of mankind when in Indian mathematics the concept of "0" was introduced. Since this point in time, one was able to use number representations by position that greatly enhanced computational abilities. However, one had also to tolerate several not immediately intuitive agreements such as $0! = 1$ for the factorial function, or $2^0 = 1$. Far from trying to measure up with this ancient achievement, we will here experiment with something similar.

When we study row- as well as column-less relations as similar borderline cases, we have certain aims in mind. Observing very big systems in full detail is principally impossible. We should not pretend to be able to snapshot the global telephone net, e.g. What we may be able to achieve is getting snapshots of rather small parts of a big system while ignoring the rest or temporarily ignoring connections with the rest. These observable parts are usually not fixed in advance, may vary over time, and observation of one and the same part may not proceed continuously and may be taken up again later.

The aim of this paper is to investigate the interplay between relations and partialities, the latter based on [8]. This article is a report on intermediate results of ongoing research. For reasons of space we cannot present all the details of the intended applications and restrict to the following hint on synoptic regions.

2 Synoptic Regions

We assume that a part of a system is identified, its connections to the rest are cut off in order to be able to observe it or to work on it. When we manage to

H. de Swart (Ed.): RAMICS 2011, LNCS 6663, pp. 314–330, 2011.

single out such a region and to handle it in full detail, we will call it a *synoptic region*. After working or observing has taken place, it is re-connected again. So we have a situation that has often been described operationally with semaphores, Petri nets, or commit/rollback structures in data bases. One may also think of a leaking hot-water pipe in a heating system to be repaired: The plumber will look for positions around the leak and will deep-freeze these points during repair. Such synoptic regions will in retrospect often overlap. Observed locally, however, every item will be set out to a *linear stream of (local) time* of belonging to varying synoptic regions.

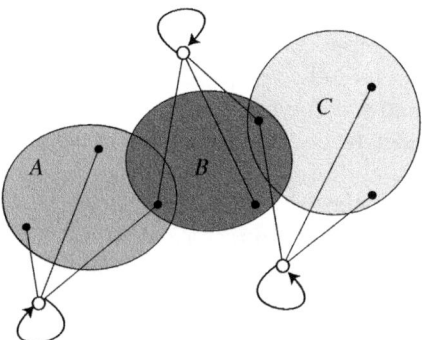

Fig. 1. Synoptic regions

The important point to observe is that the transactions taking place in these regions leave their marks when terminated so that potential later transactions in the same or in overlapping regions find these and start from such results. It is thus important that we learn to work with only temporarily cooperating regions. In particular, we have to invent some measure to cope with being non-connected.

We start giving a rather informal and naïve idea of overlapping synoptic regions. Regions A, C of Fig. 1 may work completely independently. Should also B be observed, there are many conceivable local executing sequences in the overlap parts. The loops indicate where (relational) operations may then take place and where results of such actions are left as marks in the intersections. For A, transactions in C will be *hidden transactions* in the sense of π-calculus, e.g. While in π-calculus, they are just hidden, we here aim at propagating their effects as with the — also local — atlases in algebraic topology.

With synoptic regions we aim at a deeper concept behind which allows a detailed analysis of strictness and non-strictness. *Fully observable* snapshots or transactions may only take place in synoptic regions.

But how do we manage to combine this elegantly in relation-algebraic form not using total observation? Can we maintain relation algebra even in this case of being non-connected?

It is a different topic to discuss how synoptic regions are singled out temporarily. When semaphores have been used in earlier programming concepts, it was understood that they would be handled appropriately on the compiler level. In other cases, one will use differently advanced technologies working in micro seconds as opposed to milli seconds. Scaling down observational devices is a sliding process. Concerning orders of magnitude, we started at milli, have passed micro, are currently at nano for electronic devices. Cutting out is assumed to occur in the significantly faster mode guaranteeing non-interference.

3   Relation-Algebraic Preliminaries

Since we cannot present all the prerequisites on relation algebra, we give [10], [11], [9] as a general reference. We write $R : V \longrightarrow W$ if R is a relation with source V and target W, often conceived as a subset of $V \times W$. If the sets V and W are finite and of size m and n, respectively, we may consider R as a Boolean matrix with m rows and n columns.

We assume the reader to be familiar with the basic operations on relations, namely R^{T} (*converse*), $\overline{R}$ (*negation*), $R \cup S$ (*union*), $R \cap S$ (*intersection*), and $R \,;\, S$ (*composition*), the predicate $R \subseteq S$ (*containment*), and the special relations[1] $\perp\!\!\!\perp$ (*empty relation*), $\mathbb{T}$ (*universal relation*), and $\mathbb{I}$ (*identity relation*).

We assume that a *heterogeneous relation algebra* is a structure that

— is a category with respect to composition " $;$ " and identities $\mathbb{I}$,
— has complete atomic Boolean lattices with $\cup, \cap, \,^{-}, \perp\!\!\!\perp, \mathbb{T}, \subseteq$ as morphism sets,
— obeys rules for transposition in connection with the latter two that may be stated in either one of the following two ways:

$$\text{Dedekind} \quad R\,;S \cap Q \subseteq (R \cap Q\,;S^{\mathsf{T}})\,;(S \cap R^{\mathsf{T}}\,;Q) \qquad\qquad \text{or}$$

$$\text{Schröder} \quad R\,;S \subseteq Q \iff R^{\mathsf{T}}\,;\overline{Q} \subseteq \overline{S} \iff \overline{Q}\,;S^{\mathsf{T}} \subseteq \overline{R}.$$

Residuals are often introduced via $A\,;B \subseteq C \iff A \subseteq \overline{\overline{C}\,;B^{\mathsf{T}}} =: C/B$. Intersecting such residuals with $\operatorname{syq}(R,S) := \overline{R^{\mathsf{T}}\,;\overline{S}} \cap \overline{\overline{R}^{\mathsf{T}}\,;S}$, the *symmetric quotient* $\operatorname{syq}(R,S) : W \longrightarrow Z$ of two relations $R : V \longrightarrow W$ and $S : V \longrightarrow Z$ is introduced. Given an ordering relation E and some set or vector U, one may determine the least upper bound $\operatorname{lub}_E(U)$; see, e.g., [10,11,9]. Given a relation X, it is also possible to form $\operatorname{lubR}_E(X) := \left[\operatorname{lub}(X^{\mathsf{T}})\right]^{\mathsf{T}}$, i.e., obtain the least upper bound row-wise.

We will use *membership-relations* $\varepsilon : V \longrightarrow \mathcal{P}(V)$ between a set V and its powerset $\mathcal{P}(V)$ that can be characterized algebraically via the symmetric quotient. With a membership relation the powerset ordering is easily described as $\Omega = \overline{\varepsilon^{\mathsf{T}}\,;\overline{\varepsilon}}$. Explicit examples will be provided below.

4   Two Elementary Constructions

We approach the study of synoptic regions with several construction techniques. The first construction will allow us to conceive a step that happens in a region as just one homogeneous relation. This is visualized by the schema in Fig. 2.

---

[1] Suppressing indices here.

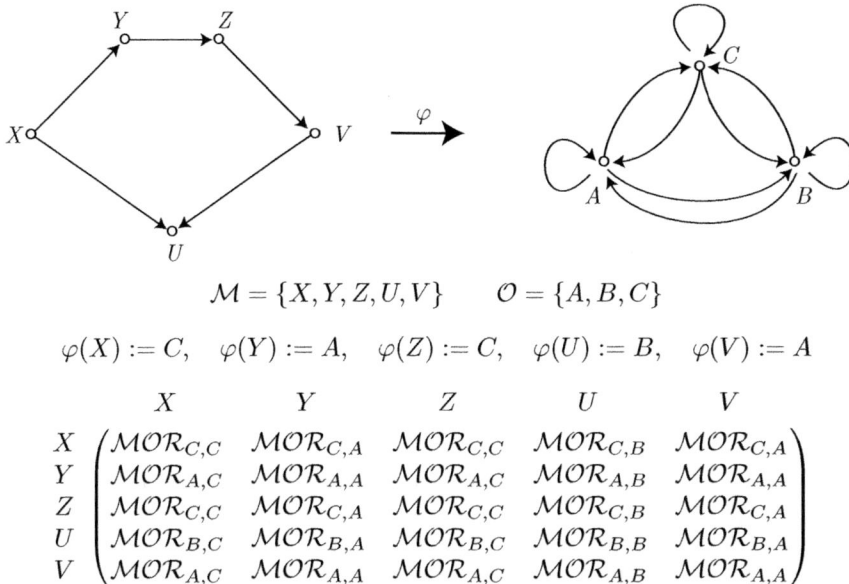

$$\mathcal{M} = \{X, Y, Z, U, V\} \qquad \mathcal{O} = \{A, B, C\}$$

$$\varphi(X) := C, \quad \varphi(Y) := A, \quad \varphi(Z) := C, \quad \varphi(U) := B, \quad \varphi(V) := A$$

$$
\begin{array}{c}
 \quad X \qquad\quad Y \qquad\quad Z \qquad\quad U \qquad\quad V \\
\begin{array}{c}
X \\ Y \\ Z \\ U \\ V
\end{array}
\left(
\begin{array}{ccccc}
\mathcal{MOR}_{C,C} & \mathcal{MOR}_{C,A} & \mathcal{MOR}_{C,C} & \mathcal{MOR}_{C,B} & \mathcal{MOR}_{C,A} \\
\mathcal{MOR}_{A,C} & \mathcal{MOR}_{A,A} & \mathcal{MOR}_{A,C} & \mathcal{MOR}_{A,B} & \mathcal{MOR}_{A,A} \\
\mathcal{MOR}_{C,C} & \mathcal{MOR}_{C,A} & \mathcal{MOR}_{C,C} & \mathcal{MOR}_{C,B} & \mathcal{MOR}_{C,A} \\
\mathcal{MOR}_{B,C} & \mathcal{MOR}_{B,A} & \mathcal{MOR}_{B,C} & \mathcal{MOR}_{B,B} & \mathcal{MOR}_{B,A} \\
\mathcal{MOR}_{A,C} & \mathcal{MOR}_{A,A} & \mathcal{MOR}_{A,C} & \mathcal{MOR}_{A,B} & \mathcal{MOR}_{A,A}
\end{array}
\right)
\end{array}
$$

Fig. 2. Homogeneous relation constructed from heterogeneous ones, using arbitrary elements of the respective morphism sets

In order to exclude any conceivable problems in such regards, everything is here assumed to be finite and non-empty. We handle the borderline cases only later.

4.1 Proposition. Let be given any heterogeneous relation algebra with object set $\mathcal{O}$ and morphism sets $\mathcal{MOR}_{o,o'}$ in the case $o, o' \in \mathcal{O}$. Assume further a mapping $\varphi : \mathcal{M} \longrightarrow \mathcal{O}$ defined on some set $\mathcal{M}$. Then the matrices of relations

$$R = \left[\psi_{m,m'}(R) \in \mathcal{MOR}_{\varphi(m),\varphi(m')} \right]_{m,m' \in \mathcal{M}}$$

or else, the matrices over $\mathcal{M} \times \mathcal{M}$ with coefficients $\psi_{m,m'}(R) \in \mathcal{MOR}_{\varphi(m),\varphi(m')}$ if $m, m' \in \mathcal{M}$, form a homogeneous relation algebra when operations are defined as

$$\left[R \sqcup_{\mathcal{M}} S \right]_{m,m'} := \psi_{m,m'}(R) \sqcup_{\mathcal{O}} \psi_{m,m'}(S)$$

$$\left[\overline{R}^{\mathcal{M}} \right]_{m,m'} := \overline{\psi_{m,m'}(R)}^{\mathcal{O}}$$

$$R \sqsubseteq_{\mathcal{M}} S \quad :\Longleftrightarrow \quad \psi_{m,m'}(R) \sqsubseteq_{\mathcal{O}} \psi_{m,m'}(S) \quad \text{for all } m, m' \in \mathcal{M}$$

$$\left[R \,{;}_{\mathcal{M}}\, S \right]_{m,m''} := \sup\nolimits_{m' \in \mathcal{M}}^{\mathcal{O}} \left\{ \psi_{m,m'}(R) \,{;}\, \psi_{m',m''}(S) \right\}$$

$$\left[R^{\mathsf{T}_{\mathcal{M}}} \right]_{m,m'} := \left(\psi_{m',m}(R) \right)^{\mathsf{T}_{\mathcal{O}}}.$$

Proof: Based on the Boolean lattices $\mathcal{MOR}_{o,o'}$, the Boolean operations are executed point-wise and, thus, form a Boolean lattice again. The monoid part is trivial as can be seen in

$$\left[(R_{;\mathcal{M}} S)_{;\mathcal{M}} T\right]_{m,m''} = \sup_{m' \in \mathcal{M}}^{\mathcal{O}} \left\{\psi_{m,m'}(R_{;\mathcal{O}} S)_{;\mathcal{O}} \psi_{m',m''}(T)\right\}$$
$$= \sup_{m' \in \mathcal{M}}^{\mathcal{O}} \left\{\sup_{m^{\circ} \in \mathcal{M}}^{\mathcal{O}} \left\{\psi_{m,m^{\circ}}(R)_{;}\psi_{m^{\circ},m'}(S)\right\}_{;}T_{\varphi(m'),\varphi(m'')}\right\}$$
$$= \ldots = \left[R_{;\mathcal{M}} (S_{;\mathcal{M}} T)\right]_{m,m''}$$

applying $\cup^{\mathcal{O}}$- and, thus, $\sup^{\mathcal{O}}$-distributivity of composition.

The task remains to prove the Schröder equivalences. For this, assume

$R_{;\mathcal{M}} S \subseteq_{\mathcal{M}} T$, i.e.,

$\left[R_{;\mathcal{M}} S\right]_{m,m''} \subseteq_{\mathcal{O}} \left[T\right]_{m,m''}$

to hold for every pair $m, m'' \in \mathcal{M}$, which means by definition of composition

$\sup_{m' \in \mathcal{M}}^{\mathcal{O}} \{\psi_{m,m'}(R) \; _{\mathcal{O}} \; \psi_{m',m''}(S)\} \subseteq_{\mathcal{O}} \left[T\right]_{m,m''}.$

Applying the definition of $\sup^{\mathcal{O}}$ and the matrix-like definition, however, this brings

$\psi_{m,m'}(R) \; _{\mathcal{O}} \; \psi_{m',m''}(S) \subseteq_{\mathcal{O}} \psi_{m,m''}(T)$

for every triple $m, m', m'' \in \mathcal{M}$, so that owing to the Schröder rule in $\mathcal{O}$

$(\psi_{m',m}(R))^{\mathsf{T}}{}_{;\mathcal{O}} \overline{\psi_{m,m''}(T)}^{\mathcal{O}} \subseteq_{\mathcal{O}} \overline{\psi_{m',m''}(S)}^{\mathcal{O}}.$

Applying the definition of the matrix operations and of the supremum, we get

$\left[R^{\mathsf{T}_{\mathcal{M}}}{}_{;\mathcal{M}} \overline{T}^{\mathcal{M}}\right]_{m',m''} \subseteq_{\mathcal{M}} \left[\overline{S}^{\mathcal{M}}\right]_{m',m''}.$ □

One may wonder why the category $\mathcal{O}$ has been used and not just one category object with morphism set $\mathbb{B}^1$ so as to obtain normal matrix coefficients $\mathbb{1}, \mathbb{0}$. We will later indeed have to resort to different coefficient types.

An 'inverse operation' of this construction is also possible:

4.2 Proposition. Let be given the algebra of homogeneous relations $2^{X \times X}$ on X and consider an equivalence Ξ. It is possible to introduce therefrom a heterogeneous relation algebra with operations directly based on the original ones as follows:

$\mathcal{O} :=$ classes of X modulo Ξ

As indicated in Fig. 3, each class D_i is injected $(\iota_i : D_i \longrightarrow X)_{1 \leq i \leq n}$ as subset into X, with $n := |X_\Xi|$.

$\mathcal{MOR}_{i,k} := \{A \mid A = \iota_i{}_; R{}_; \iota_k^{\mathsf{T}} \text{ for some } R : X \longrightarrow X\}.$

Proof: The injections constitute an n-fold direct sum, cf. [10,11,9], and correspondingly satisfy

$\iota_i{}_; \iota_i^{\mathsf{T}} = \mathbb{I}, \quad \iota_i{}_; \iota_k^{\mathsf{T}} = \mathbb{1} \text{ in the case } i \neq k, \quad \sup_i \iota_i^{\mathsf{T}}{}_; \iota_i = \mathbb{I}, \quad \text{and} \quad \Xi = \sup_i \iota_i^{\mathsf{T}}{}_; \mathbb{T}{}_; \iota_i$

Employing these formulae, the other requirements are rather trivial in view of Fig. 3.   □

Historically, researchers have preferred considering homogeneous relation algebras and introducing cylindric elements in these, e.g. The present author has

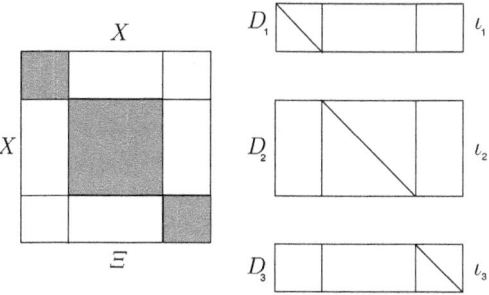

Fig. 3. Equivalence and injections of its classes

always used the other approach, i.e., starting from a heterogeneous relation algebra and considering the above homogeneous construct as a natural outcome.
For working with relations on a computer, this seems adequate. It allows domain
constructions without burdening these with non-finite models.

5 Row- Resp. Column-Less Relations

We now approach our goal from a completely different side and open another
thread. To guide our intuition, we study in Fig. 4 membership relations ε and
their corresponding powerset orderings $\Omega = \overline{\varepsilon^{\mathsf{T}}; \overline{\varepsilon}}$ for sets with 2, 1, and 0
elements.

For an empty set the set of all subsets is non-empty, which is resembled by the
above relations. The "—" and "|" shall indicate that there is no row resp. column.
There exists, however, a row as well as a column for Ω_0.

We trace the operations from $\Omega_n = \overline{\varepsilon_n^{\mathsf{T}}; \overline{\varepsilon_n}}$ down to $\Omega_0 := \overline{\varepsilon_0^{\mathsf{T}}; \overline{\varepsilon_0}}$. Source and
target are easily determined, but one will find it difficult to obtain this as a

$$\varepsilon_2 = \begin{array}{c} m \\ f \end{array}\begin{pmatrix} 0 & 1 & 0 & 1 \\ 0 & 0 & 1 & 1 \end{pmatrix}$$

$$\varepsilon_2^{\mathsf{T}} = \begin{array}{c} \{\} \\ \{male\} \\ \{female\} \\ \{m,f\} \end{array}\begin{pmatrix} 0 & 0 \\ 1 & 0 \\ 0 & 1 \\ 1 & 1 \end{pmatrix}$$

$$\Omega_2 = \begin{pmatrix} 1 & 1 & 1 & 1 \\ 0 & 1 & 0 & 1 \\ 0 & 0 & 1 & 1 \\ 0 & 0 & 0 & 1 \end{pmatrix}$$

$$\varepsilon_1 = 1\begin{pmatrix} 0 & 1 \end{pmatrix}$$

$$\varepsilon_1^{\mathsf{T}} = \begin{array}{c} \{\} \\ \{1\} \end{array}\begin{pmatrix} 0 \\ 1 \end{pmatrix}$$

$$\Omega_1 = \begin{pmatrix} 1 & 1 \\ 0 & 1 \end{pmatrix}$$

$$\varepsilon_0 = \quad - \quad \odot$$

$$\varepsilon_0^{\mathsf{T}} = \{\} \,()$$

$$\Omega_0 = \begin{pmatrix} 1 \end{pmatrix}$$

Fig. 4. Membership relation and powerset ordering for 2-, 1-, and 0-element set

matrix product when $n = 0$. It is, however, not without sense: Ω_0 is the left residual of ε_0 with itself, meaning that it indicates where a column of ε_0 is contained in a column of ε_0.

If we consider representing a Boolean matrix by an array, and an array by a list of lists, then one will observe that it is possible to represent ε_0^T as $[[]]$, namely as a matrix with one line containing no element, i.e. with no column. It is, however, impossible to represent ε_0 in this way. Even more difficult is it to include the row- and column-less relation discussed only later.

On the other hand, there exist situations in which one may wish to extend the constructions of Sect. 4 smoothly to these borderline cases. To give a first idea, we study a rather simple-minded example, investigating the interrelationship between ε and Ω simultaneously for two *disjoint* and *unrelated* sets X and Y. If we try to model such *non–connectedness* of X, Y with $\perp\!\!\!\perp$ as coefficient in ε, we get the situation of Fig. 5.

$$\varepsilon = \begin{array}{c} X \\ Y \end{array}\!\!\left(\begin{array}{cc} \varepsilon_X & \perp\!\!\!\perp_{X2^Y} \\ \perp\!\!\!\perp_{Y2^X} & \varepsilon_Y \end{array}\right) \qquad \begin{array}{c} 2^X \\ 2^Y \end{array}\!\!\left(\begin{array}{cc} \overline{\varepsilon_X^\mathsf{T}\,;\overline{\varepsilon_X}} & \overline{\varepsilon_X^\mathsf{T}\,;\mathbb{T}_{X2^Y}} \\ \overline{\varepsilon_Y^\mathsf{T}\,;\mathbb{T}_{Y2^X}} & \overline{\varepsilon_Y^\mathsf{T}\,;\overline{\varepsilon_Y}} \end{array}\right) = \Omega$$

Fig. 5. Attempting to model non-connectedness of X, Y with $\perp\!\!\!\perp$

This would be fine with Ω_X, Ω_Y in the diagonal, however with disturbing terms $\neq \perp\!\!\!\perp$ off diagonal. These in turn would corrupt every higher construct built on top of it, such as forming least upper bounds lub to get crispness or checking for continuity, or else the existential images of Sect. 6; but look ahead to the end of Sect. 9.

6   Existential Images

The result in Prop. 4.1 is immediate for relations $R \in \mathcal{MOR}_{o,o'}$ with $o, o' \in \mathcal{O}$ interpreted with nonempty sets; i.e., when all the borderline cases considered earlier are excluded. The question we try to answer next is to which extent considering existential images will allow us to extend our imagination. It will in particular be observed that some of the dubious row- or column-less relations have a reasonable counterpart as an existential image. We recall the definition and refer to [4,2,9].

6.1 Definition. Given any relation $R : X \longrightarrow Y$ together with membership relations $\varepsilon : X \longrightarrow 2^X$ and $\varepsilon' : Y \longrightarrow 2^Y$, we define its *existential image* as

$$\vartheta_R := \mathsf{syq}(R^\mathsf{T}\,;\varepsilon, \varepsilon') = \overline{\varepsilon^\mathsf{T}\,;R\,;\overline{\varepsilon'}} \cap \overline{\overline{\varepsilon^\mathsf{T}\,;R}\,;\varepsilon'}. \qquad \square$$

The existential image is known to be a (lattice-)continuous mapping with respect to the powerset orderings $\Omega = \overline{\varepsilon^{\mathsf{T}};\overline{\varepsilon}}$. It behaves nicely with respect to relational composition, being multiplicative and respecting identities:

$$\vartheta_{Q;R} = \vartheta_Q;\vartheta_R \qquad \vartheta_{\mathbb{I}_X} = \mathbb{I}_{\mathbf{2}^X}.$$

The relation R may be re-obtained from ϑ_R as $R = \overline{\varepsilon;\vartheta_R;\overline{\varepsilon'}^{\mathsf{T}}}$, (see [9]), but normally there exist many relations W satisfying $R = \overline{\varepsilon;W;\overline{\varepsilon'}^{\mathsf{T}}}$, most notably $W := \overline{\varepsilon^{\mathsf{T}};\overline{R;\overline{\varepsilon'}}}$. Furthermore, it is known that the existential image and the original relation may *simulate* each other via $\varepsilon, \varepsilon'$:

$$\varepsilon^{\mathsf{T}};R = \vartheta_R;\varepsilon'^{\mathsf{T}} \qquad \varepsilon'^{\mathsf{T}};R^{\mathsf{T}} = \vartheta_{R^{\mathsf{T}}};\varepsilon^{\mathsf{T}}.$$

As long as there are no empty rows or columns, this is well known. However, we are approaching the borderline cases in the constructions that follow. Does this smoothly extend to these? To study this, we first recall a very small but not yet borderline example in Fig. 6.

$$R = \begin{array}{r} \\ \text{male} \\ \text{female} \end{array}\!\!\begin{array}{c} 1 \\ \begin{pmatrix} \mathbf{0} \\ \mathbf{1} \end{pmatrix} \end{array} \qquad \vartheta_R = \begin{array}{r} \\ \{\} \\ \{\text{male}\} \\ \{\text{female}\} \\ \{\text{male,female}\} \end{array}\!\!\begin{array}{c} \{\}\ \{1\} \\ \begin{pmatrix} \mathbf{1} & \mathbf{0} \\ \mathbf{1} & \mathbf{0} \\ \mathbf{0} & \mathbf{1} \\ \mathbf{0} & \mathbf{1} \end{pmatrix} \end{array} \qquad \vartheta_{R^{\mathsf{T}}} = \begin{array}{r} \\ \{\} \\ \{1\} \end{array}\!\!\begin{array}{c} \{\}\ \{\text{male}\}\ \{\text{female}\}\ \{\text{m,f}\} \\ \begin{pmatrix} \mathbf{1} & \mathbf{0} & \mathbf{0} & \mathbf{0} \\ \mathbf{0} & \mathbf{0} & \mathbf{1} & \mathbf{0} \end{pmatrix} \end{array}$$

$$\varepsilon^{\mathsf{T}};R = \vartheta_R;\varepsilon'^{\mathsf{T}} = \begin{array}{r} \\ \{\} \\ \{\text{male}\} \\ \{\text{female}\} \\ \{\text{male,female}\} \end{array}\!\!\begin{array}{c} 1 \\ \begin{pmatrix} \mathbf{0} \\ \mathbf{0} \\ \mathbf{1} \\ \mathbf{1} \end{pmatrix} \end{array} \qquad \varepsilon'^{\mathsf{T}};R^{\mathsf{T}} = \vartheta_{R^{\mathsf{T}}};\varepsilon^{\mathsf{T}} = \begin{array}{r} \\ \{\} \\ \{1\} \end{array}\!\!\begin{array}{c} \text{male}\ \text{female} \\ \begin{pmatrix} \mathbf{0} & \mathbf{0} \\ \mathbf{0} & \mathbf{1} \end{pmatrix} \end{array}$$

Fig. 6. R and ϑ_R simulate each other with the memberships $\varepsilon, \varepsilon'$

$$R = \begin{array}{c} \text{male}\ \text{female} \\ (\ \cdot\ \ \cdot\ ) \end{array} \qquad \vartheta_R = \begin{array}{r} \\ \{\} \end{array}\!\!\begin{array}{c} \{\}\ \{\text{male}\}\ \{\text{female}\}\ \{\text{m,f}\} \\ (\mathbf{1}\ \ \mathbf{0}\ \ \mathbf{0}\ \ \mathbf{0}) \end{array} \qquad \vartheta_{R^{\mathsf{T}}} = \begin{array}{r} \\ \{\} \\ \{\text{male}\} \\ \{\text{female}\} \\ \{\text{male,female}\} \end{array}\!\!\begin{array}{c} \{\} \\ \begin{pmatrix} \mathbf{1} \\ \mathbf{1} \\ \mathbf{1} \\ \mathbf{1} \end{pmatrix} \end{array}$$

$$\varepsilon^{\mathsf{T}};R = \vartheta_R;\varepsilon'^{\mathsf{T}} = \begin{array}{r} \\ \{\} \end{array}\!\!\begin{array}{c} \text{male}\ \text{female} \\ (\mathbf{0}\ \ \mathbf{0}) \end{array} \qquad \varepsilon'^{\mathsf{T}};R^{\mathsf{T}} = \vartheta_{R^{\mathsf{T}}};\varepsilon^{\mathsf{T}} = \begin{array}{r} \\ \{\} \\ \{\text{male}\} \\ \{\text{female}\} \\ \{\text{male,female}\} \end{array}\!\!\begin{array}{c} | \\ \begin{pmatrix} \cdot \\ \cdot \\ \cdot \\ \cdot \end{pmatrix} \end{array}$$

Fig. 7. Simulation with existential image of a row-less relation

It is obviously interesting to ask to which extent this behaviour scales down to the row- and/or column-less relations mentioned earlier. With Fig. 7, we study a row-less relation in a similar fashion.

We have a more detailed look at $\varepsilon^{\mathsf{T}};R = \vartheta_R;\varepsilon'^{\mathsf{T}}$ of Fig. 7 in Fig. 8. While the left product looks funny, the right one is fairly normal.

$$\{\}\,(\cdot)\;;\;\begin{array}{c|cc} & \text{male} & \text{female} \\ \hline \end{array} \;=\; \begin{array}{c} \\ \{\} \end{array}\!\!\begin{array}{cc} \text{male} & \text{female} \\ (\mathbf{0} & \mathbf{0}) \end{array} \;=\; \begin{array}{c} \\ \{\} \end{array}\!\!\begin{array}{cccc} \{\} & \{\text{male}\} & \{\text{female}\} & \{\text{m,f}\} \\ (\mathbf{1} & \mathbf{0} & \mathbf{0} & \mathbf{0}) \end{array} \;;\; \begin{array}{r} \{\} \\ \{\text{male}\} \\ \{\text{female}\} \\ \{\text{male,female}\} \end{array}\!\!\begin{pmatrix} \mathbf{0} & \mathbf{0} \\ \mathbf{1} & \mathbf{0} \\ \mathbf{0} & \mathbf{1} \\ \mathbf{1} & \mathbf{1} \end{pmatrix}$$

$$\varepsilon^{\mathsf{T}};R \quad = \quad \vartheta_R;\varepsilon'^{\mathsf{T}}$$

Fig. 8. Visible ϑ_R simulating row-less relation

We observe finally the small but totally normal looking ϑ_R and $\vartheta_{R^{\mathsf{T}}}$ of a row- as well as column-less relation in Fig. 9.

$$R = \text{—}\;. \qquad \vartheta_R = \{\}\,(\mathbf{1}) \qquad \vartheta_{R^{\mathsf{T}}} = \{\}\,(\mathbf{1})$$

$$\varepsilon^{\mathsf{T}};R = \vartheta_R;\varepsilon'^{\mathsf{T}} = \emptyset\,(\cdot) \qquad \varepsilon'^{\mathsf{T}};R^{\mathsf{T}} = \vartheta_{R^{\mathsf{T}}};\varepsilon^{\mathsf{T}} = \emptyset\,(\cdot)$$

Fig. 9. A row- and column-less relation with normal-looking existential image

In all these cases, the existential image has been suited to serve as a substitute for the original relation R in as far as simulation was concerned.

7 The 1-Element Boolean Lattice and Non-connectedness

To cope with all this requires specific measures. Boolean lattices don't bring any problems when $\mathbb{B}^n$ is considered with $n > 0$. We are familiar with $\mathbb{B}^1 = \{\texttt{True},\texttt{False}\}$ or, denoted differently, $\mathbb{B}^1 = \{\,\mathbf{1}\,,\,\mathbf{0}\,\}$ and with $\mathbb{B}^2 = \{(\,\mathbf{1}\,,\,\mathbf{1}\,), (\,\mathbf{1}\,,\,\mathbf{0}\,), (\,\mathbf{0}\,,\,\mathbf{1}\,), (\,\mathbf{0}\,,\,\mathbf{0}\,)\}$. But what about the case $n = 0$ where $\mathbb{B}^0$ has cardinality 1? In the classic text by Birkhoff [3] it is not explicitly demanded that least and greatest elements be different: "A Boolean lattice ... by definition ... must contain universal bounds O and I". With coincidence $O = I$, this would normally become completely uninteresting, so that it is not mentioned more explicitly by Birkhoff that $O \neq I$ is required.

In order to use it for the borderline cases, we expressly admit the Boolean lattice $\mathbb{B}^0$ and make clear that one has to be extremely careful introducing a specific denotation for its only element $\mathbb{B}^0 = \{\Phi\}$. All Boolean operations, union, intersection, implication, negation will always result in Φ. The question to answer is whether this makes any sense. It is definitely uninteresting for its own. But when matrices with such coefficients — among others — are built, it may be expected that they serve as appropriate 'adapters', most notably when integrated in more advanced constructions. First, we take a look at a most primitive construction.

7.1 Proposition. The following is a heterogeneous relation algebra:

- Two category objects called X_0, X_1 with morphism sets as follows
    - $\mathcal{MOR}_{00} := \{\Phi_{00}\}$, $\mathcal{MOR}_{01} := \{\Phi_{01}\}$,
      $\mathcal{MOR}_{10} := \{\Phi_{10}\}$, $\mathcal{MOR}_{11} := \{\Phi_{11}\}$,
    - forming Boolean lattices that are all isomorphic to $\mathbb{B}^0$,
    - with identities Φ_{00} on X_0 and Φ_{11} on X_1 and composition defined as

;	Φ_{00}	Φ_{01}	Φ_{10}	Φ_{11}
Φ_{00}	Φ_{00}	Φ_{01}	—	—
Φ_{01}	—	—	Φ_{00}	Φ_{01}
Φ_{10}	Φ_{10}	Φ_{11}	—	—
Φ_{11}	—	—	Φ_{10}	Φ_{11}

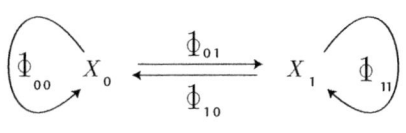

- Transposition shall result in Φ throughout — however with suitably exchanging source and target, which means exchanging indices.   □

This means two category objects, both with sets of relations on it containing just Φ. It is also possible to combine the traditional with the new and unusual relations. In the following proposition, one may wish to represent the elements of $\mathcal{MOR}_{00}$ as Boolean matrices $(\mathbb{B}^0)^{X_0 \times X_0}$ as well as the only one $\Phi_{X_0 X_1}$ of $\mathcal{MOR}_{01}$ as matrix $(\mathbb{B}^0)^{X_0 \times X_1}$. The difference is that now $\mathcal{MOR}_{11}$ has two elements $\bot, \mathbb{I}$.

7.2 Proposition. The following is a heterogeneous relation algebra:

- Two category objects called X_0, X_1 with morphism sets as follows
    - $\mathcal{MOR}_{00} := \{\Phi_{00}\}$, $\mathcal{MOR}_{01} := \{\Phi_{01}\}$,
      $\mathcal{MOR}_{10} := \{\Phi_{10}\}$, $\mathcal{MOR}_{11} := \{\bot_{11}, \mathbb{I}_{11}\}$,
    - forming Boolean lattices isomorphic to $\mathbb{B}^0, \mathbb{B}^0, \mathbb{B}^0, \mathbb{B}^1$, respectively,
    - with identities Φ_{00} on X_0 and $\mathbb{I}_{11}$ on X_1 and composition defined as

;	Φ_{00}	Φ_{01}	Φ_{10}	$\bot_{11}$	$\mathbb{I}_{11}$
Φ_{00}	Φ_{00}	Φ_{01}	—	—	—
Φ_{01}	—	—	Φ_{00}	$\bot_{01}$	$\bot_{01}$
Φ_{10}	Φ_{10}	$\bot_{11}$	—	—	—
$\bot_{11}$	—	—	Φ_{10}	$\bot_{11}$	$\bot_{11}$
$\mathbb{I}_{11}$	—	—	Φ_{10}	$\bot_{11}$	$\mathbb{I}_{11}$

– Textually, transposition shall preserve the letter throughout and exchange source and target indices.

Proof: Associativity is simple: Whenever an element Φ_{ik} is involved in composition, the result will be either $\mathbb{L}_{11}$ or the only element available, i.e., Φ_{00}, Φ_{01} or Φ_{10}, respectively. To prove the Schröder or the Dedekind rule is trivial.   □

The author is aware that the homogeneous counterparts of the tiny relation algebras mentioned are far from being new. They are at least contained in early work of Peter Jipsen with Roger Maddux, in [5], and then in [6,7]. They are here considered concerning their rôle in further constructions.

This relation algebra is certainly non-uniform — meaning that the product $\Phi_{10}\Phi_{01}$ of two universal relations is $\mathbb{L}_{11}$ which is unequal to the universal relation $\mathbb{I}_{11}$ on X_1.

Now we combine two homogeneous relation algebras into a heterogeneous one so as to have "no connection" between its parts, but in a way that facilitates the higher constructs already mentioned. This has — without success — already been attempted in Fig. 5.

7.3 Proposition. Let be given any two sets X, Y together with all the relations $R : X \longrightarrow X$ and $S : Y \longrightarrow Y$ considered as morphism sets $\mathcal{MOR}_{XX}$ and $\mathcal{MOR}_{YY}$, respectively. Define further two one-element morphism sets

$$\mathcal{MOR}_{XY} := \{\Phi_{XY}\} \text{ and } \mathcal{MOR}_{YX} := \{\Phi_{YX}\}.$$

Then the following is a heterogeneous relation algebra:

– Two category objects called X, Y with morphism sets as defined above
    — forming Boolean lattices isomorphic to $\mathbb{B}^{X \times X}$, $\mathbb{B}^{Y \times Y}$, respectively $\mathbb{B}^0$,
    — with identities $\mathbb{I}_X, \mathbb{I}_Y$ on X, Y and composition based on the agreements
$$\Phi_{XY}\Phi_{YX} = \mathbb{L}_{XX} \quad \text{and} \quad \Phi_{YX}\Phi_{XY} = \mathbb{L}_{YY}$$
$$A_{:}\Phi_{XY} = \Phi_{XY} \text{ for all relations } A : X \longrightarrow X$$
$$B_{:}\Phi_{YX} = \Phi_{YX} \text{ for all relations } B : Y \longrightarrow Y$$
$$\Phi_{XY:}C = \Phi_{XY} \text{ for all relations } C : Y \longrightarrow Y$$
$$\Phi_{YX:}D = \Phi_{YX} \text{ for all relations } D : X \longrightarrow X$$
– Transposition shall be transposition restricted to $\mathbb{B}^{X \times X}$ and $\mathbb{B}^{Y \times Y}$, together with $\Phi_{XY}^{\mathsf{T}} = \Phi_{YX}$ and $\Phi_{YX}^{\mathsf{T}} = \Phi_{XY}$.

Proof: Associativity restricted to the inside of $\mathbb{B}^{X \times X}$, or $\mathbb{B}^{Y \times Y}$, will obviously hold. Whenever Φ_{XY}, e.g., is involved, the result will necessarily be either Φ or $\mathbb{L}$. To prove the Schröder or the Dedekind rule is also trivial; either one is restricted to $\mathbb{B}^{X \times X}$ or $\mathbb{B}^{Y \times Y}$, where it holds, or one has to calculate the Dedekind as

$$R_{:}\Phi_{XY} \cap \Phi_{XY} = \Phi_{XY}$$

but also

$$(R \cap \Phi_{XY} \, \Phi_{XY}^{\mathsf{T}}) \, (\Phi_{XY} \cap R^{\mathsf{T}} \, \Phi_{XY}) = (R \cap \Phi_{XY} \, \Phi_{YX}) \, \Phi_{XY} = \Phi_{XY}. \qquad \square$$

Recalling Prop. 4.1 with $\varphi(X) = X$ and $\varphi(Y) = Y$, this may also be visualized in a homogeneous relation algebra:

$$\begin{array}{c} \qquad\quad {\scriptstyle X \quad Y} \\ \begin{array}{c} X \\ Y \end{array} \left(\begin{array}{cc} R & \Phi_{XY} \\ \Phi_{YX} & S \end{array} \right) \end{array} \; ; \; \begin{array}{c} \qquad\quad {\scriptstyle X \quad Y} \\ \left(\begin{array}{cc} R' & \Phi_{XY} \\ \Phi_{YX} & S' \end{array} \right) \end{array} = \begin{array}{c} \qquad\quad {\scriptstyle X \quad Y} \\ \left(\begin{array}{cc} R \, R' & \Phi_{XY} \\ \Phi_{YX} & S \, S' \end{array} \right) \end{array}$$

This says nothing more than that $\mathbb{B}^{X \times X}$ and $\mathbb{B}^{Y \times Y}$, considered independently, form a homogeneous relation algebra. But this independent handling integrates algebraically in the concept of a heterogeneous relation algebra with the help of the Φ.

8   Direct Sum Construction

There is another way of formulating ideas like Prop. 7.3 remembering the direct sum construct. We recall the laws of a direct sum from[10,11,9]:

$$\iota \, \iota^{\mathsf{T}} = \mathbb{I}, \quad \kappa \, \kappa^{\mathsf{T}} = \mathbb{I}, \quad \iota^{\mathsf{T}} \, \iota \cup \kappa^{\mathsf{T}} \, \kappa = \mathbb{I}, \quad \iota \, \kappa^{\mathsf{T}} = \mathbb{L},$$

that prevail also in combination with Φ. Then a homogeneous relation algebra can be constructed as follows.

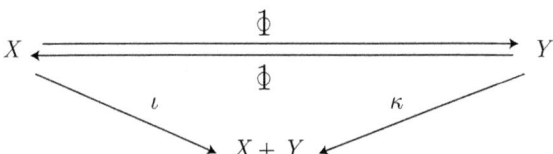

Fig. 10. Direct sum of non-connected items

8.1 Proposition. Let again be given the two sets X, Y together with all the relations $R : X \longrightarrow X$ and $S : Y \longrightarrow Y$. In addition consider the direct sum $X + Y$ of these sets. The following homogeneous relation algebra is constructed with morphisms indicated as

$$\begin{array}{c} \qquad\qquad\quad {\scriptstyle X \qquad\; Y \qquad\quad X+Y} \\ \begin{array}{c} X \\ Y \\ X + Y \end{array} \left(\begin{array}{ccc} A & \Phi_{XY} & B \, \iota \\ \Phi_{YX} & C & D \, \kappa \\ \iota^{\mathsf{T}} \, E & \kappa^{\mathsf{T}} \, F & \iota^{\mathsf{T}} \, G \, \iota \cup \kappa^{\mathsf{T}} \, H \, \kappa \end{array} \right) \end{array}$$

— For morphisms indexed 1,2, Boolean operations (here by example only for union and negation) are declared element-wise.

$$
\begin{array}{c}
\quad\quad X \quad\quad\quad\quad Y \quad\quad\quad\quad\quad\quad X+Y \\
\begin{array}{c} X \\ Y \\ X+Y \end{array}
\left(
\begin{array}{ccc}
A_1 \cup A_2 & \Phi_{XY} & (B_1 \cup B_2){;}\iota \\
\Phi_{YX} & C_1 \cup C_2 & (D_1 \cup D_2){;}\kappa \\
\iota^{\mathsf{T}}{;}(E_1 \cup E_2) & \kappa^{\mathsf{T}}{;}(F_1 \cup F_2) & \iota^{\mathsf{T}}{;}(G_1 \cup G_2){;}\iota \cup \kappa^{\mathsf{T}}{;}(H_1 \cup H_2){;}\kappa
\end{array}
\right)
\end{array}
$$

$$
\begin{array}{c}
\quad\quad X \quad\quad Y \quad\quad\quad X+Y \\
\begin{array}{c} X \\ Y \\ X+Y \end{array}
\left(
\begin{array}{ccc}
\overline{A} & \Phi_{XY} & \overline{B}{;}\iota \\
\Phi_{YX} & \overline{C} & \overline{D}{;}\kappa \\
\iota^{\mathsf{T}}{;}\overline{E} & \kappa^{\mathsf{T}}{;}\overline{F} & \iota^{\mathsf{T}}{;}\overline{G}{;}\iota \cup \kappa^{\mathsf{T}}{;}\overline{H}{;}\kappa
\end{array}
\right)
\end{array}
$$

— composition of morphisms indexed 1,2 is based solely on the laws of a direct sum, amended by $\Phi_{X,Y}{;}\kappa = \mathbb{l}_{X,X+Y}$, $\Phi_{Y,X}{;}\iota = \mathbb{l}_{Y,X+Y}$ and shall result in

$$
\begin{array}{c}
\quad\quad X \quad\quad\quad\quad Y \quad\quad\quad\quad\quad\quad X+Y \\
\begin{array}{c} X \\ Y \\ X+Y \end{array}
\left(
\begin{array}{ccc}
A_1{;}A_2 \cup B_1{;}E_2 & \Phi_{XY} & (A_1{;}B_2 \cup B_1{;}G_2){;}\iota \\
\Phi_{YX} & C_1{;}C_2 \cup D_1{;}F_2 & (C_1{;}D_2 \cup D_1{;}H_2){;}\kappa \\
\iota^{\mathsf{T}}{;}(E_1{;}A_2 \cup G_1{;}E_2) & \kappa^{\mathsf{T}}{;}(F_1{;}C_2 \cup H_1{;}F_2) & \begin{array}{l} \iota^{\mathsf{T}}{;}(E_1{;}B_2 \cup G_1{;}G_2){;}\iota \\ \cup \kappa^{\mathsf{T}}{;}(F_1{;}D_2 \cup H_1{;}H_2){;}\kappa \end{array}
\end{array}
\right)
\end{array}
$$

$$
\begin{array}{c}
\quad\quad\quad X \quad\quad\quad Y \quad\quad X+Y \\
\text{— identity} \quad
\begin{array}{c} X \\ Y \\ X+Y \end{array}
\left(
\begin{array}{ccc}
\mathbb{I}_X & \Phi_{XY} & \mathbb{l}_{X,X}{;}\iota \\
\Phi_{YX} & \mathbb{I}_Y & \mathbb{l}_{Y,Y}{;}\kappa \\
\iota^{\mathsf{T}}{;}\mathbb{l}_{X,X} & \kappa^{\mathsf{T}}{;}\mathbb{l}_{Y,Y} & \mathbb{I}_{X+Y}
\end{array}
\right)
\end{array}
$$

$$
\begin{array}{c}
\quad\quad\quad X \quad\quad Y \quad\quad\quad X+Y \\
\text{— converse} \quad
\begin{array}{c} X \\ Y \\ X+Y \end{array}
\left(
\begin{array}{ccc}
A^{\mathsf{T}} & \Phi_{XY} & E^{\mathsf{T}}{;}\iota \\
\Phi_{YX} & C^{\mathsf{T}} & F^{\mathsf{T}}{;}\kappa \\
\iota^{\mathsf{T}}{;}B^{\mathsf{T}} & \kappa^{\mathsf{T}}{;}D^{\mathsf{T}} & \iota^{\mathsf{T}}{;}G^{\mathsf{T}}{;}\iota \cup \kappa^{\mathsf{T}}{;}H^{\mathsf{T}}{;}\kappa
\end{array}
\right)
\end{array}
$$

This definition results in a homogeneous relation algebra.

Proof: The Boolean lattice properties rest mainly on $\cup$-distributivity of relational composition. The properties of Φ have already been studied earlier.

The semigroup property is relatively easy to demonstrate by executing matrix composition. It remains, thus, to convince ourselves concerning the Schröder- or Dedekind rule. There are only seven of the nine matrix positions to investigate.

$$
\begin{array}{ll}
A_1{;}A_2 \cup B_1{;}E_2 \subseteq A_3 & (1,1) \\
(A_1{;}B_2 \cup B_1{;}G_2){;}\iota \subseteq B_3{;}\iota & (1,3) \\
C_1{;}C_2 \cup D_1{;}F_2 \subseteq C_3 & (2,2) \\
(C_1{;}D_2 \cup D_1{;}H_2){;}\kappa \subseteq D_3{;}\kappa & (2,4) \\
\iota^{\mathsf{T}}{;}(E_1{;}A_2 \cup G_1{;}E_2) \subseteq \iota^{\mathsf{T}}{;}E_3 & (3,1)
\end{array}
$$

$$\kappa^\mathsf{T}{}_\mathsf{i}(F_1{}_\mathsf{i}C_2 \cup H_1{}_\mathsf{i}F_2) \subseteq \kappa^\mathsf{T}{}_\mathsf{i}F_3 \qquad (3,2)$$

$$\iota^\mathsf{T}{}_\mathsf{i}(E_1{}_\mathsf{i}B_2 \cup G_1{}_\mathsf{i}G_2){}_\mathsf{i}\iota \cup \kappa^\mathsf{T}{}_\mathsf{i}(F_1{}_\mathsf{i}D_2 \cup H_1{}_\mathsf{i}H_2){}_\mathsf{i}\kappa \subseteq \iota^\mathsf{T}{}_\mathsf{i}G_3{}_\mathsf{i}\iota \cup \kappa^\mathsf{T}{}_\mathsf{i}H_3{}_\mathsf{i}\kappa \qquad (3,3)$$

We sketch the proof of positions $(1,1)$ and $(1,3)$:

$(1,1)$	$\Longrightarrow$	$A_1{}_\mathsf{i}A_2 \subseteq A_3$	$\Longleftrightarrow$	$A_1^\mathsf{T}{}_\mathsf{i}\overline{A_3} \subseteq \overline{A_2}$	
$\iota{}_\mathsf{i}(3,1)$	$\Longrightarrow$	$E_1{}_\mathsf{i}A_2 \subseteq E_3$	$\Longleftrightarrow$	$E_1^\mathsf{T}{}_\mathsf{i}\overline{E_3} \subseteq \overline{A_2}$	
$(1,3){}_\mathsf{i}\iota^\mathsf{T}$	$\Longrightarrow$	$A_1{}_\mathsf{i}B_2 \subseteq B_3$	$\Longleftrightarrow$	$A_1^\mathsf{T}{}_\mathsf{i}\overline{B_3} \subseteq \overline{B_2}$	
$\iota{}_\mathsf{i}(3,3){}_\mathsf{i}\iota^\mathsf{T}$	$\Longrightarrow$	$E_1{}_\mathsf{i}B_2 \subseteq G_3$	$\Longleftrightarrow$	$E_1^\mathsf{T}{}_\mathsf{i}\overline{G_3} \subseteq \overline{B_2}$	$\square$

A short look at cardinalities shows that indeed something different has been built. Assume $|X| = 2$ and $|Y| = 3$. Then $|X| + |Y| + |X + Y| = 10$ giving 2^{100} relations for the normally connected direct sum. Here, we have only 2^{52}; these stem from the arbitrary choice of the four 2×2-matrices A, B, E, G and the four 3×3-matrices C, D, F, H in Prop. 8.1: $\left(2^4\right)^4 \cdot \left(2^9\right)^4 = 2^{52}$.

We add that the corresponding result for the direct product, in the axiomatic formulation

$$\pi^\mathsf{T}{}_\mathsf{i}\pi = \mathbb{I}, \quad \rho^\mathsf{T}{}_\mathsf{i}\rho = \mathbb{I}, \quad \pi{}_\mathsf{i}\pi^\mathsf{T} \cap \rho{}_\mathsf{i}\rho^\mathsf{T} = \mathbb{I}, \quad \pi^\mathsf{T}{}_\mathsf{i}\rho = \mathbb{T},$$

fails to hold in the present axiomatization. It is burdened with the unsharpness situation, namely the fact that only containment can be proved in

$$(\pi{}_\mathsf{i}R{}_\mathsf{i}\pi'^\mathsf{T} \cap \rho{}_\mathsf{i}S{}_\mathsf{i}\rho'^\mathsf{T}){}_\mathsf{i}(\pi'{}_\mathsf{i}P{}_\mathsf{i}\pi''^\mathsf{T} \cap \rho'{}_\mathsf{i}Q{}_\mathsf{i}\rho''^\mathsf{T}) \subseteq (\pi{}_\mathsf{i}R{}_\mathsf{i}P{}_\mathsf{i}\pi''^\mathsf{T} \cap \rho{}_\mathsf{i}S{}_\mathsf{i}Q{}_\mathsf{i}\rho''^\mathsf{T})$$

and not equality. There exist small finite counter examples for being unequal published in [5]. An additional problem results from

$$(\pi{}_\mathsf{i}R{}_\mathsf{i}\pi'^\mathsf{T} \cap \rho{}_\mathsf{i}S{}_\mathsf{i}\rho'^\mathsf{T}){}_\mathsf{i}\pi' = \pi{}_\mathsf{i}R \cap \rho{}_\mathsf{i}S{}_\mathsf{i}\mathbb{T} \subseteq \pi{}_\mathsf{i}R$$

$$(\pi{}_\mathsf{i}R{}_\mathsf{i}\pi'^\mathsf{T} \cap \rho{}_\mathsf{i}S{}_\mathsf{i}\rho'^\mathsf{T}){}_\mathsf{i}\rho' = \rho{}_\mathsf{i}S \cap \pi{}_\mathsf{i}R{}_\mathsf{i}\mathbb{T} \subseteq \rho{}_\mathsf{i}S,$$

meaning some sort of strictness with regard to the respective other component.

9 Lifting Relations to Partialities

Looking back at Fig. 5, we give a sketch of the Boolean lattice orderings and the lattice-continuous mappings we wish to have as diagonal blocks, avoiding the off-diagonal problems mentioned. To qualify a relation E to be a Boolean lattice, several points have to be considered; see [8]:

$$D := E \cap \overline{E}{}_\mathsf{i}\mathbb{T} \quad F := E \cap \mathbb{T}{}_\mathsf{i}\overline{E} \quad N := \overline{D^\mathsf{T}{}_\mathsf{i}D} \cap \overline{F{}_\mathsf{i}F^\mathsf{T}}$$

$$a := (\overline{\overline{E}{}_\mathsf{i}\overline{E}} \cap \overline{E}{}_\mathsf{i}\mathbb{T} \cap \mathbb{T}{}_\mathsf{i}\overline{E}){}_\mathsf{i}N \quad \text{€} := a{}_\mathsf{i}E.$$

Obviously, D it that part of the ordering E which is restricted to rows representing elements strictly above the least element. Analogously, F restricts the ordering to columns representing elements strictly below the greatest. With N, we postulate that no common upper bounds are allowed except for the greatest element and no common lower bounds except for the least. $N, a, \text{€}$ will then describe negation, atoms as part of the diagonal, and membership; see Fig. 11.

$$
\begin{array}{r|ccccccccc}
 & \{\} & \{\text{Win}\} & \{\text{Draw}\} & \{\text{Win,Draw}\} & \{\text{Loss}\} & \{\text{Win,Loss}\} & \{\text{Draw,Loss}\} & \{\text{W,D,L}\} \\
\{\} & 1 & 1 & 1 & 1 & 1 & 1 & 1 & 1 \\
\{W\} & 0 & 1 & 0 & 1 & 0 & 1 & 0 & 1 \\
\{D\} & 0 & 0 & 1 & 1 & 0 & 0 & 1 & 1 \\
\{W,D\} & 0 & 0 & 0 & 1 & 0 & 0 & 0 & 1 \\
\{L\} & 0 & 0 & 0 & 0 & 1 & 1 & 1 & 1 \\
\{W,L\} & 0 & 0 & 0 & 0 & 0 & 1 & 0 & 1 \\
\{D,L\} & 0 & 0 & 0 & 0 & 0 & 0 & 1 & 1 \\
\{W,D,L\} & 0 & 0 & 0 & 0 & 0 & 0 & 0 & 1
\end{array}
$$

$$
\begin{array}{r|cccccccc}
 & \{\} & \{\text{Win}\} & \{\text{Draw}\} & \{\text{Win,Draw}\} & \{\text{Loss}\} & \{\text{Win,Loss}\} & \{\text{Draw,Loss}\} & \{\text{W,D,L}\} \\
\{\} & 0 & 0 & 0 & 0 & 0 & 0 & 0 & 0 \\
\{W\} & 0 & 1 & 0 & 0 & 0 & 0 & 0 & 0 \\
\{D\} & 0 & 0 & 1 & 0 & 0 & 0 & 0 & 0 \\
\{W,D\} & 0 & 0 & 0 & 0 & 0 & 0 & 0 & 0 \\
\{L\} & 0 & 0 & 0 & 0 & 1 & 0 & 0 & 0 \\
\{W,L\} & 0 & 0 & 0 & 0 & 0 & 0 & 0 & 0 \\
\{D,L\} & 0 & 0 & 0 & 0 & 0 & 0 & 0 & 0 \\
\{W,D,L\} & 0 & 0 & 0 & 0 & 0 & 0 & 0 & 0
\end{array}
$$

$$
\begin{array}{r|cccccccc}
 & \{\} & \{\text{Win}\} & \{\text{Draw}\} & \{\text{Win,Draw}\} & \{\text{Loss}\} & \{\text{Win,Loss}\} & \{\text{Draw,Loss}\} & \{\text{W,D,L}\} \\
\{\} & 0 & 0 & 0 & 0 & 0 & 0 & 0 & 0 \\
\{W\} & 0 & 1 & 0 & 1 & 0 & 1 & 0 & 1 \\
\{D\} & 0 & 0 & 1 & 1 & 0 & 0 & 1 & 1 \\
\{W,D\} & 0 & 0 & 0 & 0 & 0 & 0 & 0 & 0 \\
\{L\} & 0 & 0 & 0 & 0 & 1 & 1 & 1 & 1 \\
\{W,L\} & 0 & 0 & 0 & 0 & 0 & 0 & 0 & 0 \\
\{D,L\} & 0 & 0 & 0 & 0 & 0 & 0 & 0 & 0 \\
\{W,D,L\} & 0 & 0 & 0 & 0 & 0 & 0 & 0 & 0
\end{array}
$$

Fig. 11. Lattice ordering E, atoms a, and $\text{\euro} := a{;}E$

9.1 Definition. $E : X \longrightarrow X$ is a *Boolean lattice ordering* if the constructs mentioned satisfy

– E is an ordering
– N is a bijective mapping satisfying $N{;}E = E^{\mathsf{T}}{;}N$
– $\texttt{lub}_E(\text{\euro})$ is surjective. □

Based on such a Boolean lattice E, we now recall from [8] the definition of an algebra of partialities.

9.2 Proposition. Let the Boolean lattice ordering E be given and consider the following set of lattice-continuous mappings

$$\mathcal{F} := \big\{ f \mid f \text{ a mapping that satisfies } f^{\mathsf{T}}{;}\texttt{lub}_E(R) = \texttt{lub}_E(f^{\mathsf{T}}{;}R) \big\}$$

together with operations defined as follows

$$
\begin{aligned}
&f \sqcup g := \texttt{lubR}_E(f \cup g) \qquad f \sqcap g := \texttt{glbR}_E(f \cup g)\\
&f^- := \texttt{syq}(a{;}\overline{E}{;}f^{\mathsf{T}}{;}E^{\mathsf{T}}{;}a{;}E,\, a{;}E)\\
&f \sqsubseteq g \quad :\Longleftrightarrow \quad g \subseteq f{;}E\\
&f{;}g := f{;}g \qquad \mathrm{I\!I\!I} := \mathrm{I\!I}\\
&f^\sim := \texttt{syq}(a{;}E{;}f{;}\overline{E}^{\mathsf{T}}{;}a{;}E,\, a{;}E)
\end{aligned}
$$

Then the set $\mathcal{F}$, thus equipped with relational operations, constitutes a homogeneous relation algebra. □

Following the initial idea, we can now state this result for the Boolean lattice $\mathbb{B}^0$: We have $E = (\mathbf{1}) = N$, $D = F = a = (\mathbf{0})$, i.e., no atoms and negation $N = \mathbb{T}$ equals identity as a relation. The only continuous mapping is also $f := (\mathbf{1})$ which satisfies $f^- = f$.

In general, E does not belong to the algebra and will later serve as part of an 'external arbiter' in deciding crispness, for instance. Of course, one will wish to also define heterogeneous relation algebras in this way, with orderings E in the diagonal. Fig. 5 has already demonstrated that this is more elaborate — not just in the purely technical sense. The idea thus developed is to consider a relation according to Prop. 7.3, i.e., with some Φ contained. We start with

$$E = \begin{matrix} X \\ Y \end{matrix} \begin{pmatrix} E_X & \Phi_{XY} \\ \Phi_{YX} & E_Y \end{pmatrix}$$

where it is assumed that E_X, E_Y both satisfy the requirements of Def. 9.1. Then we obtain immediately

$$\begin{matrix} X \\ Y \end{matrix} \begin{pmatrix} D_X & \Phi_{XY} \\ \Phi_{YX} & D_Y \end{pmatrix} \quad \begin{matrix} X \\ Y \end{matrix} \begin{pmatrix} F_X & \Phi_{XY} \\ \Phi_{YX} & F_Y \end{pmatrix} \quad \begin{matrix} X \\ Y \end{matrix} \begin{pmatrix} N_X & \Phi_{XY} \\ \Phi_{YX} & N_Y \end{pmatrix}$$
$$D \qquad\qquad\qquad F \qquad\qquad\qquad N$$

as opposed to what resulted in Fig. 5. Now, there is a chance to form least upper bounds in the parts of the relation simultaneously.

10 Concluding Remark

There is still work to be done. We are now in a position to work with non-connected and separately manipulable items in synoptic regions. We have not yet presented in which way crispness may be defined by the 'external arbiter' defined via the Boolean lattice ordering. In his work on Goguen categories [12], Michael Winter has elaborated that deciding crispness cannot be achieved *inside* the algebra. For this we have, thus, offered a possible means. Strict operations will no longer be continuous as the existential images are.

Acknowledgment. The author gratefully acknowledges the detailed comments and hints of the unknown referees.

References

1. Berghammer, R., Möller, B., Struth, G. (eds.): RelMiCS 2003. LNCS, vol. 3051. Springer, Heidelberg (2004)
2. Bird, R.S., de Moor, O.: Algebra of Programming. Prentice-Hall, Englewood Cliffs (1996)
3. Birkhoff, G.: Lattice Theory, vol. 25. American Mathematical Society Colloquium Publications (1940); third edition, third printing reprinted (1984)
4. de Moor, O.: Categories, Relations and Dynamic Programming. PhD thesis, Oxford University (1992)

5. Kahl, W., Schmidt, G.: Exploring (Finite) Relation Algebras With Tools Written in Haskell. Tech. Rep. 2000/02, Fakultät für Informatik, Universität der Bundeswehr München, http://ist.unibw-muenchen.de/Publications/TR/2000-02/

6. Offermann, E.: Konstruktion relationaler Kategorien. PhD thesis, Fakultät für Informatik, Universität der Bundeswehr München, Der Andere Verlag, Osnabrück (2003) ISBN 3-89959-078-3

7. Offermann, E.: On the Construction of Relational Categories. In: Berghammer and Möller [1], pp. 31–39

8. Schmidt, G.: Partiality I: Embedding Relation Algebras. Journal of Logic and Algebraic Programming 66(2), 212–238 (2006)

9. Schmidt, G.: Relational Mathematics. In: Encyclopedia of Mathematics and its Applications, vol. 132. Cambridge University Press, Cambridge (2011)

10. Schmidt, G., Ströhlein, T.: Relationen und Graphen. Mathematik für Informatiker. Springer, Heidelberg (1989)

11. Schmidt, G., Ströhlein, T.: Relations and Graphs – Discrete Mathematics for Computer Scientists. EATCS Monographs on Theoretical Computer Science. Springer, Heidelberg (1993)

12. Winter, M.: Goguen Categories – A categorical approach to L-Fuzzy relations. In: Trends in Logic, vol. 25, Springer, Heidelberg (2007)

Splitting Atoms in Relational Algebras

Prathap Siddavaatam and Michael Winter*

Department of Computer Science,
Brock University,
St. Catharines, Ontario, Canada, L2S 3A1
{ps09xy,mwinter}@brocku.ca

Abstract. Splitting atoms in a relation algebra is a common tool to generate new algebras from old ones. This includes constructing non-representable algebras from representable structures. The known method of splitting atoms does not allow that bijections different from the identity are contained in the starting algebra. This is a major drawback of that method because interesting candidates in mereotopology do contain such bijections. An ad-hoc splitting was done in those examples, and the results have been published in several papers. With this paper we want to start a thorough investigation of possible splitting methods.

1 Introduction

Region-based theories of space have been a prominent area of research in the recent years. In contrast to traditional approaches such as Euclidean geometry or general topology, region-based theories are point-free. They can be used to represent space in the context of (qualitative) spatial reasoning. This approach provides a mathematical model for how humans conceptualize our physical world.

Since the earliest work of de Laguna [4] and Whitehead [32], mereotopology has been considered for building point-free theories of space. Mereotopology combines topology, a mathematical model for connectedness and contact, and mereology, a description of the parthood relationship.

The Region Connection Calculus RCC was introduced as a formal structure for mereotopology [25]. Later, it was shown [28] that models of the RCC are isomorphic to so-called Boolean connection algebras (or Boolean contact algebras), i.e., Boolean algebras together with a binary contact relation C satisfying certain axioms. Since lattices and Boolean algebras in particular are well-known mathematical structures, this approach led to an intensive study of the properties of the RCC including several topological representation theorems [5,10,11,31].

Gotts explored in [13] how much topology can be defined by using the full first order RCC formalism. Since it is known for quite some time that the expressiveness of reasoning with basic operations on binary relations is equal to the expressive power of the three variable fragment of first order logic with at most

---

* The author gratefully acknowledge support from the Natural Sciences and Engineering Research Council of Canada.

H. de Swart (Ed.): RAMICS 2011, LNCS 6663, pp. 331–346, 2011.

binary relations [30], it seems worthwhile to use methods of relation algebras, initiated by Tarski [29], to study contact relations and to explore their expressive power with respect to topological domains. This idea led to the RCC8 composition table, i.e., a relation algebra based on eight atomic relationships between two regions [26]. Several refinements of the eight atomic relations produced new algebras up to 25 atoms [7,8,9]. Each time new relations were obtained by splitting certain atoms from the previous algebra into two new relations. It is easy to verify that there are more atoms within the current set of 25 atoms that can be split. However, computing the resulting algebra by hand becomes more and more infeasible. The aim of this paper is to develop a mechanism that can be implemented as a computer program in order to support this task.

In [2], a method for splitting atoms in a relation algebra was introduced. The authors adapted a method well known in cylindric algebra theory, originating with L. Henkin [15], which was used to obtain nonrepresentable cylindric algebras from representable ones. Their approach uses a condition of splittability on the atoms in question in order to ensure associativity of the composition operation after splitting. Unfortunately, this property is violated by all RCC tables in consideration starting with RCC11 or also known as the complemented closed disc algebra [6,7]. In this paper we are going to define a method for splitting atoms that is more general than the approach in [2]. Our definition allows also to split certain atoms if bijections different from the identity such as ECD in RCC11 are present.

Splitting atoms is only one method of generating new relation algebras from given ones. Programs computing those algebras in the finite case has been developed previously using a variety of different methods. As examples we refer to [18,23,24].

This paper is organized as follows. In the next section we want to recall some fundamentals on relation algebras. In Section 3 we present atom structure and complex algebras which are essential tools in implementing any algorithm on relation algebras. We focus on the existing theory of splitting atoms in relation algebras in Section 4. We will illustrate that mechanism in an example computed by our implementation. In Section 5 we are going to present our approach to splitting. We will provide necessary conditions so that the result of a splitting is indeed a relation algebra. Finally, we will illustrate our procedure by some examples before we close with some remarks on future research.

2   Relation Algebras

In this paper we will use the notion and basic definitions from [20]. In particular, we use the varieties NA and RA of nonassociative and associative relation algebras. We assume that the reader is familiar with the basic notions from Boolean algebras and lattice theory. For any notion used but defined here we refer to [3,14,19].

Definition 1. *A structure* $\mathfrak{A} = \langle A, +, \cdot, ^-, 0, 1, ^\smile, ;, 1' \rangle$ *of type* $\langle 2, 2, 1, 0, 0, 1, 2, 0 \rangle$ *is called a relation algebra (RA) iff it satisfies the following:*

R1. $\langle A, +, \cdot, ^-, 0, 1 \rangle$ *is a Boolean algebra.*
R2. $\langle A, ;, 1' \rangle$ *is a monoid.*
R3. *For all $x, y, z \in A$ the following formulas are equivalent:*

$$x; y \cdot z = 0 \iff \breve{x}; z \cdot y = 0 \iff z; \breve{y} \cdot x = 0.$$

We say that $\mathfrak{A}$ is a nonassociative relation algebra (NA) if $\mathfrak{A}$ is a structure satisfying all of the axioms above except associativity of the composition operation ;, i.e., **R2** is weakened by only requiring that 1' is a neutral element for composition.

We adopt the regular convention to denote the diversity element $\overline{1'}$ by 0'.

The notion of a subalgebra is as usual, and we will denote the fact that $\mathfrak{B}$ is a subalgebra of $\mathfrak{A}$ by $\mathfrak{B} \subseteq \mathfrak{A}$.

Oriented triangles can be used to visualize **R3** and its immediate consequence the so-called *cycle law*. It states that the following properties are equivalent:

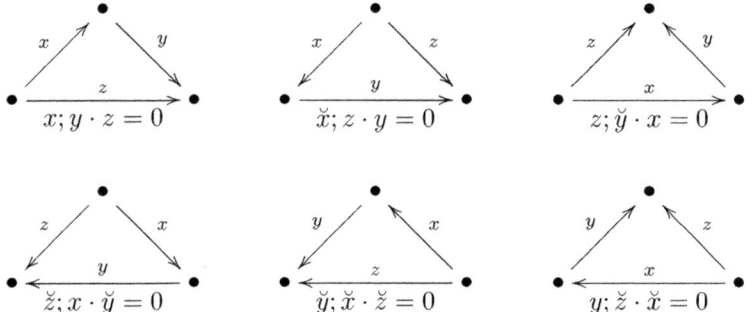

Besides the cycle law above, we will need some basic properties which are summarized in the following lemma. A proof can be found in any of [20,22,27,29,30].

Lemma 1. *Let $\mathfrak{B}$ be a relation algebra, and let $x, y, z \in \mathfrak{B}$. Then we have:*

1. $\breve{0} = 0, \breve{1} = 1, \breve{1'} = 1'$.
2. $x \leq x; \breve{x}; x$.
3. $x; (y + z) = x; y + x; z$.
4. *if $\breve{x}; x \leq 1'$, then $x; (y \cdot z) = x; y \cdot x; z$.*
5. *if $\breve{x}; x \leq 1'$, then $x; \overline{y} \leq \overline{x; y}$.*

A relation x is called univalent (or functional) iff $\breve{x}; x \leq 1'$. It is called injective if $\breve{x}$ is univalent. A univalent and injective relation is called bijective. We denote the set of all bijections (bijective relations) of a relation algebra $\mathfrak{B}$ by $\mathrm{Bij}\mathfrak{B}$. On addition, we say that x is total if $1' \leq x; \breve{x}$ and surjective if $\breve{x}$ is total. A function (or map) is a univalent and total relation.

In the next lemma we have summarized some properties of atoms in relation algebra that we will need in this paper. We will denote the set of atoms of a relation algebra $\mathfrak{B}$ by $\mathrm{At}\mathfrak{B}$.

Lemma 2. *Let $\mathfrak{B}$ be a relation algebra, and $x, y, z \in \mathrm{At}\mathfrak{B}$. Then we have:*

1. *There is an atom $i \leq 1$' with $x; i = x$.*
2. *$z \leq x; y$ iff $y \leq \breve{x}; z$ iff $x \leq z; \breve{y}$ iff $\breve{z} \leq \breve{y}; \breve{x}$ iff $\breve{y} \leq \breve{z}; x$ iff $\breve{x} \leq y; \breve{z}$.*
3. *If y is a bijection and $x; y \neq 0$, then $x; y$ is an atom.*

Again, a proof can be found in any of [17,20,21,22].

Of particular interest are integral relation algebras. They form basic building block in constructing arbitrary algebras. For details on their importance, we refer to [33,34].

Definition 2. *A relation algebra $\mathfrak{A}$ is called integral iff for all $x, y \in \mathfrak{A}$, $x; y = 0$ implies that $x = 0$ or $y = 0$.*

It is well-known that the property of being integral is equivalent to the fact that the identity is an atom of $\mathfrak{A}$. Another equivalent property is the requirement that all relations of the algebra are total, i.e., $1' \leq x; \breve{x}$.

Notice that (1) and (3) of Lemma 2 become trivial in integral relation algebras because the identity is an atom respectively every relation is total.

3   Atom Structures and Complex Algebras

In order to manipulate a finite relation algebra on the computer, atom structures are of interest. Some relation algebras are made for special purposes, e.g., see [1], and cannot be stored easily. Atom structures contain all necessary information to obtain the algebra in a smaller format, namely as relations on the set of atoms. For further details on atom structures and complex algebras not mentioned here we refer to [16,20,21,22].

Definition 3. *An atom structure $\mathfrak{At}\mathfrak{A} = \langle \mathrm{At}\mathfrak{A}, C(\mathfrak{A}), f, I(\mathfrak{A}) \rangle$ of a NA $\mathfrak{A}$ consists of a non-empty set $\mathrm{At}\mathfrak{A}$ of atoms, a unary predicate $I(\mathfrak{A}) = \{x \in \mathrm{At}\mathfrak{A} : x \leq 1'\}$, a unary function $f : \mathrm{At}\mathfrak{A} \to \mathrm{At}\mathfrak{A}$ defined by $f(x) = \breve{x}$, and a ternary relation $C(\mathfrak{A}) = \{\langle x, y, z \rangle : x, y, z \in \mathrm{At}\mathfrak{A}, x; y \geq z\}$.*

Conversely, we start from a relational structure $\mathfrak{S} = \langle U, C, f, I \rangle$, i.e., a set U together with a ternary relation C on U, a unary function $f : U \to U$, and a subset I of U. One may construct an algebra of relational type on the powerset SbU of U as follows.

Definition 4. *Given a relational structure $\mathfrak{S} = \langle U, C, f, I \rangle$ the complex algebra $\mathfrak{Cm}\mathfrak{S} = \langle SbU, \cup, \cap, ^-, \emptyset, U, ; , ^\smile, 1' \rangle$ is defined by*

$$X; Y = \{z \in U : \exists x \in X \exists y \in Y \langle x, y, z \rangle \in C\} \text{ and } \breve{X} = \{f(x) : x \in X\}.$$

The notion of a cycle is very helpful working with atom structures. Cycles basically reflect axiom **R3** or the cycle law introduced earlier. For three elements

x, y, z of a relational structure $\mathfrak{G} = \langle U, C, f, I \rangle$ we write $[x, y, z]$ for the following set of up to six triples:

$$[x, y, z] = \{\langle x, y, z \rangle, \langle \breve{x}, z, y \rangle, \langle y, \breve{z}, \breve{x} \rangle, \langle \breve{y}, \breve{x}, \breve{z} \rangle, \langle \breve{z}, x, \breve{y} \rangle, \langle z, \breve{y}, x \rangle\}.$$

$[x, y, z]$ is called a cycle. Its importance can be seen in the next theorem. A proof can be found in [21].

Theorem 1. *Let* $\mathfrak{G} = \langle U, C, f, I \rangle$ *be a relational structure consisting of a set* U *together with a ternary relation* C *on* U, *a unary function* $f : U \to U$, *and a subset* I *of* U.

1. *The following three conditions are equivalent:*
   (i) $\mathfrak{G}$ *is the atom structure of some complete atomic NA.*
   (ii) $\mathfrak{Cm}\mathfrak{G}$ *is a NA.*
   (iii) $\mathfrak{G}$ *satisfies condition (a) and (b)*
     (a) if $\langle x, y, z \rangle \in C$, *then* $\langle f(x), z, y \rangle \in C$ *and* $\langle z, f(y), x \rangle \in C$.
     (b) for all $x, y \in U$, $x = y$ *iff there is some* $w \in I$ *such that* $\langle x, w, y \rangle \in C$.
2. $\mathfrak{Cm}\mathfrak{G}$ *is a relation algebra iff* $\mathfrak{Cm}\mathfrak{G}$ *is a NA and it satisfies condition (c):*
   (c) for all $x, v, w, x, y, z \in U$, *if* $\langle v, w, x \rangle \in C$ *and* $\langle x, y, z \rangle \in C$, *then there is some* $u \in U$ *such that* $\langle w, y, u \rangle \in C$ *and* $\langle v, u, z \rangle \in C$.

From the theorem above one can easily see that condition (a) is already satisfied if C is a union of cycles. Property (c) can nicely be visualized by the following diagram.

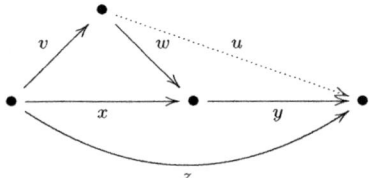

Notice that $\mathfrak{Cm}\mathfrak{G}$ is integral iff I is a singleton [22]. In an integral atom structure, i.e., an atom structure of an integral NA, it is possible to remove the cycle including the identity since property (b) determines them uniquely in this case. Therefore, we will normally only list the diversity cycles, i.e., those cycles that do not contain the identity.

4   Splitting Atoms in Relation Algebra

In [2] a method for splitting atoms in relation algebras was introduced. The method of splitting is well known in cylindric algebra theory, originating with L. Henkin et al. [15]. His work was used to obtain nonrepresentable cylindric algebras from representable ones. In this section, we want to recall the theory presented in the paper mentioned above.

Definition 5. *Let* $\mathfrak{A}$ *and* $\mathfrak{B}$ *be atomic NA's. We say that* $\mathfrak{A}$ *is obtained from* $\mathfrak{B}$ *by splitting if the following conditions are satisfied:*

S1. $\mathfrak{A} \supseteq \mathfrak{B}$.

S2. *Every atom $x \in \mathfrak{A}$ is contained in an atom $c(x) \in \mathfrak{B}$, called the cover of x.*

S3. *For all $x, y \in \mathrm{At}\mathfrak{A}$, if $x, y \leq 0'$, then*

$$x; y = \begin{cases} c(x); c(y) \cdot 0' & \text{iff } x \neq \breve{y}, \\ c(x); c(y) & \text{iff } x = \breve{y}. \end{cases}$$

Given an atomic NA $\mathfrak{B}$ one might identify one or more atoms, provide for each of those atoms a number into how many new atoms they should be split, and how many of those new atoms are supposed to be symmetric. This information can summarized by two functions η and θ mapping $\mathrm{At}\mathfrak{B}$ to cardinals. We say that $\mathfrak{A}$ is obtained from $\mathfrak{B}$ by splitting along η and θ if $\mathfrak{A}$ is obtained from $\mathfrak{B}$ by splitting and for all $x \in \mathrm{At}\mathfrak{B}$

$$\eta(x) = |\{y \in \mathrm{At}\mathfrak{A} : y \leq x, y \neq \breve{y}\}|,$$
$$\theta(x) = |\{y \in \mathrm{At}\mathfrak{A} : y \leq x, y = \breve{y}\}|.$$

In [2] the following two theorems about the existence and uniqueness of a splitting along two functions were shown.

Theorem 2. *Let $\mathfrak{A}, \mathfrak{A}'$, and $\mathfrak{B}$ be complete atomic NA's. Let η and θ be functions mapping $\mathrm{At}\mathfrak{B}$ to cardinals. If $\mathfrak{A}$ and $\mathfrak{A}'$ are obtained from $\mathfrak{B}$ by splitting along η and θ, then $\mathfrak{A}$ and $\mathfrak{A}'$ are isomorphic by an isomorphism that leaves $\mathfrak{B}$ fixed.*

Theorem 3. *Let $\mathfrak{B}$ be an atomic NA. Let η, θ be functions mapping $\mathrm{At}\mathfrak{B}$ to cardinals, and let $\alpha(x) = \eta(x) + \theta(x)$. Then we have:*

1. *There is an atomic NA $\mathfrak{A}$ obtained from $\mathfrak{B}$ by splitting along η and θ iff the following conditions hold for all $x \in \mathrm{At}\mathfrak{B}$:*
   (a) $\alpha(x) \geq 1$.
   (b) $\eta(x) = \eta(\breve{x})$.
   (c) $x \leq 1'$ implies $\eta(x) = 0$.
   (d) $x = \breve{x}$ implies $\eta(x)$ is even.
   (e) $x \neq \breve{x}$ implies $\theta(x) = 0$.
2. *Suppose $\mathfrak{A}$ is an atomic NA and obtained from $\mathfrak{B}$ by splitting along η and θ. Then $\mathfrak{A}$ is a RA iff $\mathfrak{B}$ is a RA and for all $x, y \in \mathrm{At}\mathfrak{B}$ we have; if $\alpha(x) > 1$ and $y; x \neq 0$ and $y \leq 0'$, then $y \leq y; (x; \breve{x} \cdot 0')$.*

The situation where a relation algebra and one atom that we want to split into two atoms is given is of particular interest. Therefore, we say that x is *splittable* in an atomic relation algebra $\mathfrak{B}$ if the following three conditions are satisfied:

A1. $x \leq 0'$.

A2. If $0' \geq y \in \mathrm{At}\mathfrak{B}$ and $x; y \neq 0$, then $y \leq (\breve{x}; x \cdot 0'); y$.

A3. If $0' \geq y \in \mathrm{At}\mathfrak{B}$ and $y; x \neq 0$, then $y \leq y; (x; \breve{x} \cdot 0')$.

$\mathfrak{B}$ is called splittable iff $\mathfrak{B}$ has a splittable atom.

By Theorem 3 x is splittable in $\mathfrak{B}$ iff there is a relation algebra $\mathfrak{A}$ obtained from $\mathfrak{B}$ by splitting such that $x \notin At\mathfrak{A}$. Note that if $\mathfrak{B}$ contains a functional element y below 0', i.e., $\breve{y}; y \leq 1'$ and $y \leq 0'$, then $\mathfrak{B}$ is not splittable. The relation algebra RCC11 [6,7] is integral and contains the bijective relation ECD. Therefore, no atom in RCC11 is splittable, and the theory above cannot be applied.

We want to illustrate the definition by an example. This will also introduce the way we represent integral relation algebras on the computer in order to actually compute different methods for splitting.

Example 1. Let $\mathfrak{A}$ be the relation relation algebra with $At\mathfrak{A} = \{1, 2, 3, 4, 5, 6, 7, 8\}$. In our representation the converse operation is induced by two numbers n and s, n the total number of atoms and s the number of symmetric atoms, i.e., atoms satisfying $x = \breve{x}$. As above, atoms will be represented by the numbers $1, \ldots, n$ where the first s atoms are always symmetric and the remaining $n - s$ atoms are non-symmetric. They will always come in pairs with $\breve{m} = m - 1$ if $m - s$ is even, and $\breve{m} = m+1$ otherwise. Consequently, $n-s$ must be even. In our example $n = 8$ and $s = 4$ so that $\breve{4} = 4, \breve{5} = 6$ and $\breve{6} = 5$. C is given as a list of diversity cycles. Therefore, any triple represents actually up to six triples.

$$C(\mathfrak{A}) = [\; (2,2,2), (2,2,3), (2,2,4), (2,2,5), (2,2,7), (2,3,3), (2,3,4), (2,3,5),$$
$$(2,3,7), (2,4,4), (2,4,5), (2,4,7), (2,5,5), (2,5,7), (2,7,7), (3,3,3),$$
$$(3,3,4), (3,3,5), (3,4,4), (3,4,5), (3,4,7), (3,5,5), (3,5,7), (3,7,7),$$
$$(4,4,4), (4,4,5), (4,4,7), (4,5,5), (4,5,7), (4,6,6), (4,6,8), (4,7,7),$$
$$(4,8,8), (5,5,5), (5,5,7), (5,7,7), (5,8,8), (7,7,7)]$$

This algebra is in fact the RCC8 composition table. This algebra does not contain any functions or bijections besides the identity so that the splitting mechanism can be applied. We want to split Atom 4 into two atoms. After renaming some atom in order to follow the convention on symmetric atoms followed by non-symmetric atoms in pairs, we get a new algebra $\mathfrak{B}$ with $n = 9$, $s = 5$ and the following correlation between atoms in $\mathfrak{B}$ and $\mathfrak{A}$:

Atoms of $\mathfrak{B}$	1	2	3	4	5	6	7	8	9
Atoms of $\mathfrak{A}$	1	2	3	4	4	5	6	7	8

The diversity cycles of the new algebra $\mathfrak{B}$ are as follows:

$$C(\mathfrak{B}) = [\; (2,2,2), (2,2,3), (2,2,4), (2,2,5), (2,2,6), (2,2,8), (2,3,3), (2,3,4),$$
$$(2,3,5), (2,3,6), (2,3,8), (2,4,4), (2,4,5), (2,5,5), (2,4,6), (2,5,6),$$
$$(2,4,8), (2,5,8), (2,6,6), (2,6,8), (2,8,8), (3,3,3), (3,3,4), (3,3,5),$$
$$(3,3,6), (3,4,4), (3,4,5), (3,5,5), (3,4,6), (3,5,6), (3,4,8), (3,5,8),$$
$$(3,6,6), (3,6,8), (3,8,8), (4,4,4), (4,4,5), (4,5,5), (5,5,5), (4,4,6),$$
$$(4,5,6), (5,4,6), (5,5,6), (4,4,8), (4,5,8), (5,4,8), (5,5,8), (4,6,6),$$
$$(5,6,6), (4,6,8), (5,6,8), (4,7,7), (5,7,7), (4,7,9), (5,7,9), (4,8,8),$$
$$(5,8,8), (4,9,9), (5,9,9), (6,6,6), (6,6,8), (6,8,8), (6,9,9), (8,8,8)]$$

The computation of this table took less than fraction of a second in our implementation. Computing it by hand is already tedious.

4.1   The Extension of a Relation Algebra

In this section we will make use of ordinal arithmetics. Therefore, we want to recall some of their basic properties needed throughout this paper. The definition of addition can be given inductively:

$$\alpha + 0 = \alpha,$$
$$\alpha + (\beta + 1) = (\alpha + \beta) + 1,$$

and if δ is a limit ordinal, then $\alpha + \delta = \bigcup\{\alpha + \beta : \beta < \delta\}$.

Zero is an additive identity $\alpha + 0 = 0 + \alpha = \alpha$, addition is associative $(\alpha + \beta) + \gamma = \alpha + (\beta + \gamma)$. Furthermore, ordinal addition is left-cancellative, i.e., if $\alpha + \beta = \alpha + \gamma$, then $\beta = \gamma$. This property allows to define a left subtraction for ordinals. However, right cancellation is not valid. Similar properties are shared by ordinal multiplication recursively defined by:

$$\alpha * 0 = 0,$$
$$\alpha * (\beta + 1) = (\alpha * \beta) + \alpha,$$

and if δ is a limit ordinal, then $\alpha * \delta = \bigcup\{\alpha * \beta : \beta < \delta\}$.

We have $\alpha * 0 = 0 * \alpha = 0$, multiplication is associative, 1 is an identity (or unit) $\alpha * 1 = 1 * \alpha = \alpha$, and satisfies a cancellation law, namely if $\alpha > 0$ and $\alpha * \beta = \alpha * \gamma$, then $\beta = \gamma$. As for addition right cancellation is not valid.

The notion of a splitting is too restrictive for our purposes because it does not allow the splitting of algebras that contain bijections different from the identity. The super algebra property together with the cover property seems to be sufficient. This guarantees that the composition of elements in the subalgebra can be computed by the elements of the super algebra.

Definition 6. *Let $\mathfrak{A}$ and $\mathfrak{B}$ be atomic integral RA's. We say that $\mathfrak{A}$ is an extension of $\mathfrak{B}$ if the following conditions are satisfied:*

S1. $\mathfrak{A} \supseteq \mathfrak{B}$.
S2. *Every atom $x \in \mathfrak{A}$ is contained in an atom $c(x) \in \mathfrak{B}$, called the cover of x.*

Notice that S2 is redundant if $\mathfrak{A}$ and $\mathfrak{B}$ are finite, but it is needed in the infinite case.

If the atoms in $\mathfrak{A}$ satisfy the condition imposed by to function η and θ, then we say that $\mathfrak{A}$ is an extension of $\mathfrak{B}$ along η and θ. Notice that such an extension does not have to be unique up to isomorphism. The next theorem provides necessary conditions for the existence of an extension. Notice that the proof also provides an explicit construction.

Theorem 4. *Let $\mathfrak{B}$ be a complete atomic integral RA. Let η, θ be functions mapping $\mathrm{At}\mathfrak{B}$ to cardinals, and let $\alpha(x) = \theta(x) + \eta(x)$. Then there is a complete atomic integral RA $\mathfrak{A}$ that is an extension of $\mathfrak{B}$ along η and θ if the following conditions hold for all $x, y \in \mathrm{At}\mathfrak{B}$:*

(a) $\alpha(x) \geq 1$.
(b) $\eta(x) = \eta(\breve{x})$.
(c) $x \in \text{Bij}\mathfrak{B}$ *implies* $\alpha(x) = 1$.
(d) $x = \breve{x}$ *implies* $\eta(x) =$ *even, i.e.,* $\eta(x) = 2 * \beta$ *for some ordinal* β.
(e) $x \neq \breve{x}$ *implies* $\theta(x) = 0$.
(f) $y \in \text{Bij}\mathfrak{B}$ *implies* $\alpha(x; y) = \alpha(x)$.
(g) $y \in \text{Bij}\mathfrak{B}$, $x = \breve{x}$ *and* $\eta(x) > 0$ *implies* $x; y = (x; y)^{\smallsmile}$ *and* $\theta(x) = \theta(x; y)$.
(h) $\alpha(x) > 1$, $y; x \neq 0$ *and* $y \notin \text{Bij}\mathfrak{B}$ *implies* $y \leq y; (x; \breve{x} \cap 0')$.

Proof. First we want to show that $\alpha(\breve{x}) = \alpha(x)$ since we will use this property frequently without mentioning. If $x = \breve{x}$ the assertion is trivial. If $x \neq \breve{x}$, then $\alpha(x) = \eta(x)$ follows from (e), which shows together with (b) that $\alpha(x) = \alpha(\breve{x})$.

We are going to construct a relational structure $\mathfrak{S} = \langle U, C, f, I \rangle$ so that $\mathfrak{Cm}\mathfrak{S}$ is a RA and $\mathfrak{B}$ can be embedded into $\mathfrak{Cm}\mathfrak{S}$. Therefore, notice that $\theta(x), \eta(x)$ as well as $\alpha(x) = \theta(x) + \eta(x)$ are ordinals so that we can assume $\alpha(x) = \{\beta : \beta < \alpha(x)\}, \theta(x) = \{\beta : \beta < \theta(x)\}$ and $\eta(x) = \{\beta : \theta(x) \leq \beta < \alpha(x)\}$. Let be $U = \{(x, \beta) : \beta \in \alpha(x)\}$, and define $f : U \to U$ by

$$f(x, \beta) = \begin{cases} (\breve{x}, \beta) & \text{if } x \neq \breve{x} \text{ or } \beta \in \theta(x), \\ (x, \beta + 1) & \text{if } x = \breve{x} \text{ and } \beta \in \eta(x) \text{ and } \beta = \theta(x) + 2 * \beta', \\ (x, \beta - 1) & \text{if } x = \breve{x} \text{ and } \beta \in \eta(x) \text{ and } \beta = \theta(x) + 2 * \beta' + 1. \end{cases}$$

Obviously, we have $f(f(u)) = u$ for all $u \in U$, and $f(u) = u$ iff $u = (x, \beta)$ with $x = \breve{x}$ and $\beta \in \theta(x)$. We will denote the second component of $f(x, \beta)$ by $f_2(x, \beta)$ so that $f(x, \beta) = (\breve{x}, f_2(x, \beta))$ and $\beta = f_2(\breve{x}, f_2(x, \beta))$ follows. Also notice that $f_2(x, \beta) = f_2(\breve{x}, \beta)$.

We want to show the following property:

$(*)$ If $y \in \text{Bij}\mathfrak{B}$, then $f_2(x, \beta) = f_2(x; y, \beta)$ for all $\beta \in \alpha(x) = \alpha(x; y)$.

First, suppose $f_2(x, \beta) \neq \beta$. Then $x = \breve{x}$ and $\beta \in \eta(x)$, and, hence, $\eta(x) > 0$, by the definition of f. From (g) we get $(x; y) = (x; y)^{\smallsmile}$ and $\theta(x) = \theta(x; y)$. This also implies that $\eta(x) = \eta(x; y)$ because of (f) and the left-cancellation property of ordinal addition. Consequently, we have $\beta \in \eta(x; y)$, and, hence, $f_2(x, \beta) = f_2(x; y, \beta)$ follows. Now, assume $f_2(x, \beta) = \beta$. If $f_2(x; y, \beta) \neq \beta$, then we conclude analogously to the previous case that $f_2(x, \beta) \neq \beta$ using that $\breve{y}$ is a bijection and that $x; y; \text{breve}y = x$. This is a contradiction so that $f_2(x; y, \beta) = \beta$ follows.

Notice that we also have that $y \in \text{Bij}\mathfrak{B}$ implies $f_2(x, \beta) = f_2(y; x, \beta)$ for all $\beta \in \alpha(x) = \alpha(y; x)$. This follows immediately from $\alpha(x) = \alpha(\breve{x})$ and $f_2(x, \beta) = f_2(\breve{x}, \beta)$ for all x and β.

Now let $I := \{(1', 0)\}$ and define

$$C = \bigcup \{[(x, \beta), (y, \gamma), (z, \delta)] : z \leq x; y \text{ and } x, y, z \notin \text{Bij}\mathfrak{B}\}$$

$$\cup \bigcup \{[(x, \beta), (y, 0), (z, \beta)] : z \leq x; y \text{ and } y \in \text{Bij}\mathfrak{B}\}.$$

In order to show that $\mathfrak{Cm6}$ is a NA it is sufficient to show Property (b) of Theorem 1(1)(iii) since C is defined as a union of cycles.

Suppose $(x, \beta) \in U$. Then we have $[(x, \beta), (1', 0), (x, \beta)] \subseteq C$. For the converse implication suppose $\langle (x, \beta), (1', 0), (z, \delta) \rangle \in C$. We obtain $[(x, \beta), (1', 0), (z, \delta)] = [(z, \delta), (\breve{1}', 0), (x, \beta)] \subseteq C$ so that the definition of C implies that $x = z$ and $\beta = \delta$.

In order to prove that $\mathfrak{Cm6}$ is a RA suppose $\langle (v, \beta), (w, \gamma), (x, \delta) \rangle \in C$ and $\langle (x, \delta), (y, \epsilon), (z, \rho) \rangle \in C$. We distinguish several cases:

1. $v \in \mathrm{Bij}\mathfrak{B}$: We want to show that $(\breve{v}; z, \sigma)$ with $\sigma = f_2(\breve{v}; z, f_2(z, \rho))$ is the required element, i.e., that

    $$\langle (v, 0), (\breve{v}; z, \sigma), (z, \rho) \rangle \in C \text{ and } \langle (w, \gamma), (y, \epsilon), (\breve{v}; z, \sigma) \rangle \in C.$$

    First, we have $\rho \in \alpha(z)$ which implies $f_2(z, \rho) \in \alpha(\breve{z}) = \alpha(\breve{z}; v)$ using (f) so that $\sigma \in \alpha(\breve{v}; z)$, and, hence, $(\breve{v}; z, \sigma) \in U$ follows. Furthermore, we have $v; \breve{v}; z = z$ and $w; y = \breve{v}; v; w; y \geq \breve{v}; x; y \geq \breve{v}; z$ so that the property in the definition of C on the first components of each triple is satisfied. From $[(\breve{z}, f_2(z, \rho)), (v, 0), (\breve{z}; v, f_2(z, \rho))] \subseteq C$ by definition we conclude that the first triple is clearly in C. In order to show that the second triple is also in C we distinguish four cases:

    (a) $w, y, \breve{v}; z \notin \mathrm{Bij}\mathfrak{B}$: In this case we have $[(w, \gamma), (y, \epsilon), (\breve{v}; z, \sigma)] \subseteq C$ and we are done.

    (b) $w \in \mathrm{Bij}\mathfrak{B}$: Then x is a bijection too because $x \leq v; w$ and v and w are bijections. We conclude from $\langle (x, 0), (y, \epsilon), (z, \rho) \rangle \in C$ that we have $[(\breve{y}, f_2(y, \epsilon)), (\breve{x}, 0), (\breve{z}, f_2(z, \rho))] = [(\breve{z}, f_2(z, \rho)), (x, 0), (\breve{y}, f_2(y, \epsilon))] \subseteq C$, and, hence, that $f_2(y, \epsilon) = f_2(z, \rho)$. This implies $[(\breve{z}; v, f_2(z, \rho)), (w, 0), (\breve{y}, f_2(y, \epsilon))] \subseteq C$ which shows $\langle (w, 0), (y, \epsilon), (\breve{v}; z, \sigma) \rangle \in C$.

    (c) $y \in \mathrm{Bij}\mathfrak{B}$: Then we conclude from $\langle (x, \delta), (y, 0), (z, \rho) \rangle \in C$, and, hence, $[(x, \delta), (y, 0), (z, \rho)] = [(z, \rho), (\breve{y}, 0), (x, \delta)] \subseteq C$, that $\delta = \rho$. Similarly, from $\langle (v, 0), (w, \gamma), (x, \delta) \rangle \in C$, i.e., $[(\breve{w}, f_2(w, \gamma)), (\breve{v}, 0), (\breve{x}, f_2(x, \delta))] = [(\breve{x}, f_2(x, \delta)), (v, 0), (\breve{w}, f_2(w, \gamma))] \subseteq C$, we get $f_2(w, \gamma) = f_2(x, \delta)$ since v is a bijection. We conclude

    $$
    \begin{aligned}
    \sigma &= f_2(\breve{v}; z, f_2(z, \rho)) \\
    &= f_2(\breve{v}; x; y, f_2(x; y, \delta)) && \delta = \rho \text{ and } z = x; y \\
    &= f_2(w; y, f_2(x; y, \delta)) && v; w = x \text{ and } v \text{ bijection} \\
    &= f_2(w, f_2(x, \delta)) && (*) \text{ twice} \\
    &= f_2(w, f_2(w, \gamma)) && f_2(w, \gamma) = f_2(x, \delta) \\
    &= \gamma,
    \end{aligned}
    $$

    so that $\langle (w, \gamma), (y, 0), (\breve{v}; z, \rho) \rangle \in C$ follows.

    (d) $\breve{v}; z \in \mathrm{Bij}\mathfrak{B}$: This implies $\sigma = 0$ and that $z = v; \breve{v}; z$ is also a bijection, and, hence, $\rho = 0$. From $\langle (v, 0), (w, \gamma), (x, \delta) \rangle \in C$, and, hence, $[(\breve{x}, f_2(x, \delta)), (v, 0), (\breve{w}, f_2(w, \gamma))] = [(\breve{w}, f_2(w, \gamma)), (\breve{v}, 0), (\breve{x}, f_2(x, \delta))] \subseteq C$, we get $f_2(w, \gamma) = f_2(x, \delta)$. In addition $\langle (x, \delta), (y, \epsilon), (z, 0) \rangle \in C$, and, hence, $[(\breve{x}, f_2(x, \delta)), (z, 0), (y, \epsilon)] = [(y, \epsilon), (\breve{z}, 0), (\breve{x}, f_2(x, \delta))] \subseteq C$, implies $f_2(x, \delta) = \epsilon$. Together we obtain $f_2(w, \gamma) = f_2(x, \delta) = \epsilon$ so that $[(\breve{w}, f_2(w, \gamma)), (\breve{v}; z, 0), (y, \epsilon)] \subseteq C$, i.e., $\langle (w, \gamma), (y, \epsilon), (\breve{v}; z, 0) \rangle \in C$, follows.

2. The cases w or $y \in \mathrm{Bij}\mathfrak{B}$ can be shown analogously by using the elements $(w; y, f_2(w; y, f_2(y, \epsilon)))$, and $(w; y, \gamma)$, respectively.

3. $v, w, y \notin \mathrm{Bij}\mathfrak{B}$: Since $x \leq v; w$, $z \leq x; y$ and $\mathfrak{B}$ is a RA there is an element $u \in \mathrm{At}\mathfrak{B}$ with $z \leq v; u$ and $u \leq w; y$. If $u \notin \mathrm{Bij}\mathfrak{B}$, then we choose $\sigma = f_2(v, \beta)$ if $z \in \mathrm{Bij}\mathfrak{B}$ and an arbitrary $\sigma \in \alpha(u)$ otherwise. This choice gives $[(\breve{v}, f_2(v, \beta)), (z, \rho), (u, \sigma)] \subseteq C$ which immediately implies

$$\langle (v, \beta), (u, \sigma)), (z, \rho) \rangle \in C \text{ and } \langle (w, \gamma), (y, \epsilon), (u, \sigma) \rangle \in C.$$

Now, suppose $u \in \mathrm{Bij}\mathfrak{B}$. If $\beta = \rho$ and $f_2(w, \gamma) = \epsilon$, then $[(v, \beta), (u, 0), (z, \rho)] \subseteq C$ and $[f(w, \gamma), (u, 0), (y, \epsilon)] \subseteq C$ implies the assertion. The remaining two cases are shown as follows:

(a) $\beta \neq \rho$: Then $\beta \in \alpha(v) = \alpha(v; u) = \alpha(z) \ni \rho$ because of (f) and $z = v; u$. Consequently $\alpha(\breve{z}) = \alpha(z) > 1$. In addition $\breve{x} \leq \breve{w}; \breve{v} = \breve{w}; u; \breve{u}; \breve{v} = y; \breve{z}$ shows

$$\breve{w} = \breve{w}; u; \breve{u}$$

$= y; \breve{u}$	$y = \breve{w}; u$ since $u \leq w; y$ and u bijection
$\leq y; (\breve{z}; z \cdot 0'); \breve{u}$	(h) since $y \notin \mathrm{Bij}\mathfrak{B}$, $\alpha(\breve{z}) > 1$ and $y; \breve{z} > \breve{x}$
$= \breve{w}; u; (\breve{u}; \breve{v}; v; u \cdot 0'); \breve{u}$	$y = \breve{w}; u, z = v; u$
$= \breve{w}; (\breve{v}; v \cdot u; 0'; \breve{u})$	Lemma 1
$= \breve{w}; (\breve{v}; v \cdot 0')$	Lemma 1.

This implies that there is an atom $u' \in \mathfrak{B}$ with $\breve{w} \leq \breve{w}; u'$ and $u' \leq \breve{v}; v \cdot 0'$ since otherwise $\breve{w} = 0$ would follow. In particular, $u' \notin \mathrm{Bij}\mathfrak{B}$ because otherwise $\breve{w}; u' = \breve{w}$, and, hence, $u' = 1'$ would follow, a contradiction to $u' \leq 0'$. Therefore, $u'; u$ is an atom and not a bijection. Furthermore, we have $y = \breve{w}; u \leq \breve{w}; u'; u = \breve{w}; (u'; u)$ and $u'; u \leq (\breve{v}; v \cdot 0'); u \leq \breve{v}; v; u = \breve{v}; z$ so that $u'; u \leq w; y$ and $z \leq v; (u'; u)$ follows. Consequently, we can use $u'; u$ instead of the bijection u.

(b) $f_2(w, \gamma) = \epsilon$: This case is shown analogously to the previous case.

The algebra $\mathfrak{B}$ can be embedded into $\mathfrak{C}\mathrm{m}\mathfrak{S}$ using the function $h(b) = \{(x, \beta) \mid x \in \mathrm{At}\mathfrak{B}, x \leq b, \beta \in \alpha(x)\}$ which is easy to verify. The obvious definition $c : \mathrm{At}\mathfrak{C}\mathrm{m}\mathfrak{S} \rightarrow \mathrm{At}h(\mathfrak{B})$ by $c(x, \beta) = h(x)$, i.e., $c(x, \beta) = \{(x, \beta) : \beta \in \alpha(x)\}$, shows that $\mathfrak{C}\mathrm{m}\mathfrak{S}$ is an extension of the image $h(\mathfrak{B})$ along η and θ. Finally, we obtain $\mathfrak{A}$ by replacing the image $h(\mathfrak{B})$ of $\mathfrak{B}$ in $\mathfrak{C}\mathrm{m}\mathfrak{S}$ by $\mathfrak{B}$ itself. $\square$

4.2   Further Examples

In this section we present further examples showing situations in which the new mechanism can or cannot be applied. We want to start with an example that does not satisfy the requirements of Theorem 4 and does not lead to a relation algebra after splitting.

Example 2. Let $\mathfrak{A}$ be the relation algebra with $n = 3, s = 3$ and the following diversity cycles

$$C(\mathfrak{A}) = [(2,2,3),(2,3,3)].$$

The relation algebra above is the so-called *pentagonal relation algebra* [22]. If we simply apply the splitting algorithm induced by Theorem 4 to split Atom 2, we obtain the following new structure $\mathfrak{B}$ with $n = 4, s = 4$, the following correlation between atoms in $\mathfrak{B}$ and $\mathfrak{A}$,

Atoms of $\mathfrak{B}$	1	2	3	4
Atoms of $\mathfrak{A}$	1	2	2	3

and the full composition table:

;	2	3	4
2	4	4	234
3	4	4	234
4	234	34	23

This structure is not a relation algebra since $(\{2\};\{3\});\{4\} = \{4\};\{4\} = \{2,3\} \neq \{2,3,4\} = \{2\};\{2,3,4\} = \{2\};(\{3\};\{4\})$. In fact, Property (h) of Theorem 4 is violated since

$$2;(3;3 \cdot (2+3)) = 2;((1+2) \cdot (2+3)) = 2;2 = 1 + 3 \not\geq 2.$$

We did some further investigation on this very interesting example. Actually there is no integral extension of the relation algebra $\mathfrak{A}$ along any possible pair of functions so that Atom 2 or Atom 3 (or both) splits into two atoms. However, the algebra is representable on set of five elements. All three atoms are represented by non-atomic relations which shows that all atoms can be split resulting in a simple (but not integral) relation algebra. Notice that in [12] it was shown that there are even algebras that cannot be properly embedded in any simple relation algebra.

The next example actually shows that our mechanism can be applied where regular splitting will fail.

Example 3. Let $\mathfrak{A}$ be the relation algebra with $n = 11, s = 7$ and the following diversity cycles

$$\begin{aligned}
C(\mathfrak{A}) = [\ &(8,8,8),(8,8,10),(8,9,2),(8,9,5),(8,9,3),(8,10,10),(8,11,11),\\
&(8,11,5),(8,11,2),(8,11,3),(8,5,8),(8,5,10),(8,5,5),(8,5,3),\\
&(8,5,2),(8,6,8),(8,6,5),(8,6,10),(8,6,6),(8,6,3),(8,6,4),(8,7,8),\\
&(8,7,10),(8,7,5),(8,7,6),(8,7,7),(8,3,3),(8,3,2),(8,4,3),(8,2,2),\\
&(10,10,10),(10,11,5),(10,11,3),(10,11,2),(10,5,10),(10,5,5),\\
&(10,5,3),(10,5,2),(10,6,10),(10,6,5),(10,6,3),(10,6,2),(10,7,10),\\
&(10,7,5),(10,7,6),(10,7,7),(10,7,3),(10,7,4),(10,7,2),(10,3,2),\\
&(10,4,2),(10,2,2),(5,5,5),(5,5,6),(5,5,7),(5,5,2),(5,5,3),(5,5,4),\\
&(5,6,6),(5,6,7),(5,7,7),(5,3,3),(5,3,2),(5,2,2),(6,6,6),(6,6,7),\\
&(6,7,7),(7,7,7),(3,3,3),(3,3,2),(3,2,2),(2,2,2)]
\end{aligned}$$

This algebra is actually the algebra RCC11 based on the atomic relationships 1', DC, ECN, ECD, PON, PODY, PODZ, TPP, and NTPP numbered in that order recognizing that TPP and NTPP are non-symmetric. For example, DC is Atom 2, PODZ is Atom 7, TPP is Atom 8, TPP$^{\smile}$ is Atom 9, and NTPP is Atom 10. For details on mereotopological properties of those atomic relations we refer to [7]. As mentioned earlier this mereotopological example contains a bijection different from the identity, the relation ECD or the Atom 4. No atom of this algebra can be split using the mechanism in [2]. Using our approach we can split TPP into two new atoms called TPPA and TPPB. As a consequence of that fact that TPP is non-symmetric and that the algebra contains the bijection ECD, and, hence, the Properties (b) and (f) of Theorem 4, we have also to split TPP$^{\smile}$, ECN, and PODY each into two new relations TPPA$^{\smile}$, TPPB$^{\smile}$, ECNA, ECNB, PODYA and PODYB. This yields an algebra $\mathfrak{B}$ with $n = 15, s = 9$, the following correlation between the atoms

Names in $\mathfrak{B}$	1'	DC	ECNA	ECNB	ECD	PON	PODYA	PODYB	PODZ	TPPA	TPPA$^{\smile}$	TPPB	TPPB$^{\smile}$	NTPP	NTPP$^{\smile}$
Atoms of $\mathfrak{B}$	1	2	3	4	5	6	7	8	9	10	11	12	13	14	15
Atoms of $\mathfrak{A}$	1	2	3	3	4	5	6	6	7	8	9	8	9	10	11
Names in $\mathfrak{A}$	1'	DC	ECN	ECN	ECD	PON	PODY	PODY	PODZ	TPP	TPP$^{\smile}$	TPP	TPP$^{\smile}$	NTPP	NTPP$^{\smile}$

and the following set of cycles:

$C(\mathfrak{A}) = [\,(10, 10, 10), (10, 10, 12), (10, 12, 10), (10, 12, 12), (12, 10, 10), (12, 10, 12),$
$(12, 12, 10), (12, 12, 12), (10, 10, 14), (10, 12, 14), (12, 10, 14), (12, 12, 14),$
$(10, 11, 2), (10, 13, 2), (12, 13, 2), (10, 11, 6), (10, 13, 6), (12, 13, 6),$
$(10, 11, 3), (10, 11, 4), (10, 13, 3), (10, 13, 4), (12, 13, 3), (12, 13, 4),$
$(10, 14, 14), (12, 14, 14), (10, 15, 15), (12, 15, 15), (10, 15, 6), (12, 15, 6),$
$(10, 15, 2), (12, 15, 2), (10, 15, 3), (10, 15, 4), (12, 15, 3), (12, 15, 4),$
$(10, 6, 10), (10, 6, 12), (12, 6, 12), (10, 6, 14), (12, 6, 14), (10, 6, 6),$
$(12, 6, 6), (10, 6, 3), (10, 6, 4), (12, 6, 3), (12, 6, 4), (10, 6, 2), (12, 6, 2),$
$(10, 7, 10), (10, 7, 12), (10, 8, 10), (10, 8, 12), (12, 7, 12), (12, 8, 12),$
$(10, 7, 6), (10, 8, 6), (12, 7, 6), (12, 8, 6), (10, 7, 14), (10, 8, 14), (12, 7, 14),$
$(12, 8, 14), (10, 7, 7), (10, 7, 8), (10, 8, 7), (10, 8, 8), (12, 7, 7), (12, 7, 8),$
$(12, 8, 7), (12, 8, 8), (10, 7, 3), (10, 7, 4), (10, 8, 3), (10, 8, 4), (12, 7, 3),$
$(12, 7, 4), (12, 8, 3), (12, 8, 4), (10, 7, 5), (10, 9, 10), (10, 9, 12), (12, 9, 12),$
$(10, 9, 14), (12, 9, 14), (10, 9, 6), (12, 9, 6), (10, 9, 7), (10, 9, 8), (12, 9, 7),$
$(12, 9, 8), (10, 9, 9), (12, 9, 9), (10, 3, 3), (10, 3, 4), (10, 4, 3), (10, 4, 4),$
$(12, 3, 3), (12, 3, 4), (12, 4, 3), (12, 4, 4), (10, 3, 2), (10, 4, 2), (12, 3, 2),$
$(12, 4, 2), (10, 5, 3), (10, 2, 2), (12, 2, 2), (14, 14, 14), (14, 15, 6),$
$(14, 15, 3), (14, 15, 4), (14, 15, 2), (14, 6, 14), (14, 6, 6), (14, 6, 3),$
$(14, 6, 4), (14, 6, 2), (14, 7, 14), (14, 8, 14), (14, 7, 6), (14, 8, 6), (14, 7, 3),$
$(14, 7, 4), (14, 8, 3), (14, 8, 4), (14, 7, 2), (14, 8, 2), (14, 9, 14), (14, 9, 6),$
$(14, 9, 7), (14, 9, 8), (14, 9, 9), (14, 9, 3), (14, 9, 4), (14, 9, 5), (14, 9, 2),$

344   P. Siddavaatam and M. Winter

$(14, 3, 2), (14, 4, 2), (14, 5, 2), (14, 2, 2), (6, 6, 6), (6, 6, 7), (6, 6, 8),$
$(6, 6, 9), (6, 6, 2), (6, 6, 3), (6, 6, 4), (6, 6, 5), (6, 7, 7), (6, 7, 8), (6, 8, 8),$
$(6, 7, 9), (6, 8, 9), (6, 9, 9), (6, 3, 3), (6, 3, 4), (6, 4, 4), (6, 3, 2), (6, 4, 2),$
$(6, 2, 2), (7, 7, 7), (7, 7, 8), (7, 8, 8), (8, 8, 8), (7, 7, 9), (7, 8, 9), (8, 8, 9),$
$(7, 9, 9), (8, 9, 9), (9, 9, 9), (3, 3, 3), (3, 3, 4), (3, 4, 4), (4, 4, 4), (3, 3, 2),$
$(3, 4, 2), (4, 4, 2), (3, 2, 2), (4, 2, 2), (2, 2, 2), (12, 8, 5), (12, 5, 4)]$

This splitting (and some further splittings) was already done in [7]. The main difference is that in the aforementioned paper the algebra was basically computed by hand, and the result above was obtained by a Haskell program implementing our approach.

5   Conclusion and Outlook

We consider this paper as a starting point of a variety of methods for splitting atoms in relation algebras. The very general definition of an extension provides the opportunity for this study. It will be of particular interest to characterize the different methods by additional properties. For example, the definition of a splitting together with Theorem 3 characterizes this construction precisely. We did not provide a full characterization for our method yet. We believe that our method generates a maximal extension of the given algebra along the functions η and θ within a certain class of extensions.

Another important process in generating new algebras from old ones is the removal of cycles. First of all, one may obtain further extensions of an algebra by applying Theorem 4 first, and then removing certain cycles. If Theorem 4 really result in a maximal extension, one could actually obtain all extension using this and similar approaches. Furthermore, this process has also applications in mereotopology. We want to illustrate this by an example. In the process of generating refinements of the algebra at hand, one considers an atom in one spot of the composition table and investigates whether two subcases related to the composition at hand are possible. For example, if one considers RCC11, the relation TPP of tangential proper part, one may recognize that the table suggests TPP$\leq$ECN;TPP, i.e., whenever one region is inside another region so that their borders intersect, then there is a region that is externally connected to the first and itself a tangential proper part of the second. It is not hard to see that one can construct examples as well as counterexamples for this statement. This indicates that the composition table of RCC11 is not the composition table of the concrete relations in a model of RCC. This situation can be taken as the starting point for generating a new algebra, a refinement of RCC11. This is done by first splitting TPP in RCC11 as in Example 3 obtaining an algebra with 15 atoms. In order to obtain a refinement of RCC11 we have also to remove the cycle (ECN,TPP,TPP) for one of the two copies TPPA, TPPB of TPP in the new algebra. The result of this process will produce the mereotopological algebra RCC15 which can be further refined using similar steps to RCC25. For further details about this process we refer to [7]. A mechanization of this second step is also necessary in order to further advance relational methods in qualitative spatial reasoning.

References

1. Andréka, H., Düntsch, I., Németi, I.: A non permutational integral relation algebra. Michigan Math. J. 39, 371–384 (1992)
2. Andréka, H., Maddux, R.D., Németi, I.: Splitting in relation algebras. Proc. Amer. Math. Soc. 111(4), 1085–1093 (1991)
3. Birkhoff, G.: Lattice Theory. 3rd edn., vol. XXV. American Mathematical Society Colloquium Publications (1968)
4. de Laguna, T.: Point, line, and surface, as sets of solids. J. of Philosophy 19(17), 449–461 (1922)
5. Dimov, G., Vakarelov, D.: Contact algebras and region-based theory of space: A proximity approach - I. Fundamenta Informaticae 74(2-3), 209–249 (2006)
6. Düntsch, I.: Relation algebras and their application in temporal and spatial reasoning. Artificial Intelligence Review 23, 315–357 (2005)
7. Düntsch, I., Schmidt, G., Winter, M.: A necessary relation algebra for mereotopology. Studia Logica 69, 381–409 (2001)
8. Düntsch, I., Wang, H., McCloskey, S.: Relation algebras in qualitative spatial reasoning. Fundamenta Informaticae 39(3), 229–248 (2000)
9. Düntsch, I., Wang, H., McCloskey, S.: A relation algebraic approach to the Region Connection Calculus. Theoretical Computer Science 255, 63–83 (2001)
10. Düntsch, I., Winter, M.: A Representation Theorem for Boolean Contact Algebras. Theoretical Computer Science 347, 498–512 (2005)
11. Düntsch, I., Winter, M.: Weak Contact Structures. In: MacCaull, W., Winter, M., Düntsch, I. (eds.) RelMiCS 2005. LNCS, vol. 3929, pp. 73–82. Springer, Heidelberg (2006)
12. Frias, M., Maddux, R.D.: Non-embeddable simple relation algebras. Algebra Universalis 38(2), 115–135 (1997)
13. Gotts, N.M.: Topology from a single primitive relation: Defining topological properties and relations in terms of connection. Research Report 96.23, School of Computer Studies, University of Leeds (1996)
14. Grätzer, G.: General Lattice Theory, 2nd edn. Birkhäuser, Basel (1998)
15. Henkin, L., Monk, J.D., Tarski, A.: Cylindric algebras, Part II. North-Holland, Amsterdam (1985)
16. Hirsch, R., Hodkinson, I.: Strongly representable atom structures of relation algebras. Proceedings of the Amer. Math. Soc. 130, 1819–1831 (2002)
17. Jónsson, B., Tarski, A.: Boolean algebras with operators, Part II. Amer. Journal of Mathematics 73, 127–162 (1952)
18. Kahl, W., Schmidt, G.: Exploring (Finite) Relation Algebras Using Tools Written in Haskell. Universität der Bundeswehr München, Report Nr. 2000-02 (2000)
19. Koppelberg, S.: General Theory of Boolean Algebras. In: Handbook on Boolean Algebras, vol. 1, North-Holland, Amsterdam (1989)
20. Maddux, R.D.: Some varieties containing relation algebras. Trans. Amer. Math. Soc. 272, 501–526 (1982)
21. Maddux, R.D.: Finite integral relation algebras. In: Comer, S.D. (ed.) Universal Algebra and Lattice Theory, Proceedings of a Conference held at Charleston, LNM, vol. 1149, pp. 175–197 (1985)
22. Maddux, R.D.: Relation Algebras. Studies in Logic and the Foundations of Mathematics, vol. 150. Elsevier Science, Amsterdam (2006)
23. Offermann, E.: On the Construction of Relational Categories. Presented at RelMiCS 7, Kiel (2003),
http://www.informatik.uni-kiel.de/~relmics7/submitted/

24. Offermann, E.: Konstruktion relationaler Kategorien. Dissertation, Der Andere Verlag, Osnabrück (2003) ISBN 3-89959-078-3
25. Randell, D.A., Cui, Z., Cohn, A.G.: A spatial logic based on regions and connection. In: Proc. of KR 1992: Principles of Knowledge Representation and Reasoning, pp. 165–176. Morgan Kaufmann, San Francisco (1992)
26. Randell, D.A., Cohn, A.G., Cui, Z.: Computing transitivity tables: A challenge for automated theorem provers. In: Hapur, D. (ed.) CADE 1992. LNCS (LNAI), vol. 607, pp. 786–790. Springer, Heidelberg (1992)
27. Schmidt, G., Ströhlein, T.: Relations and Graphs. Discrete Mathematics for Computer Scientists. Springer, Heidelberg (1993)
28. Stell, J.G.: Boolean connection algebras: a new approach to the region-connection calculus. Artificial Intelligence 122, 111–136 (2000)
29. Tarski, A.: On the calculus of relations. Journal of Symbolic Logic 6, 73–89 (1941)
30. Tarski, A., Givant, S.: A formalization of set theory without variables, vol. 41. Colloquium Publications, Amer. Math. Soc., Providence (1987)
31. Vakarelov, D., Dimov, G., Düntsch, I., Bennett, B.: A proximity approach to some region-based theories of space. Journal of Applied Non-classical Logics 12(3-4), 527–559 (2002)
32. Whitehead, A.N.: Process and Reality. Macmillan, Basingstoke (1929)
33. Winter, M.: Relation Algebras are Matrix Algebras over a Suitable Basis. Universität der Bundeswehr München, Report Nr. 1998-05 (1998)
34. Winter, M.: A Pseudo Representation Theorem for Various Categories of Relations. Theory and Applications of Categories 7(2), 23–37 (1998)

Relational Heterogeneity Relaxed by Subtyping

Jaap van der Woude[1] and Stef Joosten[1,2]

[1] Open Universiteit Nederland,
Postbus 2960, 6401 DL HEERLEN
[2] Ordina NV, Nieuwegein
{jaap.vanderwoude,stef.joosten}@ou.nl

Abstract. Homogeneous relation algebra is an elegant calculational framework with many applications in computing science. In one application of relation algebra, called Ampersand, heterogeneous relation algebra is used as a specification language for business processes and information systems. For this purpose a typed version of relation algebra is needed together with subtyping. This requires heterogeneous relational algebra. However, the partiality of the composition and union operators in heterogeneous relational algebra are detrimental to its manipulative power. This paper proposes a practical solution to this problem. The authors suggest to relax the partiality of the heterogeneous operators. By suitable choices this homogenisation allows for a type-based specification language, which has sufficient manipulative power.

1 Introduction

A wealth of theory has been developed for binary relations (Tarski ([9]), Schmidt and Ströhlein ([7]), Freyd and Scedrov ([4]), Backhouse et al ([1]), Brink et al ([3]), Maddux ([5]), etc), both in the homogeneous and heterogeneous versions. The theory is interesting in itself, but it has lots of nice applications too. In the current story we are concerned with the use of relations in the realm of requirements and specification. In particular we want to use the manipulative power of relation algebra to construct information systems that support business processes.

This paper is a companion to ([8]) that to a large extent introduces and motivates a method called Ampersand. Ampersand is intended for designing business processes and information systems. It features tools that generate data models and other design artifacts for constructing information systems. Ampersand derives (generates) them from a collection of business rules. The tool is used to specify real life systems and has been used successfully in industry as well as in education. Requirements engineers express business rules in a heterogeneous relation algebra. Thus, Ampersand defines business process(es) by constraints. The heterogeneous version of relation algebra provides users with a type system, which generates feedback on flawed scripts. Recently, Ampersand has been enriched with generalization, which is not discussed in ([8]). In information systems, generalization is a standard feature. It is known under various names like

H. de Swart (Ed.): RAMICS 2011, LNCS 6663, pp. 347–361, 2011.

specialization, sub- or supertyping. It manifests itself in information models via ISA relations and inheritance. We shall freely use the terms generalization and sub- or supertyping as synonyms throughout this paper. The combination of heterogeneous relation algebra with generalization has been put to good use in an improved (provisional) Ampersand compiler for ease of construction, editability and generating feedback. This paper introduces that combination and studies its consequences for Tarski's axioms. It turns out that the resulting language has the same "homogeneous" three layer structure and satisfies similar axioms.

Ampersand looks at data in an information system as a set of binary relations that "store" the data. The data is organized by a structure that resembles an ontology; it is built up of *concepts*, *relations* between concepts and *rules* governing the relations. A graph, in which concepts are represented by vertices and relations by edges, is used in Ampersand as a conceptual model. Concepts may range from concrete entities like names and numbers to complex entities like orders and abstractions like delivery-appointments. Relations vary from deterministic (functions) to partial multivalued ones. As data changes over time, the contents of these relations do too. The representation of the data in the relations in Ampersand is governed by rules that form a set of invariants. Every now and then the rules may be violated by actions that change the actual content of relations. Upon violation, the system must be brought back in a state where the invariants are satisfied.

In Ampersand every relation is given a type at define time, as in $R : A \sim B$. This declaration implies that there exist concepts A and B, and every tuple $\langle a, b \rangle \in R$ implies $a \in A$ and $b \in B$. This interpretation is specific for Ampersand and is motivated in [8]. The user can formulate rules in terms of declared relations only. Ampersand features a compiler, which deduces a computer program from these rules. The resulting program maintains a database and keeps all rules satisfied in that database. Note that the intended use is in real information systems, which implies finiteness. The possible contents of the system may be infinite, but the number of concepts and the number of basic relations in the ontology certainly will be finite.

The feedback system of the Ampersand compiler ([8]) facilitates adaptation of relations and rules. The user is signaled in cases of predictable errors, mainly based on type information. If the user provides a formula the compiler cannot assign a unique type to, there might be a mistake in the formula or a choice between overloaded term names should be made. Well-typedness arguments are straightforward in a fully typed heterogeneous relation algebra, but how should we proceed in case supertyping is allowed. E.g. let $R : A \sim B$ and $S : C \sim D$ where C is a supertype of B. The composition $R \,\mathring{;}\, S$ might make sense to the user, but it is signaled incorrect by the typing mechanism. Relaxing the stringent typing convention raises a question about the composition $(\neg R) \,\mathring{;}\, S$. Do we consider the complement with respect to the type B we get from R or should we use its supertype C resulting from S?

1.1 Goal

For the purpose of generalization (subtyping or isa-constructions), the author of an Ampersand script can specify an order between two different concepts. We denote that order by $C \preceq D$ and say that C is a subtype of D or D is a supertype of C. On the element level one may say a C-member *is a* D-member. Users of Ampersand may (sloppily) refer to concepts as types. The concepts are assumed to form a lattice with some appropriate special properties depending on the problem domain. In fact a join semilattice would suffice, but assuming finiteness we get the complete lattice structure for free. If a user does not specify any order, the resulting lattice of concepts is flat.

The lattice has extreme elements representing extreme concepts: the empty ($\perp$) and the universal one ($\top$). Ampersand calls them *nothing* and *anything*. These concepts are the ones that we try to avoid in specifying the system: nothing doesn't need rules and anything doesn't allow them within reason. Ampersand's type checker rejects any relation term that has $\perp$ or $\top$ as its type.

Although concepts may be seen as sets of elementary things (also known as atoms in Ampersand), we should not necessarily interpret their meet and join setwise. In the current research version of Ampersand with editing possibilities, for example, the meet of two concepts is the more specific of the two if the concepts are order related, but nothing if not. Another example may be the join of vector spaces. In general we assume the (join semi-)lattice structure with extremities and ignore the interpretation and specialties of the concepts until the application section where choices of the concepts play a role.

Rules are constructed using the relations occurring in the ontology and they may be transformed by way of the manipulative possibilities in relation algebra. Users of Ampersand are meant to get help from relation calculus in formulating and rewriting those rules. For this reason, we need to know which of Tarski's axioms are available as transformation rules to users. Because of the typing with concepts, the homogeneous relation calculus is only locally applicable. The heterogeneous relation calculus, however, is too restrictive in its definedness of the relevant operators. We opt for an intermediate form of relation calculus that allows for subtyping, which has much of the calculational power of the homogeneous calculus and still has the typing advantages of heterogeneous relation calculus.

1.2 Result

In this paper we propose such an intermediate relational calculus. Starting of with a heterogeneous relation calculus we use generalization in order to view the lattice and composition operators on relations as being totally defined. To that end we make the subtyping explicit by introducing embeddings ϵ as relations and use them in composition with the non-compatibly typed relations to turn them into composable or addable relations. The typed lattice and composition structure is thus slightly generalized and the reverse structure is unaltered. From that point on, all we need to do is walk through Tarski's axioms using the new total operations and check their validity.

It turns out that with this construction the homogeneous relations with the ten Tarski axioms ([5], page 21) as manipulative power can be mimicked in the heterogeneous relations, preserving almost everything except for negation and the Schröder rules. But there is an acceptable way out of that omission: division and the Dedekind rule.

For the Ampersand user, this has an implication in the use of the negation. Instead of the unary complement he gets a binary subtraction operator that enables Ampersand to infer a type specialization if necessary. By way of syntactic sugar, he may omit the left hand argument, yielding the 'look and feel' of the complement operator. The Ampersand compiler facilitates this by adding a default value for the missing argument.

The Ampersand tool provides assistance in constructing relational terms by means of a type checker. In the perception of users, the type system discriminates meaningful expressions from meaningless ones. A meaningful expression in Ampersand is a relation term, which has a unique type other than $\bot$ or $\top$. Meaningless expressions are rejected by the type checker with an informative diagnostic message. This choice for meaningless terms in the augmented Ampersand tool is captured in the concept lattice and is thus separated from the relational calculus. We consider two concepts to be unrelated if their join equals $\top$, the universe of things without discriminating properties. Composition and gathering unrelated things is considered meaningless.

2 Definitions

First we introduce the usual system of heterogeneous relations and fix the notation we use. (Such may be done via order enriched categories or via allegories (e.g. [4]) but mastery of those fundamentals of mathematics is not necessary to read this paper.) After that we propose the changes to facilitate the sub- and supertyping.

2.1 Heterogeneous Relation Calculus

We consider the category $(\mathcal{C}, \mathcal{R}el)$ of relations on the concept set $\mathcal{C}$ [1]. Following the relational habit we shall denote the categorical and the relational composition by a semicolon $\mathbin{;}$ with high precedence. The sets of morphisms $\mathcal{R}el(A, B)$, i.e. the collections of all relations between the concepts A and B form complete, chain distributive lattices and these lattices are interrelated by way of the composition $\mathbin{;}$ and the converse $^\cup$, as follows:

$\mathbin{;} : \mathcal{R}el(A, B) \times \mathcal{R}el(B, C) \longrightarrow \mathcal{R}el(A, C)$

has identities as units, is associative and universally disjunctive

$^\cup : \mathcal{R}el(A, B) \longrightarrow \mathcal{R}el(B, A)$

is an isomorphic involution (i.e. $^\cup \circ {}^\cup = id$)

---

[1] If you are not familiar with the notion of category, think of a transitive directed graph whose vertices are concepts and whose edges are relations between the vertices they connect. An edge is termed morphism or relation. Pasting edges is called composition.

and they satisfy the converse-composition layer interface of contravariant distribution

$$(R \mathbin{\fatsemi} S)^{\cup} = S^{\cup} \mathbin{\fatsemi} R^{\cup}$$

The homsets $\mathcal{R}el(A, B)$ have a top $\top_{A,B}$, the universal relation, and a bottom $\bot_{A,B}$, the empty relation. Where the context prevents ambiguous interpretation, we shall omit subscripts.

Instead of $R \in \mathcal{R}el(A, B)$ we prefer to write $R : A \sim B$ and pronounce R is a relation between A and B (which by listening from left to right differs from being a relation between B and A).

We denote the left and right types of R by $R\triangleleft (= A$ above) and $R\triangleright (= B$ above) respectively.

The meet and join are written as $\sqcap$ and $\sqcup$ and they have the usual properties except for arbitrary distributivity. Considering the set-theoretic interpretation of relations in Ampersand, it is no coincidence that they look somewhat like set-theoretic intersection and union.

Often the morphisms are even assumed to form complete, complemented lattices. Moreover, the complement is used to formulate an interface between the three layers of the relation calculus: the so called Schröder rule (but it has many names and shapes)

$$\text{equivalent are: } R \mathbin{\fatsemi} S \sqsubseteq \neg T^{\cup}, \quad T \mathbin{\fatsemi} R \sqsubseteq \neg S^{\cup} \text{ and } S \mathbin{\fatsemi} T \sqsubseteq \neg R^{\cup}$$

that connects the order, the composition and the converse. Although complementation might work well in the homogeneous case and even in the localized heterogeneous version, we do not consider it because we will run into trouble because of the intended additional subtyping and the consequential uncertainty of the complementation domain. So we should cope without negation or at least without negation in considering relations with different typing. Instead of negation we may use division (or factors) to be introduced shortly.

We replace the Schröder rule by the three-layer interface known as the modular identity ([4]) or the Dedekind rule ([6]) that partly makes up for the lack of conjunctivity of $\mathbin{\fatsemi}$ as follows:

$$R \mathbin{\fatsemi} S \sqcap T \sqsubseteq (R \sqcap T \mathbin{\fatsemi} S^{\cup}) \mathbin{\fatsemi} (S \sqcap R^{\cup} \mathbin{\fatsemi} T)$$

In the presence of complementation the Dedekind rule is equivalent to the Schröder rule.

An interesting property of many relational systems, not captured by the ten Tarski axioms, is the universal disjunctivity of the composition. I.e.

$$(\sqcup U : U \in \mathcal{U} : U) \mathbin{\fatsemi} R = (\sqcup U : U \in \mathcal{U} : U \mathbin{\fatsemi} R) \quad \text{and}$$
$$R \mathbin{\fatsemi} (\sqcup U : U \in \mathcal{U} : U) = (\sqcup U : U \in \mathcal{U} : R \mathbin{\fatsemi} U)$$

Note that we use ternary notation for quantifications and repeated operator applications, similar to the Z-notation (with colons as separators in stead of a bar and a bullet). The first part of the quantification designates the repeated operator and the dummies to be used in the terms (e.g. in the second quantification these are $\sqcup U$), the second part indicates the domain for the dummies ($U \in \mathcal{U}$) and the third part is for the terms to be quantified ($U \,\fatsemi\, R$). In section **3.3** more examples of this notation occur.

The universal disjunctivity property of relational systems is equivalent to ([2]) the existence of two Galois connections between the left and right compositions and operations that might therefore be called factorisations or divisions (the term that we shall use here):

$$R \sqsupseteq S \,\fatsemi\, T \;\equiv\; R /\!\!/ T \sqsupseteq S \quad \text{and} \quad S \,\fatsemi\, T \sqsubseteq R \;\equiv\; T \sqsubseteq S\backslash\!\!\backslash R$$

Adjointness of $\fatsemi T$ and $/\!\!/ T$ and of $S\fatsemi$ and $S\backslash\!\!\backslash$ is (in a certain sense) equivalent to universal distributivity of $\fatsemi T$ and $S\fatsemi$ respectively. The name division is illustrated by the next cancellation property:

$$R \sqsupseteq (R /\!\!/ T) \,\fatsemi\, T \quad \text{and} \quad S \,\fatsemi\, (S\backslash\!\!\backslash R) \sqsubseteq R$$

These cancellation properties show some typing restrictions in the heterogeneous setting. The composition and the inclusion should make sense, so the typing of the lefthand division is $R /\!\!/ T : R^\triangleleft \sim T^\triangleleft$ and it only exists if the right types of R and T are the same ($R^\triangleright = T^\triangleright$).

Division is strongly related to negation. Indeed, from the Schröder rule it follows for instance that $S\backslash\!\!\backslash R = \neg\,(S^\cup \fatsemi \neg R)$, and thus that $\neg\,S = S^\cup\backslash\!\!\backslash \neg I$. So, considering the reluctant use of the negation in our lattice layer, division might be a useful operation to replace some of the applications of negation. For the application in constructing constraints and business rules the division has the following set-theoretic interpretation:

$$a\langle R /\!\!/ T \rangle b \;\equiv\; (\forall\, c : b\,T\,c : a\,R\,c) \quad \text{and} \quad a\langle S\backslash\!\!\backslash R \rangle b \;\equiv\; (\forall\, c : c\,S\,a : c\,R\,b)$$

2.2 Adding Supertyping

In information modeling the notion of *ISA* relation captures inheritance of entities. It stands for embedding specialized things as things. Here we coin the term generalization or supertyping for this. Generalization or supertyping may be thought of as losing some but possibly not all information (e.g. a manager is an employee), while subtyping stands for refining or specializing the information (e.g. adding a locked drawer makes a desk more determined). In the standard heterogeneous relation calculus we cannot add the fact that manager *Baas* uses locked-drawer-desk $ldd481$ to the relation of employees using desks because the types do not match. But we do want to do so because *Baas* is an employee and $ldd481$ is a desk. What we can do is embedding subconcepts in concepts (or concepts in superconcepts) and extend that embedding to relations, so employee is a supertype of manager and locked-drawer-desk is a subtype of desk.

Having a relation between subconcepts, the question arises how to extend it to their generalizations? And the other way around: can we restrict relations between types to subtypes? Instead of the given examples we may also think of clients as special persons, appointments as generalized meetings, invitations of clients to meetings or appointments for persons and the like.

In Ampersand the concepts occurring in the ontology may be sub- or supertyped and we assume that the generalization is structured as a lattice with extremal elements ($\perp$ and $\top$). We consider the generalization as an order on the concepts, given by way of embeddings of subtypes in types. The combination of heterogeneous relation algebra and generalization is substantiated by introducing the embedding functions (ϵ) as heterogeneous relations, thus incorporating the subtyping in the relation algebra. This enables us to extend the composition and the definedness of the lattice operators. As follows:

Consider the lattice $(\mathcal{C}, \preceq)$ of concepts as a subcategory of $(\mathcal{C}, \mathcal{R}el)$ by adding an embedding relation $\epsilon_{A,B}$ for every $A \preceq B$. Clearly, we want

$$\epsilon_{A,A} = id_A \quad \text{and} \quad \epsilon_{A,B} \mathbin{\text{\small ⦂}} \epsilon_{B,C} = \epsilon_{A,C}$$

Moreover, the embeddings should be injective and functional (total and univalent) relations, hence

$$id_A = \epsilon_{A,B} \mathbin{\text{\small ⦂}} \epsilon_{A,B}^{\cup} \quad \text{and} \quad \epsilon_{A,B}^{\cup} \mathbin{\text{\small ⦂}} \epsilon_{A,B} \sqsubseteq id_B$$

We assume that the concepts, their members and their relations are such that this is possible.

Following the embedding of concepts, we can now embed the relations (the homsets) too. Assume $A \preceq C$ and $B \preceq D$, then $R : A \sim B$ is embedded as relation between the supertypes C and D by composing it with the appropriate embeddings:

$$\text{if } R : A \sim B \text{ then } \epsilon_{A,C}^{\cup} \mathbin{\text{\small ⦂}} R \mathbin{\text{\small ⦂}} \epsilon_{B,D} : C \sim D$$

Indeed, in the left concept only A-members of C are considered via $\epsilon_{A,C}^{\cup}$, while the resulting B-members are properly embedded in D.

Not only can we embed relations in supertypes, we may also restrict relations to subtypes:

$$\text{if } S : C \sim D \text{ then } \epsilon_{A,C} \mathbin{\text{\small ⦂}} S \mathbin{\text{\small ⦂}} \epsilon_{B,D}^{\cup} : A \sim B$$

Because of the embedding $\epsilon_{A,C}$ only A-members of C will occur on the left type of S, while the resulting D-members are filtered by $\epsilon_{B,D}^{\cup}$ so that only B-members are accepted.

As expected, restriction after embedding is just the identity, but embedding after restriction stays restrictive. Indeed:

$$\epsilon_{A,C} \mathbin{\text{\small ⦂}} \epsilon_{A,C}^{\cup} \mathbin{\text{\small ⦂}} R \mathbin{\text{\small ⦂}} \epsilon_{B,D} \mathbin{\text{\small ⦂}} \epsilon_{B,D}^{\cup} = id_A \mathbin{\text{\small ⦂}} R \mathbin{\text{\small ⦂}} id_B = R$$

while

$$\epsilon_{A,C}^{\cup} \mathbin{\text{\small ⦂}} \epsilon_{A,C} \mathbin{\text{\small ⦂}} S \mathbin{\text{\small ⦂}} \epsilon_{B,D}^{\cup} \mathbin{\text{\small ⦂}} \epsilon_{B,D} \sqsubseteq id_C \mathbin{\text{\small ⦂}} S \mathbin{\text{\small ⦂}} id_D = S$$

Finally, note that if two concepts, say A and B, have a common supertype, say C, we can consider the commonality within them: $\epsilon_{A,C} \,\fatsemi\, \epsilon_{B,C}^{\cup} : A \sim B$. In particular this holds for the join $A \vee B$ instead of C as well as for $\top$ instead of C.

3 Adapting the Operations

In our heterogeneous relation system we incorporated the sub- and supertyping by embeddings between refining concepts, leading to embedding and restriction of relations. The next step is to extend the inter-homset relation operators $\fatsemi$, $^{\cup}$ and the homset lattice operations $\sqcup$ and $\sqcap$, preferably retaining the interfaces between the layers.

3.1 New Composition

In the heterogeneous system, the composition is only defined if the right concept of the first and the left concept of the second argument are equal:

$$\fatsemi : (A \sim B) \times (B \sim C) \longrightarrow (A \sim C)$$

We want to relax the equality "in the middle" so, that composition is also defined in case the middle components are refining. To that end we define the extended composition by

$$\textcircled{\fatsemi} : (A \sim B) \times (C \sim D) \longrightarrow (A \sim D) \quad \text{with}$$

$$R \textcircled{\fatsemi} S \;=\; R \,\fatsemi\, \epsilon_{B,B\vee C} \,\fatsemi\, \epsilon_{C,B\vee C}^{\cup} \,\fatsemi\, S$$

There is nothing new with the new composition (i.e. $R \textcircled{\fatsemi} S \;=\; R \,\fatsemi\, S$) in case the right concept of R coincides with the left concept of S (i.e. $B = C$ in the above), so $\textcircled{\fatsemi}$ is a total extension of $\fatsemi$.

The identity is also the unit of the new composition, but even more is true: embedding and restriction are just new compositions with the appropriate identities. For the embedding (assuming $R \in A \sim B$, $A \preceq C$, $B \preceq D$):

$$\epsilon_{A,C}^{\cup} \,\fatsemi\, R \,\fatsemi\, \epsilon_{B,D} \;=\; id_C \,\fatsemi\, \epsilon_{C,C} \,\fatsemi\, \epsilon_{A,C}^{\cup} \,\fatsemi\, R \,\fatsemi\, \epsilon_{B,D} \,\fatsemi\, \epsilon_{D,D}^{\cup} \,\fatsemi\, id_D \;=\; id_C \textcircled{\fatsemi} R \textcircled{\fatsemi} id_D$$

and, similarly for the restriction (assuming $S \in C \sim D$, $A \preceq C$, $B \preceq D$):

$$\epsilon_{A,C} \,\fatsemi\, S \,\fatsemi\, \epsilon_{B,D}^{\cup} \;=\; id_A \,\fatsemi\, \epsilon_{A,C} \,\fatsemi\, \epsilon_{C,C}^{\cup} \,\fatsemi\, S \,\fatsemi\, \epsilon_{D,D} \,\fatsemi\, \epsilon_{B,D}^{\cup} \,\fatsemi\, id_B \;=\; id_A \textcircled{\fatsemi} S \textcircled{\fatsemi} id_B$$

3.2 New Order

The lattice operations are only defined per homset. If we want to join two relations with different typings, we should look for a common supertype embed the relations accordingly and join them there. Similarly we find the common workspace for the meet of two relations of different types. We have to pay the price that we do lose some information (and we'll never get that back) but the relations can be joined where originally heterogeneity wouldn't allow it. So define $\bigcup$ and $\bigcap$ by

$$\bigcup, \bigcap \; : \; (A \sim B) \times (C \sim D) \longrightarrow (A \vee C \; \sim \; B \vee D) \quad \text{with}$$

$$R\bigcup S = \epsilon^{\cup}_{A,A\vee C} \, \stretchrel{\circ}{,} \, R \, \stretchrel{\circ}{,} \, \epsilon_{B,B\vee D} \; \sqcup \; \epsilon^{\cup}_{C,A\vee C} \, \stretchrel{\circ}{,} \, S \, \stretchrel{\circ}{,} \, \epsilon_{D,B\vee D}$$

$$= id_{A\vee C} \bigcirc R \bigcirc id_{B\vee D} \; \sqcup \; id_{A\vee C} \bigcirc S \bigcirc id_{B\vee D}$$

$$R\bigcap S = \epsilon^{\cup}_{A,A\vee C} \, \stretchrel{\circ}{,} \, R \, \stretchrel{\circ}{,} \, \epsilon_{B,B\vee D} \; \sqcap \; \epsilon^{\cup}_{C,A\vee C} \, \stretchrel{\circ}{,} \, S \, \stretchrel{\circ}{,} \, \epsilon_{D,B\vee D}$$

$$= id_{A\vee C} \bigcirc R \bigcirc id_{B\vee D} \; \sqcap \; id_{A\vee C} \bigcirc S \bigcirc id_{B\vee D}$$

The new lattice operations $\bigcup$ and $\bigcap$ are total extensions of the old ones $\sqcup$ and $\sqcap$. It is readily seen that $\bigcup$ and $\bigcap$ are symmetric and associative and that they distribute. Zeros do exist for both $\bigcup$ and $\bigcap$, viz. $\top : \top \sim \top$ and $\bot : \top \sim \top$, but only $\bigcup$ allows for a unit: $\bot : \bot \sim \bot$. Therefore we don't have to hope too much for a suitable definition of a (new) negation.

Some doubt arises because of this anomaly, and quite rightly so. We can define order extensions of $\sqsubseteq$ such that $\bigcup$ or $\bigcap$ are the join or meet, but not both. Let's define for $R : A \sim B$ and $S : C \sim D$:

$$R \bigcirclesubseteq S \; \equiv \; A \preceq C \wedge B \preceq D \wedge \epsilon^{\cup}_{A,C} \, \stretchrel{\circ}{,} \, R \, \stretchrel{\circ}{,} \, \epsilon_{B,D} \sqsubseteq S$$

Then $\bigcup$ is the disjunction (join) for $\bigcirclesubseteq$ and thus we have

$$R \bigcirclesubseteq S \; \equiv \; R\bigcup S = S$$

But be aware that with this choice $\bigcap$ is only locally the meet for $\bigcirclesubseteq$. Because all unions exist, the global meet for $\bigcirclesubseteq$ does exist, but it is not an operation that is useful for our purposes. The meet would require restriction of the left and right types, but domain restriction is not a reasonable operation on relations as it suggests gain of knowledge or structure where we can only lose it by generalization. We had to forget about the specialties before we were able to intersect, so the price of this totalization is that those specialties are lost.

In an Ampersand script, the user need not worry about the supertype level. He uses $\bigcap$ as his meet operator, which is analyzed by the type checker. After checking and inferring the correct types, Ampersand works exclusively locally, so the almost-meet indeed is equivalent to $\sqcap$ where it occurs.

Note that the local complements cannot be extended to a global complement in general. Even globally complementing a local top for a nontrivial concept will not succeed. In general the notion of complement doesn't make sense in an environment with generalization, because different supertypings destroy the base of complementation. Instead we still may consider the typed complements like $\top_{A,B} \backslash R$ for relation $R : A \sim B$. This insight has influenced the implementation of Ampersand. As of the next version, the complement operator is substituted by a binary minus sign. If the left hand argument is missing, the compiler assumes $\top_{A,B}$ in which $A \sim B$ is the type of the right hand argument. The type checker deduces the type and generates an error message if it is not uniquely defined.

3.3   Interfaces

The converse doesn't need adaptation since the definedness doesn't change.

The reader may want to verify that the interface between the converse and the composition layers smoothly carries over to the converse and the new composition:

$$(R \circledS S)^{\cup} \;=\; S^{\cup} \circledS R^{\cup}$$

The interface between the composition and the lattice layer, i.e. the universal disjunctivity of the composition, is also valid for the new versions of composition and disjunction. To show that, let $\mathcal{U}$ be a collection of relations, say $U : U_\triangleleft \sim U_\triangleright$ for $U \in \mathcal{U}$. Let A and B be the joins of their left and right domain concepts respectively and let $R : C \sim D$ then

$$(\bigcirc\!\!\!\!\sqcup\, \mathcal{U}) \circledS R$$

$=$ $\{$ definition $\bigcirc\!\!\!\!\sqcup$ $\}$

$(\sqcup U : U \in \mathcal{U} : \epsilon^{\cup}_{U_\triangleleft,A} \,\mathbin{\substack{\circ\\\circ}}\, U \,\mathbin{\substack{\circ\\\circ}}\, \epsilon_{U_\triangleright,B}) \circledS R$

$=$ $\{$ definition $\circledS$ $\}$

$(\sqcup U : U \in \mathcal{U} : \epsilon^{\cup}_{U_\triangleleft,A} \,\mathbin{\substack{\circ\\\circ}}\, U \,\mathbin{\substack{\circ\\\circ}}\, \epsilon_{U_\triangleright,B}) \,\mathbin{\substack{\circ\\\circ}}\, \epsilon_{B,B\vee C} \,\mathbin{\substack{\circ\\\circ}}\, \epsilon^{\cup}_{C,B\vee C} \,\mathbin{\substack{\circ\\\circ}}\, R$

$=$ $\{$ universal distributivity of $\mathbin{\substack{\circ\\\circ}}$ $\}$

$(\sqcup U : U \in \mathcal{U} : \epsilon^{\cup}_{U_\triangleleft,A} \,\mathbin{\substack{\circ\\\circ}}\, U \,\mathbin{\substack{\circ\\\circ}}\, \epsilon_{U_\triangleright,B} \,\mathbin{\substack{\circ\\\circ}}\, \epsilon_{B,B\vee C} \,\mathbin{\substack{\circ\\\circ}}\, \epsilon^{\cup}_{C,B\vee C} \,\mathbin{\substack{\circ\\\circ}}\, R)$

$=$ $\{$ transitivity of embeddings $\}$

$(\sqcup U : U \in \mathcal{U} : \epsilon^{\cup}_{U_\triangleleft,A} \,\mathbin{\substack{\circ\\\circ}}\, U \,\mathbin{\substack{\circ\\\circ}}\, \epsilon_{U_\triangleright,B\vee C} \,\mathbin{\substack{\circ\\\circ}}\, \epsilon^{\cup}_{C,B\vee C} \,\mathbin{\substack{\circ\\\circ}}\, R)$

$=$ $\{$ exercise: $X \vee Z \sqsubseteq Y \Rightarrow \epsilon_{X,X\vee Z} \,\mathbin{\substack{\circ\\\circ}}\, \epsilon^{\cup}_{Z,X\vee Z} = \epsilon_{X,Y} \,\mathbin{\substack{\circ\\\circ}}\, \epsilon^{\cup}_{Z,Y}$ $\}$

$(\sqcup U : U \in \mathcal{U} : \epsilon^{\cup}_{U_\triangleleft,A} \,\mathbin{\substack{\circ\\\circ}}\, U \,\mathbin{\substack{\circ\\\circ}}\, \epsilon_{U_\triangleright,U_\triangleright\vee C} \,\mathbin{\substack{\circ\\\circ}}\, \epsilon^{\cup}_{C,U_\triangleright\vee C} \,\mathbin{\substack{\circ\\\circ}}\, R)$

$=$ $\{$ definition $\circledS$ $\}$

$(\sqcup U : U \in \mathcal{U} : \epsilon^{\cup}_{U_\triangleleft,A} \,\mathbin{\substack{\circ\\\circ}}\, (U \circledS R))$

$=$ $\{$ definition $\bigcirc\!\!\!\!\sqcup$ $\}$

$(\bigcirc\!\!\!\!\sqcup\, U : U \in \mathcal{U} : U \circledS R)$

Universal $\bigcirc\!\!\!\!\sqcup$-junctivity of $\circledS$ guarantees the existence of adjoint division operators, say $\oslash\!\!\!\!/$ and $\oslash\!\!\!\!\backslash$, given by

$$R \ominus S \circledS T \equiv R \oslash\!\!\!\!/ T \ominus S \quad \text{and} \quad S \circledS T \ominus R \equiv T \ominus S \oslash\!\!\!\!\backslash R$$

The typing of the new factors should follow from their construction, e.g.

$$R \oslash\!\!\!\!/ T \;=\; (\bigcirc\!\!\!\!\sqcup\, U : R \ominus U \circledS T : U)$$

leading to the typing $R \oslash\!\!\!\!/ T : R_\triangleleft \sim T_\triangleright$, provided $T_\triangleright \preceq R_\triangleright$, otherwise $R \oslash\!\!\!\!/ T = \bot_{\bot,\bot}$. Remember that R/T needed $T_\triangleright = R_\triangleright$, so the new factor is slightly more defined, with a (too) big right domain concept, but otherwise it is the same. Indeed, here we also have

$$a\langle R\oslash T\rangle b \;\equiv\; (\forall\, c : b\,T\,c : a\,R\,c) \quad \text{and} \quad a\langle S\obslash R\rangle b \;\equiv\; (\forall\, c : c\,S\,a : c\,R\,b)$$

The interface between the converse and the lattice (order) layers carries over smoothly to the converse and the new lattice(like) operations:

$$(R\obslash S)^{\cup} \;=\; R^{\cup}\obslash S^{\cup} \quad \text{and} \quad (R\oslash S)^{\cup} \;=\; R^{\cup}\oslash S^{\cup}$$

After introducing the total versions of the composition and the disjunction (order) we harvested the majority of the Tarski rules:

– We do have a suitably rich order structure, be it that we have an operation that in several respects looks like the meet but it isn't. It does have a reasonable interpretation and sufficient manipulative possibilities to keep it (more evidence below).
– The extended composition structure still forms a monoid
– The converse didn't even change.
– All interfaces between the layers are preserved except for Schröders rule that doesn't make sense without complementation.

The interface between all three layers is still lacking. Our caution with respect to the negation directed our interest towards the modular law ([4]). An inconvenience for the asymmetric modular law is the unfortunate typing, but this may be corrected by choosing the slightly more complicated but symmetric version called the Dedekind rule ([6]). (We gratefully thank the referee for suggesting the symmetric version and pointing at the typing advantage.) A severe drawback is the fact that the meet-like operator occurring in this new Dedekind rule is only locally the meet. So, globally, the Dedekind rule does not capture the boundary for conjunctivity as it did for the original relational calculus. Nevertheless we are happy with it because it still gives the manipulative power comparable to the original rule and it serves a purpose in mimicking the Tarski-rules for our new operations in the heterogeneous setting with generalization. Indeed, the Ampersand user, though working in a heterogeneous setting, is allowed to transform his expressions with this rule as if his environment is homogeneous, leaving the reponsibility for the syntactic correctness to the type-checker.

So we will show that

$$R\mathbin{_9}S \sqcap T \;\sqsubseteq\; (R \sqcap T\mathbin{_9}S^{\cup})\mathbin{_9}(S \sqcap R^{\cup}\mathbin{_9}T)$$

Let $R : A \sim B$, $S : P \sim Q$ and $T : X \sim Y$, then the two composed relations on the righthand side are

$$\epsilon^{\cup}_{A,A\vee X} \mathbin{_9} R \mathbin{_9} \epsilon_{B,B\vee P} \;\sqcap\; \epsilon^{\cup}_{X,A\vee X} \mathbin{_9} T \mathbin{_9} \epsilon_{Y,Q\vee Y} \mathbin{_9} \epsilon^{\cup}_{Q,Q\vee Y} \mathbin{_9} S^{\cup} \mathbin{_9} \epsilon_{P,B\vee P}$$

and

$$\epsilon^{\cup}_{P,B\vee P} \mathbin{_9} S \mathbin{_9} \epsilon_{Q,Q\vee Y} \;\sqcap\; \epsilon^{\cup}_{B,B\vee P} \mathbin{_9} R^{\cup} \mathbin{_9} \epsilon_{A,A\vee X} \mathbin{_9} \epsilon^{\cup}_{X,A\vee X} \mathbin{_9} T \mathbin{_9} \epsilon_{Y,Q\vee Y}$$

The supertyped relations $\epsilon^{\cup} \,\mathring{,}\, U \,\mathring{,}\, \epsilon$ occur in these expressions also in conversed version with the same embeddings, so we may abbreviate them as R', S', T' and their converses in the calculation below

$$(R \ominus T \oslash S^{\cup}) \oslash (S \ominus R^{\cup} \oslash T)$$
$=$ { above with abbreviation }
$$(R' \sqcap T' \,\mathring{,}\, S'^{\cup}) \oslash (S' \sqcap R'^{\cup} \,\mathring{,}\, T')$$
$=$ { intermediate types are the same $(B \vee P)$, so the composition is local }
$$(R' \sqcap T' \,\mathring{,}\, S'^{\cup}) \,\mathring{,}\, (S' \sqcap R'^{\cup} \,\mathring{,}\, T')$$
$\sqsupseteq$ { Dedekind in the original relation calculus }
$$R' \,\mathring{,}\, S' \sqcap T$$
$=$ { unfold abbreviation }
$$\epsilon^{\cup}_{A,A \vee X} \,\mathring{,}\, R \,\mathring{,}\, \epsilon_{B,B \vee P} \,\mathring{,}\, \epsilon^{\cup}_{P,B \vee P} \,\mathring{,}\, S \,\mathring{,}\, \epsilon_{Q,Q \vee Y} \sqcap \epsilon^{\cup}_{X,A \vee X} \,\mathring{,}\, T \,\mathring{,}\, \epsilon_{Y,Q \vee Y}$$
$=$ { new operators }
$$R \oslash S \ominus T$$

This concludes the rendering of almost all Tarski axioms from the homogeneous relations in our adaptation of the heterogeneous relations.

3.4 Example

A very small example to illustrate some of the above notions is about business rules for a company that sells ordered items to clients and bills them either by crediting them or, to evade taxes, by direct payment. So he needs two kinds of bills modeled by subtypes, the black bills and the white ones. The mini-ontology is:

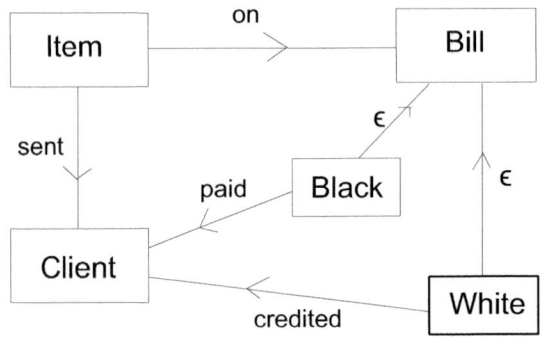

The following two business rules show the generalized composition and union and show the use of division that cannot completely do without type restriction.

– An item on a bill that is paid for or credited is sent to the client.

$$\text{on} \oslash (\text{paid} \ominus \text{credited}) \sqsubseteq \text{sent}$$

– A black bill must be paid by the client when all items on it are sent to him. For a translation we reformulate it into predicate calculus and transform it to a rule in terms of relations: for arbitrary bill β and client γ

$$(\forall i : i\langle\text{on}\rangle\beta : i\langle\text{sent}\rangle\gamma) \;\Rightarrow\; \beta\langle\text{paid}\rangle\gamma$$
$$\equiv \quad \{ \text{ division } \}$$
$$\beta\langle\text{on}\backslash\text{sent}\rangle\gamma \;\Rightarrow\; \beta\langle\text{paid}\rangle\gamma$$
$$\approx \quad \{ \text{ forall bills and clients } \}$$
$$\text{on}\backslash\text{sent} \sqsubseteq \text{paid}$$
$$\approx \quad \{ \text{ don't mind the subtypes, but for a domain of black bills } \}$$
$$\mathit{id}_{\text{Black}} \,\textcircled{\scriptsize ;}\, (\text{on}\,\textcircled{\backslash}\,\text{sent}) \,\textcircled{\scriptsize \copyright}\, \text{paid}$$

4 Application in Ampersand

Ampersand is a tool for constructing an information system from formally given business rules. These rules are defined as relation terms over a finite set of relations. Rules and relations are introduced (declared) by the user in a suitable syntax. The system assists its users (business rule engineers as well as students) in constructing relevant relational terms and rules. To that end a syntactic calculus of relations based on the ten Tarski axioms is enriched with typing arguments to filter out the evidently meaningless terms. The totalization of the heterogeneous relations above represents the first stage in constructing Ampersand relation terms. Relation terms may be composed, joined, intersected and reversed without bothering too much about the definedness of the operators. The necessity of that carelessness stems from the polymorphic naming as well as from the use of sub- and supertypes (overloading of relations and atoms, see [8] section 3.3). E.g. if some director is combined with a specific desk it should not primarily concern us whether he uses his desk as a manager, an employee or as a person, we may not want to make that typing distinction when constructing our desk-usage rule now. (Although it may be relevant later if the quality of the drawer locks is at stake.) So we may discuss many relations all called "uses" between every subtype of "person" and every subtype of "desk". The totality of the new operators allows that carelessness.

A user of Ampersand experiences that he needs not worry about types until the type checker detects an ambiguity (or omission). The language contains a mechanism by which the user can add type information to relation terms, in order to disambiguate his expressions and satisfy the type checker. Since Ampersand will not do anything until the script of the user is correct, (s)he experiences the semantics of the operations in the second stage only.

The second stage starts with the introduction of a suitable concept lattice, the generalizations in which are declared initially. We assume that that two concepts are completely independent if they have $\top$ as join. We don't have a relevant supertype of numbers and employees, of currencies and blood pressures, of amino acids and scientific journals. This is used in judging parts of relational terms:

- A composition with coarsest intermediate concept equal to $\top$ is certainly empty, since the glued left and right types are completely independent.
- A disjunction of relations with one of the domain concepts $\top$ combines independent things and is considered to be undesirable, we don't mix unmixable atoms.

The types of all domain concepts are calculated by the type checker and the instances of the above anomalies are detected and used to inform the user of probable mistakes. The precision of catching mistakes depends heavily on the strength of the concept lattice as a discriminator of (ir)relevancies. But it may give hints even if independence is not equivalent to joining up to $\top$.

The Ampersand user, or the information architect in the preceding modeling process, may define the dependencies and independencies in the subtyping order corresponding to her own needs. E.g. in recent provisional Ampersand applications the choice for trivial concept conjunctions was made, in older ones the flat lattice resulted. It is up to the user.

In the current Ampersand version negation poses a problem. If we have a relation in ordinary and supertype version, what should we do with the negation? Are we dealing with the complement in the ordinary type or does the complement in the supertype play a role. This is especially relevant in case the relation is composed with terms related to the supertype. In fact, that problem was the incentive for the investigations in this paper. The situation can now be significantly improved by adopting division while replacing the Schröder rules by the Dedekind rule and suitably restricting the $\top$-side of the division.

Negation is undoubtedly a natural element in specification of real world phenomena. In the industrial practice, negation is typically used in conjunction with other operators that allow the compiler to replace it. In case it is used explicitly by the user, the type checker helps. If the use of negation is such that an ambiguity arises, the user can fix this error by adding the intended type information. Theoretically we are not too much astray, since division and negation are related by a combination of the Schröder rule (though replaced by the Dedekind rule) and the complement of the identity, which exists in Ampersand. With this combination of theoretical and practical typing assistance Ampersand offers a version of relation calculus to its users that has appeared to be more than acceptable in practice.

5   Conclusion

We have enriched the heterogeneous relations with the notion of supertyping. With a rather natural construction the partial operations (composition and disjunction) are extended to total operations, so that much of the manipulative power of the homogeneous relations is regained for the heterogeneous relations. The result is a relational framework that explains why the current typing arguments in the business rule tool Ampersand do stand to reason. It also raises the question what we can do about the role of the negation, a partial answer to which may be found in the use of the division. The design advantage and the educational value of the Dedekind rule and the divisions could has not been judged, since these novelties are not yet taught in relation to Ampersand. The use of negation in constructing business rules needs to be investigated further.

References

1. Aarts, C., Backhouse, R., Hoogendijk, P., Voermans, T., van der Woude, J.: A relational theory of datatypes (1992), STOP summer school notes, http://www.cs.nott.ac.uk/~rcb/papers/abstract.html#book
2. Backhouse, R.: Pair algebras and Galois connections. Information Processing Letters 67(4), 169–176 (1978)
3. Brink, C., Kahl, W., Schmidt, G. (eds.): Relational methods in computer science. Advances in computing. Springer, Heidelberg (1997)
4. Freyd, P., Scedrov, A.: Categories, allegories. North-Holland Publishing Co., Amsterdam (1990)
5. Maddux, R.: Relation algebras. Elsevier, Amsterdam (2006)
6. Riguet, J.: Relations binaires, fermetures, correspondances de Galois. Bull. Soc. Math. France 76, 114–155 (1948)
7. Schmidt, G., Ströhlein, T.: Relationen und Grafen. Springer, Heidelberg (1988)
8. Michels, G., Joosten, S., van der Woude, J., Joosten, S.: Ampersand: Applying Relation Algebra in Practice. In: de Swart, H. (ed.) RAMICS 2011. LNCS, vol. 6663, pp. 280–293. Springer, Heidelberg (2011)
9. Tarski, A.: On the calculus of relations. Journal of Symbolic Logic 6(3), 73–89 (1941)

Author Index